New Aspects of
High-Energy
Proton-Proton Collisions

ETTORE MAJORANA
INTERNATIONAL SCIENCE SERIES
Series Editor:
Antonino Zichichi
European Physical Society
Geneva, Switzerland

(PHYSICAL SCIENCES)

Recent volumes in the series:

A Continuation Order Plan is available for this series. A continuation order will bring delivery of each new volume immediately upon publication. Volumes are billed only upon actual shipment. For further information please contact the publisher.

New Aspects of High-Energy Proton-Proton Collisions

Edited by

A. Ali

DESY
Hamburg, Federal Republic of Germany

Plenum Press • New York and London

Library of Congress Cataloging in Publication Data

INFN Eloisatron Project Workshop on New Aspects of Very High-Energy Proton–Proton Collisions (1987: Erice, Sicily)
  New aspects of high-energy proton–proton collisions / edited by A. Ali.
    p.   cm.—(Ettore Majorana international science series. Physical sciences; 39)
  "Proceedings of the Fourth INFN Eloisatron Project Workshop on New Aspects of Very High-Energy Proton–Proton Collisions, held May 31–June 7, 1987, in Erice, Italy"—T.p. verso.
  Includes bibliographies and index.
  ISBN-13: 978-1-4615-9542-7   e-ISBN-13: 978-1-4615-9540-3
  DOI: 10.1007/ 978-1-4615-9540-3
    1. Proton–Proton interactions—Congresses. 2. Proton–antiproton interactions—Congresses. 3. Hadron interactions—Congresses. 4. INFN Eloisatron Project—Congresses. I. Ali, A. (Ahmed) II. Title. III. Series: Ettore Majorana international science series. Physical sciences; v. 39.
  QC793.5.P728I54 1989   88-37962
  539.7′212—dc19   CIP

Proceedings of the Fourth INFN ELOISATRON Project Workshop on New Aspects of Very High-Energy Proton–Proton Collisions, held May 31–June 7, 1987, in Erice, Italy

© 1988 Plenum Press, New York
Softcover reprint of the hardcover 1st edition 1988
A Division of Plenum Publishing Corporation
233 Spring Street, New York, N.Y. 10013

Preface

ELOISATRON (Eurasiatic Long Intersecting Storage Accelerator) is the name of a research and development project in the field of high energy physics, approved and funded by the Instituto Nazionale di Fisica Nucleare INFN in Italy. The main objective of the project is to conduct research and development studies to promote the construction of a $(100 + 100)$ TeV proton-proton collider in Europe.

The present volume contains the proceedings of the 4th INFN ELOISATRON project workshop, held on the topic: New Aspects of High-Energy Proton-Proton Collisions. The workshop took place at the Centro Internazionale di Cultura Scientifica "Ettore Majorana" (CCSEM), Erice-Trapani, Sicily, Italy, in the period May 31–June 7, 1987. This was the first workshop in this series which concentrated on physics issues in proton-proton collisions with 1–100 TeV beams; the earlier three INFN ELOISATRON workshops, held at Erice during 1986 and 1987, had mostly dealt with technical issues related to the accelerator and detector aspects of high energy hadron colliders. The present workshop was supported by the Italian Ministry of Education, the Italian Ministry of Scientific and Technological Research, the Sicilian Regional Government and the Ettore Majorana Centre for Scientific Culture.

With the successful operation of the CERN Superconducting antiproton-proton Synchrotron (S$\bar{p}$pS), resulting in the discoveries of the vector bosons W and Z and providing evidence for new aspects of flavour mixings, the interest in very high energy proton beams as probes of fundamental phenomena in nature has mounted worldwide. It is now generally felt that high intensity proton beams in the energy range 1–100 TeV provide trustworthy tools, which may eventually pave the way to a qualitatively new understanding of some of the outstanding problems in particle physics. The promise that future proton colliders hold and the concomitant technical issues that have to be resolved on the way deserve dedicated studies, since both the stakes and the price tags in this pursuit are very high.

The physics and expected experimental signatures related to $\mathcal{O}(1$ TeV$)$ parton-parton centre of mass energy have received wide attention in recent years. Hard collisions in this energy region are expected to provide clues to possible new physics, for example related to the nature of spontaneous symmetry breaking, possible sub-structures of the present elementary particles and/or supersymmetry. These issues are taken up once again in a number of papers published in this volume. Several theoretical strategies are reviewed and further developed by including detailed estimates of a large variety of final states, involving electroweak gauge bosons, leptons, quarks, gluons and hadrons, expected to result from collisions of 1–100 TeV proton beams. Existing experimental results at lower energies have been used to constrain extrapolations to the ELOISATRON energies. A sharp profile of these final states is needed to devise experimental strategies and build adequate detectors. In addition, problems associated with experimentation with very high energy proton beams and extraction of useful signals for anticipated new physics in proton-proton collisions from the background processes are also discussed.

It should be mentioned that earlier studies in this direction were undertaken at Lausanne and Geneva (Switzerland) in 1984, and at La Thuile (Italy) and Geneva in 1987, under the auspices of the European Organization for Nuclear Research (CERN) and the European Committee for Future Accelerators (ECFA). A series of studies dedicated to physics and technology of the so-called Superconducting Super Collider (SSC) project in U.S.A. has been conducted at the summer institutes held regularly since 1982 at Snowmass, Colorado. Workshops concerning various aspects of the SSC project have also been held at Berkeley, Los Angeles and Madison-Wisconsin in the last two years. The interest in high energy proton collisions in the Soviet Union is reflected in the proceedings of the workshop on the experimental programme at the UNK accelerator at Serpukhov, published in early 1988.

The present volume has many aspects not hitherto discussed in detail elsewhere, including an updated description of the ELOISATRON project, its aims and objectives. The projected ELOISATRON energy is much higher than that of the SSC. This feature demands assessment of the physics potential and reach of the ELOISATRON machine in a number of test case studies. Comparisons with the corresponding reaches of the proposed machines like the Large Hadron Collider (LHC) at CERN and the SSC are also of great interest. In particular, the choice of optimum beam energy, needed to understand the physics related to the anticipated $\mathcal{O}(1\ \text{TeV})$ mass scale, is a crucial and open issue. A beginning in this direction is made in several papers published here, which take up comparisons of physics reach in selected benchmark processes for various proton beam energy options. We hope that subsequent publications in the present series of studies, as well as independent investigations elsewhere, will take a closer look at this question.

It is worth remarking that an intermediate 10% ELOISATRON machine will have to be built to test the technical feasibilty of the full scale project. This is a particularly important issue in the European context, since a 10% ELOISATRON machine will easily fit in the LEP tunnel at CERN. This aspect as well as the long term interests of the European high energy physics community, which go beyond this century, deserve both scientific discussion and public debate. We hope that this volume is a constructive contribution in this direction.

As the director of the 4th INFN ELOISATRON project workshop, I would like to express my sincere thanks to the authors, speakers and participants, who contributed to the scientific programme and these proceedings. Unfortunately, the talks of L. DiLella, R.K. Ellis, A. De Rujula, J. Kirkby, J. Sculli, V. Soergel and G. Wolf were not available for publications in this volume. I would like to specially thank the director of the INFN ELOISATRON project, A. Zichichi, and the staff of the CCSEM, Erice, for providing all possible logistic and scientific support. In particular, the help of Alberto Gabriele, Jerry Pilarski, Pinola Savalli and Guido Torelli is gratefully acknowledged. Finally, I would like to thank Helga Laudien and Zbigniew Jakubowski of DESY, and my friend Angelika von Amelunxen for their help with the organization of the workshop and the preparation of the manuscript.

Ahmed Ali
Editor

Hamburg, August 24, 1988

Contents

9. Physics of Superheavy Onia
Ken-ichi Hikasa

10. New trigger System for High Luminosity Collisions
A. Contin

11. Particle Detectors Based on the New Multichannel Photomultipliers
R. Meunier

12. Beauty Physics at the Ultrahigh Energies of the ELOISATRON
B. Cox

21. Non-Accelerator Physics at $\sqrt{s} \cong 40$ TeV
T.K. Gaisser

22. Cosmic Accelerators: a New Era of Cosmic Ray Astrophysics and Particle Physics
F. Halzen

THE ELOISATRON PROJECT

EURASIATIC LONG INTERSECTING STORAGE ACCELERATOR

A. Zichichi

CERN

Geneva, Switzerland

1. INTRODUCTION

In the last forty years, our understanding of Nature and its Laws has undergone a dramatic development. The "old" elementary particles (leptons, mesons, baryons) have been replaced by "new" elementary particles (quarks, gluons, leptons, $W^{\pm}$, Z^0).

Where can we expect to further go? To the fundamental constituents of the elementary particles (leptonic quarks, preons, rishons) and to the supersymmetric particles. There is an immense amount of phenomena to be discovered. We know the list and it is a very long one.

The energy level has been and will continue to be a key factor in the progress of our knowledge towards the final objective of the unification of all the fundamental forces of Nature.

1.1 The position of Italy

Italy is not a superpower. Instead of trying to do many things and not doing them very well it is considered far better to do a few things and do them excellently.

Authentic innovations stem from the discovery of new fundamental Laws of Nature. And this is a field where Italy has engaged its intellectual forces.

This is the reason why the unprecedented scientific development in this advanced field of Modern Science did not find the Country unprepared. Italy has nowadays reached an outstanding position in this basic field of Science and Technology.

1.2 The main objectives

Following the above considerations Italy, through the CIPE (Comitato Interministeriale per la Programmazione Economica) has undertaken a preliminary step of financing the ELOISATRON Project in the 5-years INFN plan 1984-1988. The main objective being to promote the construction of the largest subnuclear machine in the world, in collaboration with those Countries which share its interest in a field so vital for scientific, cultural and technological progress.

Such a machine is identified as a (100+100) TeV proton proton Collider. The implementation of this project would in fact constitute a unique instrument for expanding the frontiers of our present knowledge.

1

Many fundamental scientific problems - like, for example, the family, the hierarchy, the proliferation and the compositness - would be opened towards new horizons.

Furthermore the construction of the Collider would provide for technological development in several leading sectors of the so-called High Technology, i.e. the technology of the year 2000. These leading sectors are:

> **cryogenics;**
> **superconductivity;**
> **powerful magnetic fields;**
> **high vacuum;**
> **ultra fast electronics;**
> **supercomputers;**
> **new particle detectors.**

Finally, it should not be forgotten that ELOISATRON is a scientific project whose goals are Peace and civil progress. For this to be real, a new trend of Scientific and Technological solidarity between the NORTH and the SOUTH is needed. As shown by the list of Authors, the ELOISATRON Project has stimulated a tremendous interest among scientists from developing Countries. Only if peaceful high technology is made competitive, victory over war technology will become possible. Science for Peace can only become a reality through concrete projects.

1.3 The logic

In order to build the largest accelerator in the world it was felt that the most effective approach would be as follows.

First of all to develop the required technologies. Here the ELOISATRON Project has chosen to follow a totally new approach to the problem.

It was, in fact, recognized that amongst the various technologies needed for the construction and the experimental exploitation of such a machine,there are for example some, accelerating techniques or particle detection techniques, which are best developed in Research Laboratories and others like, for example, the construction of superconducting magnets, which are much better pursued by Industry.

Accordingly, it has been considered essential for the ELOISATRON Project to promote, at a very early stage, the interest of those industries where the most basic components of the machine could be constructed. Next, in order to establish the basis for such a project to become real, the creation of a large area of collaboration, the world over, was of vital importance. In fact the project cannot be isolated from the scientific and cultural environment. It must promote a stronger and wider interaction among all the scientists interested.

1.4 The basic steps

The basic steps of the ELOISATRON Project are shown in Table 1.1.

So far the time schedule has been followed as aspected. At present we are fully engaged in the R & D (LAA) project for new Detectors to be used at the MultiTeV Supercollider such as the 10% ELOISATRON model. The next step is to start the construction of the 10% model. The last step is at the level of conceptual design. Intensive R&D in fundamental technologies is going to be an integral part of this step.

All the steps towards ELOISATRON are discussed in more detail in the following.

Table 1.1 - Steps towards ELOISATRON

1979 - CS Project (Superconducting Cyclotron) resumed.

1981 - Start of the CS construction (Ansaldo - LMI - Zanon).
The HERA Project is presented to the INFN Council.

1982 - First Meeting of the HERA designers with Ansaldo - LMI - Zanon.

1985 - The construction in Italy of the superconducting magnets (prototypes) for HERA begins.

1986 - R&D for Detectors - LAA Project is approved and financed.

1987 - R&D for LSCM (Long SuperConducting Magnet) prototypes - 10% ELOISATRON Model - starts at Ansaldo.

1988 - The construction of the HERA magnets is expected to be completed.

1989 - Magnet construction for the 10% ELOISATRON Model is expected to start.

- The Full Scale ELOISATRON PROJECT:
  - 1st Conceptual Design
  - 2nd R & D in fundamental technologies
  - 3rd Construction of the Superconducting Long Magnets.

1.4.1 - The Superconducting Cyclotron (CS). In 1979 INFN took up again the project of building in Italy a Superconducting Cyclotron. Besides the scientific importance for the studies in Nuclear Physics, the aim of the CS Project was to develop around it a scientific undertaking of high technology affecting three Italian industries: one specialised in magnets (ANSALDO), another in superconducting cables (LMI) and a third one in cryogenics (ZANON).

This first step was crowned with success. In 1981 the construction of the CS began.

1.4.2 The superconducting magnets for HERA. The following year the second step started with the participation of Italy in the construction of the most delicate part of the German Project HERA (Hadron-Electron Ring Accelerator): that is, the superconducting magnets for the 820 GeV proton ring at DESY, Hamburg. Also this step has been successfully completed. In fact Italy is building 50% of the superconducting magnets for HERA. The series production of about 300 dipoles has already started and will be completed by 1988.

1.4.3 The 10% ELOISATRON Model. After completion of the CS and HERA phases a general test at the 10% level of the Full-scale ELOISATRON Project is needed.

This test must be made at the minimum cost and in the most efficient way using the best structures existing today.

It should be noticed that in order to construct this machine only moderate R&D of the magnets developed by the Italian industry for the HERA machine is needed.

1.4.4 Research and Development for Detectors (LAA). Experience with the LEP and HERA programmes has shown that the design and construction of the required complex apparatus needed in present accelerators is an undertaking of the same order of magnitude as the construction of the machine itself.

Moreover, considering the event complexity and the severe experimental conditions involved in a multi-TeV hadron Collider, it has been recognized that with present-day techniques nobody would know how to perform an experiment at such a machine.

Notice that this situation is completely reversed for electron Colliders. In this case the detector doesn't present relevant difficulties, but the technology to build the machine is unknown.

In parallel to the development of the 10% ELOISATRON Model it was therefore deemed necessary to start the LAA Project: an intensive R&D programme for new experimental techniques. The goal is to prove, on the basis of prototypes, the feasibility of essential components for a detector capable of operating at the next generation of Colliders, the first one being the 10% ELOISATRON Model. Special attention must therefore be paid to radiation hardness, rate capability, momentum resolution and hermeticity of such a detector assembly.

Bearing this in mind, the LAA Project covers the following most important components:

High precision tracking devices
Calorimetry
Large area devices for muon detection
Leading particle detection
Data acquisition and analysis
Theoretical QCD calculations and MonteCarlo
simulations
Very high magnetic fields
Superconductivity at high temperature.

At present a number of projects in the various fields listed above has been approved and financed by the LAA Scientific and Technical Advisory Board. They are under way.

1.4.5 The Full-scale ELOISATRON Project. In addition to the problems related to the machine itself two most basic problems had to be considered. Where to build ELOISATRON and how to build it.

The Civil Engineering problems have been studied in specialized Working Groups and possible solutions have been found. A detailed study of the machine lattice and of the full-scale project with its executive plans have not been attempted since it was felt that this must constitute an intrinsic part of the effort put in the 10% test model.

However, existing studies of projects at the (10-20)% scale (LHC,SSC) and the studies for the full-scale ELOISATRON performed so far, allow to conclude, as discussed later, that such a project has no intrinsic prohibitive features.

1.5 The structure

According to the principles discussed above the structure of the ELOISATRON Project is, at present, as follows.

The Project is subdivided in a number of Working Groups to which scientists from various Research Laboratories, Universities and Institutions participate together with several Italian industries.

Table 1.2 shows the Working Groups and the scientists, scientific institutions or industries in charge of organizing the activity in the various domains.

Table 1.2 - ELOISATRON Working Groups

1	-	**Superconducting Magnets**	-	*Ansaldo, LMI, Zanon*
2	-	**Refrigeration Power Plants**	-	*Zanon*
3	-	**Civil Engineering**	-	*L. Valeriani, Grandi Lavori*
4	-	**Geophysical Studies**	-	*E. Boschi, ING*
5	-	**Machine Workshops** *(Puglisi, Torelli, Villa)*	-	*K. Johnsen*
6	-	**Experimental Areas**	-	*H. Wenninger*
7	-	**R&D for Detectors (LAA) Workshops** *(G. Charpak)*	-	*A. Zichichi*
8	-	**Supercomputers** (QCD and Data Handling)	-	*T.D. Lee*
9	-	**Physics - Experimental**	-	*S.C.C. Ting*
10	-	**Physics - Theory Workshops** *(A. Ali)*	-	*A. Salam*

The results of the studies performed are then presented and discussed by the scientific community in Specialised Workshops.

The needed research and development of prototypes is carried on, either in Laboratories, in the framework of the LAA programme, or, in industries, in the framework of concrete projects apt to develop the required technologies at the industrial level.

The crucial point in the ELOISATRON Project is that the location of the main Lab is where all detailed R&D is going on. And this means that ANSALDO, LMI, ZANON are basic labs of the Project.

2. THE NEED FOR ELOISATRON

2.1 A new era

A new era in modern scientific thought has opened up. New ideas, new concepts and new phenomena make physics of just twenty years ago seem as old as millennia.

Einstein's quadridimensional space-time seemed to be a conquest beyond which none would be able to go. This, however, seems to be an incredibly narrow outlook for two reasons, both fundamental: the <u>number</u> of dimensions and the <u>property</u> of those dimensions.

No one had thought, before the sixties, that there could exist space-time dimensions with "fermionic" properties. Those of Einstein are "bosonic". This is how the new concept of Superspace was born, and with it, Superparticles and Supermatter. In Table 2.1 a synthesis is given of the present state of our knowledge of Matter - quarks, leptons, photons, strong and weak gluons, Higgs particles - and Supermatter.

The world in which we live and the matter we are made of could have their roots in a Superspace with many more than four "bosonic" dimensions plus the corresponding "fermionic" ones. And this is not all.

The concept of "point" that has held its position for centuries and centuries, falls by the wayside. In its place there is a Superstring or a membrane: i.e. either a unidimensional or a multidimensional entity with a non point-like structure in the "still to be definitely identified" pluridimensional Superspace. Fig. 2.1 shows the present status of our knowledge.
In this extraordinary progress towards the unification of all the fundamental forces of Nature, the winning parameter has been so far, and will certainly remain, the energy level.

Table 2.1 - Present status of World and Superworld

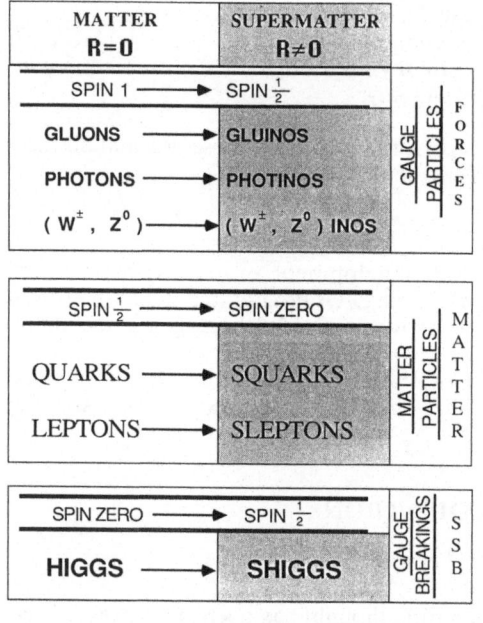

exception : NEUTRINO ➔ SNEUTRINO

SBS = Spontaneous supersymmetric breaking

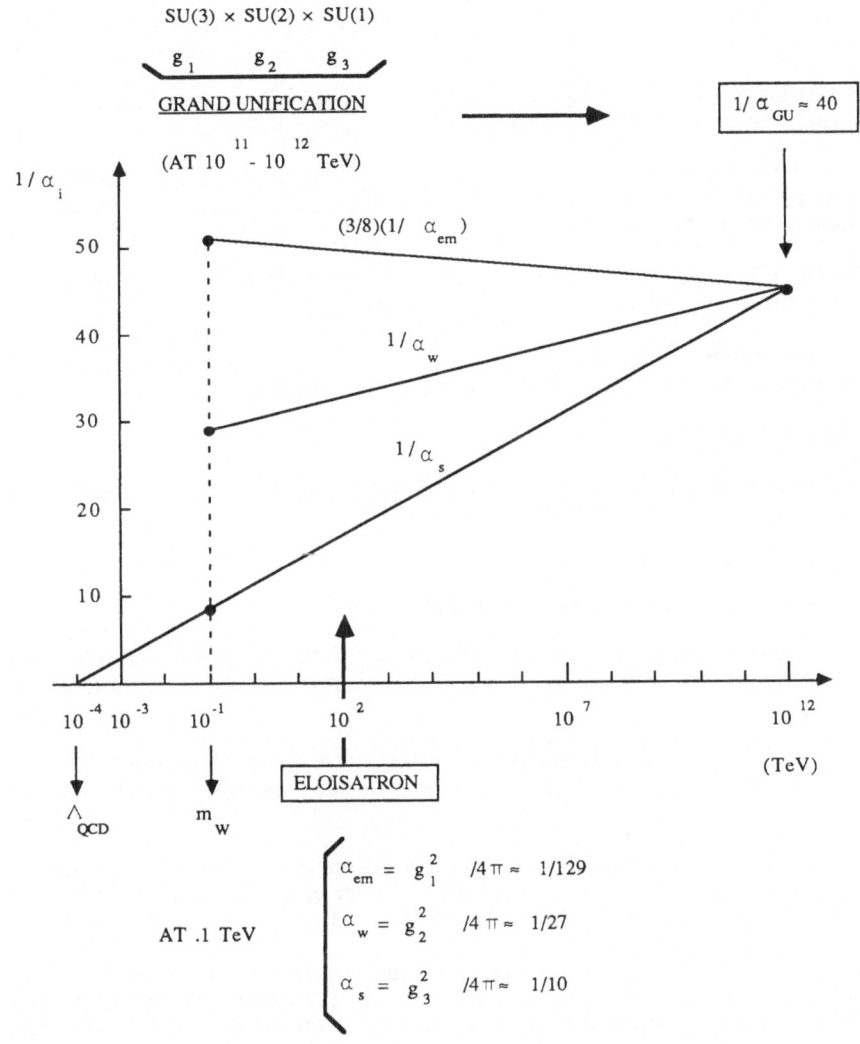

$$SU(3) \times SU(2) \times SU(1)$$

$$g_1 \qquad g_2 \qquad g_3$$

GRAND UNIFICATION

$$(AT \ 10^{11} - 10^{12} \ TeV)$$

$$1/\alpha_{GU} \approx 40$$

$1/\alpha_i$

$(3/8)(1/\alpha_{em})$

$1/\alpha_w$

$1/\alpha_s$

ELOISATRON

(TeV)

Λ_{QCD}

m_W

AT .1 TeV

$$\alpha_{em} = g_1^2 \ /4\pi \approx 1/129$$

$$\alpha_w = g_2^2 \ /4\pi \approx 1/27$$

$$\alpha_s = g_3^2 \ /4\pi \approx 1/10$$

Fig. 2.1 - All the fundamental forces of nature should be generated by a unique force. The convergence of the three fundamental constants g_1, g_2, g_3 is the basis for GUT.

2.2 Why start now to discuss a multi-TeV machine

Given that by 1990 both LEP and HERA will be operating, and if, as many of us believe, Subnuclear Physics is to be pursued with vigour and without a break of many years, then it is appropriate to start work now on a new hadronic machine in the 10^2 TeV range.

2.3   From new ideas to reality

Some people might say that it would be better to wait for new ideas. But past experience has shown that to wait is not a wise choice.
In fact, a long time is needed to transform new ideas into reality. Two examples should suffice:

i)   Superconducting high-field magnets, first proposed in 1961, became "reality" in 1986 (the Tevatron): 25 years were required.

ii)   Collective field accelerators were proposed by Veksler, Budker and Fainberg in 1956: a third of century later, there is still no practical design for a high-energy machine based on these ideas.

In this context it is important to notice that the ELOISATRON design is based on extrapolation from known facts and technologies. Nevertheless,

- new acceleration techniques,
- new superconductivity technologies (at high temperatures),

should be encouraged for the full-scale Project.

The 10% test model should not wait for these results. It should consolidate the validity of the extrapolation of present day technologies.

2.4   The desert and the lesson from the past

At the end of the seventies, a fashionable approach to the extreme energies was the "desert", i.e. nothing should exist in the energy range from a few 10^{-1} TeV up to about 10^{12} TeV.

The "desert" would be a serious obstacle if there were no problems whatsoever in this field of extreme energies. The high energy limit of our present knowledge has two frontiers.

One is in the domain of experimental physics. At present the known limits on the inverse radii of all known leptons (e, m, t) are of the order of 10^{-1} TeV. This means that point-like structures of the particles existing at present, are already in the multi-TeV range.

The other is in the theoretical domain. Here we have the family and the hierarchy problems, which, together with the proliferation of the Higgs sector and of its associated large number of parameters, make the multi-TeV range overcrowded with problems to be understood.

But even if everything looked "perfect", we should not forget the lesson we can learn from past experience. Other "deserts" were believed to be present in all history of Physics and none of them has over resulted to be there.

In the last 400 years, for at least 5 times the following statement was made: "There is nothing to discover, we have really understood everything". This happened after the discoveries of

1 - Galilei [Unification of Rectilinear and Circular motion]
2 - Newton [Universality of Gravitational forces: $F = k\ (m_1 \cdot m_2)/r_{12}^2$]
3 - Maxwell [Unification of Electricity, Magnetism and Optics]
4 - Plank, Einstein [Quantum ($h \neq 0$) and Relativistic ($c \neq \infty$) Physics]
5 - Fermi [Three "matter" particles (p, n, e) and three "glues" (π, γ, g)]

For five times a "desert" was predicted in the world of Physics and the lesson we can derive from the past is that this desert is our enemy. One example of desert not mentioned in the above list, in spite of being the most famous one, is the one predicted by Lord Kelvin in 1897: "There is nothing to be discovered in Physics now, all that remains to be done is more precise measurements". A few months later J.J. Thomson discovered the electron.

The last example in the list is the most recent case: when Enrico Fermi was convinced that, with the discovery of the "nuclear glue" - the π meson - there was nothing more to be searched for, in order to understand the microscopic world. Since then a continuous confirmation for the lack of any predicted or unpredicted desert has characterized the development of modern physics. It is probably appropriate to review this in more detail.

30 GeV PS (CERN, BNL)

Original motivations for the construction of this accelerator:

- (πp) (pp) scattering, phase-shift analysis
- test of isospin (I) and time reversal (T) invariances

Discoveries:

- new particle states $\Rightarrow$ SU(3) (Gell-Mann, Ne'eman)
- (ω–ϕ) mixing angle
- (e^+e^-), ($\mu^+\mu^-$) production in hadronic interactions
- electromagnetic structure of the proton in the time-like region
- antideuteron
- $\nu_e \neq \nu_\mu$
- $\nu_\mu \neq \nu_\mu$
- C, P, CP, T violation
- weak neutral current
- J particle state

SLAC

Original aims:

- electromagnetic form factors of nucleons
- electromagnetic transition form factors (N-N*)
- checks of Quantum ElectroDynamics (QED)

Discoveries:

- point-like structures inside the proton
- parity violation in purely electromagnetic interactions

ADONE

Conventional motivations:

- checks of QED and radiative corrections
- (μe) electromagnetic equivalence
- study of the tails of the vector mesons

Unconventional proposals:

- search for heavy leptons via acoplanar (μe) pair techniques
- search for leptonic quarks

Unexpected discovery:

- the ratio (hadronic cross-section / muonic cross-section) greater than theoretically predicted

SPEAR and DORIS

Discoveries:
- heavy lepton
- J/Ψ family of new particles
- open charm states

PETRA and PEP

Discoveries:

- evidence for gluons (expected)

ISR

The ISR was a special case. In fact, both the choice of the energy range and the technical performances of the machine were optimal. However the Physics results obtained were not as outstanding as expected since the quality of the experimental setups used was not adequate. This points out the importance of a dedicated R&D programme for experimental setups, as the present LAA Project, before the construction of any new machine is undertaken.

400 GeV SPS (CERN, FERMILAB)

Discoveries:

- Υ states (unexpected)

SppS (540 GeV)

Discoveries:

- $W^{\pm}$, Z^0 (expected)

TO CONCLUDE, the lesson from the past can be expressed in the following points:

1 - the same effort must be devoted to experimental set-ups and to machine construction (ISR docet);

2 - watch the energy gaps:

<pre>
       ADONE   SPEAR   PETRA
      |————————————|——|————————————|
            3 GeV 9   10 GeV
            (J/Ψ)   (Υ)
</pre>

3 - when unexpected discoveries are made, the expected ones appear as DESERT OF IMAGINATION.

This is why at present, the need for a multi-TeV machine and the effort to build such a machine are justified.

3. FROM THEORY TO DOWN-TO-EARTH PHYSICS

3.1 The great problems of MultiTeV Physics.

The theoretical goals outlined above can be expressed in terms of the following basic questions:

> do new heavier quarks and leptons exist?
> do supersymmetric partners exist?
> are there other intermediate bosons?
> do the Higgs bosons exist?
> are quarks and leptons composite?
> would some unexpected exotic process occur?

In addition to this impressive list of new physics, we should not forget the straigthforward extrapolation expected from present knowledge.

3.2 Extrapolation from present knowledge

In order to go from theory to down-to-earth physics, i.e in order to be able to make experiments, it is necessary to study which are the expected observable features of events produced by physics at multi-TeV energies and derive from these the main characteristics of the experimental apparatus. Rare phenomena, for instance, suggest that the highest luminosity should be aimed at. Let us consider, for example, 10^{33} cm^{-2}s ^{1}. Furthermore, as discussed later, if only one event per bunch crossing is desired, a short separation time between bunches (<100 ns) is needed. This implies specific performances from the experimental setups (see Section 4.2).

Let us then review some of the expected physics features in the multi-Tev range.

The present QCD knowledge on hard parton interactions predicts that at multi-TeV energies the machine would essentially be have as a "broad band gluon-gluon collider". By extrapolating CERN ISR results at $\sqrt{s}$ = 62 GeV, it was possible to predict the average charged particle multiplicity and the multiplicity distribution at the CERN pp Collider ($\sqrt{s}$=540 GeV), in good agreement, as shown in Fig.s 3.1 and 3.2, with experimental results. It should be noticed that to do this it is necessary to consider not the nominal energy $\sqrt{s}$, but the effective energy available for particle production once the "leading" particle effect - see Fig. 3.3 - is taken into account, i.e. once the energy carried away by the "leading" particles is subtracted from $\sqrt{s}$. The same procedure at $\sqrt{s}$=20 TeV gives an average charged multiplicity of 100 with important fluctuations up to multiplicities greater than 200, as shown in Fig. 3.4.

Hard parton interactions will provide quarks and gluons which fragment into jets. They will also presumably produce new massive states decaying into intermediate bosons, leptons, quarks and gluons, therefore predominantly into jets. Jets and leptons are expected to be fundamental instruments to extract interesting rare events from the standard parton scattering processes. Moreover, "missing jets" - i.e. missing energy/momentum - would provide signatures for events with undetected particles such as neutrinos or photinos, or other "inos".

It should be noticed that high p_T jets must not only be considered as a test for QCD, or as a background to cope with, when hunting for more interesting new phenomena: they could reveal if quarks are composite. The cross-section behaviour with p_T should in fact be modified by the effect of a deeper contact interaction caused by a superstrong binding force.

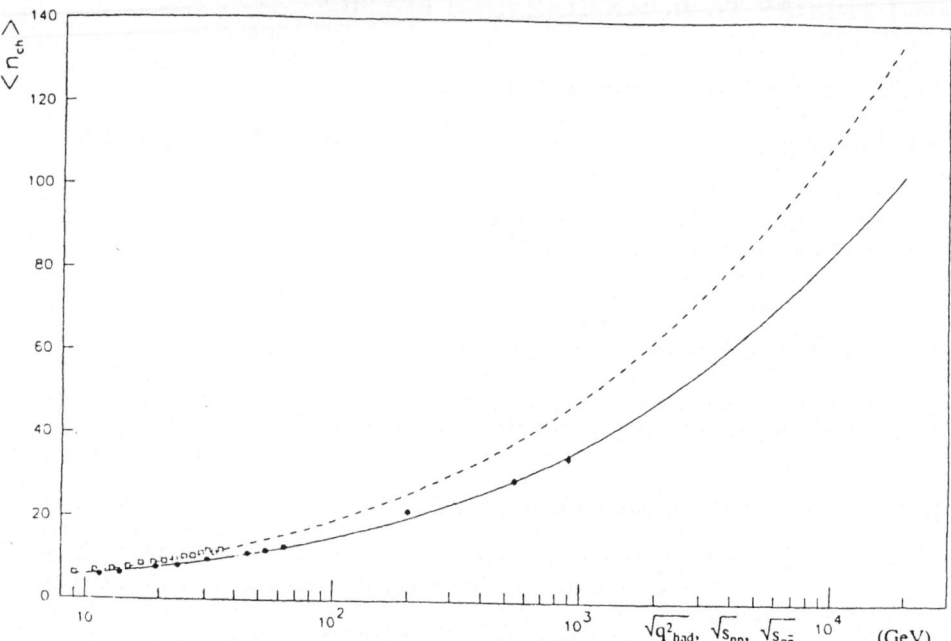

Fig. 3.1 - The average particle multiplicities as a function of $\sqrt{q^2_{had}}$ (dashed line) and $\sqrt{s}$ (full line). The dependence of the average particle multiplicity on $\sqrt{s}$ is derived from its dependence on $\sqrt{q^2_{had}}$.

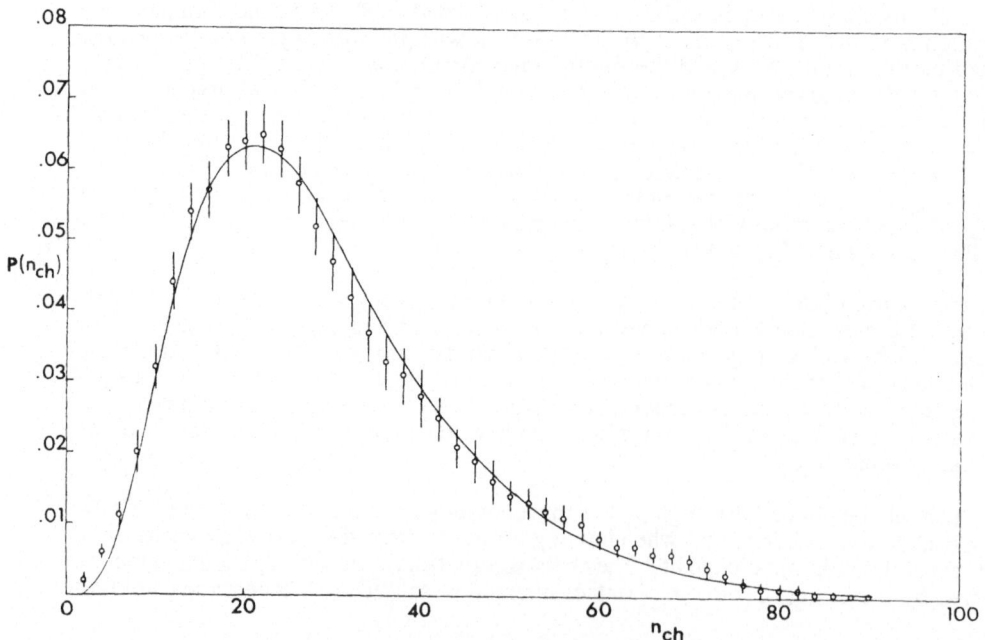

Fig. 3.2 - The charged particle multiplicity distribution measured by the UA5 collaboration at $\sqrt{s}=540$ GeV. The curve is the prediction with the BCF method of taking into account the leading effect.

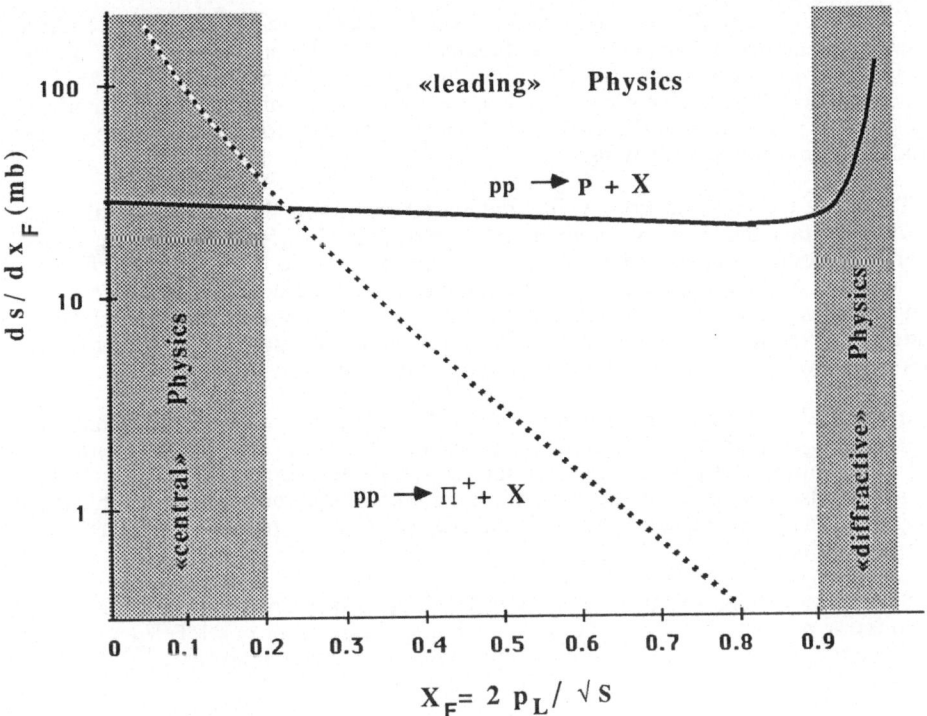

Fig. 3.3 - Protons and pions x_F distributions in (pp) interactions. Central, leading and diffractive regions are also shown.

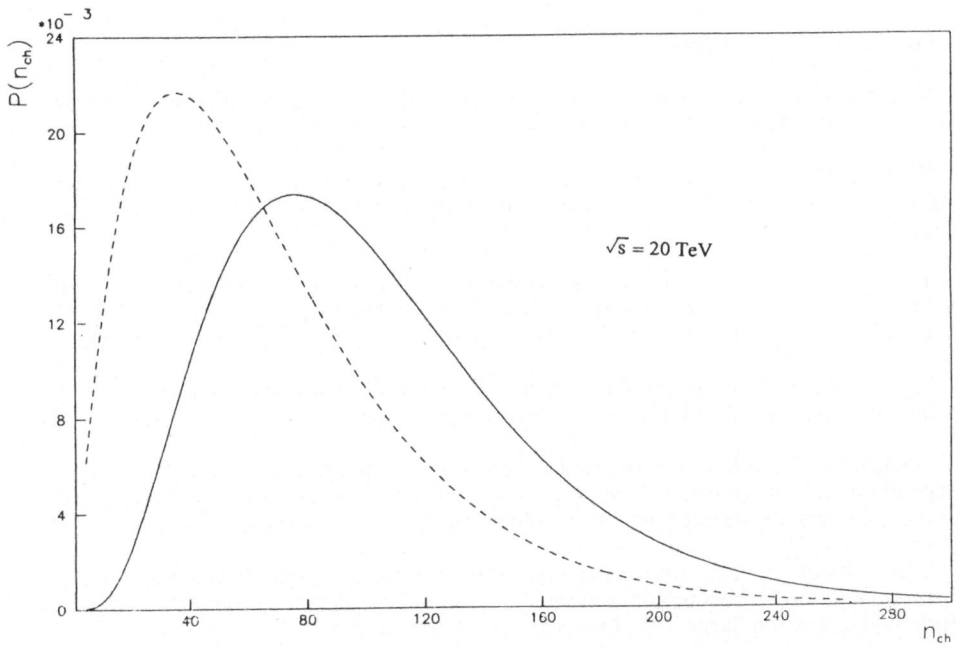

Fig. 3.4 - The charged particle multiplicity distribution predicted at $\sqrt{s}$=20 TeV following the UA5 method (dashed line) and the BCF method (full line).

Concerning heavy flavour production, it is not unlike to predict that a large hadron Collider will be a good old and new quarks factory. The calculations, mostly based on the fusion mechanism, show that at $\sqrt{s}=20$ TeV heavy quarks with masses up to 400 GeV could be observed. It must also be recalled that the fusion mechanism underestimated the rate of c and b production at the CERN ISR and failed to reproduce the shape of the longitudinal momentum distributions.

Therefore the cross-section evaluations based on this model have to be taken as lower limits. In fact, the "leading" production observed at the ISR for Λ^+_c and Λ^0_b should be taken into account. In addition, the "leading" mechanism could be much more efficient in the detection of very heavy quarks. In fact, the semileptonic decay of baryonic states carrying the heavy flavour is expected to produce a distinctive pattern: the sign and the p_T dependences of the lepton charge are directly correlated to the **"up-like"** or **"down-like"** nature of the new heavy quarks and to their masses, respectively.

In general, the signatures for new heavy flavour production would consist of multi-jet events or, if their semileptonic decay is considered, of jets + lepton (within or nearby a jet, with high p_T with respect to the jet axis) + missing energy. Multi-lepton events should also be relevant in this kind of search. Substantial rates up to a few hundred GeV leptonic energy are expected from c, b and t decays. Leptons from heavier quarks will be even more energetic.

SUSY production, for instance pp -> ggX, with g -> $q\bar{q}\gamma$, could be investigated via a huge missing p_T signature due to the non-interacting photino. For $m_g<1$ TeV and $\sqrt{s}=20$ TeV, the cross-section could exceed 0.1 pb, giving 10 events/day at 10^{33} cm^{-2}s^{-1} luminosity.

Higgs production through WW fusion could occur at the 1 pb level for $m_H=400$ GeV and $\sqrt{s}=20$ TeV. Assuming H -> WW decay, the 4-jet invariant mass could be a valuable tool to detect the signal.

Another feature of present-day knowledge to be extrapolated in the field of "flavour physics", is the structure of the multi-quark states, still to be discovered. There are hundred of these "particles" as shown in Fig.s 3.5 - 3.8.

3.3 - The observable effects

The impressive list of phenomena to be searched for, in terms of down-to-earth Physics, corresponds to the following experimental effects to be studied:

- Short-lived hadrons should be tagged by secondary vertex identification. Adequate tracking power is needed. The ability to define points in space will be of considerable advantage.

- Long-lived hadrons must be distinguished from leptons and photons and their energy precisely measured. The high multiplicity, jet-like events produced in multi-TeV collisions calls for high granularity detectors, capable of reconstructing the jet directions.

- Electrons and photons should be identified, even if produced inside jets. Their energy should be measured with the highest possible resolution.

- Neutrinos and other non-interacting particles can be "identified" as energy unbalance in the fully reconstructed event. Hermeticity will be an essential attribute of any apparatus, since most of the expected "new" physics will produce missing energy.

- Muons should be identified in the largest possible solid angle. Their charge should be measured up to the highest possible momentum, in order to detect charge asymmetries. Therefore, high precision, large area, low cost devices are in order.

- Leading protons can be identified by their momentum. Energies up to the maximum attainable should be measured with good precision, in the smallest possible angle with respect to the beam direction.

On the basis of past experience discussed in Section 2.4, we have to add to this impressive list the other one: "the unexpected". So far, as we have already said, "when unexpected discoveries are made, the expected ones appear as a DESERT OF IMAGINATION".

The ELOISATRON Project is before us with an unprecedented challenge: can we think of something to be discovered, which is more exciting than all the items in the above list? Needless to say that this list is really impressive. Could someone imagine a "new" breakthrough which is not part of our supervast horizon? This is the goal for multi-TeV physics.

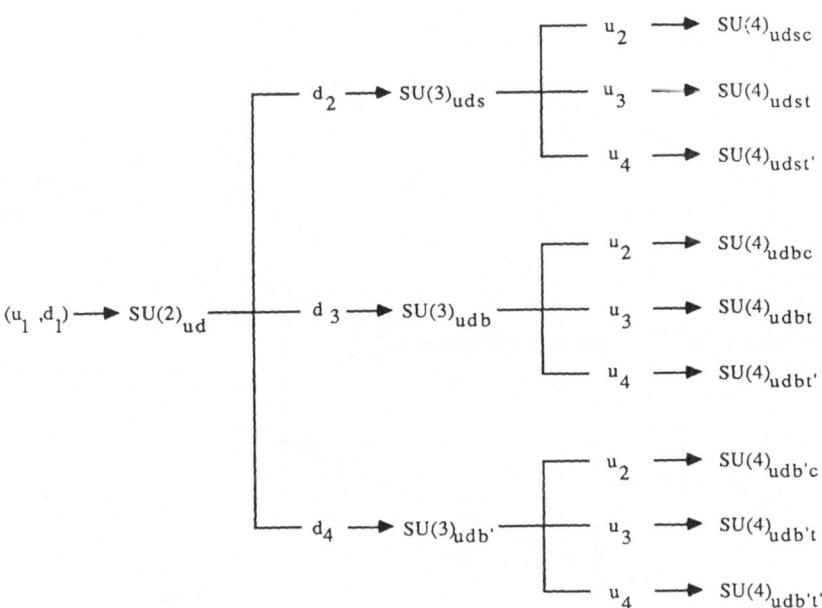

$$\begin{pmatrix} u & c & t & t' \\ d & s & b & b' \end{pmatrix} \equiv \begin{pmatrix} u_1 & u_2 & u_3 & u_4 \\ d_1 & d_2 & d_3 & d_4 \end{pmatrix}$$

Fig. 3.5 - $SU(3)_{uds}$ repeated with b and b' replacing s, and $SU(4)_{udsc}$ repeated with t and t' replacing c. Notice the above equivalence between old and new notation for flavours.

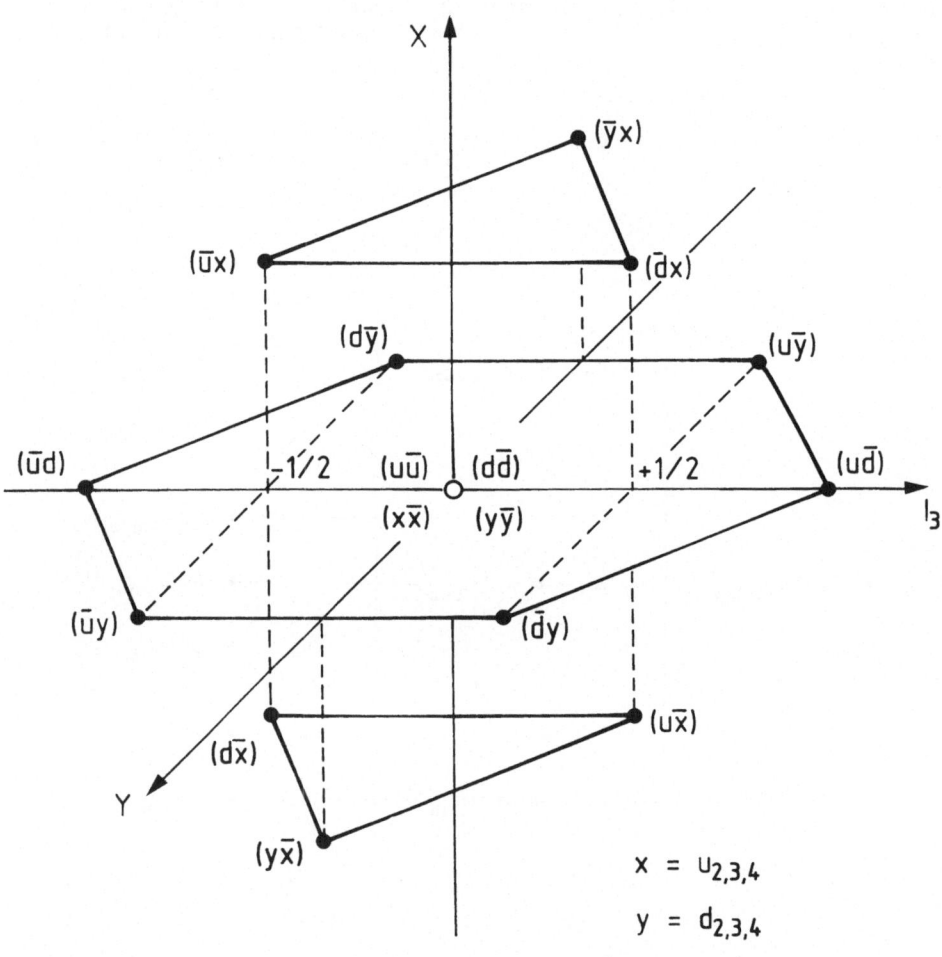

Fig. 3.6 - The SU(4)$_{udxy}$ structure for $J^P = 0^-$ and $J^P = 1^-$ meson states.

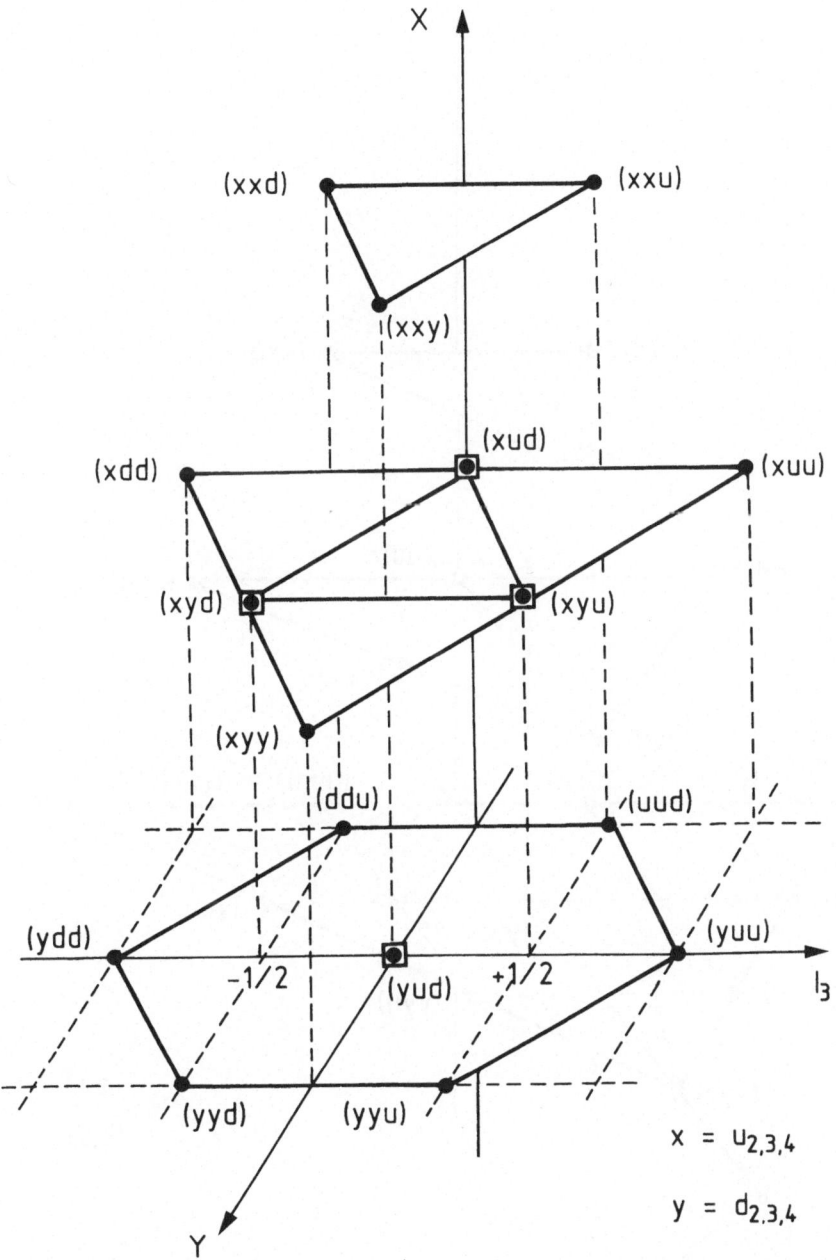

Fig. 3.7 - The SU(4)$_{udxy}$ structure for the $J^P = (1/2)^+$ baryon states.

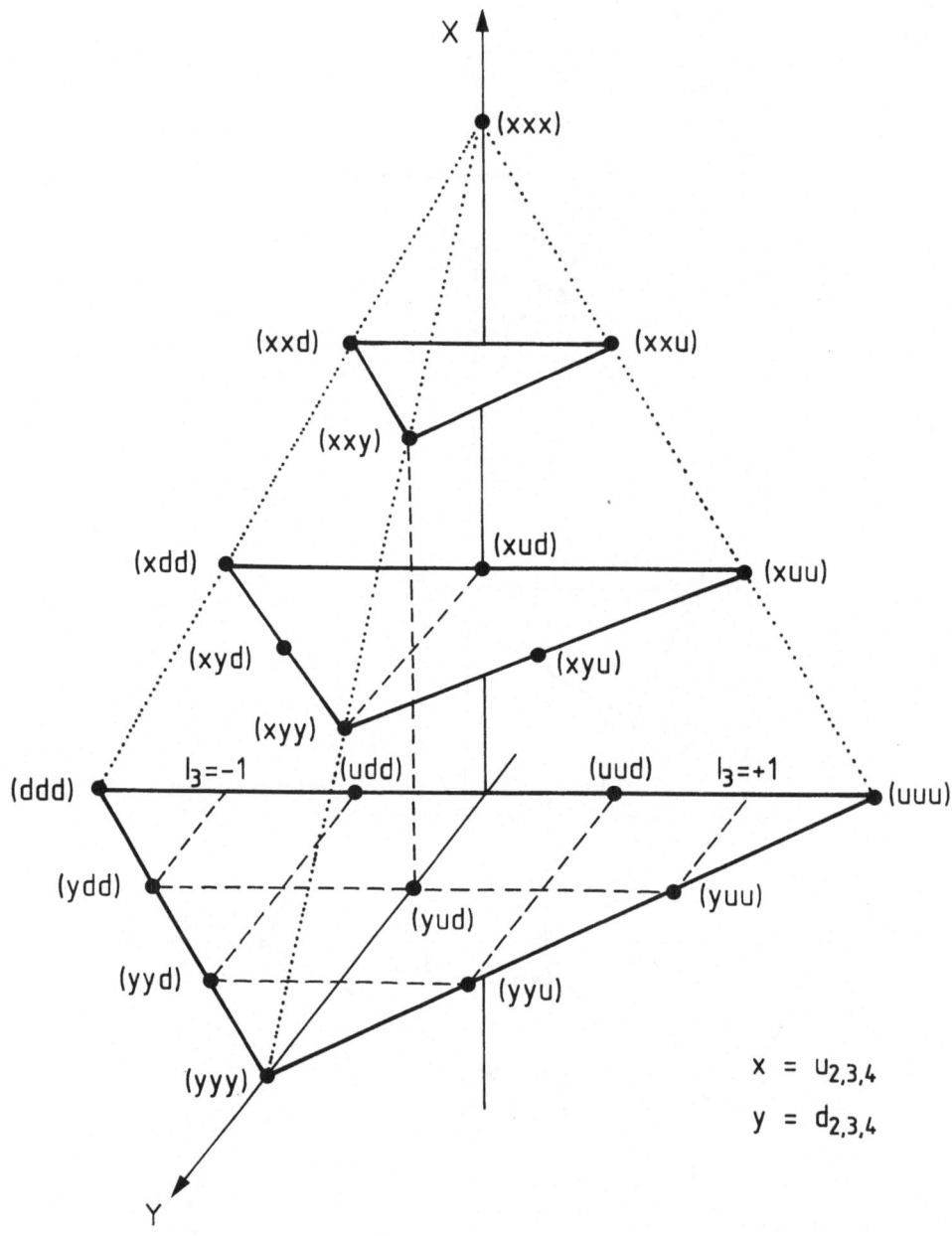

Fig. 3.8 - The SU(4)$_{udxy}$ structure for the J^P = (3/2)$^+$ baryon states.

4. THE ELOISATRON LAB

Since the very beginning, the conception of the ELOISATRON Project was based on a new way of tackling the problem of big projects. This new way is based on the following points:

1 - Do not start to build a new Laboratory if the work needed can be done in existing structures.

2 - Do involve the Industry in the implementation of the project. And this also means that the R&D work should be done in the laboratories of the industries in close connection with the physicists.

4.1 Hardware for the machine: ANSALDO, LMI, ZANON

At present the ELOISATRON Lab, as far as Machine Components are concerned, is housed in three places:

- at Genoa (ANSALDO)
- at Florence (LMI)
- at Vicenza (ZANON)

In these three places R&D work is being accomplished:

1 - at ANSALDO (Genoa): Superconducting technology to build powerful and high precision magnets is being studied .
2 - at LMI (Florence): Superconducting cables technology is going on .
3 - at ZANON (Schio-Vicenza): Powerful cryogenic plants and specific cryogenic problems are being studied .

Instead of having hundreds of people working in a new Laboratory, the ELOISATRON Project has hundreds of people working in these industries, each one highly specialized for the key work needed to build ELOISATRON.

Note that these industries are not only engaged in pure R&D work. They have been and are fully engaged in the implementation of the various phases of the ELOISATRON main steps: at present their major involvement is in the construction of the HERA superconducting dipoles and in the collaboration with the LAA Project.

In the following we will report briefly on the work being done by ANSALDO, LMI and ZANON, in order to give a closer view of the status of this ELOISATRON Lab.

As mentioned before, one of the main points of the ELOISATRON Project has been to stimulate the Italian industry to acquire the technologies neeeded for the implementation of the Project. One of these technologies, which has a special key function, is the series production of superconducting magnets.

Ansaldo has specialized in the design and contruction of the magnets, LMI (La Metalli Industriale) in the production of the superconducting cables and Zanon in the cryogenics.

The main characteristics of the superconducting dipoles that these industries are at present building are those of the 820 GeV proton ring of HERA. For completeness and in order to give to the reader a close view with the real performance of these ELOISATRON Labs we report in Table 4.1 the effective data for these dipoles.

Table 4.1 - HERA Superconducting Dipole Magnets

DESIGNED FIELD   6 Tesla

DESIGNED STORED ENERGY   756 kJ   (without yoke)

TYPE OF WINDING   2 Dipole coils

DESIGN NOMINAL CURRENT   4992 A

CABLE INSULATION 1 overlapped (58%) taping of kapton tape (0.025 mm thickness) + glass fiber tape (0.13 mm thickness) impregnated with B-stage epoxy resin wrapped around the cable with = a 3 mm gap between adjacent turns.

CONDUCTOR Rutherford-type superconducting cable built up of 24 individual twisted wires of 0.84 mm diameter; total dimension 10x(1.28;1.67) mm; trapezoidal shape.

TYPE OF COOLING   Helium bath

COIL WEIGHT   250 kg approx. (without collars)

The production capability is 250 dipoles per year. This is just a starting point. The production capability could be easily doubled or increased even more, if needed.

Fig.s 4.1 through 4.11 show several phases of the manufacturing of the HERA Superconducting Dipoles at the Ansaldo plant.

Fig.s 4.12 through 4.18 show some of the process equipment and facilities needed for the large-scale production of superconducting wires and cables at LMI.

The total length of cable produced for the HERA magnets is about 500 km with a production rate of 2 km per day. The critical current measured on a short sample, taken from the first 20 km produced, is 9010 Amps at 5.5 T, 4.6 °K.

Fig.s 4.19 through 4.22 show the cross section of various superconducting wires produced at LMI and the cable used for the HERA magnets.

Finally, Fig.s 4.23 and 4.24 show a cryostat in the vacuum test room and a welding test for the He shield of the HERA dipoles at the Zanon plant.

The **conclusive remark** is that this part of the **ELOISATRON Lab** is **fully operative** and **very efficient**. And, as discussed in the following Section, this is not all.

Fig. 4.1 - Winding machine (Ansaldo)

Fig. 4.2 Winding machine (Ansaldo)

Fig. 4.3 Pressing machine for coils and collars (Ansaldo)

Fig. 4.4   Pressing machine for coils and collars Ansaldo)

Fig. 4.5 Pressing machine for packing magnetic half-yokes (Ansaldo)

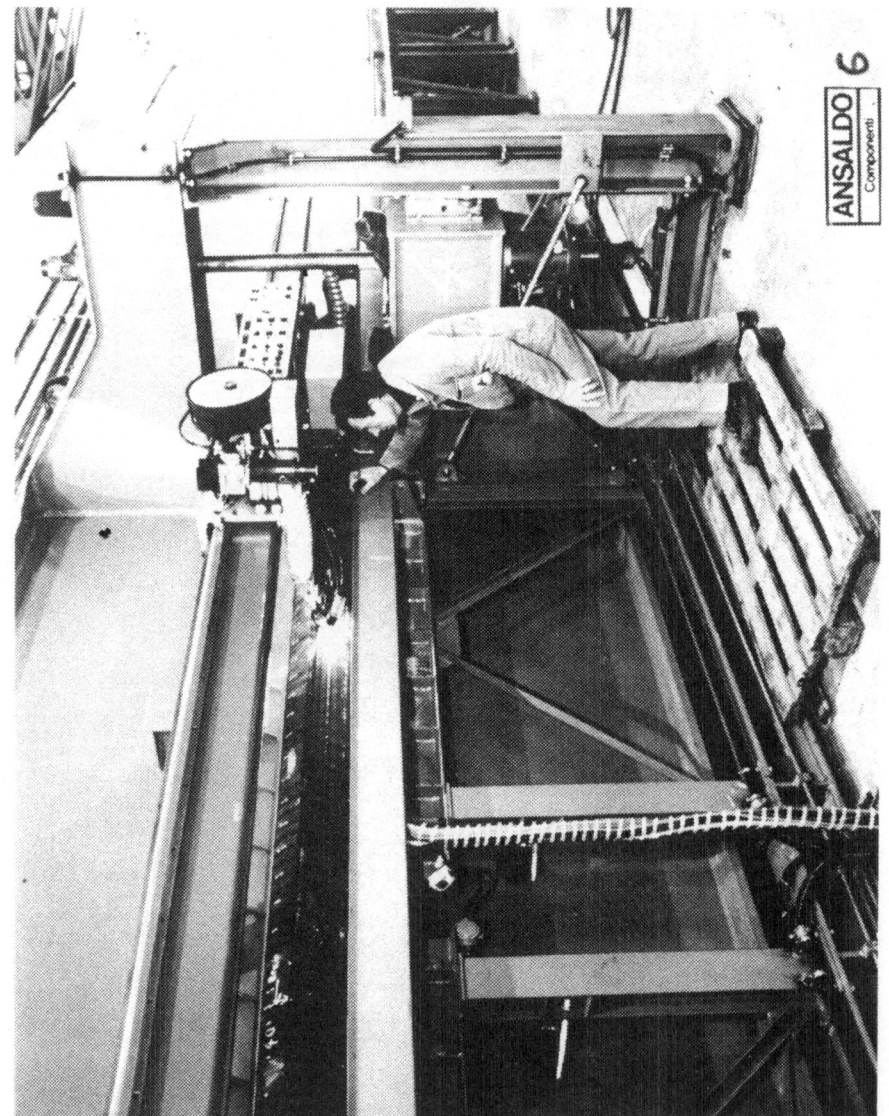

Fig. 4.6   Pressing machine to solder the magnetic yokes (Ansaldo)

Fig. 4.7   Manufacturing the ferromagnetic yokes (Ansaldo)

Fig. 4.8 Polymerized half yokes ready for assembly (Ansaldo)

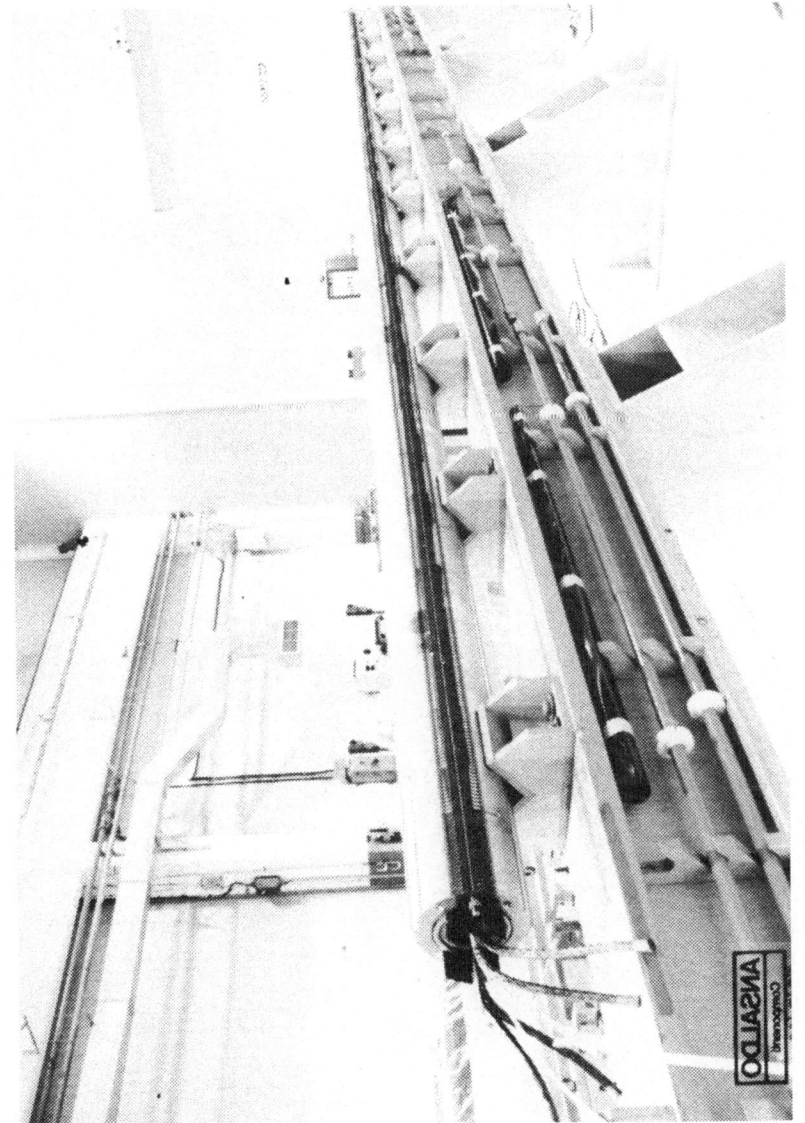

Fig. 4.9 Complete winding with collars ready for magnetic field measurements (Ansaldo)

Fig. 4.10 Coils with collars: detail of one end with the electrical connections (Ansaldo)

Fig. 4.11 Complete magnet before the introduction into the cryostat (Ansaldo)

Fig. 4.12 Continuous casting plant for the production of high purity copper used as the superconductor stabilizer (LMI)

Fig. 4.13 - Chemical surface treatment of the Nb-Ti rods (LMI)

Fig. 4.14 The 4000 tons extrusion press which provides the first step in the metalwork of the Cu/Nb-Ti composite billet (LMI)

Fig. 4.15 Set of draw-benches for drawing from large to small rod diameters (LMI)

Fig. 4.16   Multiple drawing bench for wire drawing (LMI)

Fig. 4.17   Assembly of the required number of superconducting strands into cables
according to customer's specifications (LMI)

Fig. 4.18　Overall view of the new workshop for the final phases of the superconductors production process (LMI)

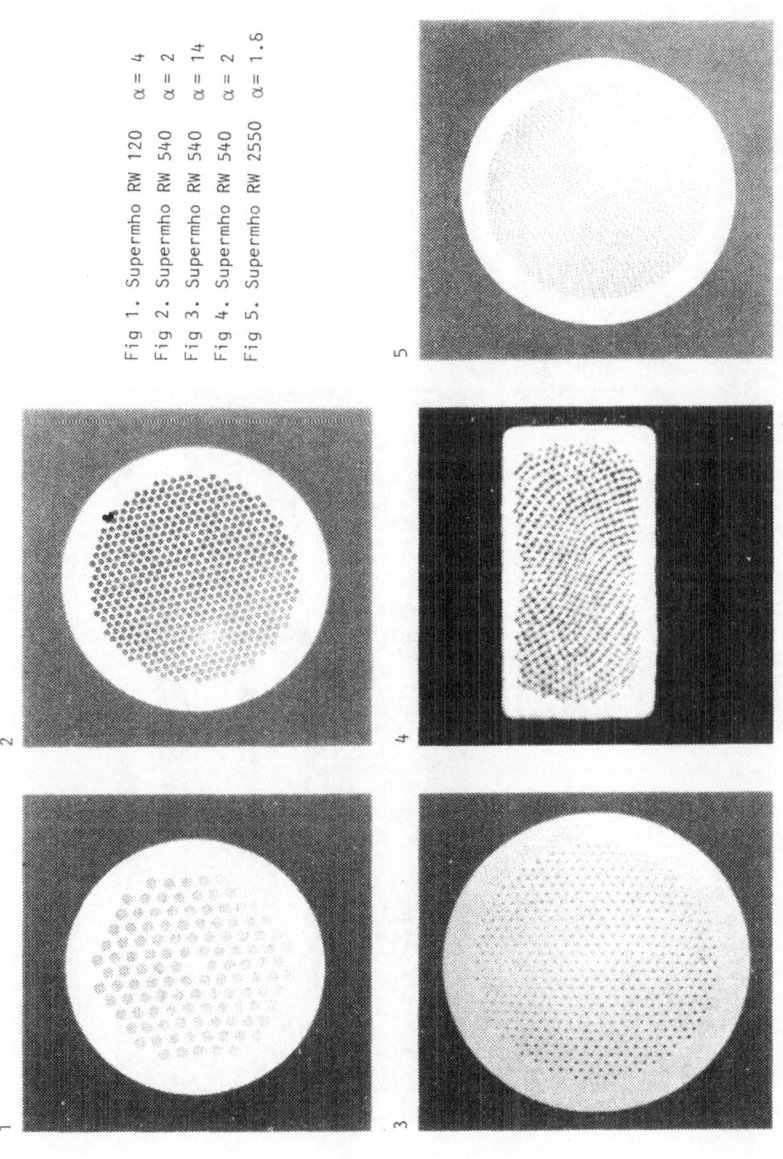

Fig 1. Supermho RW 120 $\alpha = 4$
Fig 2. Supermho RW 540 $\alpha = 2$
Fig 3. Supermho RW 540 $\alpha = 14$
Fig 4. Supermho RW 540 $\alpha = 2$
Fig 5. Supermho RW 2550 $\alpha = 1.6$

CS wire

HERA wire

Fig. 4.19 - Cross section of various superconducting wires (LMI)

39

CS cable

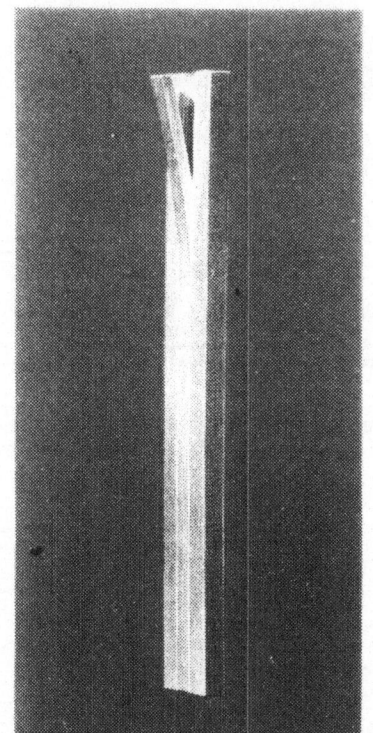

HERA cable

A 167 mm
C 128 mm
B D 10.00 mm

Fig. 4.20 - Superconducting cables (LMI)

Fig. 4.21 - The HERA cable (LMI)

Fig. 4.22   A coil of the superconducting dipole of the HERA proton ring (LMI)

Fig. 4.23 - A cryostat in the vacuum test room (Zanon)

Fig. 4.24 Welding tests for the He shield of the HERA dipoles (Zanon)

4.2 R&D for Detectors (LAA Project)

If a Supercollider would be there, no physicist would know how to do experiments. A multi-TeV accelerator with luminosities in the 10^{32}-10^{33} cm^{-2}s^{-1} range opens a totally unexplored field in detector technology. An intensive R&D programme for new experimental techniques is needed in order to proceed along the ELOISATRON line, and this is the main goal of the LAA Project. In the following we report briefly which are the main features of this programme.

In order to study the new and rare phenomena of multi-TeV physics it is necessary to aim at the highest possible luminosity (10^{33} cm^{-2}s^{-1}). As far as the detectors are concerned, the limiting luminositiy is:

$$L = <n> / (t_b \cdot \sigma_{pp})$$

where
- $<n>$ is the maximum average number of events per bunch crossing that the experiment can tolerate
- t_b is the minimum time between bunch crossings
- s_{pp} is the total (pp) cross-section.

The average number of events per bunch crossing $<n>$ must be one if the missing energy is to be used as a signature in event selection and analysis. Note that however $<n>=1$ still corresponds to a probability of 0.26 to have more than one event per crossing. The total (pp) cross-section at the 10% ELOISATRON Model energies can be derived from theoretical extrapolations to be at the level of 10^{-25} cm^2 (100 mb).

With this cross-section value and $<n>=1$, a luminosity of 10^{32} cm^{-2}s^{-1} can only be reached by having a very short bunch spacing, in the 100 ns($=10^{-7}$ s) range. The lower limit on t_b is determined by the time response of the detectors used by the experiment and by the occupancy of the detector elements. Correspondingly, the interaction rate would be of the order of 10^7 Hz.

A higher luminosity, 10^{33} cm^{-2}s^{-1}, with many events per bunch crossing, could be dedicated to rare phenomena with cross-sections

- $\sigma = 10^{-33}$ cm^2 = 1 nb $\quad\Rightarrow\quad$ production rate = 1 event/s
- $\sigma = 10^{-37}$ cm^2 = 0.1 pb $\quad\Rightarrow\quad$ production rate = 1 event/day.

Complex trigger configurations are needed to take data both with an acceptable rate and with a high efficiency.

Hard parton interactions will provide quarks and gluons which fragment into jets. They will also presumably produce new massive states decaying into intermediate bosons, leptons, quarks and gluons, therefore predominantly into jets.

The formidable task of particle identification in a high rate, high energy and high multiplicity environment, where precision measurements are needed, both in time and space, call for a dedicated R&D in order to solve the many problems still ahead of us.

The presently-available technologies for particle detection have to be pushed to their limits, new applications of old technologies have to be invented, and new technologies have to be developed.

For these purposes to be achieved the LAA Project must cover the following items:

- high-precision tracking;
- calorimetry;
- large area devices for muon detection;
- leading particle detection;
- application specific integrated circuits (ASIC);
- data acquisition and analysis;
- supercomputers and Monte-Carlo simulations;

- very high magnetic fields;
- superconductivity at high temperature.

Let us briefly review each one separately.

High precision tracking

i) Scintillating fibre microtracker

The goal is to investigate and improve the properties of narrow scintillating fibres (light yield and attenuation, radiation hardness), and to design the microtracker itself (assembly of multi-fibre bundles in shell structures), together with high-speed optoelectronics read-out chain.

ii) Vertex detector based on multidrift gas proportional tubes

The idea here is to develop, build and test a fast, triggerable and modular detector made of hexagonal thin carbon fibre tubes (30mm diameter, 70 signal wires per tube), together with its dedicated Large Scale Integrated (LSI) circuits for the on-tube electronics.

iii) GaAs microstrip detector

The properties of GaAs in term of performances and radiation hardness up to the Mrad level will be investigated in order to prove the feasibility of GaAs microstrips with integrated front-end electronics to be used, instead of Silicon microstrips, in future high-precision tracking devices.

Calorimetry

i) Electromagnetic calorimeter based on BaF_2 scintillator

A prototype of calorimeter with BaF_2 scintillators and photosensitive wire chambers for shower containment up to 100 GeV will be built. An exceptional time resolution, obtainable by detecting the fast component emitted from the BaF_2 crystals in the UV region, and very good position accuracy, due to the application of photosensitive wire chambers, will be tested.

ii) "Spaghetti" hadron calorimeter

The goal is to build a prototype of hadron calorimeter made of "spaghetti", i.e. scintillating plastic fibres, as active material and lead absorber. This calorimeter is of particular interest for its compactness and hermeticity. Its uniformity of response and expected compensation capability (e/h=1) will be tested.

Large area devices

i) Comprehensive study of muon detection in multi-TeV hadron colliders.
   Particular emphasis will be devoted to the construction problems. The goals are:

- construction of plastic chambers (LST) with new design read-out strips for higher accuracy and higher rate capability;
- study of alternative solutions such as plastic tubes filled with liquid scintillator and viewed by photomultiplier tubes;
- study of high-precision drift chambers;
- feasibility study of high field magnets of large dimensions.

ii) High precision alignment study

The development of an alignment and calibration system for large drift chamber detectors will be performed. The goal is the construction of precision tools to measure angles, lengths, straight lines and planarity with a precision of 20 mm over 5m (UV Lasers). Test drift chambers, using "cool" gases and newly developed electronics, will be constructed.

Leading particle detection

The idea is to integrate a detector design with an advanced accelerator design in order to detect and measure particles produced at very low angles with respect to the beams. The problems of producing radiation hard electronics and detectors will be investigated, together with mechanical problems. The goal is to produce a prototype microstrip system small enough to contain its associated electronics within 30 mm around the detectors. The required radiation resistance is a few Mrad.

Application specific integrated circuits

The development of front-end and read-out electronics will be performed. Microelectronics, consisting of highly integrated circuits in which detection, amplification and data processing are made, will be developed. The goals are:

- low noise amplifier design;
- fast signal comparator design;
- fast analogue pipelining, using CCD;
- fast on-chip digitization, using Sigma-Delta modulation;
- process compatibility development: electronics in Silicon detectors;
- evaluation and improvement of radiation hardness of electronics.

Data acquisition and analysis

Data acquisition components and new data acquisition architectures in both hardware and software, via several small and coordinate projects, will be developed. Particular emphasis will be put on the problems of second level trigger and experiment monitoring.

Supercomputers and Monte-Carlo simulations

The idea here is to develop and build a supercomputer for QCD dynamics lattice calculations. This supercomputer would be more powerful than the presently available CRAY. A programme chain, using our best knowledge on QCD, with particular emphasis on Heavy Flavour Physics in future high energy hadron colliders, will also be implemented.

Very High Magnetic Fields

The final dimensions of an LAA-like Detector to be operated in a multi-TeV Collider will depend critically on the value of the Magnetic Field in the VERTEX volume.

At present, if no progress is made in this area, the dimensions of the LAA-like Detector would be of the order of 30 meters along the beam and 20 meters transverse, for a total weight of about 30 ktons.

To gain an order of magnitude in the VERTEX Magnetic Field is of great value to reduce these dimensions. This is why the study of large volume Magnetic Fields is an essential part of the LAA Project.

To set up at CERN a laboratory for these studies would be extremely interesting, but very expensive.

However, such a laboratory exists in one of the most powerful and advanced Italian industries: ANSALDO. We are in close collaboration with this industry and therefore this item of the LAA Project will be developed along this line of close collaboration with ANSALDO.

It should be noticed that progress in the maximum value of B to be achieved for the vertex volume, implies progress in precision measurement in the VERTEX Detector. The two progresses are strongly coupled and this is a further reason to have R&D for High Magnetic Fields in Large Volumes as an integral part of the LAA Project.

SuperConductivity at High Temperature

Superconductivity at high temperature has two possible applications in the field of High Energy Physics. One is in the area of detector technologies; the other is in the production of superpowerful magnetic fields.

To be in close contact with those centres where such research is pursued is the purpose of the LAA Project. On the other hand, an Italian industry, LMI, is at the forefront of "classical" superconductivity, and it is now engaged in the field of SuperConductivity at High Temperature.

SuperConductivity at High Temperature will be followed by the LAA Project, with the purpose of establishing a series of contacts with the various research centres in the World and to stimulate close collaboration between European experts and CERN.

4.3 Geophysical studies

The problem of finding a site for such a large machine as ELOISATRON with appropriate geophysical characteristics has been investigated by ING (Istituto Nazionale di Geofisica) in a 1985 study. The results of this study are summarized in Fig. 4.25. At least seven sites, in Italy, are suitable for the ELOISATRON ring.

4.4 Civil Engineering studies

The civil engineering problems have been investigated in a specialized Working Group. Depending on the flatness of the ground, the machine could be built partly as prefabricated tract (Fig. 4.26) and partly in a tunnel (Fig. 4.27). The cost of this solution has been estimated to be not higher than other possible solutions, such as the pipe-line, once all needed items are included. Fig. 4.28 show the schematic plan view of an experimental hall.

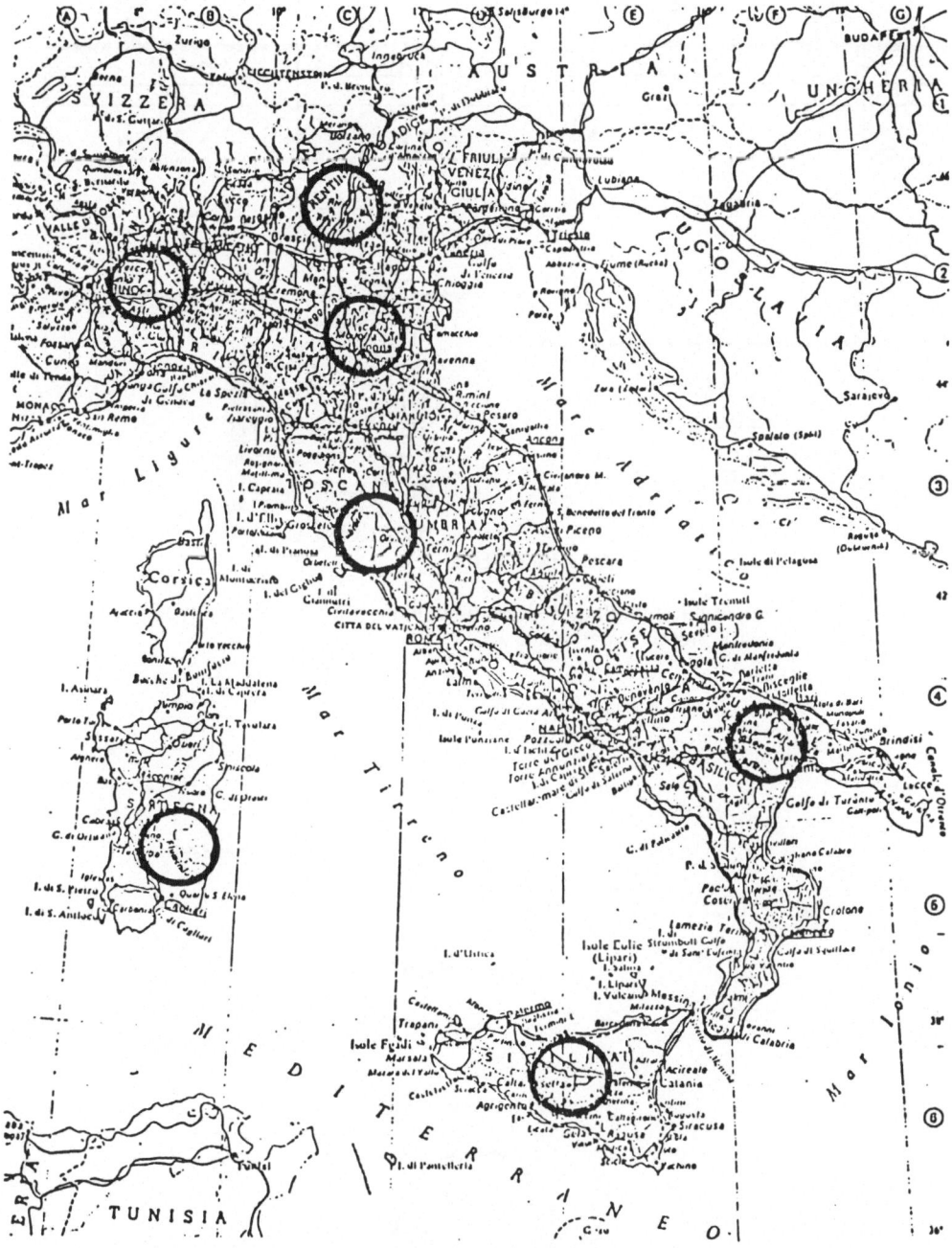

Fig. 4.25 - Possible sites for the large ELOISATRON ring. [Study made by the Italian National Institute for Geophysics (ING), 1985]

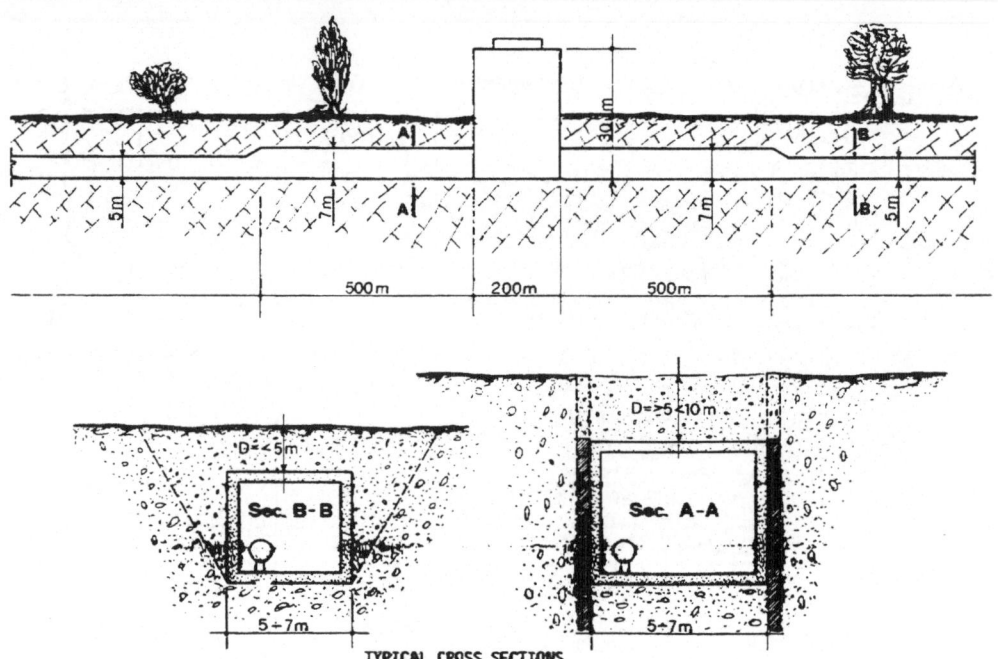

TYPICAL CROSS SECTIONS

Fig. 4.26 - The schematic longitudinal section of an experimental hall and typical cross sections for prefabricated tracts.

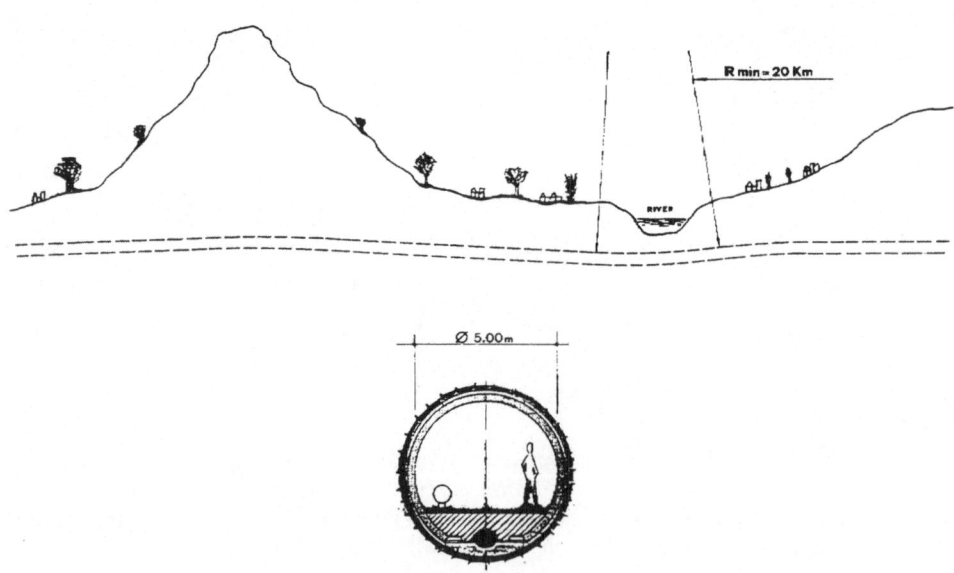

Fig. 4.27 - A typical longitudinal cross section of the ELOISATRON tunnel.

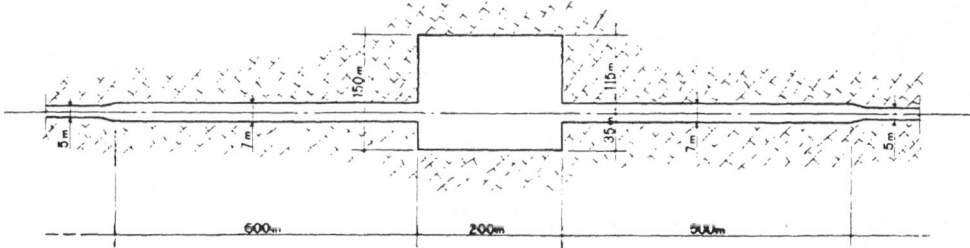

Fig. 4.28 - The schematic plan view of an experimental hall.

4.5 Workshops

Four Workshops have been held at the Ettore Majorana Centre for Scientific Culture in Erice in the years 1986-1987 as listed below:

◊   Seminar on New Techniques for Future Accelerators;
◊   Workshop on Vertex Detectors: State of the Art and Perspectives;
◊   3rd Workshop - New Techniques for Future Accelerators:
       RF and Microwave Systems;
◊   4th Workshop - Very High Energy Proton-Proton Physics.

Other workshops are scheduled.

4.6 The 10% Model

The 10% Model is a necessary step towards ELOISATRON. Its main data are:

◊ **1600 Dipoles;   length ≈ 20 m, B $^{\underline{a}}$ 8 Tesla**
◊ **400 Quadrupoles**
◊ **Total length (magnetic) ≈ 20 km**
◊ **Estimated cost : 1000 GLiras**

Notice that in order to build this machine only a moderate development of the technologies presently available to Italian industry is required. In fact, as mentioned above, series production of 10 m long, 7T dipoles is already possible.

If the 10% ELOISATRON Model is to be ready without a long break between its start up and the end of the LEP and HERA operation, its implementation has to be done in such a way that the most efficient structures and facilities must be incorporated, wherever they are.

One obvious possibility, but by no means the only one, would be to build the 10% ELOISATRON Model in the LEP-tunnel at CERN. Fig. 4.29 shows that there would be no conflict between the LEP programme and the 10% ELOISATRON Model. Other possibilities should be investigated because the crucial point is that, no matter where the 10% ELOISATRON Model is going to be built, its implementation has to be linked with the full scale project.

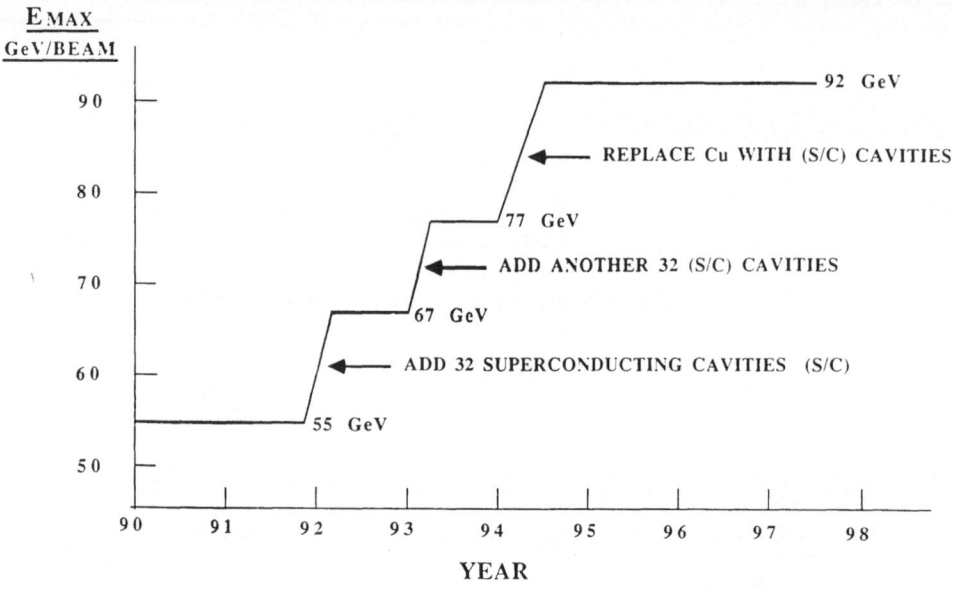

Fig. 4.29 - Possible time evolution of the LEP programme.

5. CONCEPTUAL DESIGN OF THE FULL SCALE ELOISATRON PROJECT

5.1. Introduction

The main part of the ELOISATRON Project is a 2x100 TeV proton-proton collider of the maximum possible luminosity. With a circumference of 300 km the bending field of the collider will have to be about 10 T. The total bending length will in that case be 209.6 km/ring. The rest of the circumference will be used for focusing quadrupoles, correcting magnets and other auxiliary magnets,acceleration cavities, beam instrumentation, injection and extraction systems, and, most important of all, the insertions for the interaction regions. It is proposed to group these in two major groups, each group occupying about 15 km. The facility will thus have the shape of a race track similar to the arrangement already proposed during the LSR study about 10 years ago and also adopted in the SSC design. However, the long sides will contain some bending, so they are not quite straight. Within each group it could be envisaged to have three interaction points, as indicated in Fig. 5.1. However, other solutions should be analysed as well, for instance having many interaction points like a string of pearls with many detectors almost continuously along the 15 km.

The main rings will be fed from a cascade of synchrotrons (most likely three in succession) which again will be fed from a linear accelerator. This is also shown schematically in Fig. 5.1. It should be noted that the last of the injector synchrotrons is, in this sketch, assumed to be placed in the same tunnel as the main rings, and therefore does not appear as a separate ring. Other arrangements are possible as will be mentioned in Section 5.4.

In the following are some comments on the main subsystems.

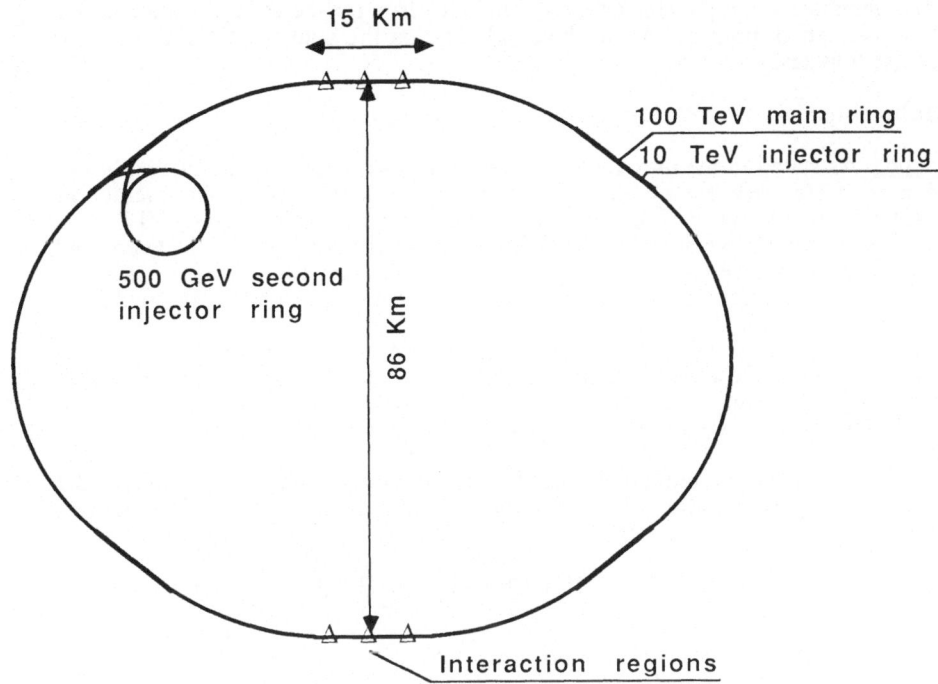

Fig. 5.1 - Possible ELOISATRON layout

5.2. Insertions

The main purpose of the experimental insertions is to transform the beam properties such that the crossing regions become very intense event sources and still maintain accessibility for detectors and otherwise make the facility flexible and convenient to use.

These are often contradictory requirements, and compromises must be found, perhaps leading to different solutions for different interaction regions. The final arrangement will depend much on the detector development over the next decade or so, for instance, on how beam-focusing systems can be integrated into detector components. It seems, however, likely that under any circumstances interaction regions of very high luminosity will be wanted, leading to the requirement of the strongest possible focusing near the interaction points. For instance, a β of about 1m or less will be required to get luminosities above 10^{33}cm^{-2}s^{-1}.

In order to get an idea whether this is feasible and practicable simple scaling from the corresponding LHC arrangement can be performed. Only the triplets at either side of the interaction point will be considered since these are the most critical elements. The scaling gives a 75 m long triplet consisting of four 15 m long quadrupoles with 5 m between each. The gradient of the quadrupoles became 350 T/m. The free space from the interaction point to the first quadrupole is 30 m and the system is symmetric about the interaction point. With such an arrangement the b at the crossing point can be made 1.25 m and the maximum b in the triplet becames 8250 m.

This illustrates that a β of the order of 1 m is realistic for the interaction point, but much study is needed, both on the machine side and on the detector side, to arrive at optimum insertion designs.

5.3. Main rings

5.3.1. Lattice. As stated in the introduction the main rings should provide two colliding proton beams of 100 TeV each with the highest feasible luminosity in the interaction regions. A total circumference of 300 km will require a bending field of about 10 T. This seems a safe projection of what can be obtained on an industrial scale in some years, and 10 T is therefore the assumed figure for the present study. If in the future higher fields will be proven feasible and practical, the total circumference can be correspondingly smaller.

Considerable further analysis and computation is needed to arrive at an optimum lattice for the machine. However, simple scaling from existing designs of similar smaller projects give a good starting point for such analysis. An example of such scaling is given in Table 5.1. The numbers are for one ring only.

In this example the number of dipoles per period is fairly high, and if the construction techniques will allow about 20m long magnets, the more convenient number of 8 dipoles per period could be adopted.

TABLE 5.1 - LATTICE PARAMETERS

Length of period	200 m
Phase advance per period	$\pi/3$
Betatron wavelength	1200 m
Bending angle per normal period	4.7 mrad
Number of normal periods	<1332
Number of quads per period	2
Effective length of each quad	13.6 m
Total nº of quadrupoles	2664
Maximum dipole field	10 T
Number of dipoles per normal period	12
Effective dipole length	13.1 m
Total nº of dipoles	15984
Bending radius	33356 m

5.3.2. Performance evaluation and performance-related parameters. Once the energy has been fixed, the next most important parameter for the ELOISATRON Project is the luminosity given by

$$L = N_p^2 f_b / (4\pi\sigma^2) = N_p^2 f_b \gamma / (\beta^* \epsilon)$$

Here N_p is the number of particles per bunch, f_b is the average bunch frequency, σ is the r.m.s. beam radius at the crossing points, and ϵ is the normalized emittance defined by

$$\epsilon = 4\pi\gamma\sigma^2 / \beta^*$$

where β^* is the beta value at the crossing point.

These equations illustrate where to put emphasis to get high luminosities. There are, however, a few limitations; the most important ones for the design can be listed as follows:

i)   The detectors have difficulty in discriminating between events if the bunches cross more frequently than once every 25 nsec. This is not a hard limit, and may go down. However, for the time being this is taken as giving the following limit on the bunch frequency

$$f_b \leq (25 \times 10^{-9} \text{ s})^{-1} = 40 \text{ MHz}$$

or 7.5 m between bunches.

ii)   The beam-beam tune shift parameter must be kept smaller than a given number. This parameter is given by

$$\xi = N_p r_p / \epsilon$$

where r_p is the classical proton radius. For pp interactions the total tune shift over all interaction regions should be less than 0.01, which gives

$$\xi < 0.01/6 = 0.00167$$

for the case of six interaction regions.

iii)   It is in practice very difficult to make β^* very small at very high energies. For SSC at 20 TeV it is assumed that β^* can be made 0.5 m. For 100 TeV we assume optimistically that the smallest we can make it is $\beta^* \sim 1m$.

iv)   Synchrotron radiation at 100 TeV becomes a serious load on the cryogenic system of the magnets. It does not seem reasonable to assume more than say 2 W/m of the vacuum pipe, and even in that case it will be very desirable to try to trap much of this on a separately cooled screen inside the vacuum chamber.

v)   The normalized emittance from present day proton accelerators is $5\pi \times 10^{-6}$ or larger. However, since iv) above will be imposing a limit on the circulating current far below the capability of an accelerator, it is assumed that the emittance can be correspondingly reduced to say

$$\epsilon \approx 0.75\pi \times 10^{-6} \, m$$

It is difficult to put all these conditions into simple mathematical formulae. It should also be remembered that some of the conditions are rather flexible and may change as the various technical solutions are being studied. Examples, however, have been worked out and some iterations have resulted in some tentative numbers as presented in Table 5.2.

TABLE 5.2 - SOME TENTATIVE PERFORMANCE PARAMETERS

FOR ELOISATRON

Energy per beam	100 TeV
Number of bunches	39600 per beam
b-value at interaction point	1.25 m
Normalized emittance	$0.75\pi \times 10^{-6} \, m$
r.m.s. beam radius at inter. point	$1.25 \times 10^{-6} \, m$
Circulating current	16.43 mA
Particles per bunch	2.56×10^9
Beam-beam tune shift (with 6 active crossings)	1.67×10^{-3}
Bunch spacing	$25 \times 10^{-9} \, s$
Stored beam energy	$1.623 \times 10^9 \, J$
Luminosity	$0.91 \times 10^{33} \, cm^{-2}s^{-1}$
Energy loss per turn due to synchrotron radiation	23.34 MeV
Radiated power (per beam)	385 kW
Power per unit length of one beam	1.89 W/m
Transverse em. damping time	1.2 h

The most important parameter is, as already stated, the luminosity, and one might speculate a little on how this can be increased without violating the other constraints. One simple way would be to by-pass two out of the six interaction regions during the highest luminosity runs. This would permit a reduction of emittance (if possible) to $0.5\pi \times 10^{-6}$ m and an increase in luminosity by a factor 1.5 to 1.4×10^{33} cm^{-2}s^{-1}. One might also speculate about increasing the circulating current, but this can only be done after the handling of the synchrotron radiation is better known (perhaps as a later improvement programme). Further, how to dump 100 TeV protons with stored energies in the gigajoules range must be studied.

5.3.3. Aperture. The physical aperture of an accelerator constitutes a hard limit to the beam in this respect that it cannot be changed once the machine has been built. For this reason the aperture requirement must be estimated on the basis of conservative assumptions.

In the next chapter arguments will be given for an injection energy of 10 TeV. However, since it is equally likely that the final choice may be about 5 TeV the latter figure will be assumed for the aperture estimates.

In section 5.3.2 a normalized emittance of $0.75\pi \times 10^{-6}$ m was assumed for performance estimates. This may, however, be difficult to achieve, in particular in the early operational phase. For aperture determinations we therefore assume the more commonly chosen figure of $5\pi \times 10^{-6}$ m.

If it is assumed that no physics will be done at energies below 2x20 TeV, the beam should not need more room than 4 σ at injection. This would give 8 σ at 20 TeV, and therefore good beam conditions from this energy and upwards.

The closed-orbit deviations will depend on the quality of the magnets, the beam observation and the correction system. In the lack of more detailed information LEP values will be scaled. Results from this kind of evaluation are given in Table 5.3.

TABLE 5.3 - APERTURE EVALUATIONS

Max. beam radius (4σ) at injection	2.59 mm
Corrected closed-orbit error	± 4 mm
Radial extent of bucket	± 0.17 mm
Needed vertical aperture	± 6.6 mm
Needed horizontal aperture	± 6.8 mm
Aperture of vacuum chamber	± 7.5 mm
Aperture of coil	± 12.5 mm

Note should be taken of the at present big uncertainty in the space needed between the vacuum chamber and the coil. This space will be needed for vacuum chamber wall, correction windings, and possibly insulation etc. to keep the vacuum chamber at 4.6 °K while the coil must be at 1.8 °K in case niobium-titanium coils are chosen.

If we want to have space for 8σ of this beam in the insertion triplets (pg. 95), their aperture must be made larger than in the rest of the machine. However, a more realistic scenario is to assume that the low-b insertion will only be fully activated when an emittance of $0.75\pi \times 10^{-6}$ m has been achieved. This will then give $8\sigma=2.5$ mm in the maximum β point at 20 TeV, i.e. the same aplerture can be assumed in the insertion triplets as in the rest of the machine.

5.3.4 Comments on some of the technical components

5.3.4.1 - Magnets. Some remarks can be made about various possibilities:

i)   One can use niobium - titanium as winding material. By going to 1.8 °K one can probably reach 8 - 10 T. However, the highest possible field may not correspond to a cost optimum, and this will eventually have to be studied.

ii)   If Nb_3Sn is chosen 10 - 12 T may be reached at normal liquid helium temperature. Much development on metallurgical processes for wire production and on the thermal treatment of the insulated windings is needed to arrive at a satisfactory design. Such work goes on at many places in the world, Europe included.

iii)   With the dramatic development recently of "warm" superconducting materials, there is hope that such materials will become available for accelerator magnets in the future. This must be followed closely. It must however also be remembered that such solutions will introduce new problems. For instance, the advantage of cryogenic pumping is lost and distributed pumping of very small transvere dimensions will have to be developed.

iv)   Two possibilities exist for the magnetic circuits for the two rings. The conventional approach is to make the two rings magnetically independent, i.e. each magnet has its own return yoke. Another approach is to let each magnet gap take the return flux from its neighbour. This is called the "two-in-one" approach. This has two advantages. The first one is that space is saved. The LHC has adopted this version in order to arrive at a design that will fit in the LEP tunnel. The second advantage is that this solution leads to some savings in superconducting material, steel and cryostat.

The main disadvantage of the "two-in-one" solution is the restriction it imposes on operational flexibility. It becomes difficult to tune and operate the rings separately. This has led the SSC Design Group to choose the independent rings. The same is recommended for the ELOISATRON, in particular, since the savings mentioned above are very small compared to the cost of the whole project.

However, it should be mentioned that it may still be desirable to put the two magnets in the same cryostat. This will introduce a little operational dependance, but much less than if they are magnetically coupled.

5.3.4.2 - Radiofrequency systems. The radiofrequency system is needed for acceleration, replacement of synchrotron radiation losses, and keeping the beam stable and longitudinally bunched. Very likely a 200 MHz system based on the HERA development will be appropriate.

5.4. Injection

In order to obtain an estimate for the aperture requirements an injection energy of 5 TeV was assumed. There are, however, reasons to investigate the possibility of a higher injection energy, for instance 10 TeV. The main advantage comes from the effect of the

persistent currents in the magnet windings of the main ring. These effects become more serious the larger the ratio between maximum and minimum field. A ratio of twenty is known to be bothersome. A ratio of ten will ease the situation.

The injector will be a very low duty cycle machine. Since it seems wasteful to keep such a machine at low temperature all the time, it is suggested to make the injector, or rather the whole injector chain, with normal magnets. With such a choice it is natural to study the possibility of having the injector in the same tunnel as the main ring, as this would require about 1 T bending magnets, a natural field for conventional magnets. Since this machine cannot be permitted to go through the interaction regions, by-passes will have to be provided. The savings in the civil engineering may nevertheless be substantial and justify this approach. This accelerator needs not have fast-cycling capability.

The next ring in the injection chain can have an energy of about one twentieth of the 10 TeV injector, i.e. about 500 GeV. The ring may therefore be rather similar to the SPS at CERN. Its injector, at 25 GeV, may be rather similar to the CERN PS or the Brookhaven AGS, but the whole chain with more modern designs. The injection chain has been sketched in Fig. 5.2.

The above remarks about the injection system are possibly based on unduly conservative assumptions. If the development of "warm" superconductors leads to feasible and economical magnet designs for temperatures at or above liquid nitrogen the remarks about the relative advantages of normal conducting magnets for the injector chain will no longer be valid. In such a case the 5-10 TeV injector ring will be taken out of the main-ring tunnel and made as a separate, relatively small, injector ring. The other lower energy injectors will also be made of superconducting magnets. However, even the most superficial design of such a system is premature with present-day knowledge. This may, however, change fast.

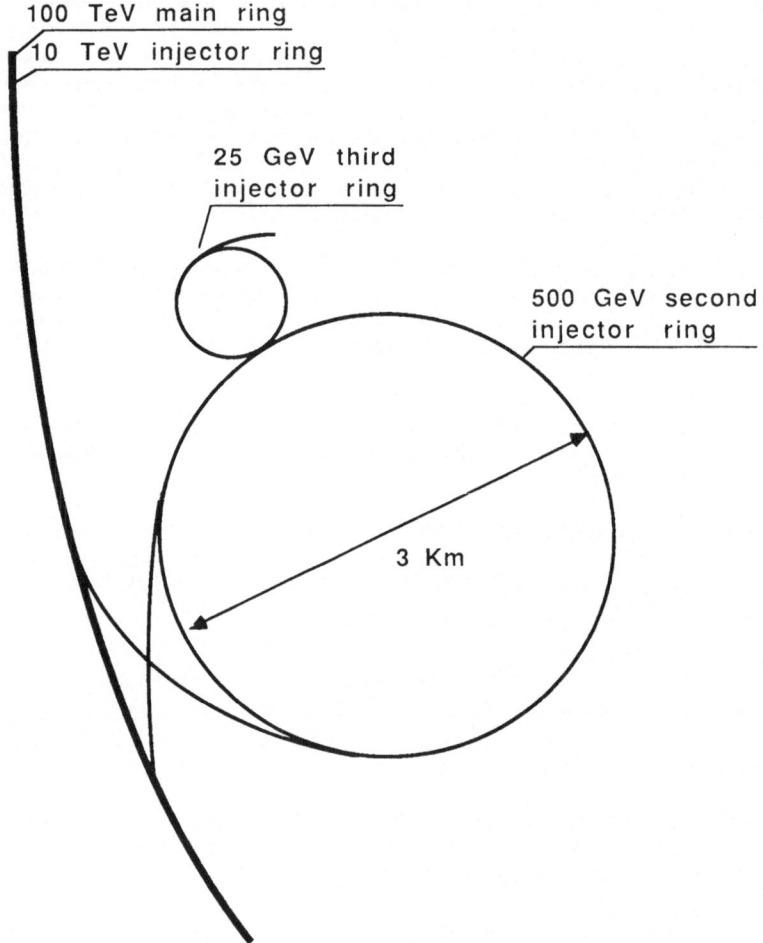

Fig. 5.2 - ELOISATRON injection.

5.5 - FURTHER COMMENTS

Two comments are needed in connection with the problem of the full scale ELOISATRON project.

i) R&D for superconductivity at high temperature

As mentioned before, superconductivity at high temperature presents possible applications of extreme interest for High Energy Physics both in the area of detector technologies and in the production of superpowerful magnetic fields.

An intensive R&D programme in this field is therefore needed to ascertain the practical relevance of high temperature superconductivity for the full-scale ELOISATRON Project.

ii) Executive design of the full-scale ELOISATRON

A detailed study of the full-scale ELOISATRON is still to be done. This will be better done, in fact, once the possible implications of the high temperature superconductivity on the full-scale design will be more fully understood.

However, it is important to notice that all the studies performed so far at the conceptual design level, point to the fact that the full-scale Eloisatron is feasible by extrapolating present technologies with an intensive R&D.

Therefore, there is no point in spending a large amount of money (10 GLiras) for theoretical studies of the full-scale Project now. The R&D needed for it must be, on the contrary, an intrinsic part of the effort put in the 10% ELOISATRON Model.

6. WHERE THE ELOISATRON PROJECT HAS BEEN PRESENTED AND DISCUSSED

The ELOISATRON Project and its different phases of development have been presented and discussed in many national and international forums, the world over.

A relevant factor in this context is the presentation of the Project to the Italian Governement via the CIPE (Interministerial Committee for Economical Planning), which has undertaken a preliminary step of financing the ELOISATRON Project in the 5-years INFN plan (1984-1988) with the sum of 50 GL (~ 75 MUS$).
In the following the scientific presentations are listed:

-	**VERSAILLES Summit Meeting (France)**	**June 1982**
-	**SANREMO (Italy)** Nobel and Galileo Celebrations	**2 May 1983**
-	**ROME (Italy)** Nobel and Galileo Celebrations	**9 May 1983**
-	**CASTIGLIONE DELLA PESCAIA (Italy)** 2nd INFN Meeting on Advanced Detectors	**2 June 1983**
-	**ERICE (Italy)** Ettore Majorana Centre for Scientific Culture International School of Subnuclear Physics 21st Course	**8 August 1983**
-	**COMO (Italy)** 3rd International Conference on Physics in Collisions	**2 September 1983**
-	**WASHINGTON (DC,USA)** National Science Foundation Working Group Meeting on High Energy Physics	**3 October 1983**
-	**SALERNO (Italy)** University of Salerno	**4 December 1984**
-	**POTENZA (Italy)** University of Potenza	**3 March 1985**

- **BOLOGNA (Italy)** 14 March 1985
  University of Bologna

- **LIVERMORE (California, USA)** 9 April 1985
  Lawrence Livermore National Laboratory

- **BOULDER (Colorado,USA)** 11 April 1985
  University of Colorado - Conference on
  World Affairs

- **CASTELGANDOLFO (Italy)** 10 July 1985
  INFN National Conference

- **ERICE (Italy)** 13 August 1985
  Ettore Majorana Centre for Scientific Culrure
  International School of Subnuclear Physics
  23rd Course

- **ERICE (Italy)** 3 September 1985
  Ettore Majorana Centre for Scientific Culture
  International Conference on Advanced Solid
  Earth Geophysics

- **TRIESTE (Italy)** 8 October 1985
  21st Congress of the SIF
  (Italian Physical Society)

- **GENEVA (Switzerland)** 11 November 1985
  CERN - Senior Staff Consultative Committee

- **STRASBOURG (France)** 20 November 1985
  Council of Europe - Parliamentary Assembly
  Committe on Science and Technology

- **BOLOGNA (Italy)** 15 December 1985
  Academy of Sciences - Invited Plenary Address for
  the opening of the academic year

- **MOSCOW (USSR)** 29 January 1986
  Presidium of the USSR Academy of Sciences

- **CAGLIARI (Italy)** 15 March 1986
  University of Cagliari

- **GENEVA (Switzerland)** 19 June 1986
  CERN - RECFA

- **ERICE (Italy)** 12 August 1986
  Ettore Majorana Centre for Scientific Culture
  International School of Subnuclear Physics
  24th Course

- **ERICE (Italy)** 23 September 1986
  Ettore Majorana Centre for Scientific Culture
  INFN ELOISATRON Project
  Workshop on Vertex Detectors

- **BEIJING (China)** 18 October 1986
  Academia Sinica and University of Beijing

-	CATANIA (Italy) INFN National Laboratory for Southern Italy	5 November 1986
-	MOSCOW (USSR) Presidium of the USSR Academy of Sciences	15 January 1987
-	ROME (Italy) INFN Symposium	9 June 1987

7.   CONCLUSIONS

Past experience and present knowledge call for a big jump in the Energy level to be achieved in Subnuclear Physics.

The most effective approach towards this goal is the ELOISATRON Project which, as shown in table 1-1 (page 10), has already realized a number of important steps.

Let us summarize a few basic achievements:

i)   a powerful collaboration between Industry and Research Laboratories.

ii)   the first series production (at an industrial level) of technically highly developed superconducting magnets.

iii)   an intensive R & D programme for detectors (LAA) to be used in MultiTeV Colliders.

The next step is the 10% ELOISATRON Model, whose final goal is the most powerful superaccelerator in the world: (100+100) TeV.

The activity being pursued by the Italian Industries (Ansaldo,LMI, Zanon, Laben), in close collaboration with the most advanced Research Centers, will permit the involvement of Italy in the next big step towards the MultiTeV SuperCollider and its Physics achievements. In this contest it should be emphasized that a few more steps in the superconducting magnet are required in order to achieve the best performances with present technologies. In fact we have been able to reach a level of technological accomplishment which is higher than the one aimed at. In other words, the 10% ELOISATRON model can be started practically with our present level of technological goals, already achieved (i.e.superconducting magnets 10 metres long with 8 Tesla).

During the construction of the 10% model an intensive R & D programme should be conducted so as to follow the spectacular developments of high temperature Superconductivity. However, even if this new, high technologically advanced field, does not produce effective results in the next five years, the R & D using present-day technologies would grant the feasibility of the full-scale ELOISATRON Project. In fact, 10 Tesla magnets are certainly within reach if the R & D programme in standard Superconductivity is carried out in the next few years.

The ELOISATRON Project, realized along the outlined programme of collaboration between Research Centres and Industries together with the effective participation of Third World Scientists, represents the greatest challenge for Subnuclear Physics to push forward our present knowledge in Science and Technology with a highly qualified component of International Solidarity. This is very much needed in order to halt the present trend of a continuously increasing gap between the NORTH and the SOUTH.

CRITERIA FOR STEPS TO HIGHER ENERGIES

Herwig Schopper
CERN
1211 Geneva 23
Switzerland

I.   INTRODUCTION

In discussing or proposing a new generation of high energy accelerators a number of arguments have to be taken into account. Physics
motivation should, of course, be the main criteria, but machine technology
is important and the development of detectors is getting an increasing
weight.   Financial boundaries cannot be neglected and political, ecologi-
cal and sociological issues might also play a certain influence.

II. PHYSICS MOTIVATION

Physics questions are, of course, of major importance in deciding on
a new accelerator. Theoretical guidance is needed and different predictions from various theories may be put to experimental test. However,
looking back in history, one notices that very often physics questions
considered in the proposal for a particular accelerator were not answered
by this machine, but on the contrary many machines have discovered new
phenomena which were not mentioned at all in the original proposal.

Two machines have certainly accomplished the task for which they were
built. The BEVATRON at Berkeley was constructed with the aim to observe
the antiproton, and the SPS at CERN was converted into a proton-antiproton
collider in order to discover the W and the Z particles.   Some examples
for phenomena which were found unexpectedly at proton machines are the
muon neutrino and the neutral currents at the AGS in Brookhaven and the PS
at CERN, the large transverse momentum dependence indicating partons
inside the protons at the ISR, the EMC effect at the SPS, the Y particle
at Fermilab. The discovery of partons by inelastic electron scattering at
SLAC was as unexpected as the discovery of the Charm-quark at the AGS and
SPEAR.   Gluons have been discovered at PETRA, although they were never
mentioned in the original proposals and more recently B-mixing was
established at DORIS.

In particular predictions of particle masses are vague or even wrong,
which was true for the Charm-, Beauty- and Top-quark, and the exceptions
are the W and Z^0 particles.

Most reliably seemed to be the limits imposed by violation of
unitarity. New physics was predicted for weak interactions below the
Fermi limit at 250 GeV, and indeed the W and Z^0 were found well below that
limit. For the electroweak interaction such a limit should exist around 1

TeV and one might well expect that the mechanism of symmetry breaking, e.g. the Higgs mass could be below this limit.

In conclusion one might state that the lesson we have learned from the past is that guidelines by theory are important, but that one should not rely entirely on them. Indeed, the most exciting issue is to discover new phenomena not predicted by theory in a newly accessible energy range. Nature is very imaginative and always has surprises at hand.

One should also not forget to test the validity of quantum mechanics and relativity. Going from dimensions of 1 cm to 10^{-8} cm forced us to replace classical physics by quantum mechanics. Now we are exploring dimensions of 10^{-16} cm, and therefore it might not be excluded that quantum mechanics might have to be modified. The same might be true for relativity.

III. MACHINE TECHNOLOGY

Two parameters are essential for the discovery potentiality of an accelerator: maximum energy and luminosity. The technical limitations for these two parameters are different for proton and electron machines.

Proton Machines

The maximum energy of a proton machine is given by the product of radius times magnetic field $E \sim R.B$.

Using superconducting magnets and in particular the new warm superconductors, it might be possible to push B up to field strengths of more than 10 Tesla. At the moment, however, this kind of field seems to be difficult to produce since mechanical stresses become a serious problem. Hence, an increase of the radius of the machine might be inevitable. Apart from the cost this poses no principal technical problem.

A limit to the present design of proton machines might be approached at about 100 TeV, where synchrotron radiation becomes nonnegligible for beam dynamics. For superconducting magnets and the associated cryogenics synchrotron radiation might pose some problems even at lower energies, but this could be taken care of by a proper design of the magnet structure. Above 100 TeV synchrotron radiation losses will influence the beam dynamics, and the design principles for proton machines will be similar to those of present electron machines. At the Accelerator Conference at Oxford in 1982 Björken considered an admittedly virtual machine of 500 TeV beam energy, with 170 km radius and a luminosity of 10^{33} cm^{-2}s^{-1}.

Possible future technical developments should also have the objective to produce smaller apertures and perhaps a machine which is not in a plane but follows the terrain. This could maybe be achieved by a truly cybernetic accelerator.

In conclusion it seems that proton colliders in the energy range of 40 to 100 TeV per beam are technically feasible and even beyond these energies no principal technical barriers seem to exist.

Electron Machines

In circular electron colliders the maximum energy is not determined by the bending power of the magnetic field, but by synchrotron radiation losses, which increase with the fourth power of the energy. To compensate these energy losses an electric field is necessary which has to be

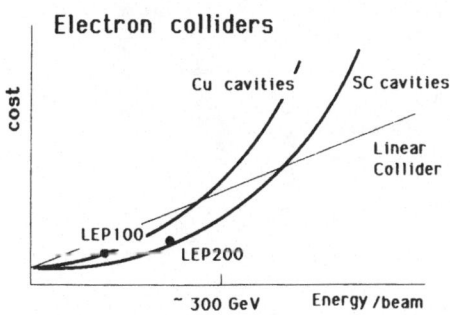

Fig. 1   Cost of linear and circular electron colliders versus energy
         (schematically)

produced by a RF system whose cost will increase with the eighth power of
the beam. It is well known that for an optimized machine both the bending
radius$_2$ and the length of the RF accelerating structure has to increase
with E^2 which implies also that the total cost will increase approximately
with E^2.

    The advantage of circular machines, of course, lies in the fact that
particles which did not produce a collision at the first encounter meet
very often per second thus increasing considerably the chance of a
collision. Luminosities therefore can be relatively high.

    For linear accelerators the cost increases obviously proportional to
the maximum energy apart from a small base constant cost. If the cost is
plotted as function of the energy, one finds a cross-over point for the
cost of a linear collider with that of a circular machine (Fig. 1).   If
one uses copper cavities in a circular machine this cross-over is
somewhere between 200 and 300 GeV, whereas it is higher when one uses
superconducting cavities.   Many technical problems have to be solved
before a linear collider could be constructed.

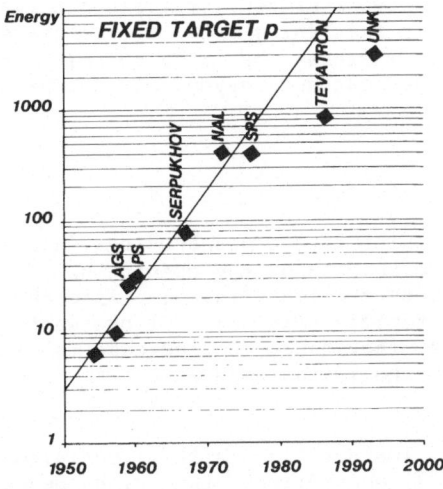

Fig. 2   Livingston plot for fixed target proton accelerators

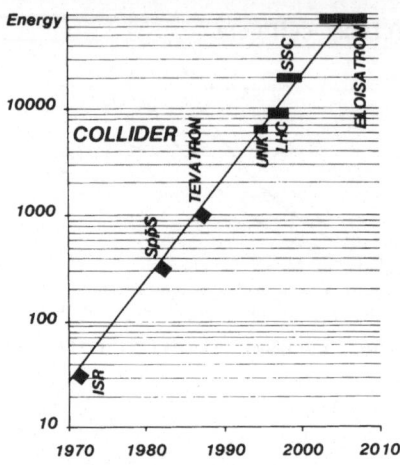

Fig. 3   Livingston plot for proton colliders

The obvious disadvantage of a linear accelerator is that the particles have only one chance to produce a collision.   Therefore, high beam currents are required. The production of a well defined beam focus with dimensions between microns and angstroms and the prevention of a disruption of the beams might be the most difficult task to solve.

A machine with an energy of about 500 GeV per beam might seem to be realistic in the not too distant future. A linear collider with beam energies in the TeV range would have to be a later step.

Energy versus Time

In the famous Livingston-plot the maximum energy achieved with accelerators is plotted as function of time. In figures 2, 3 and 4 this plot is given separately for proton colliders, proton fixed target machines, and e^+e^- storage rings. For proton colliders (Fig. 3) one can notice an increase of energy by a factor of about ten every ten years, and this trend could continue in the future with projects like LHC, SSC and ELOISATRON for two more powers of ten. For fixed target proton machines (Fig. 2) one sees a certain flattening off. e^+e^- storage rings (Fig. 4) have achieved an enery increase by a factor of 7 to 8 about every 10 years. This trend could be continued beyond LEP 200 with colliding linear accelerators.

IV.   DETECTOR TECHNOLOGY

As is well known, point like cross sections fall with $1/E^2$ although some enhancement due to resonances or the opening-up of threshholds like WW or ZZ increase the cross section (Fig. 5). Also the cross section for Jet production decreases rapidly (Fig. 6). If one wants to observe hard collisions, it will be inevitable to increase the luminosity. In the ideal case the luminosity should increase with E^2 in order to obtain a certain number of events.

If one plots the luminosities either achieved or planned for different machines (Fig. 7) one notices that by no means did the luminosity increase proportionally to E^2.

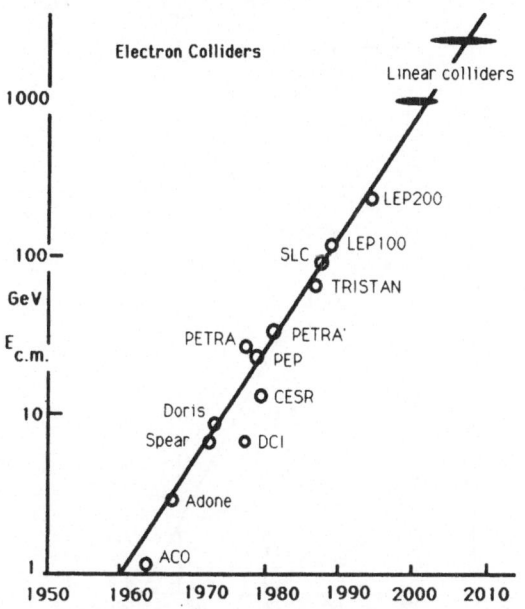

Fig. 4   Livingston plot for electron colliders

But even if the luminosities foreseen for LHC and SSC of the order of 10^{33} or 10^{34} $cm^{-2} s^{-1}$ can be achieved, detectors will have a hard time to cope with the large counting rates and high multiplicities. In addition it might be necessary to introduce single bunch crossing luminosities of 10^{28} $cm^{-2} s^{-1}$ which result in many hundred events per crossings.

As a result the development of new detection techniques might be a more difficult and more challenging task than the development of accelerators. Completely new ideas and techniques are needed. Maybe one will

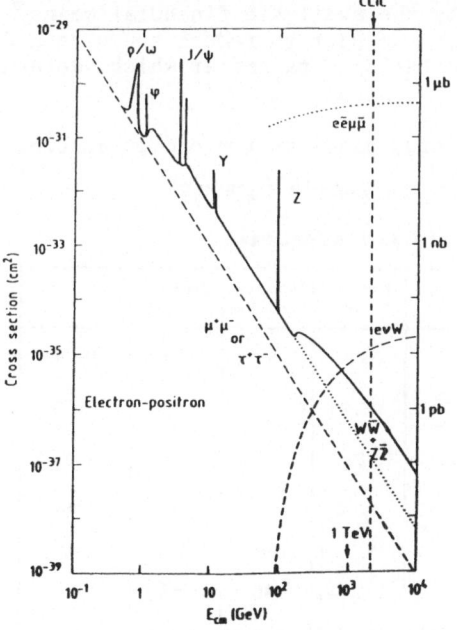

Fig. 5   Pointlike cross section

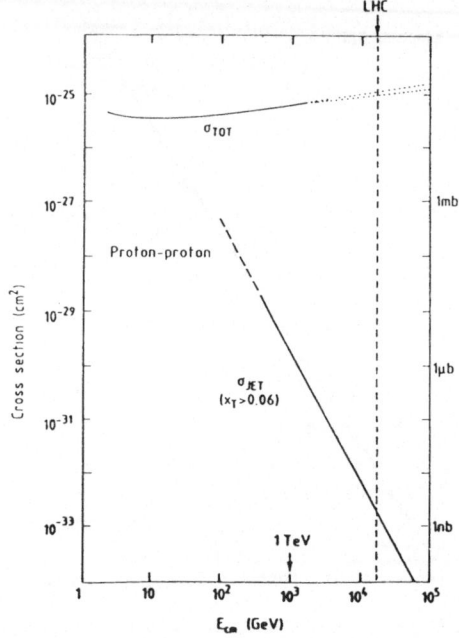

Fig. 6   Jet production cross section

have to give up tracking of individual particles, and even the detection
of muons behind thick absorbers might become difficult. Detector development is therefore as important as new accelerator technologies.

V.   FINANCIAL BOUNDARIES

The highest energies one can achieve are not only limited by technical problems, but also by the available financial means.   Hence, a big
effort has been made in the past to reduce the size and with it the cost
of accelerators. A characteristic parameter which determines to a large

Table 1   Energy gradient for proton accelerators

Energy/length = gradient

Proton synchrotrons

Name	Energy GeV	Gradient GeV/km	Magnet	Year
Bevatron	6.1	51	Fe	1954
PS , AGS	26, 31	40	Fe	1959
SPS, NAL	500	70	Fe	1976/72
Tevatron	1000	160	4.4 T	1986
UNK	3 TeV	140	6 T?	1992?
SSC	20 TeV	240	6.6 T	
LHC	9 TeV	330	10 T	

Table 2. Energy gradient for linear electron accelerators

Electron machines

Linear accelerators

Name	Energy GeV	Gradient GeV/km	Year
LBL	1.2 (2.3)	5.2	1959
SLAC	22	7.3	1966
KEK	2.5	7.8	1981
SLC	50	17	1987

extent the cost efficiency of an accelerator is its maximum energy divided by its length (its circomference for a circular accelerator). This gradient is shown in Table 1 for successive generations of proton machines. One can notice that going from the BEVATRON to the LHC, a factor of 7 has been gained for the gradient. This is to a large extent due to the transition from iron to superconducting magnets.

In Table 2 the gradient for electron linear accelerators is shown. They reflect the development from 1959 to 1987 during which time an improvement factor of more than 3 was achieved.

For circular electron machines it is rather the quantity E^2/length which is characterizing the efficiency of the accelerator since for such machines the radius goes approximately with the square of the maximum energy. Table 3 shows that an improvement factor of about 15 was obtained going from ADONE to LEP 200.

Comparing the absolute cost of different accelerator generations is difficult since the ways of presenting the budget are quite different in different laboratories. I tried to estimate the total cost of the accelerator including its infrastructure, but without experiments and the infrastructure for experiments. The result for fixed target proton machines is shown in Fig. 8. If one takes the cost per GeV one notices that going from the BEVATRON to the Fermilab or SPS machine one has gained about a factor of 20. In Fig. 9. the corresponding figures are shown for hadron colliders where an impressive gain of about a factor of 100 can be noticed comparing the cost per GeV of these different machines.

Table 3. Energy gradient for circular electron accelerators

Circular machines

Name	E /beam	E^2/length	Year
Adone	1.5	22.5	1969
Spear Doris	5	18	1972
Petra PEP	23	230	1977
LEP200	100	332	1989

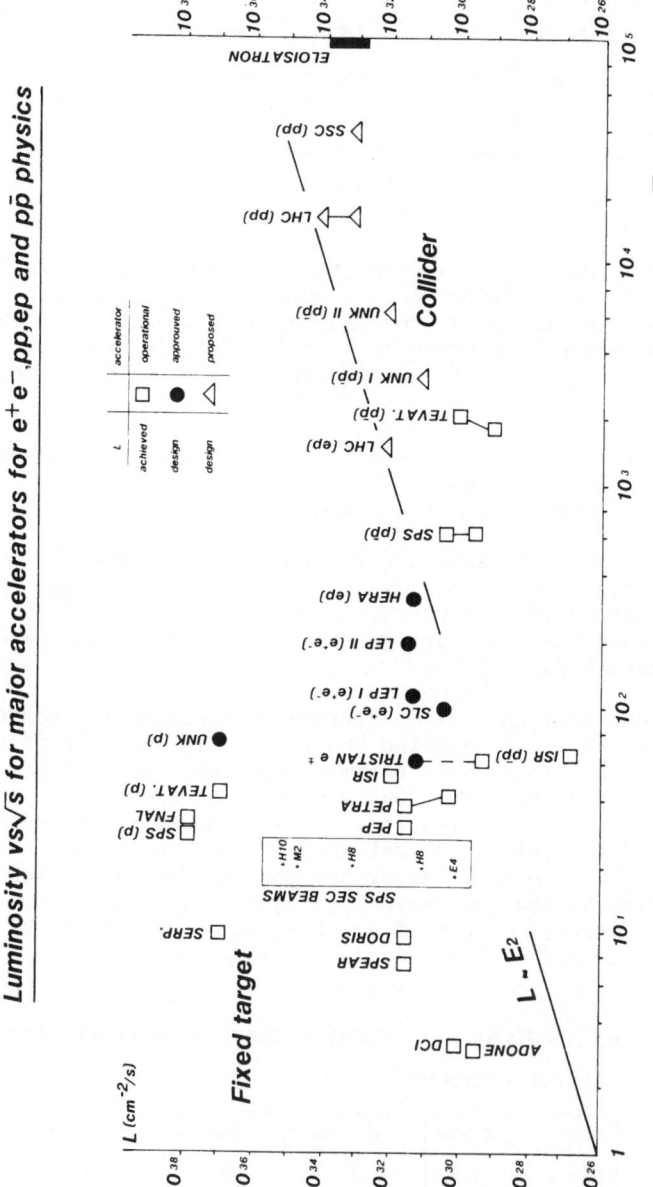

Fig. 7 Luminosities achieved for different kinds of accelerators versus energy

Much effort is usually spent to reduce the price for the machine components themselves like magnets, RF accelerating cavities, controls etc. However, very often the importance of civil engineering and infrastructure like power distribution are neglected. Indeed, considering different projects I found that on the average about 45% of the project cost went into civil engineering and infrastructure (Fig. 10). As a consequence, a comparable effort should be made to reduce these costs, whereas a reduction of the cost of the accelerator components alone will not suffice.

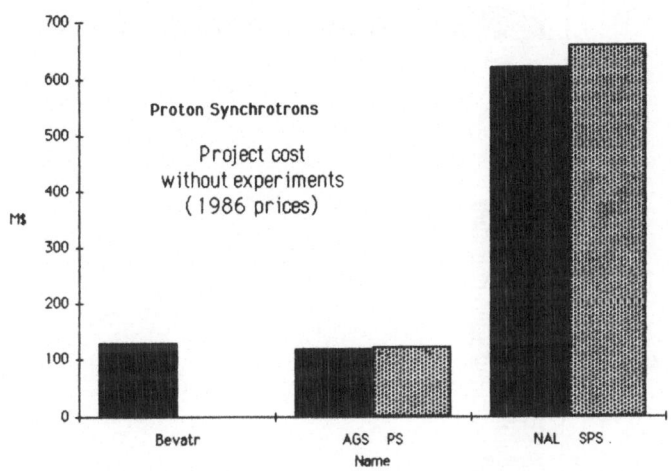

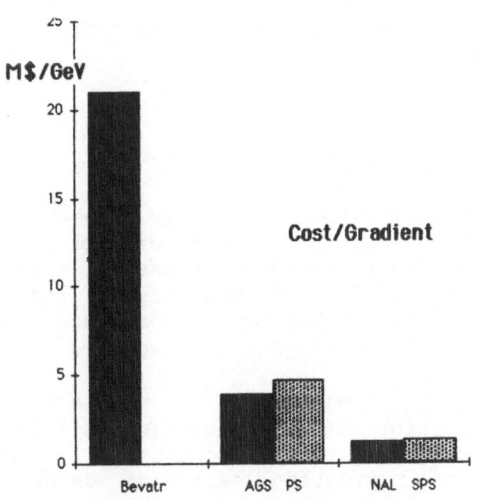

Fig. 8   Cost of fixed target machine
         (without experiments and experimental infrastructure)
         a) total cost   b) cost/GeV

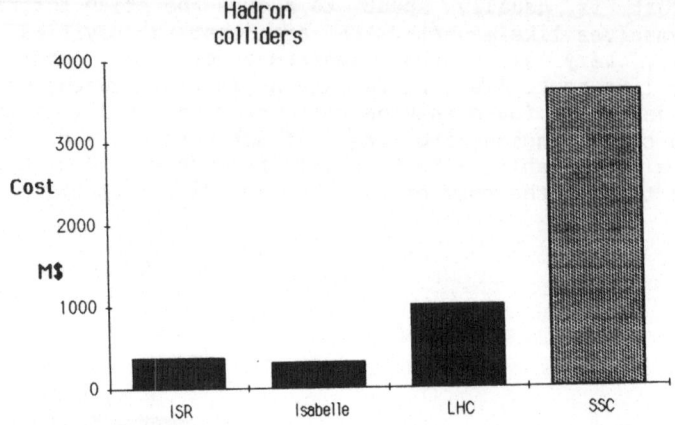

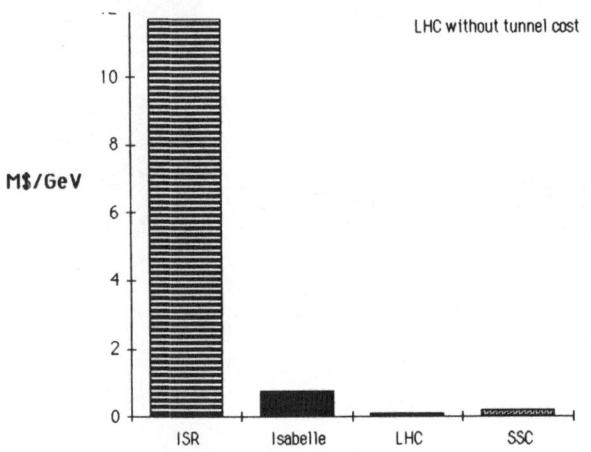

Fig. 9   Cost of hadron colliders
(without experiments and experimental infrastructure)
a) total cost   b) cost/GeV

VI.   ECOLOGICAL, SOCIOLOGICAL AND POLITICAL ASPECTS

The interference with the environment becomes an ever more important topic. Nowadays even deserts belong to the best protected regions. I believe, therefore, that the construction of tunnels for large accelerators by the cut and fill method will not be possible in the future. The largest part of the accelerator will have to be housed in underground tunnels. This unfortunately makes civil engineering more expensive.

Working conditions and safety regulations for people working for the construction of the accelerator but also operating it later become more rigorous. Such problems are in particular important for machines using liquid helium or experiments employing explosive gases. Ventilation systems and safety protections become a non negligible part of the project cost.

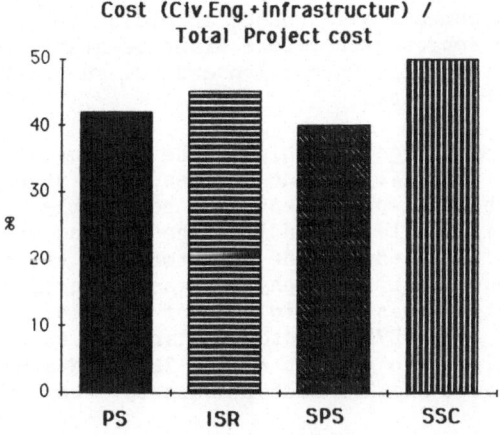

Cost (Civ.Eng.+Infrastructur) /
Total Project cost

Fig. 10   Cost of civil engineering and infrastructure
relative to total project cost

Sometimes the length of time to construct an accelerator and to
prepare the experiments is considered as a major sociological difficulty.
However, I believe that these periods have not essentially changed over
the last two decades. It still takes five to six years from the first discussion of a project to its authorisation and about the same time for the
construction of the machine. For students, of course, it seems very hard
to bridge such long times without producing physics. However, it should
be accepted that developing accelerators and detectors implies as
interesting physics as the analysis of data.

The size of collaborations involving several hundred people is
sometimes considered as a natural barrier for future developments. This
poses certainly a sociological problem, and requires the development of
new sociological structures. Of course, inside such a large group not
everybody can interact with everybody, but a certain distribution of
responsibilities must evolve. The experience with present large
collaborations shows that even in these large groups excellent young
people have a very good chance to distinguish themselves, get to be known
and make a career.

With very large and costly machines, such as those considered for the
next generation of accelerators, a duplication in different parts in the
world is of course excluded. Already the present generation of accelerators under construction corresponds quite well to a complementary world-
wide programme. This, however, was not achieved in any systematic way, and
I believe that coordinating bodies, where a world-wide programme can be
discussed both by physicists and politicians are missing. It is my strong
opinion that an extension of international collaboration, which works
already so well for experiments (planning and financing) will have to be
extended to accelerators and should replace nationalistic "accelerator
races".

VII. CONCLUDING REMARKS

Taking all these arguments into account for the planning and the
decision on future accelerators is, of course, difficult and the weight

of the different arguments will change according to the project and the country. A different approach therefore might be simply to look back into the past to see which steps were taken and how successful they had been from the physics point of view.

For proton fixed target machines the energy steps for different machines are shown in Fig. 11. As one notices, the steps were about a factor of 3 and these steps seemed to be reasonable, since the task of building the accelerator could be handled, the funding could be obtained, and also programmes related to the detectors and the analysis could be coped with and last not least, good physics was produced. If one looks at proton colliders, there was a big step of a factor 10 between the SPS collider and the ISR. The TEVATRON provides another factor 3 with respect to the SPS collider. With respect to the TEVATRON a hadron collider in the LEP tunnel would give another factor close to ten, whereas the SSC

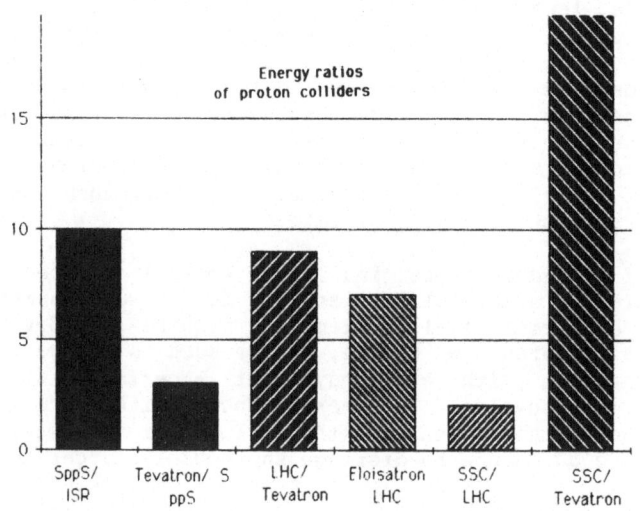

Fig. 11   Energy ratios of successive proton accelerator generations

would present a step of a factor of 20. Compared to the experience in the past this appears to be a very large step. Of course, one might find this fact attractifve, since it presents an interesting challenge. However, my opinion is that a step from TEVATRON to a hadron collider in the LEP tunnel of about a factor 10 would be more reasonable, followed by another step to a collider with an energy above 40 GeV per beam. Such a procedure would seem justified not only because of the global arguments given here, but also based on the arguments given above concerning the physics, detector technology and accelerator technology. Such a machine could either be an upgraded SSC or the ELOISATRON.

THE LEP PROJECT AT CERN

J.J. Thresher

CERN,
Geneva, Switzerland

1. INTRODUCTION

Electron–positron physics and LEP may seem to be somewhat out of place at this meeting, where the theme is physics with very high energy proton–proton colliders, such as the ELOISATRON. However, the platform on which the physics with future machines will be built — be they hadron–hadron, electron–proton, or electron–positron colliders — will surely have as a major component the output of LEP. The detectors, even with the relatively low rates at which they will operate, will have something to contribute to the evolution of future detectors, while the physics, with the hints of activity (or otherwise) in the multi-TeV region, and with the basic information that will emerge about the Standard Model and anything else that shows up within the mass range up to 200 GeV, cannot fail to have an impact.

Accordingly, in this talk I shall discuss, briefly, the status of LEP, the detectors being built for it, the physics goals as I see them, and finally the way the programme may evolve during the 1990s.

2. THE STATUS OF LEP

LEP is a large electron–positron storage ring designed to operate at energies up to 100 GeV per beam at least. The upper limit set by its magnets is close to 120 GeV per beam. Some of the more important basic parameters are given in Table 1.

The machine is being built in a tunnel, 27 km in circumference, which lies at depths varying from 50 m to 170 m below ground level in a region between the Jura mountains and Geneva airport (see Fig. 1). When it first comes into operation it will be equipped with sufficient RF power to accelerate the electrons and positrons up to 55 GeV — sufficient, that is, to span the Z^0 peak comfortably. When operating at the Z^0 peak at its design luminosity of $1.6 \times 10^{31} \, \text{cm}^{-2} \, \text{s}^{-1}$, it will be producing some 20,000 Z^0's per day at each of the four intersection regions where the detectors will be located.

One of the important features of LEP is that it will use two of CERN's existing accelerators, namely the Proton Synchrotron (PS) and the Super Proton Synchrotron (SPS), operating in tandem, to make up the injector system. For this purpose the PS is being equipped with two electron linacs. One is a high-current, 200 MeV machine producing an output of 2.5 A for use as an electron-to-positron converter. The aim is to produce 12 mA of positrons in a form suitable for

Table 1
LEP design parameters

Beam energy (GeV)	55	95
Circumference (km)	26.66	26.66
Dipole field (T)	0.0645	0.1114
Injection energy (GeV)	20	20
R.F. frequency (MHz)	352	352
Dist. between supercond. quads (m)	± 3.5	± 3.5
r.m.s. bunch length (mm)	17.2	13.9
r.m.s. beam radii: σ_x (μm)	255	209
σ_y (μm)	15.3	10.8
Bunch spacing (μs)	22	22
Nom. luminosity (cm^{-2} s^{-1})	1.6×10^{31}	2.7×10^{31}
Beam lifetime (h)	6	5
r.m.s. energy spread	0.92×10^{-3}	2.06×10^{-3}
Current (4 bunches) (mA)	3	3
Synchr. rad. loss per turn (GeV)	0.263	2.303

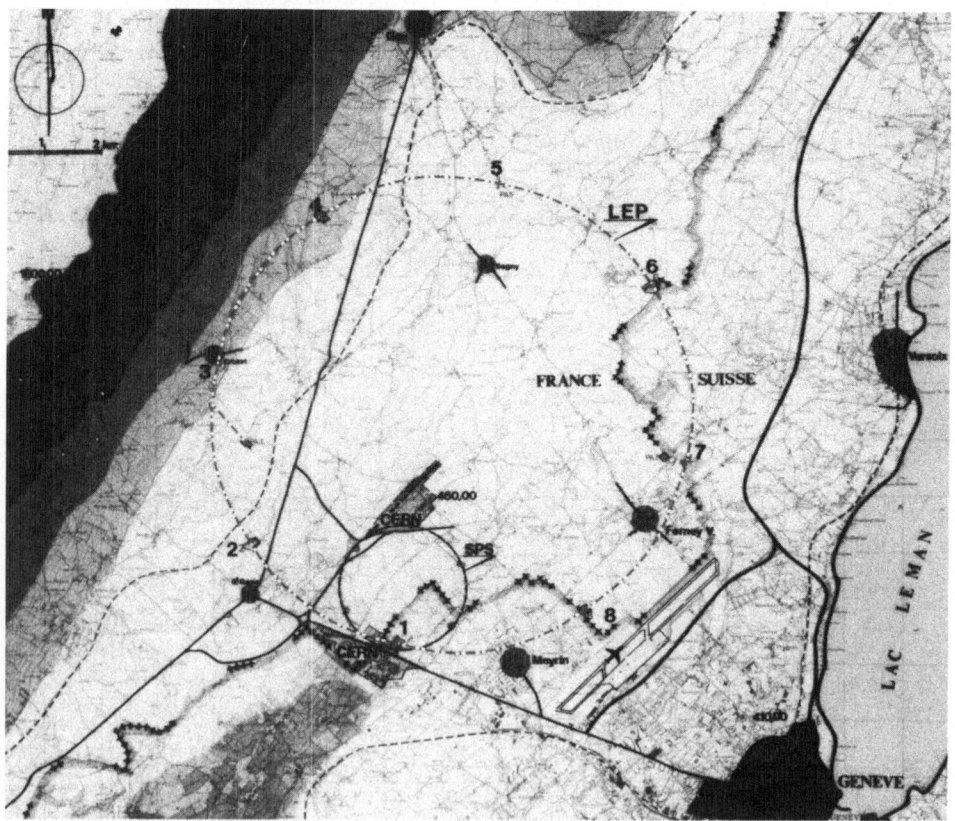

Fig. 1. Map of the region in which LEP is located

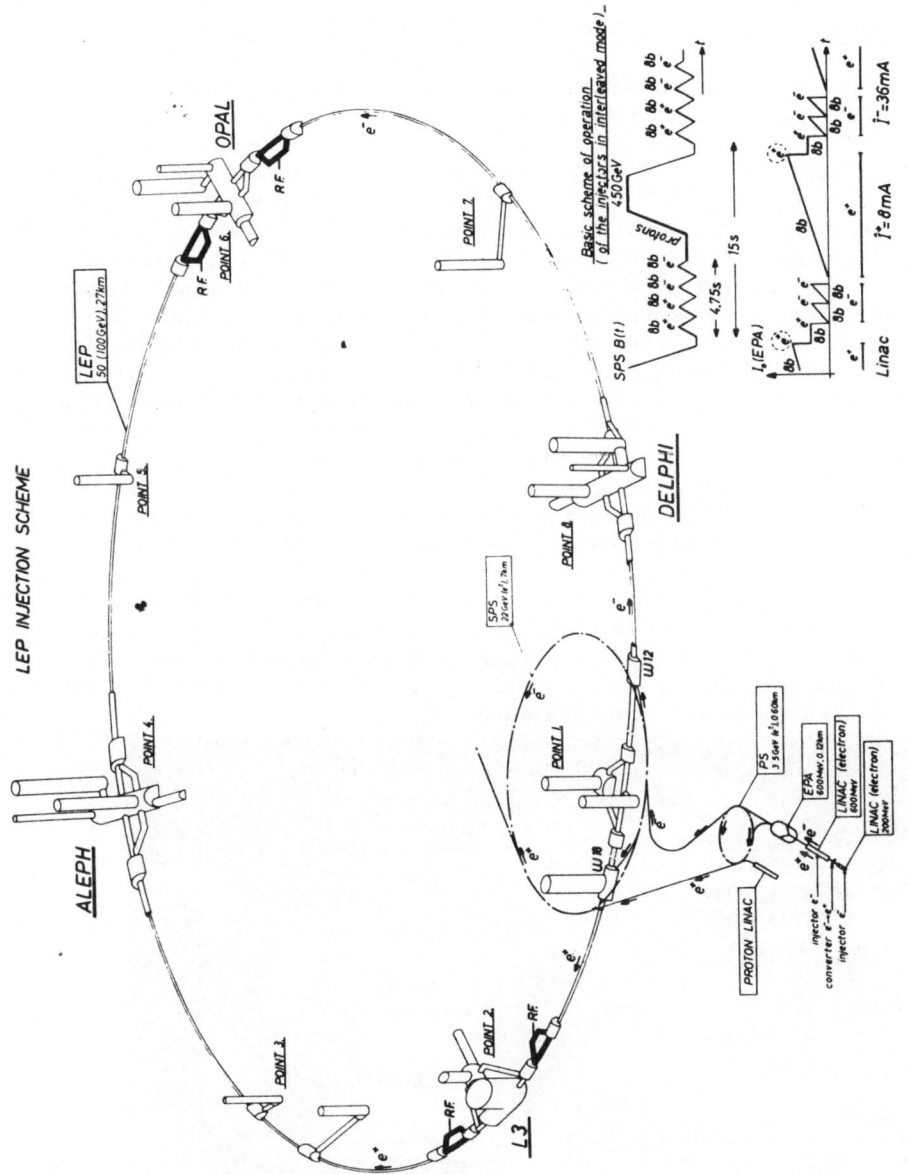

Fig. 2. The injector system for LEP

subsequent acceleration. The second linac will be used to accelerate these positrons, and also electrons, up to an energy of 600 MeV. They will then be fed into an electron–positron accumulator ring (EPA) before being transferred, first to the PS for acceleration to 3.5 GeV, and then on to the SPS where they will be accelerated to 20 GeV, prior to injection into LEP. The full scheme is illustrated in Fig. 2.

It is planned to have the injector system ready well in advance of the completion of LEP itself. So far, both linacs have been fully tested; electrons and positrons have been stored in the EPA with good lifetimes and have been accelerated to 3.5 GeV in the PS. In September 1987, positrons will be transferred to the SPS and accelerated to about 13 GeV (the maximum that the existing SPS RF system will allow). Acceleration to the full injection energy of 20 GeV using special single-cell standing-wave RF units is planned for mid-1988.

As regards the construction of LEP itself, I shall comment first on the status of the tunnel. This is being built in two main sections. The larger section, 23 km long, is in good-quality rock (molasse) and has been excavated with three large tunnelling machines. It is now complete, the machines have been removed, and only the concrete lining remains to be finished. The other, smaller, section is under the Jura mountains, where the excavation has to be done by normal blasting methods since the nature and quality of the rock (mainly limestone) is not suited to the use of tunnelling machines. Also, from the beginning there were known to be geological faults in this region, with the likelihood of striking underground water at high pressure. Despite elaborate precautions and careful planning there have indeed been serious difficulties with flooding which have delayed completion of this part of the tunnel by over a year. At one stage, water at a pressure of 10 atmospheres was flowing into the tunnel at a rate in excess of 100 l/s. Fortunately, a satisfactory method of proceeding in the face of such difficulties has been established and steady progress, at a rate of about 10 m a week, is now being made, with the expectation that the remaining 250 m to be excavated will be complete by the end of 1987 or early in 1988.

Other aspects of the civil engineering are proceeding as fast as progress with excavation will allow. Well over half the concreting and much of the infrastructure — such as ventilation, monorail transport system, elevators, electrical services — is complete. Installation of the machine has also started (see Fig. 3), and it is planned to have at least the first octant ready for a full injection test with 20 GeV positrons from the SPS in July 1988.

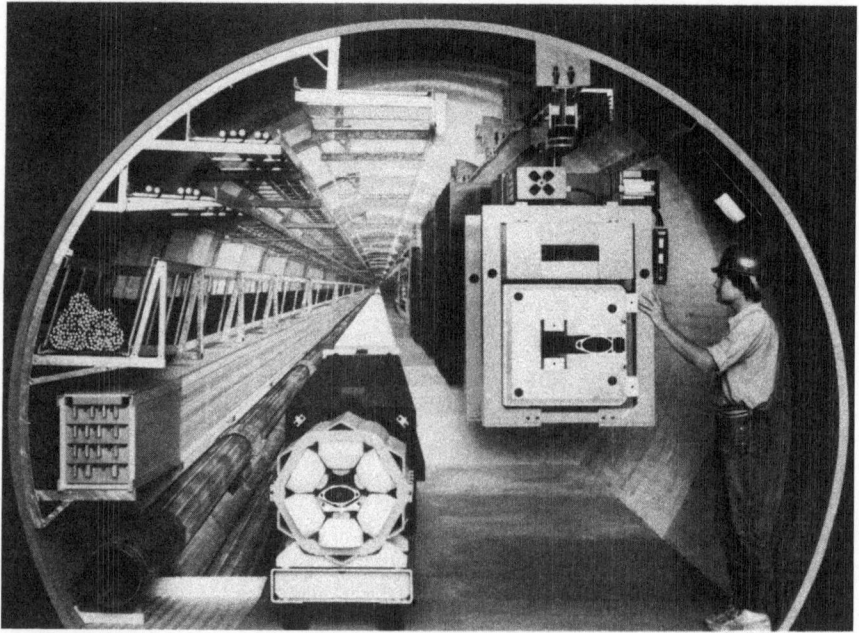

Fig. 3. A model of the LEP tunnel showing the method of installation of machine components

Machine components are being delivered and tested on schedule. In all, 3400 dipoles, 760 quadrupoles, 512 sextupoles, and 630 correcting dipoles are required, plus components for the RF system with its 128 cavities and for the vacuum system. In each case well over 80% has been delivered.

LEP will be provided with four large underground experimental areas to house the four experiments that have been approved. An artist's impression of a LEP experimental hall and associated surface buildings is shown in Fig. 4. Here the main problem will be to ensure that these areas, complete with cranes and general infrastructure, can be made available in time for the final installation and assembly of the detectors to be finished before LEP starts up in 1989. This is particularly important for the L3 detector, which cannot be moved and so has to be installed in the beam in its final position without interference with the machine construction or commissioning. The other three detectors will be mounted on rails, and can be assembled in a 'garage' to one side of the LEP ring and moved into position when required. The situation is summarized in Table 2. A picture of the L3 experimental hall is shown in Fig. 5.

Fig. 4.
Artist's impression of a LEP experimental zone

Table 2
Installation schedule for the LEP experiments

Experiment Location Depth below surface	L3 Point 2 50 m	ALEPH Point 4 145 m	OPAL Point 6 105 m	DELPHI Point 8 105 m
Civil eng. complete	31/01/87	16/11/87	15/02/88	07/12/87
Cranes operational: underground surface	12/08/87 07/12/87	16/10/87 01/02/88	15/02/88 01/05/88	22/12/87 15/04/88
Start of installation	17/08/87	15/02/88	01/06/88	18/04/88

Fig. 5. The L3 experimental hall during construction

3. THE DETECTORS FOR LEP

The four LEP detectors cannot be discussed in any detail in this brief review. I shall therefore restrict myself to making some general comments and to mentioning just a few of the many interesting features which characterize each of them [1].

Inevitably, all four detectors are designed to have cylindrical geometry. Also they all incorporate solenoidal magnetic fields, central tracking chambers, electromagnetic and hadron calorimeters, muon detectors, and a high degree of hermeticity. All four make use of multilevel trigger systems which aim to reduce the data-taking rates to 1–2 Hz. The first-level triggers operate within the 22 μs gap between successive bunch crossings.

Most of these elements are, of course, essential ingredients in all modern detectors used at colliders. At LEP the techniques chosen vary considerably from one detector to another and depend very much on the particular aspects of the physics which the respective collaborations wish to emphasize. Together the four detectors provide a powerful combination of systems with which to study the many physics problems confronting particle physics today. An added strength is that LEP itself offers the experimentalists a clearly defined initial state of well-known energy and momentum. This helps to give significance to effects in which only small numbers of events may be involved.

3.1 ALEPH

The ALEPH detector (see Fig. 6) uses a large superconducting solenoid (Fig. 7), 6.4 m long and 5.3 m in diameter, to provide a magnetic field of 1.5 T. Inside the solenoid there will be a microvertex detector immediately surrounding the beam pipe, followed by an inner tracking

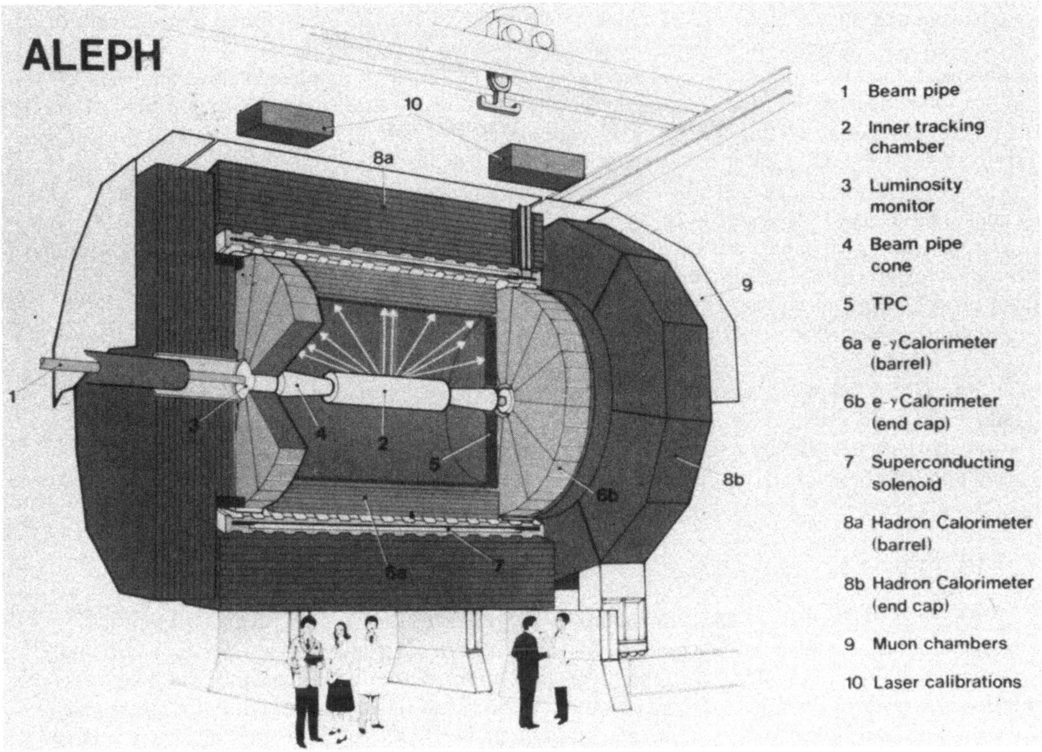
ALEPH
10
8a
9
1
3
4
2
5
6b
8b
6a
7
1 Beam pipe
2 Inner tracking chamber
3 Luminosity monitor
4 Beam pipe cone
5 TPC
6a e-γ Calorimeter (barrel)
6b e-γ Calorimeter (end cap)
7 Superconducting solenoid
8a Hadron Calorimeter (barrel)
8b Hadron Calorimeter (end cap)
9 Muon chambers
10 Laser calibrations

Fig. 6. The ALEPH detector

AIMANT SUPRACONDUCTEUR 65T

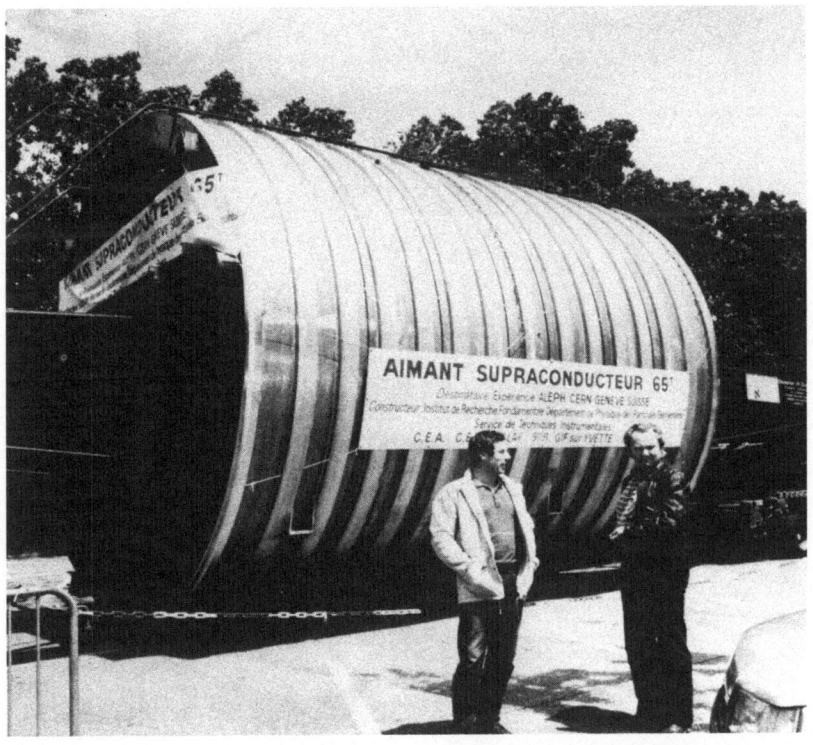

Fig. 7. The ALEPH superconducting solenoid

chamber used primarily for triggering, a large time projection chamber (TPC), and then an electromagnetic (e.m.) calorimeter. The solenoid is enclosed in an instrumented iron return yoke in which layers of streamer tubes are interleaved with 5 cm thick iron plates for use as a hadron calorimeter and as part of the muon detection system. The latter is completed by two double layers of streamer tubes outside the yoke to detect and measure the direction of muons which escape from the iron. An interesting feature of the e.m. and hadron calorimeters for ALEPH is that both are built up with pad readout in towers pointing towards the interaction vertex. The same basic system is used for the barrel and the end-caps. The e.m. calorimeter is designed for good granularity. It has 45 layers of lead, 33 of which are 2 mm thick and 12 (the outer layers) 4 mm thick. Between each plane is a layer of wire chambers filled with a gas mixture of xenon (80%) and carbon dioxide (20%). In all, there are 48,000 towers in the barrel section and 24,000 in the end-caps. The spatial resolution will be $\pm 2\text{-}3$ mm for locating the shower centre, and the energy resolution is expected to be $\pm 18\%/\sqrt{E}$.

The TPC is 4.8 m long and 3.6 m in diameter. There are approximately 3000 wires and 20,000 pads on each end-plate, which is built in 18 sections, arranged to have radially stepped boundaries. An unconventional pad structure built up in concentric circles helps to avoid deterioration in resolution for high-momentum tracks by ensuring that they make only small angles with respect to the pad axis.

3.2 DELPHI

The DELPHI detector, uniquely amongst the LEP detectors, places particular emphasis on hadron identification over a wide momentum range. For this purpose it uses a complex system of ring-imaging Cherenkov (RICH) counters covering most of the full solid angle. In many other respects the detector is rather similar to that of ALEPH. As shown in Fig. 8 it has a microvertex detector, an inner detector for triggering, a TPC, and an e.m. calorimeter lying inside a large superconducting solenoid which provides a magnetic field of 1.2 T. The RICH detectors are also located inside the solenoid; the TPC is therefore made much smaller than in ALEPH in order to accommodate them. The iron return yoke around the solenoid is instrumented with streamer tubes

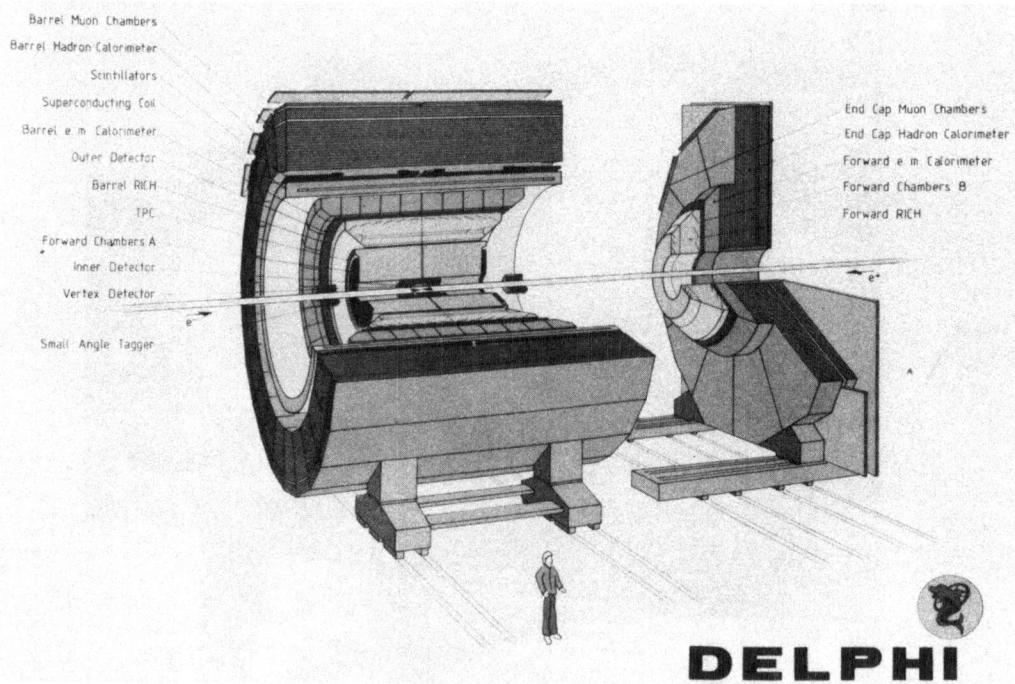

Fig. 8. The DELPHI detector

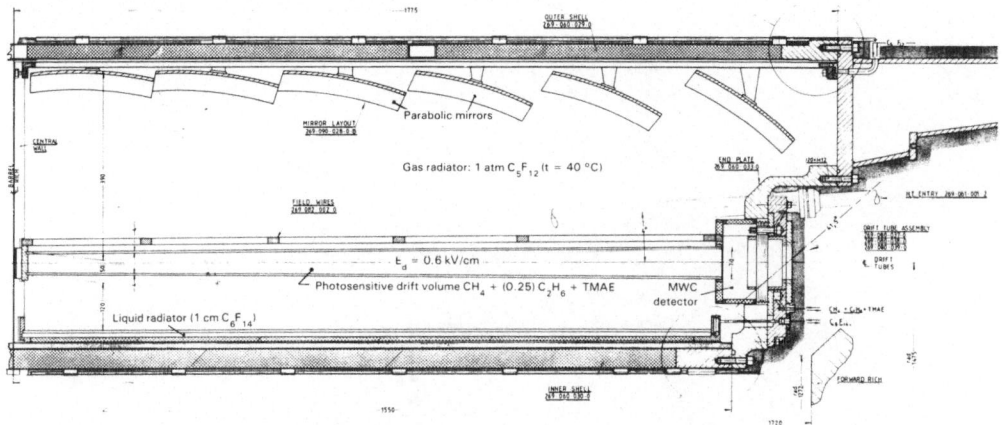

Fig. 9. Cross-section through the beam axis of the DELPHI Barrel RICH

for hadron calorimetry; muon detection is done with two planes of drift chambers, one inside the return yoke and the other on the outside.

The RICH detectors make use of liquid and gas radiators, cleverly arranged so that a single readout system can be used for both. They provide π, K, and p identification for momenta from 0.3 to 25 GeV/c [2]. The principle of the scheme (Fig. 9) is as follows. The inner cells are filled with a liquid Freon and are 1 cm thick in the radial direction. Cherenkov light, produced in a well-defined cone when particles of sufficient velocity pass through a cell, enters a system of drift tubes filled with gas containing a small amount of TMAE. The TMAE converts the Cherenkov photons into electrons by photoionization, within a distance of less than 3 cm from the inner walls of the drift tubes. The electrons are then made to drift along the tubes, in a direction parallel to the magnetic field, to a set of anode wires backed by cathode strips, which together give the coordinates of the conversion point of each photon. The region outside the drift tubes is filled with a Freon gas. Cherenkov light produced in the gas is reflected, by a set of parabolic mirrors, back into the drift-tubes where a ring image is formed. The photons are converted in the outer 3 cm of the drift-tube gas, i.e. well separated in the radial direction from the conversion region of the inner-cell photons. Thus the two sets of electron rings produced by Cherenkov light from the inner and outer radiators can be distinguished from one another in a simple way.

3.3 L3

The most obvious feature that distinguishes the L3 detector from the other three at LEP is its size. All the detector elements lie within an octagonally shaped aluminium coil having an inner diameter of 11.86 m and producing a field of 0.5 T. A picture of one of the coil elements on its mounting frame is shown in Fig. 10.

The main design aim is to enable the energies of muons, electrons, and photons to be measured with a resolution $\Delta E/E = \pm 1\%$. For electrons and photons this is done in a compact e.m. calorimeter consisting of tapered crystals of bismuth germanium oxide (BGO), each viewed by a photodiode. There will be 7680 crystals in the barrel section of the calorimeter. Each crystal will measure 2 cm $\times$ 2 cm at the front face and will be 24 cm ($22X_0$) in length. They will be slotted into a honeycomb structure made of thin carbon-fibre wafers to give the pointing geometry required. Calibration of the first half-barrel is under way and is giving excellent results for the energy resolution, namely $\pm 0.5\%$ at 50 GeV, $\pm 1.5\%$ at 2 GeV, and $\pm 6\%$ at 100 MeV.

A hadron calorimeter surrounds the BGO array. It will be built up of layers of depleted uranium and wire chambers to provide a fine-grained system, 4λ thick. Layers of copper,

Fig. 10. One of the coil elements for the L3 detector on its mounting frame

instrumented with streamer tubes, are placed outside the uranium section to give a further 2λ of material to complete the absorption of hadrons.

Muons which filter through this system (7λ including the BGO) are detected (still within the magnetic field) in a three-layer system of wire chambers, with a separation of ~ 1.5 m between successive layers to give a good measurement of the muon momentum. The system is made up of 16 modules assembled in two octagonal sections which can be precisely aligned to ± 40 μm. Tests carried out on one of the modules have shown that muons can be detected with spatial resolution of ± 150 μm in each layer, well within the design specification.

Charged-particle tracking is done in a time expansion chamber (TEC) located at the centre of the detector inside the BGO array. The TEC is only 1 m in diameter in order to keep to a minimum the volume of BGO which surrounds it, but it aims to measure tracks to ± 40 μm with a two-track resolution of 500 μm in the r, φ plane.

As can be seen in Fig. 11, the TEC, the BGO array, the hadron calorimeter, and the forward muon chambers are all located inside a massive support tube, 32 m long and 4.5 m in diameter; it

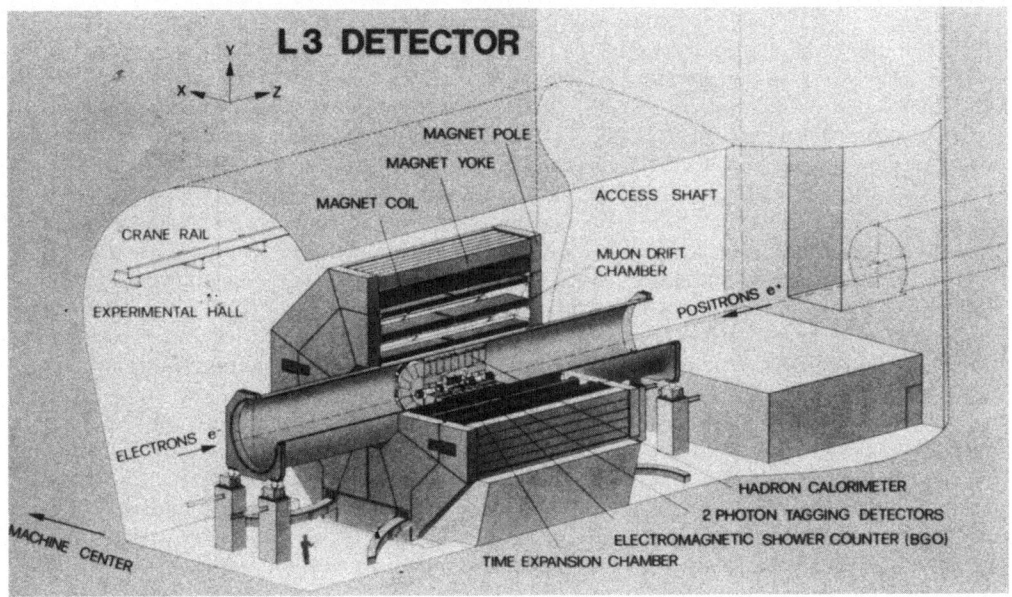

Fig. 11. The L3 detector

weighs 300 tons and also supports the muon detectors on the outside, so making all the detectors mechanically independent of the magnet.

3.4 OPAL

The OPAL detector is illustrated in Fig. 12. It is built up of elements based on well-proven techniques used mainly at PETRA but also at the SPS at CERN. Often referred to as the safe conventional detector at LEP, it nevertheless has many features which on closer inspection would seem to make such a description somewhat naïve.

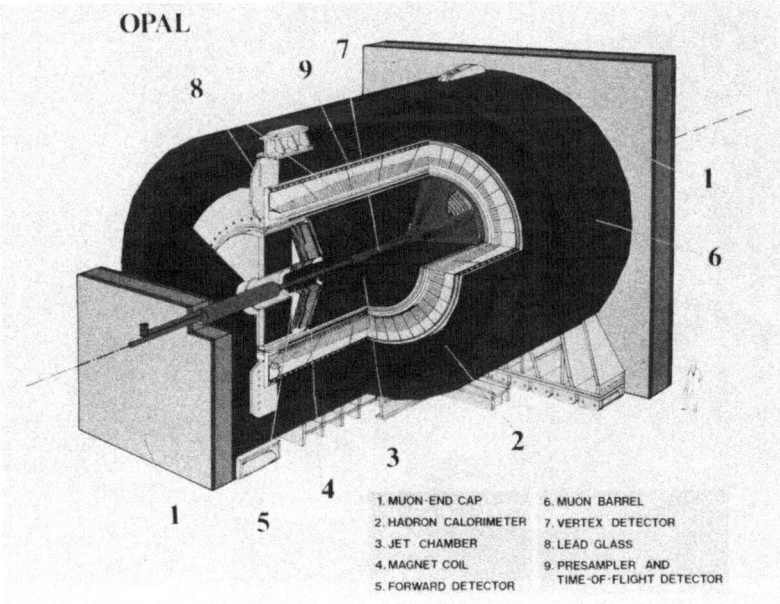

1. MUON END CAP	6. MUON BARREL
2. HADRON CALORIMETER	7. VERTEX DETECTOR
3. JET CHAMBER	8. LEAD GLASS
4. MAGNET COIL	9. PRESAMPLER AND
5. FORWARD DETECTOR	TIME-OF-FLIGHT DETECTOR

Fig. 12. The OPAL detector

There are three basic tracking devices which are located inside a 0.4 T solenoidal magnetic field produced by a water-cooled aluminium coil, 4.5 m in diameter and 6.5 m long. First there is an inner detector used for triggering, rather like in ALEPH and DELPHI, but having a substantially better r, ϕ resolution of ± 30 μm which enables it to be used as a vertex detector; it is thus an essential part of the tracking system. It is surrounded by a jet chamber which constitutes the main tracking device, and which is modelled on the one used for the JADE detector at PETRA. In essence it is a large multicell drift chamber containing 3840 sense wires and is filled with a mixture of argon (90%), methane (8%), and isobutane (2%) at a pressure of 4 atmospheres. Up to about 150 ionization measurements can be made along the track of each charged particle, giving good dE/dx resolution and π–e separation up to 7 GeV/c. The r, ϕ resolution has been checked in a full-scale prototype, and has confirmed the design figures of ± 110–150 μm. Although this jet chamber also gives information on the z-coordinates of the tracks, this is improved by a system of drift chambers immediately surrounding the jet chamber, which give a $\Delta z/z$ of ± 300 μm.

The e.m. calorimeter in OPAL is built up of 12,000 lead-glass blocks, each about $24X_0$ long. Of these, 9600 make up the barrel section and are specially shaped to give pointing geometry. They are located outside the coil, which is $1.5X_0$ thick, and are viewed by photomultipliers. The end-caps, consisting of 1200 blocks each, are in the full (axial) field of the magnet and are viewed by specially developed, low-noise vacuum phototriodes. Much of the e.m. calorimeter has already been calibrated; the energy resolution obtained is $\Delta E/E \approx \pm 5\%/\sqrt{E}$. Part of the barrel section being prepared for calibration in a test beam is shown in Fig. 13. Immediately in front of the

Fig. 13. A half-ring of the barrel section of the OPAL electromagnetic calorimeter being prepared for calibration in a test beam.

lead-glass there is a presampler to provide information on showers which develop in the magnet coil or in the end-plates of the jet chamber. There is also a set of time-of-flight counters in the same region.

The magnet yoke is instrumented with limited streamer tubes to provide hadron calorimetry, and muon identification is done by means of a system of drift chambers arranged in four layers outside the return yoke. For the barrel section, the chambers employ a neatly devised cathode readout system with segmented diamond- and triangular-shaped pads on the top and bottom of each chamber; this enables the z-coordinate to be determined to ± 1 mm [3].

4. PHYSICS GOALS

The physics about which we feel most confident will be centred on studies at and around the Z^0 pole. The field is rich and, with each experiment aiming to collect some 10^7 events in this region, high-precision measurements in many decay channels will be possible, with good sensitivity to small effects. A key measurement will be the determination of the Z^0 mass. For this an accurate measurement of the LEP beam energy is required; the plan is to do this to an accuracy of $\pm 1 \times 10^{-4}$ by measuring the spin precession frequency of the $e^{\pm}$ beams in LEP, assuming, as seems likely, that they develop an adequate degree of polarization. In this case the error on the measured mass of the Z^0 could be as low as $\Delta m_Z = \pm 20$ MeV/c^2, giving an error on $\sin^2\theta_w$ of ± 0.0004, at which level the uncertainty in the radiative corrections dominates. If the spin-precession method cannot be used, it should be possible to achieve Δm_Z of ± 50 MeV/c^2.

A measurement of the width to ± 20 MeV/c^2 will also be feasible. Remembering that the partial width $\Gamma_{\nu\bar{\nu}} \approx 170$ MeV/c^2, a limit could be placed on the number of neutrino types, and evidence of new channels may emerge. Neutrino counting could also be done by studying the reaction $e^+e^- \rightarrow \nu\bar{\nu}\gamma$ just above the Z^0 while scanning across the Z^0 peak. Another obvious but important piece of physics will be to check that the partial widths $\Gamma_{e^+e^-}$, $\Gamma_{\mu^+\mu^-}$, and $\Gamma_{\tau^+\tau^-}$ are all equal, where an accuracy of $\pm 1\%$ could be achieved.

High on the list of priorities when LEP starts up will be to search for the Higgs boson H^0. Theory has little to offer as regards the value of the mass of the Higgs. If it does indeed exist, and if it has a mass below that of the Z^0, then decay channels such as

$$Z^0 \rightarrow H^0\ell^+\ell^- \ (\text{BR} \approx 10^{-5}\text{-}10^{-7} \text{ for } m_H = 20\text{-}70 \text{ GeV}/c^2)$$

and

$$Z^0 \rightarrow H^0\gamma \ (\text{very low BR})$$

offer ways of discovering it. Rate limitations may restrict the range of detectable masses to $m_H < 40$ GeV/c^2. For higher masses up to $m_H \approx 80$ GeV/c^2, it will probably be preferable to wait for the energy of LEP to be increased, and to look for the reaction

$$e^+e^- \rightarrow Z^0H^0$$

just below the W^+W^- threshold. Here the cross-section is estimated to lie in the range 1-10 pb, depending on the value of m_H and on the incoming energy of the $e^{\pm}$ beams. The most promising H^0 decay channel should be $H^0 \rightarrow b\bar{b}$, leading to two jets plus the decay products of the Z^0 in the final state; for example:

$$
\begin{array}{l}
e^+e^- \rightarrow Z^0H^0 \\
\qquad\quad \mid \ \ \overset{\textstyle\llcorner}{} \rightarrow b\bar{b} \ \ \Big\} \\
\qquad\quad \llcorner\!\!\!\!\!\!\!\!\rule{0pt}{0pt} \longrightarrow q\bar{q} \ \Big\}
\end{array}
\quad \text{four jets (73\%)},
$$

$$e^+e^- \rightarrow Z^0H^0$$
$$\hookrightarrow b\bar{b} \;\Big\}$$
$$\rightarrow \nu\bar{\nu}$$

two jets + missing mass (18%),

$$e^+e^- \rightarrow Z^0H^0$$
$$\hookrightarrow b\bar{b}$$
$$\rightarrow e^+e^-$$
$$\rightarrow \mu^+\mu^-$$
$$\rightarrow \tau^+\tau^-$$

two jets + $\ell^+\ell^-$ (6%).

The search for the Higgs will be no easy matter. Rates will be low, backgrounds could be high, and there could be further complications if the top-quark can be produced within the LEP energy range.

When the energy of LEP approaches its full design value of about 100 GeV per beam, operation above W^+W^- threshold becomes possible. This will provide a unique opportunity to study the three-boson couplings γWW and ZWW which underpin the electroweak theory. The reaction $e^+e^- \rightarrow W^+W^-$ is dominated by the three diagrams shown below. Large cancellations between the amplitudes are expected, so measurements of this reaction provide sensitive tests of the theory.

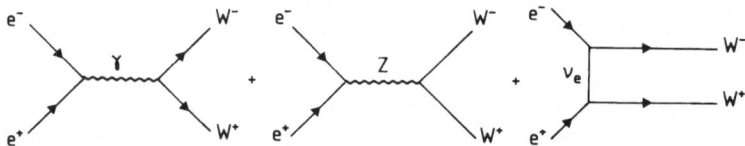

It will be important to measure the mass of the W once the energy of LEP is high enough. There are various possible approaches to this: for example, measuring the excitation function for producing W^+W^- pairs; looking at the two-jet invariant mass from $W \rightarrow q\bar{q}$ decays; or studying the $W \rightarrow \nu e$ invariant mass. Each of these could give $\Delta m_W \approx \pm 100$ MeV/c^2, so a final experimental uncertainty of $\Delta m_W \approx \pm 50$ MeV/c^2 seems achievable.

The higher LEP energies would also seem to be needed in the search for the t-quark, assuming that it will not already have been discovered in experiments at the $p\bar{p}$ colliders at CERN and Fermilab. Predictions of its mass cover a wide range, but there are good reasons to expect it to lie below 250 GeV/c^2. The lower limit from the UA1 experiment is 55 GeV/c^2.

The above gives a selection of some of the physics aims of the LEP experiments [4]. Clearly there are many other interesting studies to be made at LEP: for example, measurements of the longitudinal and transverse polarization of the W's; forward–backward asymmetries; search for supersymmetric particles; and so on. There can be few accelerators built to date where the richness of the physics return has been better assured from the start.

5. EVOLUTION OF THE LEP PROGRAM

In this last section of my report I should like to say something about how I see the LEP programme evolving in the coming years.

During the first half of 1989 we expect to see the completion of the installation of the LEP machine and the four experiments. The second half of 1989 will be very largely a running-in phase for both the machine and the experiments. By the end of that year we might hope to be approaching a peak luminosity of about 10^{31} cm^{-2} s^{-1} and, optimistically perhaps, to have had some 1000 hours of useful operation for the experiments with an integrated luminosity $\int L \, dt$ reaching 5 pb^{-1}.

In 1990 and 1991 it would seem reasonable to operate LEP for about 3600 hours in each year. Assuming a peak luminosity of $L_{peak} = 1.5 \times 10^{31}$ cm^{-2} s^{-1} and reasonable efficiency factors to allow for the time to fill the ring, for the lifetime of each fill, and for overall operating efficiency, an integrated luminosity $\int L \, dt = 100$ pb^{-1} per year for each experiment should be possible. Much, if not all, of the running will doubtless be at or around the energy of the Z^0 peak, but in any case no higher than 55 GeV per beam, which is the maximum beam energy expected with the klystrons that will be installed when LEP starts up. Thus by the end of 1991, each experiment could in principle have accumulated close to 10^7 Z^0 events.

From 1992 onwards, I would foresee LEP continuing to operate for about 3600 hours per year. A steady increase in energy is planned, first by adding superconducting cavities to the 128 copper cavities already installed, and then finally replacing the latter with more superconducting cavities as illustrated in Fig. 14. The aim would be to reach, in this way, an energy significantly above the W^+W^- production threshold of 82 GeV per beam — 92 GeV per beam would seem to be possible — by 1994. At these higher energies, L_{peak} should increase to about 2.5×10^{31} cm^{-2} s^{-1} and $\int L \, dt$ should reach 170 pb^{-1} per year. Of course, the precise schedule of operation will depend on how the subject evolves during these years; whether the Higgs boson or the t-quark are discovered, or whether something quite unexpected emerges. Also there is a possibility that the beam in LEP may develop substantial levels of transverse polarization. If spin rotators are added to the machine in order to transform the polarization of the beams from the transverse to the longitudinal direction, some very stringent tests of the Standard Model become possible by running at the Z^0 peak. All of which will keep CERN and its LEP user community very busy until at least the turn of the century.

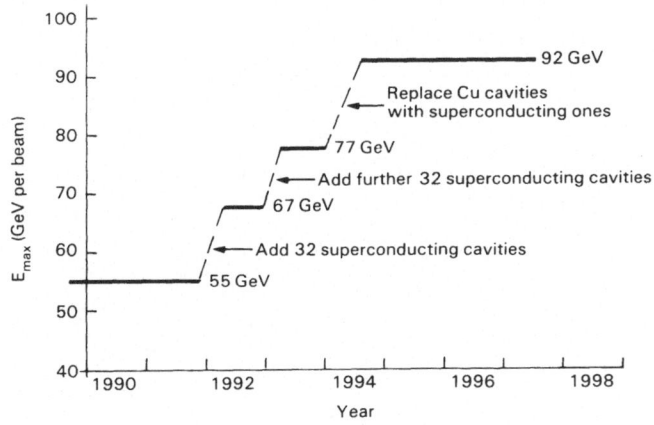

Fig. 14. Evolution of the LEP program

REFERENCES

[1] A more detailed report has been given by J.V. Allaby, Ann. New York Acad. Sci. **461**:447 (1986).
[2] R. Arnold et al., CRN Report CRN/HE 87-01 (1987), to be published in Nuclear Instruments and Methods.
[3] J. Allison et al., Nucl. Instrum. Methods **A236**:284 (1985).
[4] For a full treatment, see:
J. Ellis and R. Peccei (eds.), Physics at LEP (CERN 86-02, Geneva, 1986).
A. Böhm and W. Hoogland (eds.), Proc. ECFA Workshop on LEP 200, Aachen, 1986 (CERN 87-08/ECFA 87-108, Geneva, 1987).

PHYSICS AT FUTURE COLLIDERS

Mario Greco

INFN-Laboratori Nazionali di Frascati
P.O. Box 13, 00044 Frascati, Italy

ABSTRACT

The physics potential of next generation colliders is briefly reviewed, with emphasis on the CERN Large Hadron Collider (LHC) and e^+e^- Linear Collider (CLIC). Results are based on realistic calculations from several studies, especially the 1987 La Thuile-CERN Workshop on Physics at Future Accelerators.

Much of activity has been recently devoted to explore the physics possibilities of next generation hadron and lepton colliders, in the TeV region[1-3]. Indeed the success of the standard $SU(3) \otimes SU(2) \otimes U(1)$ gauge theory of strong and electroweak interactions and, at the same time, its well known limitations suggest that the frontier of new physics has to be expected in the energy range up to the order of a TeV in the parton-parton centre of mass. Although one may look forward to major discoveries from experiments at present or next coming machines as the $Sp\bar{p}S$ collider, the Tevatron, the SLC, LEP and Hera, it is unlikely that some specific issues of and beyond the Standard Model, and foremost the problem of electroweak symmetry breaking, will be settled there.

Over the past few years several studies[1] have been undertaken in the USA to explore the physics possibilities of a 40 TeV c.m. energy Superconducting pp Supercollider (SSC). More recently the discovery potentials of a Large Hadron Collider

(LHC) in the LEP tunnel, at 16 TeV c.m. energy, with an e-p option at $\sqrt{s} \approx 1.8$ TeV, and of an e^+e^- linear collider (CLIC) at $\sqrt{s} \cong 2$ TeV have been investigated at CERN[4]. In these studies a comparison has been made of the physics interest and the feasibility of experiments at the three types of particle colliders, with nominal luminosities $L = 10^{33} (\rightarrow 10^{34})$ $cm^{-2}s^{-1}$ for the pp and e^+e^- options.

In the present talk I will review the main physics points emerging from the La Thuile-CERN Workshop. More detailed considerations, as well as specific questions on experimental instrumentation and machine performances, can be found in the Workshop's Proceedings[4].

Spontaneous symmetry breaking in the minimal Standard Model[5] leads to the existence of a single neutral Higgs boson. As well known, there is no precise prediction for its mass. One can, however, use arguments of self-consistency to get qualitative lower and upper bounds which usually place m_H in the range (10-1000) GeV, although there is no argument that m_H cannot be several TeV. For $m_H \gtrsim 1$ TeV, where weak interactions become strong, one has $\Gamma_H \approx m_H^3/2$, where Γ_H and m_H are in TeV units, and the particle identification becomes subtle. There is however hope to find the Higgs at LEP in the channels ee$\rightarrow$ZH or $\theta(t\bar{t}) \rightarrow H\gamma$, provided the toponium θ is in the LEP range[7]. At future colliders experiments will be designed to detect the Higgs boson with $m_H \leq O(1$ TeV$)$ and either it will be discovered or the emergence of some new physics will be found.

Higgs production in pp collisions occurs via two dominant mechanisms, the gg and WW(ZZ) fusion (see Fig. 1)

FIG. 1

The first[2,8] dominates for low m_H, and is very sensitive to the spectrum of the heavy particles which could contribute in the fermion loop, in particular to m_{top}. The latter can be computed[9,10] quite unambigously and dominates for large m_H. A large value of m_{top}, which is presently suggested by the large mixing effects observed[11] in the $B\bar{B}$ system, enhances sizeably the gg mechanism, as shown in Fig.2 [12]. Clearly the WW cross section, shown in Fig.3 [10] for various beam energies, represents a minimal reference value for future experiments, the actual H production rate depending critically upon m_{top}. The question of the actual detection of the Higgs, according to the value of its mass and the corresponding main decay modes, and the critical problem of distinguishing the signal from the QCD background, will be discussed later.

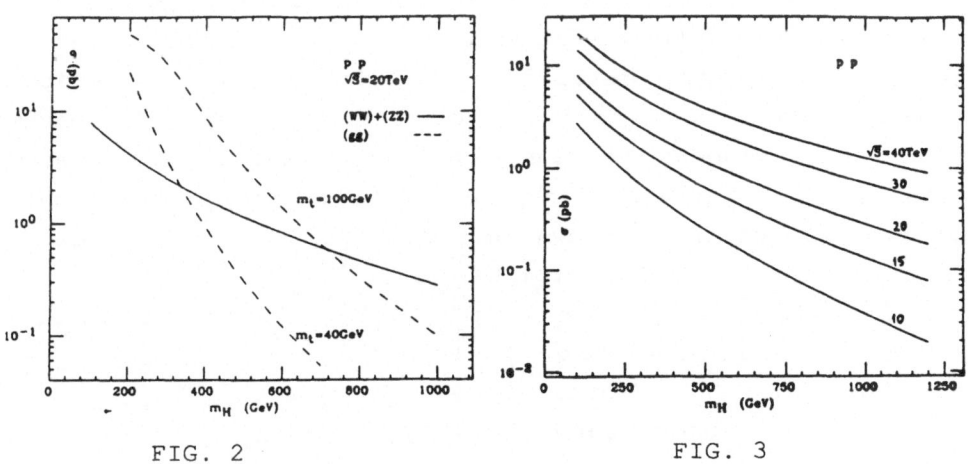

FIG. 2   FIG. 3

In e^+e^- annihilation it is important to note that all cross sections can be computed precisely and they fall into two categories, according to their energy dependence. The first includes all processes that are mediated by an s-channel γ or Z and fall like s^{-1}, as the $\mu^+\mu^-$ cross section $\sigma_{point} \approx 10^{-1}x$ pb/s(TeV^2). A typical example is the reaction $e^+e^- \to Z \to ZH$, which is dominant at LEP energies[7]. The cross section for $e^+e^- \to W^+W^-$ at high energy is $\pi\alpha_W^2 \ln(s/M_W^2)/2s$, with $\alpha_W \equiv \alpha/\sin^2\theta_W$, and also falls. On the other hand the cross sections which proceed through $\gamma\gamma$, WW or ZZ exchanges are scaled by M_W^{-2} and increase logarithmically with s, as for example $\sigma(e^+e^- \to \nu\bar{\nu}H) \approx \alpha_W^3 \ln(s/M_H^2)/16M_W^2$. They are shown in Fig. 4 [12,13].

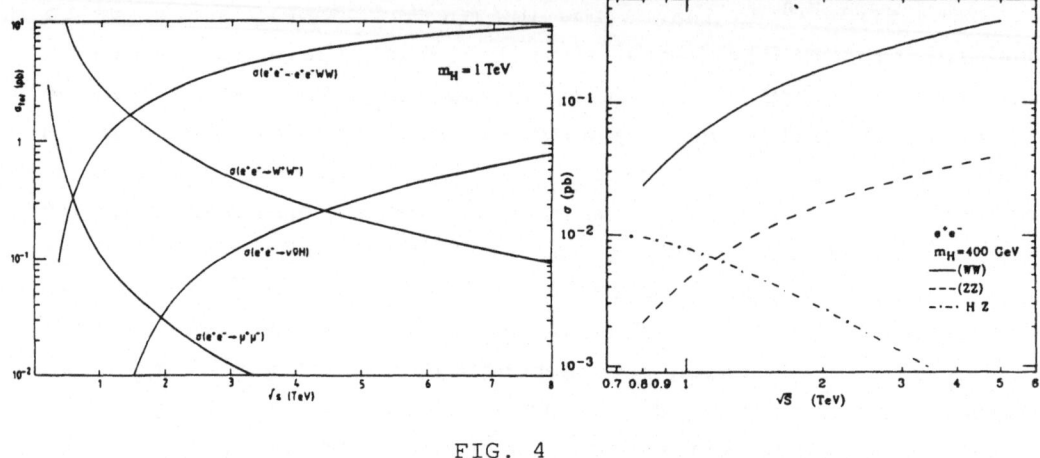

FIG. 4

The WW fusion mechanism has been studied estensively, particularly when $M_H > 2M_W$. Previous estimates[9] in the real W approximation, where one calculates $WW \rightarrow WW$ for real W's, folding thereafter with the W structure functions - similarly to the Weizsäcker-Williams approximation - have been recently implemented by exact calculations[10]. In particular, exact expressions have been derived for the differential cross sections with respect to the Higgs trimomentum, which are relevant for the background separation. For example the Higgs p_T distribution $d\sigma/dp_T$[10], shown in Fig. 5, is peaked near $p_T \approx m_W$ for all values of $\sqrt{s}$ and m_H of interest.

This feature plays an important role in the separation of the Higgs signal from the $\gamma\gamma$ background ($\gamma\gamma \rightarrow WW$) which is concentrated at $p_T \approx 0$.

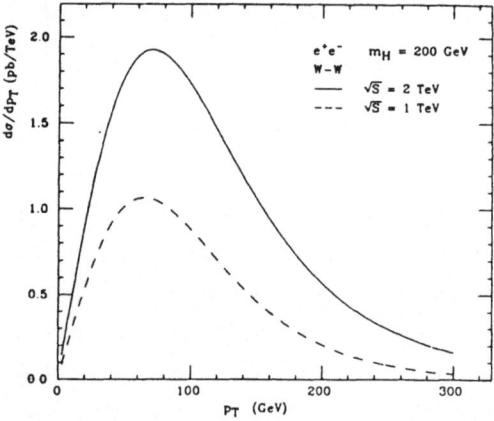

FIG. 5

In case of a "light" Higgs ($M_Z < m_H < 2M_W$), when the dominant decay mode is into heavy quarks $H \rightarrow Q\bar{Q}$, ($Q = b/t$ for $m_H </> 2m_t$), e^+e^- annihilation is practically the only viable mechanism for observing the Higgs. Indeed the peaking of the p_T-distribution as $p_T \approx m_W$, mentioned above, allows a clear separation of $H \rightarrow Q\bar{Q}$ from the $Q\bar{Q}$ pairs produced through $\gamma\gamma$ and γW fusion - which is the most relevant background - peaked at $p_T \approx 0$. This is shown in Fig. 6. On the contrary the detection is almost impossible in pp collisions, because of the overwhelming QCD background from $gg \rightarrow Q\bar{Q}$[14]. Some rare decay mode ($H \rightarrow \gamma\gamma$, $\tau^+\tau^-$) have been envisaged to overcome the problem, particularly at SSC energies, but the detection limitations are very severe. Therefore an e^+e^- collider is the ideal facility for the discovering of a "light" Higgs.

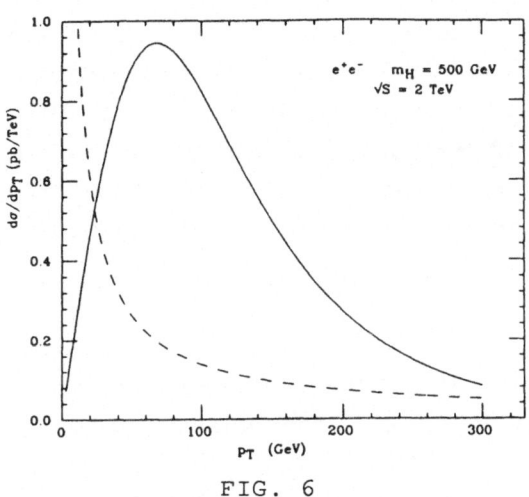

FIG. 6

When $m_H > 2m_{W,Z}$, as well known, the main decay modes are $H \rightarrow WW, ZZ$ with $\Gamma(H \rightarrow WW)/\Gamma(H \rightarrow ZZ) = 2$, with $\Gamma H \approx m_H^3/2$ for $m_H \gg 2m_W$. In pp collisions the production cross sections are comfortably large, as discussed above and explicitly shown in Fig. 3. The main problem is the background and here the value of the energy plays an important role. Indeed, for SSC, one can clearly use the fully leptonic decay mode $H \rightarrow ZZ \rightarrow 2l\bar{l}$, reaching a discovery limit of $m_H \lesssim 1.2$ TeV. At LHC, on the contrary, this decay mode is not enough and a limit of only $m_H \lesssim 0.6$ can be obtained through the channel $H \rightarrow Z \rightarrow \nu\bar{\nu} + e\bar{e}$ ($\mu\bar{\mu}$). On the other hand, the QCD background in the mixed channels Wjj or Zjj or even the ZZ production with subsequent decay $t \rightarrow b+W$, when $m_t > m_W$, are extremely hard to beat. The only hope of detecting heavy Higgs in the hadronic final states is restricted to the possibility of realizing an efficient system of tagging the outgoing quarks, after the emission of the W's or Z's producing the Higgs by fusion.

In e^+e^- annihilation the detection of the W's or Z's in the jet-jet mode, possible because of the absence of serious background problems, compensates the lower absolute production rates. The luminosity is the only crucial parameter which fixes the upper bounds of the discovery limits. Indeed, for $L=10^{33}(10^{34})cm^{-2}s^{-1}$ one gets $m_H \lesssim 0.7$ (1.1) TeV. Luminosity is also crucial for the detection of possible charged Higgs bosons, which are almost an impossible task for pp machines. Indeed "normal" charged Higgs, i.e. those expected in a natural supersymmetric extension of the Standard Model, do not couple as $H^{\pm}W^{\mp}Z$ or $H^{\pm}W^{\mp}\gamma$ at the tree level. Therefore the production rates in pp collisions are quite low and impossible to compete with the QCD background. In e^+e^- annihilation H^+H^- pairs can be produced via γ or Z exchange or via $\gamma\gamma$, and if $m_{H^{\pm}}$ is not to close to m_W – the reactions $e\bar{e} \rightarrow W^+W^-$ or $\gamma\gamma W^+W^-$ are otherwise a serious problem – one can reach a limit[12] $m_{H^{\pm}} \lesssim 0.8\ E_{beam}$ for $L \geq 10^{33}$ $cm^{-2}s^{-1}$.

The production and detection of heavy quarks and leptons of sequential type in the S.M. has been studied extensively and does not pose any problem. The heavy leptons (quarks) are expected to decay into W and $v(q')$ and can be revealed up to about 0.5 TeV (0.8 TeV) and 0.8 TeV (0.8 E_{beam}) in LHC and CLIC respectively[12,17]. These upper limits are slightly higher for SSC.

The study of the physics possibilities of LHC and CLIC in a framework going beyond the Standard Model has been confined[18] to a few topics – Supersymmetry, Additional Neutral Z' Bosons, Compositness and Leptoquarks – also for complementarity to previous studies[1,3]. Some general detector characteristics have been assumed, which include energy and momentum resolutions, granularity and angular coverage of the detector. The analysis performed by the study groups[18] cover a variety of cases which is hard to summarize shortly. We will just mention here a few illustrative examples, giving emphasis, in particular, to physics possibilities of CLIC.

The LHC machine can only produce sparticles with masses 0(1) TeV with observable cross sections if they are strongly interacting, i.e. squarks or gluinos. Furthermore, because of the high and difficult background situation for jets, a number of different topological cuts have to be applied, leading[18] to a bound of about 1 TeV for the sparticle masses. The SSC could reach out to ≈ 1.5 TeV. In Fig. 7 we show a typical distribution[19] for the process $pp \to \tilde{g}\tilde{q}$ assuming $m_{\tilde{g},\tilde{q}} \cong 300$ GeV and the decay modes $\tilde{g} \to \tilde{q}\bar{q}, \tilde{q} \to q\tilde{\gamma}$. Similarly in Table I a comparison of the expected SUSY signal/QCD background is reported[18,19].

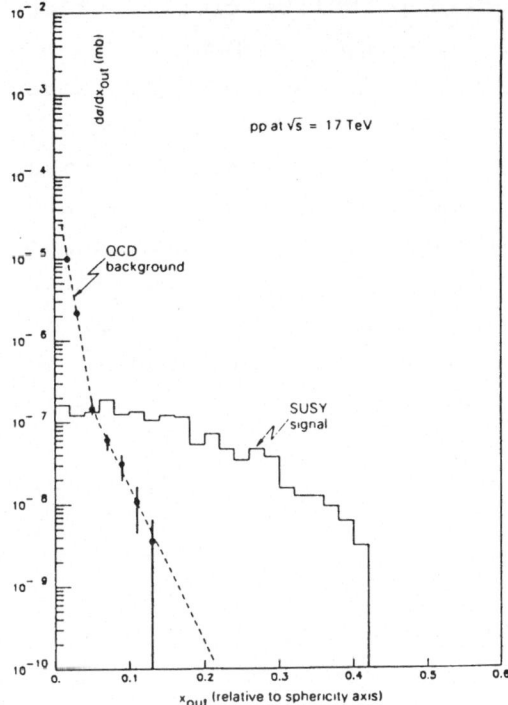

FIG. 7

TABLE I

Sparticle masses (GeV)		σ (mb)	Signal/background ratio (% of signal retained)			
$m_{\tilde{g}}$	$m_{\tilde{q}}$		No cut	$\not{E}_T > 200$ GeV	$x_E > 0.24$	$x_{out} > 0.08$
210	483	0.59×10^{-5}	0.21	13 (7.6%)	8.7 (6%)	94 (29%)
315	285	1.84×10^{-6}	0.06	21 (42%)	14 (31%)	58 (50%)
350	805	0.47×10^{-6}	0.21	5 (27%)	4 (11%)	17 (32%)
525	475	1.3×10^{-7}	0.06	3.3 (63%)	4.2 (41%)	9 (54%)
700	1610	0.74×10^{-8}	3.4×10^{-3}	0.27 (65%)	0.1 (12%)	0.54 (40%)
1050	950	1.92×10^{-7}	8.4×10^{-2}	0.75 (80%)	0.1 (52%)	0.2 (68%)

Supersymmetric events are in general much cleaner at CLIC, where all heavy, electroweakly interacting sparticles, including sleptons can be produced. The cross sections are generally small, since for producing a generic spin-0 $s\bar{s}$ one has

$$R \equiv \sigma(e^+e^- \to \gamma \to s\bar{s})/\sigma(e^+e^- \to \gamma \to \mu\bar{\mu}) = 1/4Q_s^2 N_c \beta^3,$$

where $\beta = p/E$, Q_s is the charge of s and N_c is the number of colours it has (one for sleptons, three for squarks). The standard pointlike cross section $\sigma(e^+e^-\rightarrow\gamma\rightarrow\mu\bar{\mu}) = 4\pi\alpha^2/3s = 87$ fb/$[s(TeV^2)]$ gives 220 $\mu\bar{\mu}$ pairs per year at $\sqrt{s}=2$ TeV, for $L = 10^{33}$ $cm^{-2}s^{-1}$. Some typical cross sections[18] for $e\bar{e} \rightarrow$ sparticles are shown in Fig. 8, including Z and γ exchanges. To do some reasonable physics one needs $L = 4\times10^{33}$ $cm^{-2}s^{-1}$ at $\sqrt{s} = 2$ TeV, corresponding to $\sim10^3$ $\mu\bar{\mu}$/year.

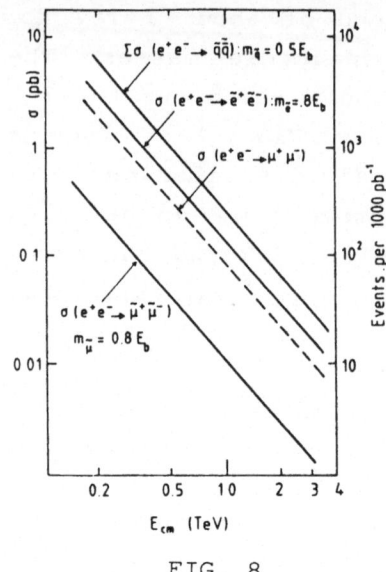

FIG. 8

The characteristic signature of supersymmetric particle pair-production is events with missing energy, missing transverse and total momentum, and dijet or dilepton final states which are acollinear and acoplanar. A typical example[20] is shown in Fig. 9, where the signal for $e^+e^-\rightarrow\mu^+\mu^-$ is compared to the background. In general, it can be concluded that CLIC, with an integrated luminosity of about 50 fb^{-1} at $\sqrt{s} = 2$ TeV, would allow detection of almost all sparticles with electroweak couplings in the mass range of 500 to 850 GeV.

Many different possible additional neutral gauge bosons Z' occur in different models. Some have been considered in previous studies[1]. Superstring inspired models[21] include extra U(1) subgroups, in addition to the Standard Model SU(3)⊗SU(2)⊗U(1) and therefore predict one or two new Z' with masses ≲ 1 TeV. With a certain ambiguity on the models and the corresponding Z' couplings, some illustrative examples[22] for the production cross-sections and the total widths of the Z' are shown in Fig. 10 for pp collisions. One can clearly probe for a high-mass Z' at the LHC, but CLIC could be used as a Z' factory to study its properties in great detail. This example explicitly shows the complementarity of the two machines.

A concise and schematic summary[12,18] of the discovery potentials of the two machines, including also some limits obtainable with the LHC ep option, are finally shown in Tables

II-III. It is clear that LHC represents the most natural continuation of the actual physics programmes at CERN and Fermilab, but it cannot compete with the physics expectations from SSC. On the other hand a future high energy e^+e^- collider, with a luminosity L $\gtrsim$ 1CO^{33} cm^{-2}s^{-1}, could offer additional exiting results which are certainly required for a full elucidation of new phenomena in the TeV energy region.

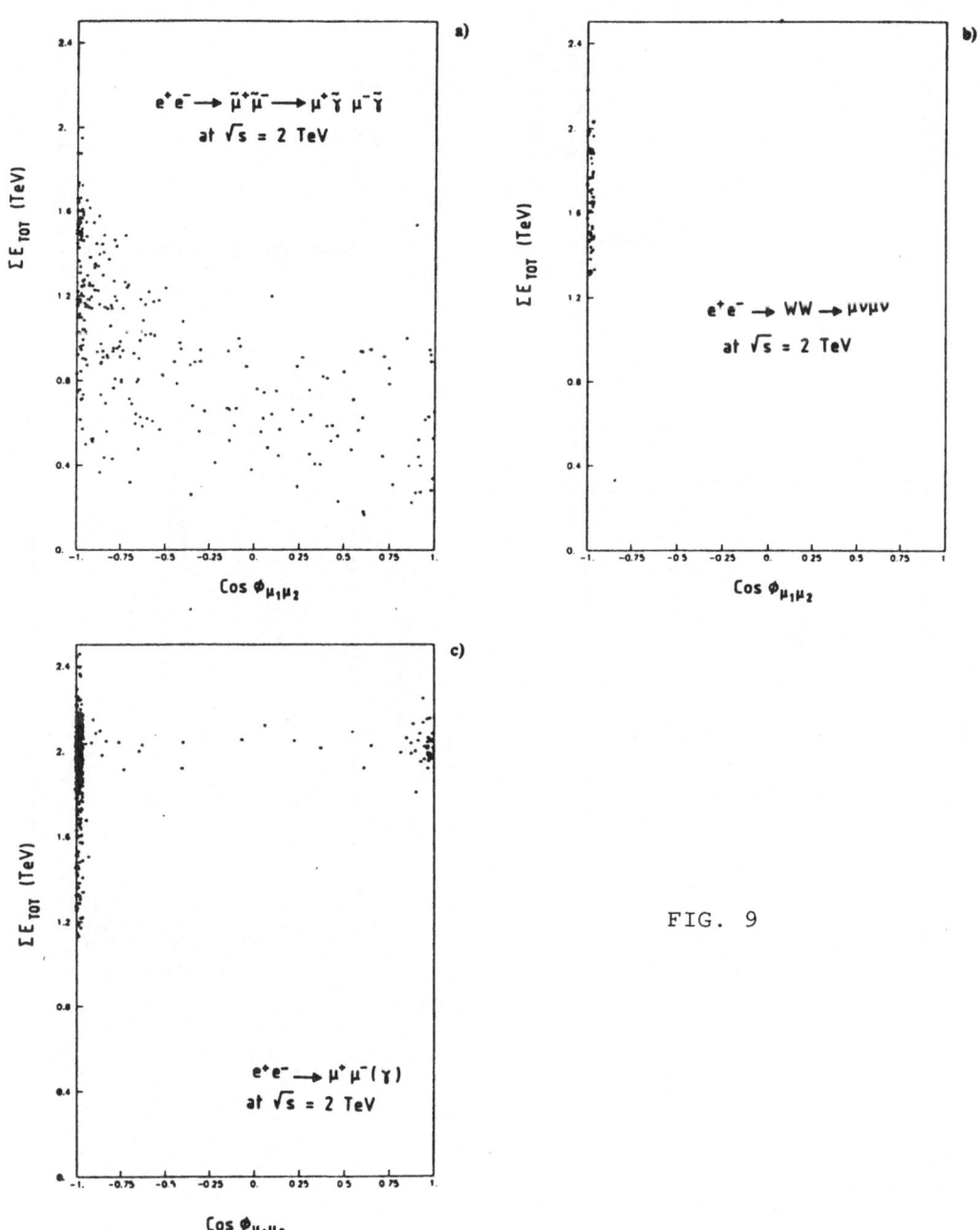

FIG. 9

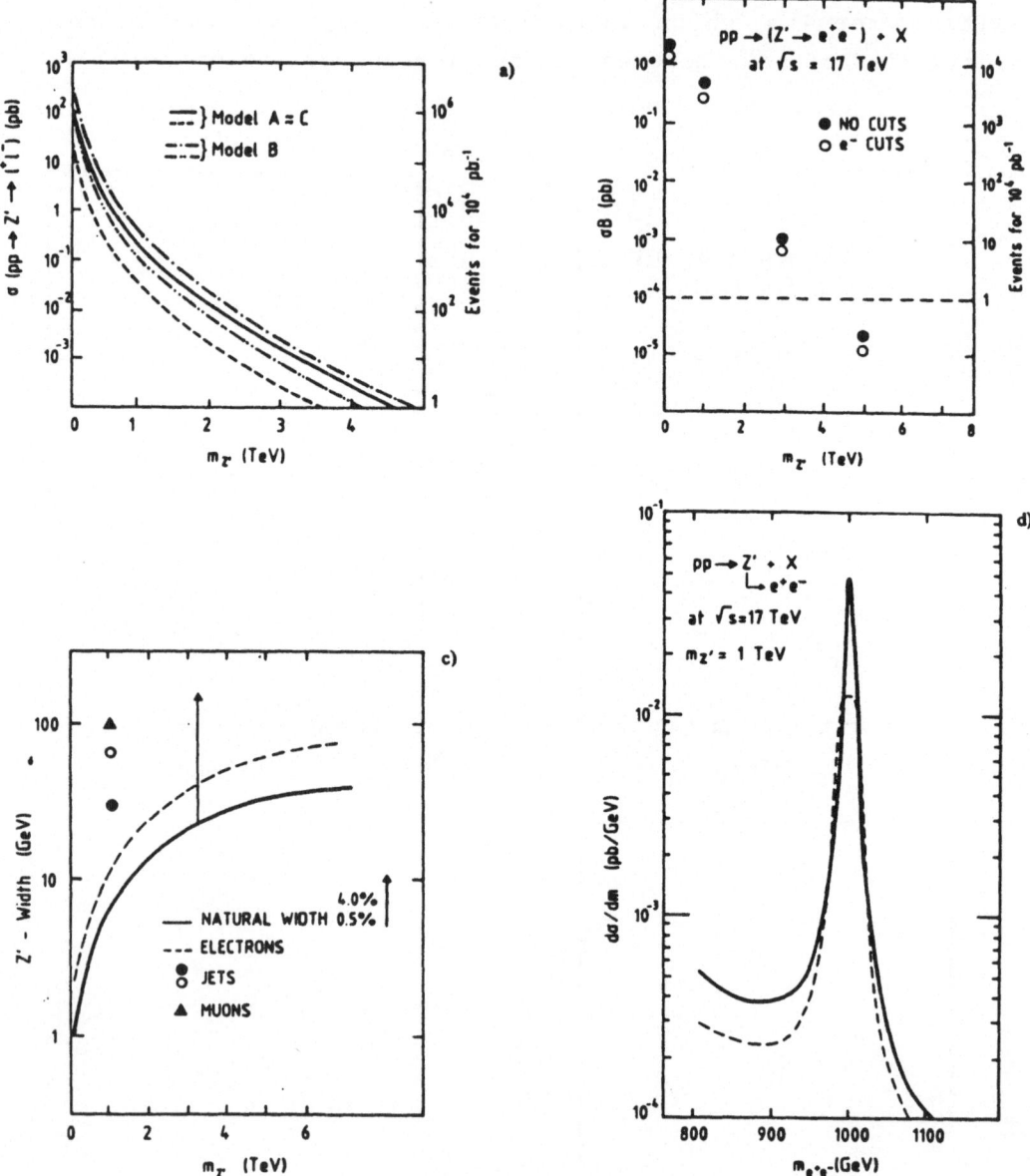

FIG. 10

TABLE II

	CLIC	LHC
Intermediate-mass Higgs: $m_Z < m_H < 200$ GeV $H \to Q\bar{Q}$ (Good up to $m_H < 300$ GeV)	Yes $\sqrt{s} \approx 1$ TeV also good ($L \approx 10^{32}$ cm^{-2} s$^{-1} \to$ marginal, $L \approx 10^{33}$ OK)	No (SSC: No)
Heavy Higgs: $m_H > 200$ GeV $H \to WW$ $H \to ZZ$	Yes $H \to 4$ jets, $m_H < 0.6$–0.8 TeV (If $L = 10^{34}$ cm^{-2} s^{-1}, $m_H < 1$–1.2 TeV: luminosity crucial)	Yes $H \to ZZ \to$ $\to \nu\nu + e^+e^-, \mu^+\mu^-$, $m_H < 0.6$ TeV ($m_H < 1$ TeV with quark tagging), SSC: $m_H < 1$–1.2 TeV, $\sqrt{s}$ crucial.
Charged Higgs: $H^+ \to t\bar{b}$	Difficult: $\sqrt{s} = 2$ TeV 60 ev. per year $\sqrt{s} = 1$ TeV 250 ev. per year. May be possible for $2m_W < m_H < 0.8E_{beam}$, Large m_H better; Luminosity crucial.	No (SSC: No)
Heavy leptons: $L \to \nu W$	$m_L < 0.8E_{beam}$ Possible $\sqrt{s} = 1$ TeV: better S/B	Possible $m_L < 0.5$ TeV (SSC: 0.7 TeV)
Heavy u,d quarks: $Q \to qW$	Yes (easy) $m_Q < 0.8E_{beam}$ Large m_Q better	6 Jets: No $4j + \ell\nu$: Promising $m_Q < 0.8$ TeV (SSC: 1 TeV)

TABLE III

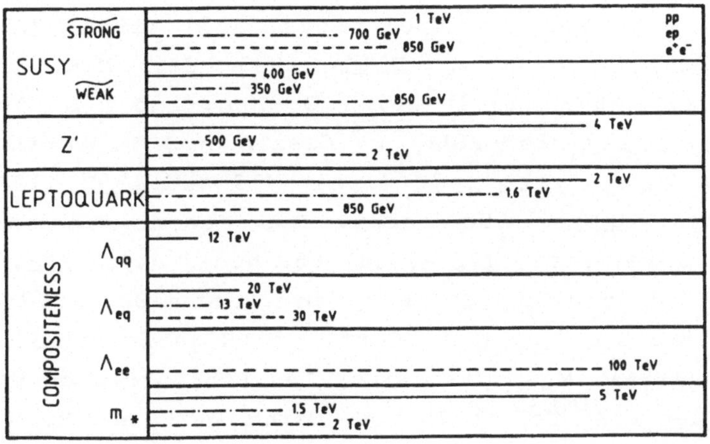

REFERENCES

1. Proc. 1982 Summer Study on "Elementary Particle Physics and Future Facilities", Snowmass, Colo., (1982), eds. R. Donaldson, R. Gustafson and F. Paige (AIP, New York, 1983); Proc. 1984 Summer Study on the "Design and Utilization of the Superconducting Super Collider", Snowmass, Colo., (1984), eds. R. Donaldson and J. Morfin (AIP, New York, 1985); Proc. 1986 Summer Study on the "Physics of the Superconducting Super Collider", Snowmass, Colo., (1986), eds. R. Donaldson and J. Marx, in preparation; Supercollider Physics, Proc. Oregon Workshop on "Super High Energy Physics", Eugene, Oregon (1985), ed. D.E. Soper (World Scientific, Singapore, 1986).

2. E. Eichten, I. Hinchliffe, K. Lane and C. Quigg, Rev. Mod. Phys. $\underline{56}$ (1984) 579.

3. Proc. ECFA-CERN Workshop on a "Large Hadron Collider in the LEP Tunnel", Lausanne and CERN (1984) ed. M. Jacob (ECFA 84/85, CERN 84-10, Geneva, 1984).

4. Proc. of the Workshop on "Physics at Future Accelerators", La Thuile-CERN (1987), CERN 87-07.

5. S.L. Glashow, Nucl. Phys. $\underline{22}$ (1961) 579; S. Weinberg, Phys. Rev. Lett. $\underline{19}$ (1967) 1264; A. Salam, in Proc. of the 8th Nobel Symposium, ed. N. Svartholm (Almquist and Wiksell, Stockholm 1968) 367; S.L. Glashow, J. Iliopoulos, L. Maiani, Phys. Rev. $\underline{D2}$ (1970) 1285, G. 't Hooft, Nucl. Phys. $\underline{B33}$ (1971) 173 and $\underline{B35}$ (1971) 167.

6. M. Veltman, Acta Phys. Pol. $\underline{B8}$ (1977) 475; B.W. Lee, C. Quigg and H.B. Thacker, Phys. Rev. $\underline{D16}$ (1979) 1519; J.J. Van der Bij and M. Veltman, Nucl. Phys. $\underline{B231}$ (1984) 205; J.J. Van der Bij, Nucl. Phys. $\underline{B161}$ (1985) 341; M.B. Einhorn, Nucl. Phys. $\underline{264B}$ (1984) 75; P.Q. Hung and H.B. Thacker, Phys. Rev. $\underline{D31}$ (1985) 2866; R. Casalbuoni, D. Dominici and R. Gatto, Phys. Lett. $\underline{147B}$ (1984) 419 and $\underline{155B}$ (1985) 95, Nucl. Phys. $\underline{B282}$ (1987) 235; M.S. Chanowitz and M.K. Gaillard, Phys. Lett. $\underline{142}$ (1984) 85, Nucl. Phys. $\underline{B261}$ (1985) 379; M.S. Chanowitz, Berkeley preprint LBL-21973 (1986) presented at the 23rd Int. Conf. on "High Energy Physics", Berkeley (1986); M.S. Chanowitz, Les Rencontres de Physique de la Vallée d'Aoste, La Thuile, March 1987, ed. by M. Greco (Editions Frontières).

7. For reviews of LEP/SLC Physics, see for example: Proceedings of the "1978 LEP Summer Study", CERN 79-01 (1979); "Cornell Z^0 Theory Workshop", February 1981; M.E. Peskin and S.H. Tye editors; SLC Workshop, SLAC Report 247 (1982); Workshop on "Radiative Corrections in $SU(2)_L xU(1)$", Trieste 1983, B.W. Lynn and J.F. Wheater editors; Conference on "Tests of Electroweak Theories", Trieste 1985, B.W. Lynn and C. Verzegnassi editors; Physics of LEP, ed. by J. Ellis and R. Peccei, CERN 86-02 (1986).

8. H.M. Georgi et al., Phys. Rev. Lett. <u>40</u> (1978) 692.

9. R. Cahn and S. Dawson, Phys. Lett. 136B (1984) 196; E <u>138B</u> (1984) 464. See also D. Jones and S. Petcov, Phys. Lett. <u>84B</u> (1979) 440 and Z. Hioki, S. Midorikawa, and H. Nishiura, Prog. Theor. Phys. <u>69</u> (1983) 1484. For the effective *W* approximation see M. Chanowitz and M. Gaillard, Phys. Lett. <u>136B</u> (1984) 196; S. Dawson, Nucl. Phys. <u>B29</u> (1985) 42; G. Kane, W. Repko, and W. Rolnick, Phys. Lett. <u>148B</u> (1984) 367.

10. G. Altarelli, B. Mele and F. Pitolli, Nucl. Phys. <u>B287</u> (1987) 285.

11. ARGUS Collaboration, D.G. Cassel at the Rencontres de Physique de la Vallée d'Aoste, La Thuile, March 1987, ed. by M. Greco (Editions Frontières).

12. G. Altarelli et al., The Standard Theory Group, ref. 4.

13. G.L. Kane and J.J.G. Scanio, CERN-TH 4532/86 (1986).

14. M. Greco, Phys. Lett. <u>156B</u> (1985) 109; J.F. Gunion, Univ. California Davis preprint UCD-86-39 (1986).

15. J.F. Gunion et al., Phys. Rev. <u>D34</u> (1986) 181.

16. J.F. Gunion, Z. Kunszt and M. Soldate, Phys. Lett. <u>163B</u> (1985) 389, <u>E168B</u> (1986) 427; W.J. Stirling, R. Kleiss and S.D. Ellis, Phys. Lett. <u>163B</u> (1985) 261; J.F. Gunion and M. Soldate, Phys. Rev. <u>D34</u> (1986) 826.

17. D. Froidevanx, ref. 4.

18. J. Ellis and F. Pauss, Beyond the Standard Model, ref. 4.

19. A. Savoy-Navarro and N. Zaganidis, ref. 4. See also S. Dawson and A. Savoy-Navarro in Proc. Snowmass '84, ref. 1.

20. C. Dionisi and M. Dittmar, ref. 4.

21. For reviews see J. Ellis, preprint CERN-TH 4439/86 (1986; H.P. Nilles, preprint CERN-TH 4444/86 (1986); L.E. Ibáñez, preprint CERN-TH 4459/86 (1986).

22. P. Bagnaia, ref. 4.

HIGGS PRODUCTION IN HIGH ENERGY HADRON COLLISIONS

Zoltán Kunszt

Institute of Theoretical Physics
ETH, Höngg, Zürich
and
ITP Santa Barbara
Santa Barbara, California, 93106

I. INTRODUCTION

The standard model is generally accepted as the theory of the strong, electromagnetic and weak interactions. However, there are important missing ingredients and loose ends which can only be answered if new facilities will come into operation. In particular, the top quark and the Higgs boson have to be yet discovered. The experimental determination of the values of their masses will provide fundamental information on the validity of the Higgs mechanism and on the range of validity of the standard model.

The top quark mass will be measured in the next round of collider experiments or the standard model with three flavours can not be valid[1]. The present lower limit on the top mass is about 40GeV, while the measured values of the masses of the W and Z bosons provide an upper limit ≈ 200 GeV. It is expected that at the CERN $Sp\bar{p}S$ and at the Tevatron of Fermilab either the top will be discovered or the allowed region of its mass will shrink to a small interval $\approx 120 - 150$ GeV. The precision measurements at the Z-pole will determine the top mass indirectly from radiative corrections with an accuracy of ≈ 5 GeV.

The Higgs search appears to be more difficult. At present we have no direct or indirect experimental evidence on the existence of the Higgs particle. The theory is unable to predict its mass. The Higgs mass can be anywhere between a few GeV and a TeV. Low energy experiments constraint the value of the Higgs mass rather weakly[2]. High precision data of nuclear reactions ($0^+ - 0^+$ transitions, angular correlations, neutron scattering), and of the decays $\pi^+ \to e^+ \nu \phi$, $K^+ \to \pi^+ \phi$ and $\eta\prime \to \eta\phi$ can exclude only the mass range $\approx 0 - 400$ MeV. Recent analyses[3] of the $\Upsilon \to H \gamma$ decay have concluded that the Higgs mass cannot be in the mass range $0.5 - 3.0$ GeV. The only region not covered is a small mass interval around the kaon mass. At LEP-I, a mass limit $M_H > 60$ GeV will be attained while at LEP-II the lower mass limit of ≈ 85 GeV can be established [2]. However, we have one fundamental information on the Higgs sector. The measured value of the ρ parameter[1] $\rho = m_W^2/m_Z^2/cos^2\theta_W$ is equal to 1 within the 5 percent experimental error ($\rho \approx 1.03 \pm 0.05$), a value predicted by the standard model with one or more doublets . The data support the minimal Higgs structure. There are probably just one or two doublets (supersymmetry needs two). It is likely that Nature does not prefer the proliferation of Higgs bosons .

There is a theoretical upper bound on the mass of the Higgs if we require that the renormalizable perturbative description remain valid. We cannot move the Higgs mass to infinity while keeping the Z and W mass values and the gauge coupling fixed since this limit correspond to the non-renormalizable theory of the massive gauge bosons. So there must be an upper limit on the Higgs mass. Considering the breakdown of tree level unitarity in WW scattering Veltman[4], and Lee, Thacker and Quigg[5] obtained the upper bound $M_H < 1$ TeV. Therefore either the Higgs boson mass is less that ≈ 1 TeV or the standard model breaks down at an energy scale $\Lambda \approx 1$ TeV.

This conclusion can be further substantiated by arguments using the one loop renormalization group equation for the coupling constants of the Higgs self-interaction λ appearing in the effective potential

$$V(\phi) = \lambda(\phi^\dagger \phi - v^2/2)^2. \tag{1.1}$$

The mass of the W-boson is $m_W = gv/2$ and in the tree approximation λ is directly related to the Higgs mass

$$\lambda = \frac{M_H^2}{2v^2} = \frac{M_H^2}{8m_W^2}g^2, \tag{1.2}$$

where g is the SU(2) gauge coupling constant. The one loop renormalization group equations have been obtained in the general case by Cheng, Eichten and Li[6]. In the case when the Higgs mass is large the contributions from the light fermions and gauge couplings can be neglected and we obtain

$$\frac{d\lambda}{dt} = \frac{3}{2\pi^2}\lambda^2 \tag{1.3}$$

where $t = ln(\mu)$ and μ is the running scale. This equation tells us that the four component ϕ^4 theory is "infrared free", i.e. for any arbitrary positive coupling the theory becomes free at asymptotically large distances (in QCD the theory becomes free at asymptotically small distances):

$$\frac{1}{\lambda(\mu)} = \frac{1}{\lambda(\Lambda)} + \frac{3}{2\pi^2}ln(\frac{\Lambda}{\mu}) \tag{1.4}$$

This is the intuitive argument for the triviality of the ϕ^4 field theory first suggested by Kogut and Wilson[7]. That the ϕ^4 theory is indeed trivial has been shown by Aizemann and Fröhlich[8] using non-perturbative, exact inequalities. Recently, non-perturbative studies of the lattice ϕ^4 field theory have shown[9] that the one loop perturbative solution to the renormalization equations is a good approximation up to scales of $\mu \approx \frac{1}{2}\Lambda$. Although the effect of the gauge couplings has been neglected these arguments suggest that if the mass of the Higgs boson is relatively heavy[2] the standard model must be considered as an effective field theory with a cut-off. The numerical value of the cut-off is related to the Higgs mass and to the top mass. The interesting feature of these arguments is that they are based on the mathematical properties of the model. They are used within the framework of the standard model without making any reference to grand unification, to quadratic divergences, or to the gauge hiearchy problem. However, since the standard model does not include gravity we know that its range of validity can be extended at most up to the Planck mass $M_P = 10^{19} GeV$ which gives the physical limit for the energy scale where the break down of the standard model must occur.

---

[2]If the renormalized Higgs self coupling and the Yukawa couplings are not too large at the Fermi scale the standard model may have non-trivial continuum limit[10]. Physically this is irrelevant since the Planck mass gives a physical cut-off.

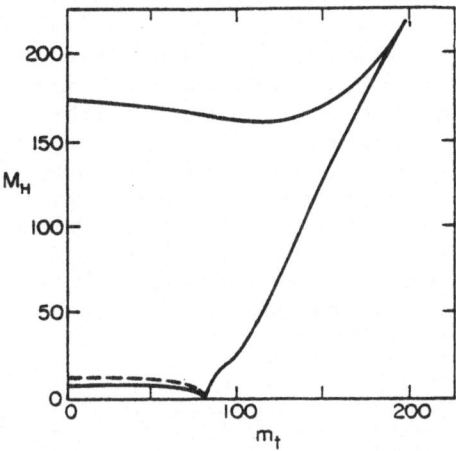

Figure 1. Allowed region of the Higgs mass M_H and the top mass m_t obtained in ref.[11] assuming the validity of perturbation theory up to the unification scale 10^{15}GeV. The shaded area gives the experimentally excluded regions.

The requirement that the standard model remains valid up to the Planck scale will put constraints on the mass values of the heavy particles. In fig.1 we show the allowed region of the top mass and the Higgs mass obtained by Cabibbo, Maiani, Parisi and Petronzio[11] requiring that the standard model remains valid up to the grand unification scale $\Lambda \approx 10^{15}$GeV (the dependence of the allowed region on Λ above this scale is very weak). With the measurement of the value of the Higgs mass, we can obtain the value of the cut-off and hence the energy scale where the the standard model should break down. The scale obtained by these arguments are smaller than the unitarity limit. Tentatively we can conclude that if the Higgs mass value is above ≈ 200 GeV the cut-off Λ must be substantially below the Planck-scale. Increasing the Higgs mass above 200 GeV the cutoff value decreases exponentially

$$\Lambda < m_W e^{\frac{16\pi^2 m_W^2}{3M_H^2 g^2}} = m_W e^{(\frac{940}{M_H})^2} \tag{1.5}$$

where the Higgs mass is measured in GeV units. If the Higgs mass is above $\approx 500-600$ GeV the scale of new physics must be around 1 TeV. This is the so called "no loose scenario". If future accelerators will be capable of discovering the Higgs boson up to $\approx 600 - 800$ GeV then either the Higgs boson will be discovered or we shall find signatures of new physics. This new physics signature will be given either by production of new particles (supersymmetery, technicolor, WW bound states etc.) or by strong scattering of longitudinally polarized W and Z bosons.

In section 2 we shall discuss the scattering of longitudinal gauge bosons, in section 3 the main results obtained in LHC and SSC studies are reviewed. Finally in section 4 we briefly comment on the Higgs search potential of "post SSC" machines.

II. SCATTERING OF LONGITUDINAL GAUGE BOSONS

In general, the scattering of longitudinally polarized gauge bosons are the key processes in the study of the electroweak symmetry breaking mechanism[4]. In these reactions the effects which motivated the introduction of the Higgs boson will show up in leading order. Without the contributions of the Higgs boson, the cross-section of these processes will grow and violate the unitarity limit. The Higgs boson appears as an s-channel resonance with a coupling strength to longitudinal gauge bosons given by (1.2). This feature can be easily understood with the help of the equivalence theorem[12,13] which says that at energies large compared to m_W the longitudinal components of the gauge bosons W_L and Z_L are essentially the Goldstone bosons arising from the spontaneous symmetry breaking in the scalar sector in $g = 0$ limit. It has been recently shown[14] that the equivalence theorem becomes an identity in the axial gauge. The key point in the argument is that in the axial gauge a space like unit vector n_μ is introduced to define the gauge and the expressions of the gauge boson propagators greatly simplify when they are expanded in terms of the parameter $m_W^2/(nq)^2$ where q denotes the four momentum of the propagator. In the axial gauge we have no ghost particles and in this limit the propagator becomes diagonal in the transverse and longitudinal degrees of freedom. Furthermore, if $M_H \gg m_W$, the couplings of these longitudinal gauge bosons (see fig.2) become the same as the scalar (pseudo-Goldstone) bosons of the original Lagrangian (before the Higgs mechanism was implemented). In the limit that $M_H^2, s, -t \gg m_W^2$ the amplitudes of $W_L W_L$ scattering can be easily obtained[5,15,14]

$$A(W_L^+ W_L^- \to W_L^+ W_L^-) = i\frac{g^2}{4}\frac{M_H^4}{m_W^2}\left(\frac{1}{s - M_H^2 + iM_H\Gamma_H} + \frac{1}{t - M_H^2 + iM_H\Gamma_H} + \frac{2}{M_H^2}\right)$$

$$(2.1)$$

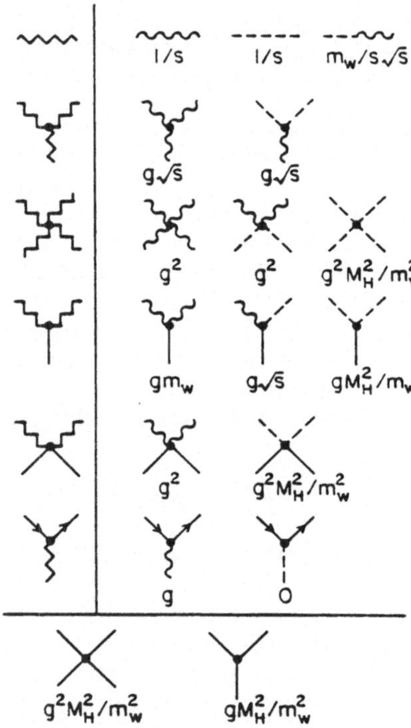

Figure 2. Feynman rules for gauge boson and Higgs couplings in the axial gauge[14].

The regime of the strongly interacting $W_L W_L$ sector is expected to emerge at $m_H > 0.8\ TeV$. The theoretical analysis of this regime requires at least a partial understanding of two questions. First, we have to analyze whether the effective W-approximation[16], which gives the theoretical framework to relate the machine luminosity to the longitudinal gauge boson luminosity remains valid in the case of strongly interacting gauge bosons. The question has been positively answered recently [14]. The next important question concerns whether we can estimate reliably the cross-sections of WW scattering at $\sqrt{s} > 0.8\ TeV$. With some assumptions concerning s-channel resonances, Chanowitz and Gaillard[12] has shown that the answer is affirmative also to this question. The key observation is that the longitudinal gauge bosons are the Goldstone bosons of the Higgs sector, therefore, the soft boson theorems (similar to soft pion theorems) determine the magnitude of the cross-section in the region around 1 TeV centre of mass energy[12,17].

The amplitude of $W^+W^- \to Z^0Z^0$ scattering in the limit when $M_H^2 \gg \hat{s} \gg m_W^2$ for example (see eq.(2.2)) has the form[3]

$$A(W^+W^- \to Z^0Z^0) = ig^2\frac{\hat{s}}{4m_W^2} \qquad (2.2)$$

In this picture the Higgs boson appears (similarly to the σ resonance in $\pi\pi$ scattering) around ≈ 1 TeV. Even if a light Higgs particle will be found the experimental study of the scattering of longitudinal gauge bosons will remain a crucial test of the electroweak symmetry breaking mechanism. At the LHC[18] and the SSC[19−20] the existence of the strongly interacting Higgs sector can be established but larger energy is needed to study its properties.

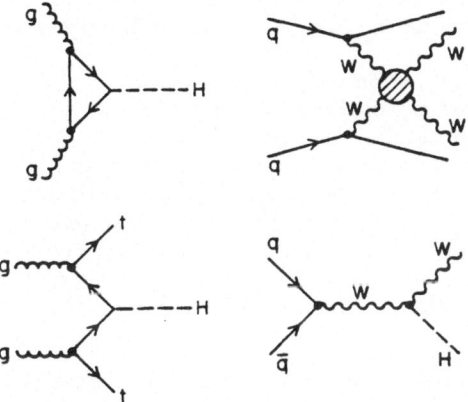

Figure 3. Illustrative Feynman diagrams for the four main Higgs production mechanisms in hadron collisions.

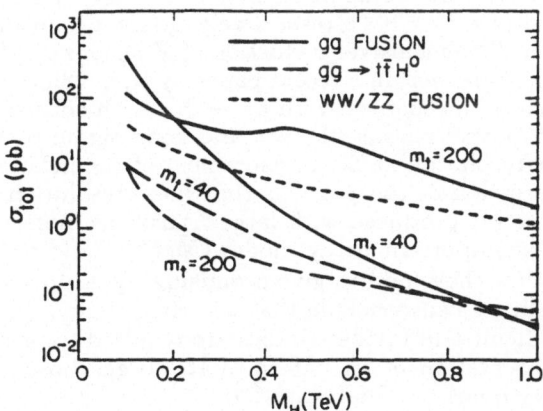

Figure 4. Higgs production cross sections at $\sqrt{s} = 40$ TeV via gluon fusion, associated production with top quark and WW/ZZ fusion at two different top mass values $m_t = 40\text{GeV}$ and $m_t = 100\text{GeV}$[25].

---

[3]The s-wave $\pi^+\pi^- \to \pi^0\pi^0$ scattering amplitude has the same form but m_W^2/g^2 is replaced by $4\pi F_\pi$

It is an exciting prospect that the next generation of proton-proton collider the LHC ($\sqrt{s} = 16.5\ TeV, \mathcal{L} \approx 10^{33} cm^{-2} s^{-1}$) and the SSC ($\sqrt{s} = 40\ TeV$), $\mathcal{L} \approx 10^{33} cm^{-2} s^{-1}$) and perhaps an $e^+ e^-$ colliders like the recently discussed Cern Linear Collider (CLIC) ($\sqrt{s} = 2\ TeV, \mathcal{L} \approx 10^{32} cm^{-2} s^{-1}$), will reach the necessary energies where decisive experimental information can be gained on the existence of the Higgs boson. The Higgs mass range can be explored up to $\mathcal{O}(1\ TeV)$, a value close to the unitarity limit. At recent workshop studies for LHC, SSC and CLIC one of the main goals was to establish the feasibility of Higgs search up to the 1 TeV regime.

Let us summarize the most significant new results of these studies (see the recent reports by G. Altarelli, by D. Froideveaux [18] at the La Thuile Workshop (1987, LHC, CLIC) and by J. Gunion et al.[19] at the Madison Worshop (1987,SSC) and R. Cahn et.al.[20] at the Berkeley Workshop(1987,SSC).

One can naturally devide the discussions of the Higgs search (after LEP-II) into three mass regions

Intermediate mass region $m_W < m_H < 2 m_W$

Heavy Higgs mass region $2 m_W < m_H < 0.8\ TeV$

Strong interaction region $m_H > 0.8\ TeV$.

There are four main production mechanisms for hadron colliders : gluon-gluon fusion[21], WW/ZZ fusion[16,22], associated production[23] (fig.3) with heavy quarks and associated production with weak gauge bosons[24].[4] The last two mechanisms give small contributions. In the heavy Higgs region the competition between the first two mechanism depends on the Higgs mass . In the intermediate region the gluon gluon fusion, while in the strong interaction region the WW fusion is the dominating production mechanism (see fig.4)[25].

The intermediate mass region. The production rate is rather large, the difficulty of Higgs search is in the huge jet background. In this regime the decay into a heavy quark pair is the dominant mode. Therefore, the tentative conclusion has been drawn in earlier studies that the observation of the 'intermediate' Higgs boson at hadron colliders may not be feasible[23,26]. However, the prospects for observing the intermediate Higgs boson improve significantly in the heavy top quark case favoured by recent Argus data[27] on $B^0 - \bar{B}^0$ mixing. If $m_t > m_H/2$, the dominant decay mode is the decay into $b\bar{b}$ and the branching ratios into $\tau^+ \tau^-$ and $\gamma\gamma$ final states increases up to $\approx 4 \times 10^{-1}$ and $\approx (1-5) \times 10^{-4}$, respectively, giving promising rates and good signatures for detection. The inclusive production of Z-bosons to $\tau^+ \tau^-$ decay gives a formidable background. However, in a recent paper by R.K. Ellis et al.[28] it has been shown that the production mechanism with $gg \to Hg$ with subsequent decay of the Higgs boson into $\tau^+ \tau^-$ gives measurable rate and good signal to background ratio assuming $\approx 10\%$ mass resolution in the measurement of the invariant mass of the τ pairs. The main observation of the paper is that mass resolution improves significantly if the Higgs boson is produced with large transverse momentum. A detailed study of the three most important decay modes ($b\bar{b}, \tau^+ \tau^-$ and $\gamma\gamma$) by J.F. Gunion et al.[29] has revealed that they tend to give promising signatures in complementary mass intervals. Overall, we can conclude that the Higgs search in the intermediate mass region is very difficult and further studies are required to prove its feasibility.

At CLIC, the observation of an 'intermediate' Higgs boson should not be a problem even via heavy quark jet final states[30].

Heavy Higgs mass region. In this region the dominant modes are the decays into gauge boson pairs. The decay $H \to Z\bar{Z}$ with both Z's decaying to e's or μ's,

---

[4] For $e^+ e^-$ colliders only the WW/ZZ fusion mechanism is relevant.

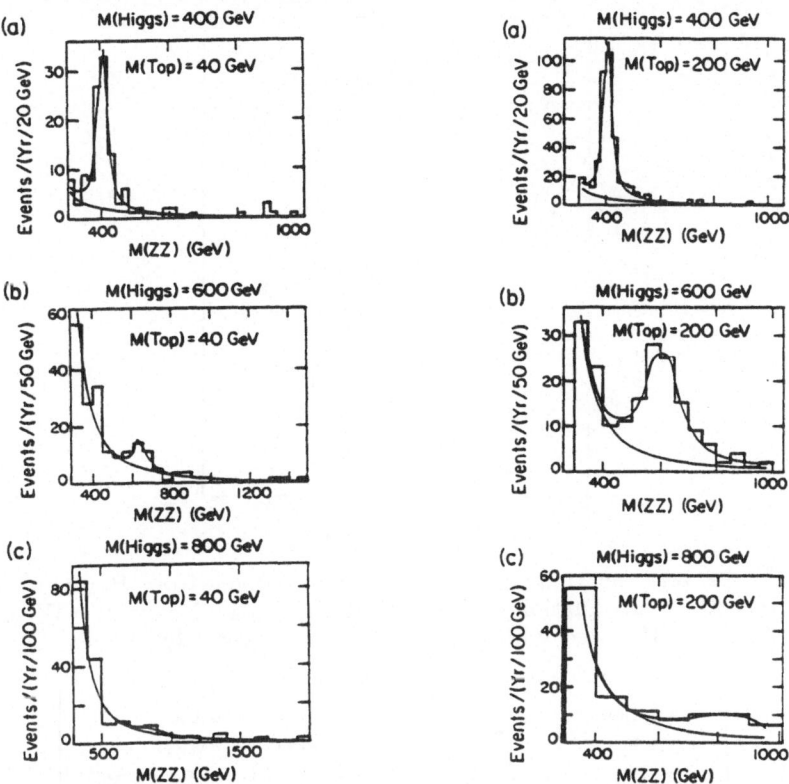

Figure 5. The ZZ invariant mass distributions arising from Higgs decay and from the background process $q\bar{q} \to ZZ$. The branching ratios when both Z decay into ee or $\mu\mu$ are included in the normalization. The curves have been calculated for integrated luminosity of 10^{40} with rapidity $|\eta_Z| < 1.5$. The distributions are shown for Higgs masses $400, 600, 800$ GeV at two values of the top mass $m_t = 20$ GeV and 200 GeV (figs. (a)-(f)).

gives an unmistakable signature of Higgs production. The background is given by the continuum production of Z-pairs and it is small. Figure 5 summarizes the main result. The conclusion again depends on the value of the top mass. Assuming $m_t = 200$GeV the discovery limits are $m_H \approx 0.6\ TeV$ at the LHC and $m_H \approx 0.8\ TeV$ at the SSC. With $m_t = 40$GeV these limits decrease to $m_H \approx 0.3\ TeV$ and $m_H \approx 0.6\ TeV$, respectively. Studies are in progress at CERN and at SSC workshops to consider a higher luminosity option, $\mathcal{O}(10^5 pb^{-1})$, with special high resolution lepton detectors. The aim is to push the discovery limit to larger Higgs mass values.

The decay mode $H \to WW$ with one W decaying into leptons and with the other decaying into jets is overwhelmed with the huge QCD background of $W + jets$ production[31]. In fig.7 we show invariant mass distributions for the Higgs production and the QCD background at LHC energies. When the Higgs boson is produced by WW fusion mechanism it is produced in association with forward-backward jets at small angles and transverse momentum of the $\mathcal{O}(m_W/2)$. Therefore the very large QCD background is not irreducible, therefore one may try to further supress the background by tagging these forward jets.[32] This question has been recently considered by Kleiss and Stirling[33] for LHC. They have calulated in a certain approximation the QCD process $qq \to Wjjjj$ as a source of background to the WW fusion Higgs production with tagged 'forward jets'. Requiring two high energy jets

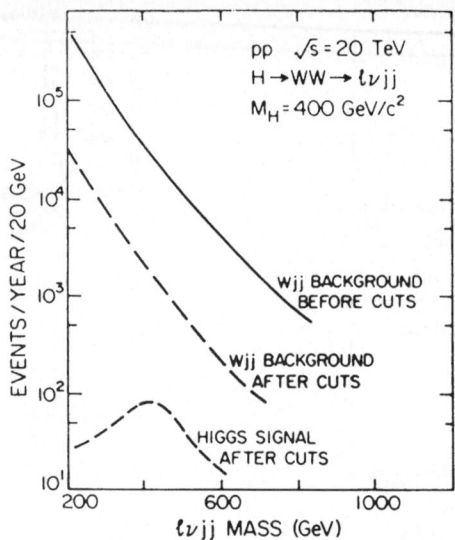

Figure 6. $W + 2\ jets$ invariant mass distributions arising from Higgs production and direct QCD production at LHC when the Higgs decays into W pair and subseqently the W's decay into $l\nu$ and $2\ jets$, respectively[18].

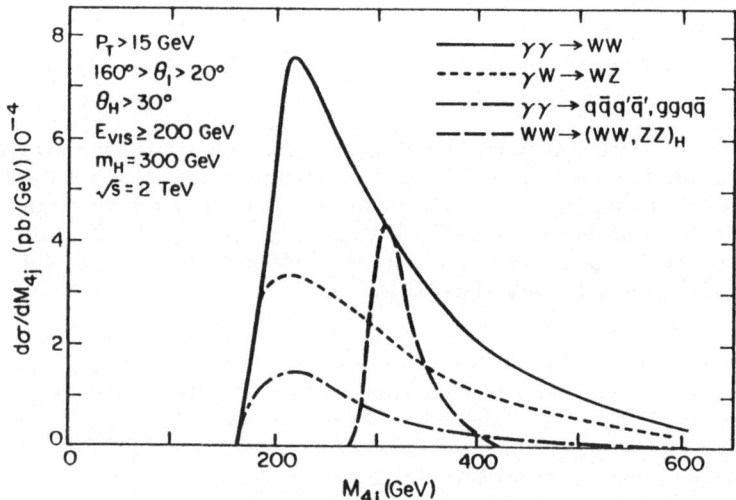

Figure 7. Invariant mass distributions of four jet final states at CLIC arising from Higgs production and background processes[35] $\gamma\gamma \rightarrow WW, \gamma W \rightarrow WZ,$ $\gamma\gamma q\bar{q}q'\bar{q}', ggq\bar{q}$ with $M_H = 300$ GeV and with polar angle cut $30^0 < \theta < 150^0$. It has been required that the jets have total visible energy $E > 300$GeV, transverse momentum $p_T > 15$GeV and they are separated with an opening angle $\theta_{jj} > 30^0$.

in the forward region in the rapidity interval $3 < \eta_{jet} < 5$ and two jets in the central region $|\eta_{jets}| < 2$ in invariant mass interval, which reconstructs the W-mass with 5% accuracy, they still find about 3-5 times larger background than signal. They also investigated an asymmetry cut requiring that the energies of the central jets differ at most 50%. In this way at LHC they obtain 1:1 signal to background ratio with ≈ 30

events. Since the event structure is rather complex, clearly a more realistic study is needed before final conclusion can be reached. The success of the method depends also on the actual value of the top mass. Quark tagging is an important question since this is the only hope to exploit the advantage of the larger branching ratio of this decay mode. It is important to have an independent second method to establish the Higgs signal.

The third useful decay mode of the heavy Higgs boson is provided by the decay $H \to ZZ$ with one Z decaying into $\nu\bar\nu$ and with the other Z decaying into e or μ pairs. The branching ratio of this channel is also relatively large. However, it has a large background from three different sources . There is $q\bar q \to ZZ$ continuum production of Z pairs as well as continuum production of WZ pairs when the charged lepton of the $W \to l\nu$ decay remains undetected. The most difficult background, however, comes from the contribution of $Z + jet(s)$ when one of the jets remains undetected giving large unbalanced transverse momentum and jet activity similar to the signal. Detection of this signal requires central production of the lepton pairs and crack-free detectors. While this is an exacting demand, estimates indicate that the observation of the Higgs particle in this channel should be feasible. However, it is unlikely that this decay mode could be used to extend the discovery limit to higher values than the value obtained in the case when both Z bosons decay to charged lepton pairs.

At CLIC[18], heavy Higgs production[34] can be best observed in four jets final states. The background processes $\gamma\gamma \to W^+W^-, \gamma W \to WZ, \gamma\gamma \to$ four QCD jets, e^+e^-, etc. have all been calculated[18,35]. They are significant, but their suppression does not cause any special problem (see fig.7). At CLIC, the discovery limit (assuming luminosity of $\mathcal{L} \approx 10^{32}cm^{-2}s^{-1}$) is about $m_H < 0.7\ TeV$

Strong interaction region. From the discussion of the heavy Higgs region, it is clear that with the present design parameters of LHC (even of SSC) we can obtain only indicative experimental information on the strong interaction region. A detailed exploration requires higher energy. If the Higgs mass is large $m_H > 0.8\ TeV$ we can not observe it as a resonance peak since the width becomes comparable with the mass value. In this case one should establish the existence of the strongly interacting Higgs sector by finding an excess of events in final states with pairs of gauge bosons, above the $q\bar q \to WW/ZZ$ or $q\bar q \to WZ$ background. To explore the detailed properties of the strongly interacting gauge bosons we need higher energies than $\sqrt{s} = 40\ TeV$ at hadron colliders and more than $\sqrt{s} = 2$ TeV at e^+e^- colliders. The physics of this region has been recently explored in refs.[12,17,36].

IV. HIGGS SEARCH AT ELOISATRON

It has been established in LHC studies by considering parton luminosity functions[37] that if we want to reach higher discovery limits it is more efficient to increase the energy than the luminosity. Let us assume that Eloisatron means a proton proton collider at collision energy $\sqrt{s} = 100$ TeV. It is useful to define an effective parton luminosity as

$$\frac{d\mathcal{L}}{dM} = \int_\tau^1 \frac{dx}{x} F_{a/A}(x) F_{b/B}(\frac{\tau}{x}), \tag{4.1}$$

where the function $F_{a/A}(x)$ denotes the number density of parton a in particle A, $\tau = m^2/s$ and $\sqrt{s}$ is the collision energy in the centre-of-mass frame. The number density of a longitudinal W boson in the quark is

$$F_{W_L/q}(x) = \frac{\alpha}{4sin(\theta_W)} \frac{(1-x)}{x} \tag{4.2}$$

where $sin(\theta_W)$ denotes the Weinberg angle. The number density of the longitudinal W-boson in the proton is obtained by the folding

$$f(x)_{W_L/p} = \Sigma_j \int_x^1 \frac{dz}{z} q_j(z) F_{W_L/q_j}\left(\frac{x}{z}\right) \tag{4.3}$$

where $q_j(z)$ is the number density function of quarks ($j = u, d, s, ..$) in the proton. In fig.8 we show the numerical evaluation of the ratios of the qq, gg and $W_L W_L$ luminosity functions caluclated by Stirling[38]. The Eloisatron and LHC luminosities are normalized to the SSC luminosities. We can see that the luminosity improvement going from SSC to Eloisatron is about the same as the improvement going from LHC to SSC. The partonic cross section is expected to be rather smooth between $1-2$ TeV. Unitarization effects should start to damp the increase of the partonic cross section in this region[17]. Multiplying the cross section values of fig.4 with the (≈ 6 times) increased luminosity values given in fig.8 we can see that at Eliosatron the sensitivity to the strong interaction region can be extended up to about 2 TeV. The background problem is expected to improve in this region. The hadrons arising from W-decay and boosted up to 1 TeV energy will give a very narrow jet. The QCD jets of > 1 TeV energy will have larger invariant mass and angular spread. Therefore, we expect that jet spectroscopy will improve in this region. This can be tested by the study of Z-production. In the strong interaction region, the interaction of WZ scattering will also be strong and the WZ final states provide better signal than the WW final states.

It is remarkable that the luminosity curve increases rather significantly in the mass region $M > 3$ TeV. This feature the curve is important in search for new heavy Z-bosons and heavy (fourth generation) quarks. Since the mass generation mechanism of the fermions is not necessarily given by the Higgs mechanism the search for heavy, almost degenerate fourth generation quarks is important in our attempt to obtain decisive experimental tests of the electroweak symmetry breaking mechanism[39].

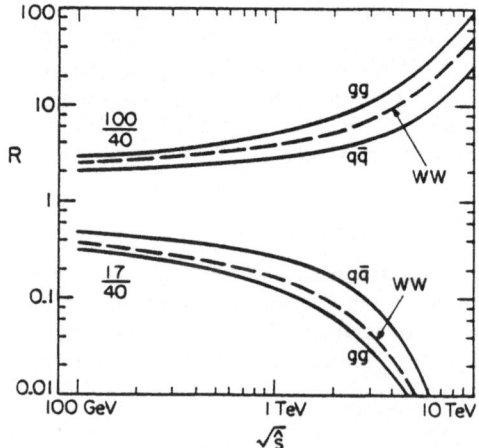

Figure 8. Ratios of $q\bar{q}q, gg, W_L W_L$ luminosities in pp collisions at LHC, SSC and Eloisatron energies[38].

Acknowledgement. I would like to thank the Worskhop Director Professor A. Ali and the "Ettore Majorana" Centre for Scientific Culture, Erice, Sicily for organizing a most stimulating and enjoyable meeting.

REFERENCES

[1]   For a recent review see:
      P. Darriulat, Proceedings of the "International Conference on High Energy
      Physics", Uppsala, Sweden, 1987 ,ed. O. Botner and refernces therein.

[2]   Physics at LEP, CERN 86-02 (1986) edited by J. Ellis and R. Peccei and ref-
      erences therein.

[3]   J. Lee-Franzini *et al.*Columbia Univ. preprint (1987) and *Phys. Rev.* **D35**
      (1987) 2883

[4]   M. Veltman, *Acta Phys. Pol.* **124B** (1977) 475

[5]   B. W. Lee, C. Quigg and H. B. Thacker, *Phys. Rev.* **16** (1977) 1519

[6]   T.P Cheng, E. Eichten, and L.F. Li *Phys. Rev.* **D9** (1974) 2259

[7]   K.G. Wilson and J.Kogut, *Phys. Rep.* **12** (1974) 75

[8]   M. Aizemann *Phys. Rev. Lett.* **47** (1981) 1 ,
      J. Fröhlich *Nucl. Phys.* **B200** (1982) 281

[9]   M. Lüscher and P.Weisz, *Nucl. Phys.* **B290** (1987) 25 ; J. Kuti,L. Lin
      and Y. Shen, preprint UCSD/PTH 87-18 (1987); P. Hasenfratz and J. Nager,
      preprint BUTP-86/20 (1986); A. Hasenfratz and T.Neuhaus, preprint FSU-
      SCRI 87-29 (1987);   W. Langguth and I. Montvay, preprint DESY 87-020
      (1987)

[10] C. Wetterich, DESY preprint, DESY 87-154 (1987)

[11] L. Maiani, G. Parisi and R. Petronzio, *Nucl. Phys.* **B136** (1978) 115 ;
      N. Cabibbo, L. Maiani, G. Parisi and R. Petronzio, *Nucl. Phys.* **B179** (1978)
      295

[12] M. Chanowitz and M. Gaillard, *Nucl. Phys.* **B261** (1985) 379

[13] J. Cornwall, D. Levin and M. Tiktopoulos, *Phys. Rev.* **D10** (1974) 1145

[14] Z. Kunszt and D. Soper, *Nucl. Phys.* **B296** (1988) 253

[15] M.J. Duncan, G.L. Kane and W.W. Repko, *Nucl. Phys.* **b272** (1986) 517

[16] D.R.T. Jones and S.T. Petkov, *Phys. Lett.* **136B** (1979) 440 ; R.N. Cahn and
      S. Dawson, *Phys. Lett.* **136B** (1984) 196 ; M.S. Chanowitz and M.K. Gaillard,
      Phys. Lett. **142B** (1984) 85 ; G.L. Kane, W.W. Repko and W.B. Rolnick,
      Phys. Lett. **148B** (1984) 367

[17] M. Chanowitz, H. Georgi and M. Golden, *Phys. Rev. Lett.* **57** (1986) 2344

[18] G. Altarelli, Proceedings of the Workshop on Physics at Future Accelerators
      La Thuile (Italy) 7-13 January 1987 and refs. therein; D. Froidevaux, *ibid.*

[19] J.F. Gunion *et al.*,*Proceedings of the 1986 Snowmass Workshop*, ed. R. Don-
      aldson and J. Marx

[20] R.N. Cahn *et al.*, LBL preprint LBL-24497 (1987) to appear in *Proceedings of the
      1987 Berkeley Workshop on "Experiments, Detectors and Experimental Aeras
      for SSC"*; see also V. Barger, T. Hahn and R.F.N. Phillips, MAD/PH/368
      (1987).

[21] H. Georgi et.al. *Phys. Rev. Lett.* **40** (1978) 692

[22] R.Cahn, *Nucl. Phys.* **B255** (1985) 341 ;
      G. Altarelli, B. Mele and F. Pitolli *Nucl. Phys.* **B287** (1987) 205 .

[23] Z. Kunszt, *Nucl. Phys.* **B247** (1984) 339

[24] E. Eichten, I. Hinchliffe, K. Lane, and C. Quigg, *Rev. Mod. Phys.* **56** (1984) 247 and Errata, *Rev. Mod. Phys.* **58** (1986) 1065

[25] J.F. Gunion *et al.*, *Nucl. Phys.* **B294** (1987) 621

[26] J. F. Gunion, P. Kalyniak, M. Soldate, and P. Galison, *Phys. Rev.* **34** (1986) 101

[27] H. Albrecht *et al. Phys. Lett.* **192B** (1987) 245

[28] R. K. Ellis *et al.*, Fermilab preprint Pub-87/100-T (1987)

[29] J. F. Gunion, G. Kane and J. Wudka University of California preprint UCD-87-28 (1987)

[30] G. Altarelli and E. Franco, *Mod. Phys. Lett.* **A1** (1986) 517

[31] W.J. Stirling, R. Kleiss and S.D. Ellis, *Phys. Lett.* **B163** (1985) 261 ; J.F. Gunion, Z. Kunszt and M. Soldate, *Phys. Lett.* **163B** (1985) 389

[32] R. Cahn, S.D. Ellis, R. Kleiss and W.J. Stirling, *Phys. Rev.* **D35** (1987) 1626

[33] R. Kleiss and W.J. Stirling, *Phys. Lett.* **200B** (1988) 193

[34] G. Kane and J. Scanio *Nucl. Phys.* **B299** (1987) 221 ; J.F. Gunion and A. Tofighi-Naki, *Phys. Rev.* **D36** (1987) 2671 .

[35] Z. Kunszt and W.J. Stirling presented at the LaThuile Workshop, unpublished;

[36] M. Bento and C.H. Llewellyn Smith *Nucl. Phys.* **B289** (1987) 36

[37] Z. Kunszt, Proceedings of the Workshop on Physics at Future Accelerators, La Thuile, CERN 87-07, Vol.I p.22;

[38] W.J. Stirling contribution to this Workshop

[39] T. Appelquist and M.S. Chanowitz, LBL perprint, LBL-23951 (1987)

SUPERSYMMETRIC PARTICLE SEARCHERS AT FUTURE HADRON COLLIDERS

Queen Mary College, London University

R.Batley

Abstract

The production and detection of supersymmetric particles at future high energy hadron colliders is reviewed, with the emphasis on searches for squarks and gluinos at the LHC. The usual assumption that squarks and gluinos decay directly to a light photino ($\tilde{q}{\rightarrow}q\tilde{\gamma}$ and $\tilde{g}{\rightarrow}q\bar{q}\tilde{\gamma}$) gives the signature of large missing transverse energy plus two or more high p_T jets. Background to this signature from top quark production and from $Z^0 \rightarrow \nu\nu$ decays will be examined in some detail. Recent work on squark and gluino searches at very high LHC luminosity ($\mathcal{L}{=}5\mathrm{x}10^{34}$ cm^{-2}s^{-1}) and on more complex squark and gluino signatures involving $W^\pm$ and Z^0 bosons is summarised.

1 Introduction

The next generation of high-energy proton-proton colliders (the LHC at $\sqrt{s}{=}17$ TeV or the SSC at $\sqrt{s}{=}40$ TeV) will probably be the first machines to probe parton-parton collisions at energies of order 1 TeV. Such energies may be essential, and should be sufficient, to obtain insight into some of the outstanding problems of the standard model, particularly those associated with the symmetry breaking sector and the origin of elementary particle masses. If supersymmetry is to play a rôle in resolving these problems, then supersymmetric particles should be found with masses below $\sim$1 TeV/c^2.

The detection of supersymmetric particles at high energy hadron colliders has therefore received a lot of attention, and studies of the more interesting signals and their backgrounds have been undertaken at various LHC [1,2] and SSC [3,4] workshops. In this report, we attempt to summarise the results of these studies, though, for reasons of familiarity, there will be an emphasis on work carried out for LHC energies.

In the next section, we present an overview of possible signatures for supersymmetry at hadron colliders, and give a brief summary of supersymmetric particle searches carried out at the SPS p$\bar{\text{p}}$ Collider. In section 3, we shall consider squark and gluino searches via the "classical" signature of large missing transverse energy plus jets, concentrating on work carried out in connection with the LHC La Thuile workshop. Recent extensions of this work to consider the case of squark and gluino searches at very high luminosity ($\mathcal{L}{=}5\mathrm{x}10^{34}$ cm^{-2}s^{-1}) will be described. The more general case in which the squark or gluino decays first into a $\widetilde{W}$ or $\widetilde{Z}$ will also be discussed. Finally, in section 4, we consider the electroweak sector and searches for selectrons and winos.

2 Supersymmetric Particles Signatures

2.1 Sparticle production in pp collisions

The minimal particle content of any supersymmetry model is shown in Table 1 [5]. Associated with each quark, lepton, gauge boson or Higgs boson of the standard model is a new particle with the same quantum numbers as the original, but differing by 1/2 unit of spin. The left- and right-handed components of each known quark or lepton give rise to two spin zero squarks $\tilde{q}_L$ and $\tilde{q}_R$ and two spin zero sleptons $\tilde{l}_L$ and $\tilde{l}_R$. At least two Higgs doublets are needed to give masses to all the particles, and additional neutral and charged Higgs bosons are a feature of supersymmetric models.

Table 1. Minimal supersymmetric particle spectrum

	spin 0	spin $\frac{1}{2}$ weak eigenstates	spin $\frac{1}{2}$ mass eigenstates	spin 1
Matter multiplets	$\tilde{l}_L, \tilde{l}_R$ $\tilde{q}_L, \tilde{q}_R$	l q		
Field		$\tilde{g}$		g
	$H^\pm$	$\widetilde{W}$ $\widetilde{H}^\pm$	$\widetilde{W_1^\pm}, \widetilde{W_2^\pm}$ (charginos)	$W^\pm$
Multiplets		$\tilde{\gamma}$		γ
	Z H_1^0, H_2^0	$\tilde{Z}$ $\widetilde{H_1^0}, \widetilde{H_2^0}$	$\widetilde{Z_1^0}, \widetilde{Z_2^0}, \widetilde{Z_3^0}, \widetilde{Z_4^0}$ (neutralinos)	Z^0

In general, mixing will occur between the supersymmetric partners of the gauge and Higgs bosons. Thus, the photino, the zino and the neutral higgsino's (the weak eigenstates) can mix to produce, in order of increasing mass, the mass eigenstates $\widetilde{Z_1^0}$, $\widetilde{Z_2^0}$, $\widetilde{Z_3^0}$ and $\widetilde{Z_4^0}$. Similarly, winos and charged higgsinos can mix to give the mass eigenstates $\widetilde{W_1^\pm}$ and $\widetilde{W_2^\pm}$.

Most supersymmetric models contain a new conserved multiplicative quantum number known as R-parity. The known particles have an R-parity of +1 while their supersymmetric partners have R-parity equal to −1. The consequences of R-parity conservation are (1) supersymmetric particles must be pair produced, (2) each supersymmetric particle must decay into a lighter supersymmetric particle, except that (3) the lightest supersymmetric particle (LSP) must be stable. The latter has a very small interaction cross section and will normally escape detection, so that missing energy appears as the standard signature of supersymmetry. Most analyses assume that the LSP is the lightest neutralino $\widetilde{Z_1^0}$, and that it is a pure photino.

A systematic survey of supersymmetric particle production mechanisms in high energy hadron-hadron collisions has been carried out by EHLQ[6], and matrix elements for most processes of interest can be found in ref. [7]. Some examples of sparticle cross sections for SSC energies ($\sqrt{s}=40$ TeV) are shown in figure 1, as a function of the sparticle mass. The largest cross sections are those for (strongly interacting) squarks and gluinos: pp→$\tilde{g}\tilde{g}$+X, pp→$\tilde{q}\tilde{g}$+X and pp→$\tilde{q}\tilde{q}$+X.

Squark (gluino) production involves the exchange of a virtual gluino (squark), e.g:

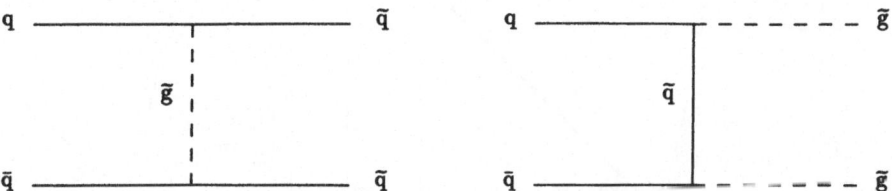

Therefore squark (gluino) cross sections depend on the assumed gluino (squark) mass. The cross sections shown in figure 1 were computed assuming all sparticle masses to be equal. Also shown for comparison in figure 1 are the cross sections for some standard model processes, namely production of $W^{\pm}$, Z^0, and heavy quarks (top). Unfortunately, some of these cross sections are also large and backgrounds to most signals for supersymmetry are potentially severe.

Squarks and gluinos should be copiously produced at the SSC and the LHC. For a total integrated luminosity of 10^{40} cm^{-2} $= 10$ fb^{-1} for example (corresponding to 1-2 years running at a luminosity of 10^{33} cm^{-2}s^{-1}), we expect over 10^7 (10^6) produced gluino pairs at the SSC

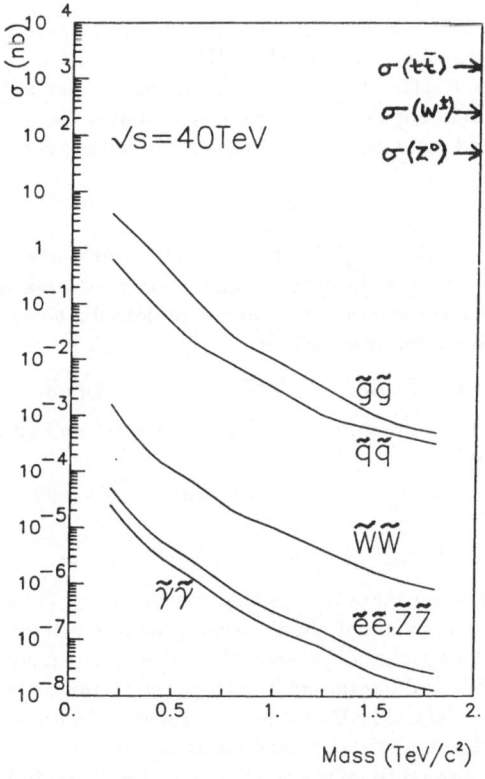

Figure 1. Supersymmetric particle production cross sections as a function of sparticle mass for pp collisions at $\sqrt{s} = 40$ TeV. Also shown for comparison are the cross sections for $W^{\pm}$ and heavy flavour (top) production.

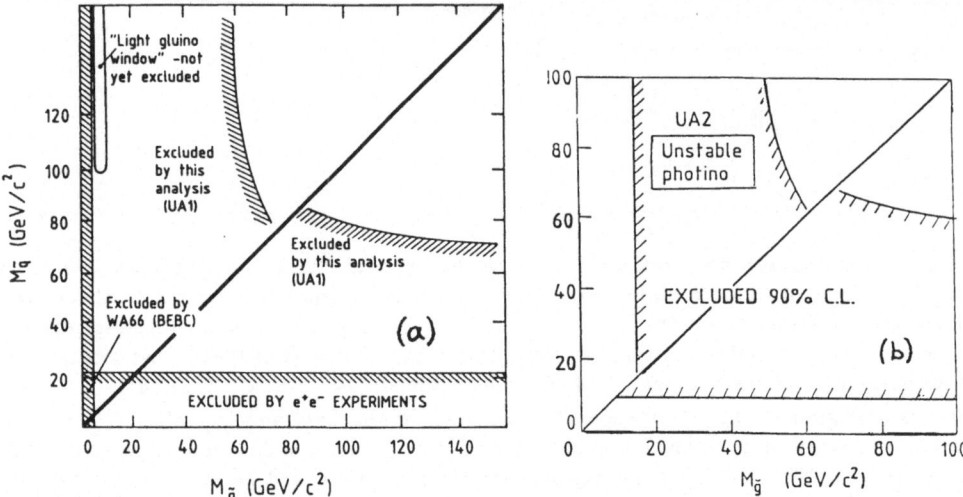

Figure 2. Current limits on squark and gluino masses from UA1 and UA2. The UA1 analysis assumes a light stable photino, while the UA2 result assumes $\tilde{\gamma} \to \gamma + \tilde{H}$.

(LHC) for a gluino mass of 400 GeV/c^2, and over $\sim 10^4$ gluino pairs for gluino masses as high as 1.5 TeV/c^2 (1 TeV/c^2). Production rates at the SSC are typically a factor ~ 10-15 higher than those at the LHC for masses of 500-1000 GeV/c^2. Cross sections for potential backgrounds such as Z^0 or heavy flavour production generally increase by a smaller factor, and we might anticipate that the signal to background ratio would improve with increasing energy.

2.2 Current searches

The best current limits on squark and gluino masses come from recent analyses of SPS Collider data by the UA1 [8] and UA2 [9] experiments. These analyses serve to illustrate many of the features of searches at higher energies and it is useful to describe them here. The UA1 analysis considered the following production mechanisms:

$$\text{pp} \to \bar{q}\bar{q} + X, \qquad \text{pp} \to \bar{q}\tilde{g} + X, \qquad \text{pp} \to \tilde{g}\tilde{g} + X, \qquad \text{pp} \to \tilde{\gamma}\bar{q} + X, \qquad \text{pp} \to \tilde{\gamma}\tilde{g} + X.$$

The LSP was assumed to be a light, stable photino. The squark and gluino decay modes depend on which of them is the heavier:

$$m_{\tilde{q}} > m_{\tilde{g}}: \qquad \bar{q} \to q\tilde{g} \text{ or } \bar{q} \to q\tilde{\gamma} \qquad \tilde{g} \to q\bar{q}\tilde{\gamma}$$

$$m_{\tilde{g}} > m_{\tilde{q}}: \qquad \bar{q} \to q\tilde{\gamma} \qquad\qquad\qquad \tilde{g} \to \bar{q}\tilde{q}$$

The final state therefore contains large missing transverse energy (from the two photinos) and at least two quark jets. The squark and gluino decay products will generally be found at large angles to the original squark or gluino direction. Thus we expect squark and gluino events to be rather isotropic, whereas potential backgrounds such as semi-leptonic decay of heavy quarks are characterised by back-to-back jets. The UA1 limits are shown in figure 2(a). Gluino masses below 53 GeV/c^2 are excluded independently of the assumed squark mass, while squark masses below 45 GeV/c^2 are excluded independently of the gluino mass. The mass limits improve when $m_{\tilde{q}} \simeq m_{\tilde{g}}$ (i.e. near the diagonal of figure 2(a)), mainly because the total squark and gluino production cross section is much larger when the squark and gluino masses are comparable. Finally, these limits are valid for photino masses up to ~ 25 GeV/c^2.

UA2 report somewhat weaker limits for similar model assumptions, but have, in addition, considered the possibility of an unstable photino decaying to a photon plus a higgsino ($\tilde{\gamma} \to \gamma + \tilde{H}$), the higgsino being the LSP and escaping as missing energy. The signature here is two photons (one from each photino decay) plus two or more jets [9]. In this case, they obtain mass limits (figure 2(b)) comparable to those of figure 2(a).

UA1 and UA2 have also looked for evidence of selectron and wino production from $W^{\pm}$ or Z^0 decay. UA1 have searched for the decay $W \to \tilde{e}\bar{\nu}$, which would produce a single high p_T electron plus large missing E_T. UA2 have looked for the decays

$$Z^0 \to (\tilde{e}^+ \to e^+\tilde{\gamma}) + (\tilde{e}^- \to e^-\tilde{\gamma}) \quad \text{and} \quad Z^0 \to (\widetilde{W}^+ \to e^+\bar{\nu}) + (\widetilde{W}^- \to e^-\bar{\nu})$$

In both cases, the signature is two high p_T opposite sign non-coplanar electrons plus large missing E_T, a signature which should also be important at higher energies. The limits obtained depend on the assumed photino mass, but reach up to $\sim$30-40 GeV/c^2 in each case. [1]

In the near future, experiments at the Tevatron Collider should be sensitive to squark and gluino masses up to 150-200 GeV/c^2 [11]. The best limits on slepton and wino or zino masses will probably come from LEP II [12] or from HERA [13].

2.3 Signatures at Future Hadron Colliders

2.3.1 Squarks and gluinos

If they are kinematically allowed, the strong decay modes $\tilde{q} \to q\tilde{g}$ and $\tilde{g} \to \tilde{q}q$ will always dominate. If not, the usual assumption is that squarks and gluinos will decay directly to the LSP (usually a photino): $\tilde{q} \to q\tilde{\gamma}$ and $\tilde{g} \to q\bar{q}\tilde{\gamma}$. This results in the "classical" signature (used by UA1 and UA2) of large missing transverse energy plus two or more high E_T jets.

Though this signature should remain valid at LHC and SSC energies, it has been empha- sised recently that, for larger squark or gluino masses ($>$200-300 GeV/c^2), direct decays to the LSP generally become less important [14,15], being overtaken instead by decays to the heavier neutralinos ($\widetilde{Z_i^0}$) and charginos ($\widetilde{W_i^{\pm}}$):

$$\tilde{q} \to q'\widetilde{W_i^{\pm}}, \quad \tilde{q} \to q\widetilde{Z_i^0} \qquad \text{and} \qquad \tilde{g} \to q\bar{q}'\widetilde{W_i^{\pm}}, \quad \tilde{g} \to q\bar{q}\widetilde{Z_i^0}.$$

The allowed squark and gluino decay modes and their branching ratios depend on details of the sparticle mass spectrum and on their couplings to the gauginos. Although couplings involving the *weak* eigenstates are fixed by supersymmetry, couplings to the *mass* eigenstates are not. Squark and gluino signatures therefore become highly model dependent and require a complete analysis of mixing in the higgsino and gaugino sectors.

The $\widetilde{W_i^{\pm}}$'s and $\widetilde{Z_i^0}$'s from the squark or gluino decay will in turn decay into lighter charginos and neutralinos, and eventually into the LSP. These decays will often include $W^{\pm}$ or Z^0 bosons or neutral or charged Higgs bosons amongst their decay products:

$$\widetilde{Z_i^0} \to \widetilde{Z_j^0} + Z^0 \qquad \widetilde{Z_i^0} \to \widetilde{Z_j^0} + H_k^0 \qquad \widetilde{W_i^{\pm}} \to \widetilde{W_j^{\pm}} + Z^0 \qquad \widetilde{W_i^{\pm}} \to \widetilde{Z_j^0} + H^{\pm}$$

$$\widetilde{Z_i^0} \to \widetilde{W_j^{\pm}} + W^{\pm} \qquad \widetilde{Z_i^0} \to \widetilde{W_j^{\pm}} + H^{\mp} \qquad \widetilde{W_i^{\pm}} \to \widetilde{Z_j^0} + W^{\pm} \qquad \widetilde{W_i^{\pm}} \to \widetilde{W_j^{\pm}} + H_k^0$$

Given that these cascade decays generally occur for *both* of the produced gluinos or squarks, a bewildering variety of final states, often containing $W^{\pm}$ or Z^0 bosons, can emerge at high mass. The final states will generally be quite complex and will contain several jets. Further, since the LSP's are produced at the end of a long decay chain, the missing E_T distribution will generally be softer than that for the direct decays $\tilde{q} \to q\tilde{\gamma}$ and $\tilde{g} \to q\bar{q}\tilde{\gamma}$. Final states containing Z^0's probably offer the best promise of being relatively free from background.

---

[1]For completeness, we should mention that the best limits on selectron and wino masses ($m_{\tilde{e}} > 58$ GeV/c^2 and $m_{\widetilde{W}} > 61$ GeV/c^2, assuming $m_{\tilde{\gamma}} \sim 0$) come from the ASP experiment [10].

2.3.2 Selectrons, winos etc

Production cross sections for electroweak sparticles such as selectrons, winos and zinos are small relative to those for squarks or gluinos of the same mass. Decay modes involving charged leptons, e.g.

$$\widetilde{W} \rightarrow (W \rightarrow l\nu) + \widetilde{\gamma} \quad \text{or} \quad \widetilde{e} \rightarrow e\widetilde{\gamma}$$

should produce the cleanest signatures. The hadronic decay modes produce the same final state as $\widetilde{q} \rightarrow q\widetilde{\gamma}$ and $\widetilde{g} \rightarrow q\bar{q}\widetilde{\gamma}$ decays (i.e. large missing E_T plus jets), but with much smaller rate. Wino and zino cross sections are largest in the case of associated production with **squarks or gluinos**, e.g. pp$\rightarrow \widetilde{q}\widetilde{W}$+X or pp$\rightarrow \widetilde{g}\widetilde{W}$+X. This gives rise to events containing missing E_T, charged leptons and jets. The process pp$\rightarrow \widetilde{\gamma}\widetilde{W}$ has a smaller cross section (down by roughly α/α_s), but may give a cleaner signature. Selectron and wino detection will be discussed further in section 4.

3 Squarks and Gluinos

Because of the large production cross sections involved, searches for squarks and gluinos have received a lot of attention at various SSC and LHC workshops [16,19,20,17,18,15]. These studies have concentrated almost exclusively on direct decays to the LSP (usually a photino): $\widetilde{q} \rightarrow q\widetilde{\gamma}$ and $\widetilde{g} \rightarrow q\bar{q}\widetilde{\gamma}$. In this section, we shall focus on some work carried out for LHC energies at the La Thuile workshop, for both "low" ($\mathcal{L}=10^{33}$ cm^{-2}s^{-1}) and "high" ($\mathcal{L}=5$x10^{34} cm^{-2}s^{-1}) luminosity running. We shall also briefly consider the more complex squark and gluino signatures associated with cascade decays via heavy gauginos.

3.1 LHC Studies (La Thuile)

3.1.1 Signals and Backgrounds

The La Thuile study [17] considered the following squark and gluino production channels and decay modes:

$$(1) \ \text{pp} \rightarrow \widetilde{q}\bar{\widetilde{q}} + X \quad (m_{\widetilde{q}} \ll m_{\widetilde{g}}) \qquad\qquad (2) \ \text{pp} \rightarrow \widetilde{g}\widetilde{g} + X \quad (m_{\widetilde{g}} \ll m_{\widetilde{q}})$$
$$\begin{array}{l} \longmapsto q\widetilde{\gamma} \\ \longrightarrow q\widetilde{\gamma} \end{array} \qquad\qquad\qquad \begin{array}{l} \longmapsto q\bar{q}\widetilde{\gamma} \\ \longrightarrow q\bar{q}\widetilde{\gamma} \end{array}$$

Thus, for squark production, the final state consists of two high p_T jets, and large missing energy from the two photinos. For gluino production, the final state consists of four quark jets and large missing energy, again from two photinos. The lightest supersymmetric particle was assumed to be a massless photino, and all squark flavours were assumed to have the same mass $m_{\widetilde{q}}$.

A variety of standard model processes can also produce events with large missing E_T in association with jets. Those considered in detail were:

1. heavy flavour production (pp$\rightarrow$c+X, pp$\rightarrow$b+X or pp$\rightarrow$t+X) followed by semi-leptonic decay e.g. t$\rightarrow$beν. Top quark masses of m_t=40 GeV and m_t=200 GeV were considered;

2. W$\rightarrow$eν, W$\rightarrow\mu\nu$ and W$\rightarrow\tau\nu$ decays, where the W$^\pm$ is produced with large transverse momentum (pp$\rightarrow$ W$^\pm$+jets);

3. Z$^0\rightarrow\nu\nu$ or Z$^0\rightarrow\tau^+\tau^-$ decay of a high p_T Z^0 (pp$\rightarrow$ Z^0+jets);

4. pair production of W$^\pm$ and Z^0 bosons: pp$\rightarrow$W$^\pm$W$^\pm$+X, $\rightarrow$ W$^\pm$Z^0+X and $\rightarrow$ Z^{0}Z^0+X. Cross sections here are small, and we do not expect a large background contribution from these processes.

Possible background contributions from W$^\pm$ and Z^0 decays into heavy flavours (W$^\pm\rightarrow$c$\bar{\text{s}}$,t$\bar{\text{b}}$ and Z$^0\rightarrow$c$\bar{\text{c}}$,b$\bar{\text{b}}$,t$\bar{\text{t}}$) were not evaluated.

Quantitative studies of both the signal (squark and gluino) and the standard model background were made with version 5.25 of the ISAJET Monte Carlo program [21]. The ISAJET program generates a complete pp interaction, including the "underlying event". For QCD jet production, the hard scatter in ISAJET is computed using only the leading order 2→2 matrix element (e.g. gg→gg). Similarly, $W^\pm$ and Z^0 production is computed to leading order in perturbative QCD (e.g. pp→q+$W^\pm$+X). However, gluon bremsstrahlung from initial and final state partons (including bremsstrahlung from squarks and gluinos) is included in ISAJET, so 2→3 processes (e.g. pp→$W^\pm$+multijets) can occur. In what follows, it will be useful to distinguish between "direct" heavy flavour production, where the heavy quark Q (=c,b or t) is produced directly in the hard scatter, and "indirect" heavy flavour production where a $Q\overline{Q}$ pair is produced via gluon splitting g→$Q\overline{Q}$ in the parton cascade:

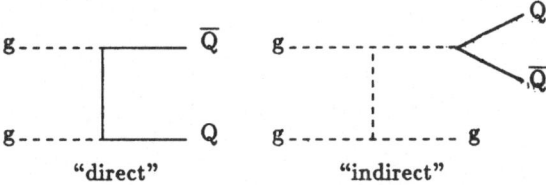

A major problem with such analyses is that of generating adequate Monte Carlo statistics in a reasonable amount of computer time. This problem is particularly severe for the "indirect" heavy flavour background, since, not only does the p_T spectrum of the parton produced in the 2→2 hard scatter fall very steeply with increasing parton p_T, but also, for a given p_T, the missing E_T distribution falls very steeply with increasing missing E_T. For QCD heavy flavour production, events were generated in 17 bins of p_T^{hard} (i.e. the p_T of the outgoing partons produced in the 2→2 hard scattering process) between 100 GeV/c and 2.5 TeV/c. To reduce the amount of computer time used, the parton evolution was carried out many times for each generated hard scatter, and the parton hadronisation was carried out many times for each parton evolution. For indirect heavy flavour production, the *effective* number of events generated in each p_T^{hard} bin was approximately 500,000 though only ∼1000 events were actually output to tape. Even these statistics were inadequate to study fully all possible cuts and selections.

The total cross sections obtained from ISAJET for the background processes listed above are summarised in the first column of Table 2. More relevant for background calculations at large missing E_T are cross sections at high p_T^{jet} and at high p_T^W or p_T^Z. Cross sections for p_T^{hard} (i.e. the p_T of the outgoing jet or the p_T of the $W^\pm$ or Z^0 boson) greater than 500 GeV/c are given in the second column of Table 2. For W→$l\nu$ or Z^0→$\nu\nu$ decays for example, large missing E_T necessarily implies that the $W^\pm$ or Z^0 be produced with large transverse momentum. The computed background rates will therefore be sensitive to knowledge of the high p_T $W^\pm$ and Z^0 cross sections, and to the topology of the high p_T events. In the UA1 analysis, the poorly known cross section for high p_T $W^\pm$ and Z^0 production was found to be a major source of sytematic error in the analysis of the large missing E_T data.

The distribution of missing transverse energy (i.e the vector sum of the two photinos and any neutrinos which might also have been produced) for squarks and gluinos is shown in figures 3(a) and 3(b). The missing E_T distribution for squarks is somewhat harder than that for gluinos since the two body decay $\tilde{q}$→$q\tilde{\gamma}$ gives higher transverse momentum photinos than the three body decay $\tilde{g}$→$q\bar{q}\tilde{\gamma}$. The inclusive missing E_T distribution (computed from the vector sum of the neutrino transverse momenta) from all the above background processes combined is shown in figure 3 (points with error bars). The error bars represent the statistical errors resulting from the Monte Carlo generation. Although the squark signal rises slightly above the background at large missing E_T, the gluino signal always remains below the background.

It is interesting to look at the relative importance of the different background processes as a function of the missing transverse energy, and to study the dependence of the QCD background on the top quark mass, and the relative importance of "direct" (2→2 processes such as gg→$Q\overline{Q}$) and "indirect" (2→3 processes such as gg→$Q\overline{Q}$g) heavy flavour production. Figure 4(a) shows again the total missing E_T distribution for the background (points with error bars) and also the

Table 2. Cross sections for standard model processes

Process	σ(total)	$\sigma(p_T > 500)$	$\sigma(E_T^{miss} > 500)$ inclusive (fb)	e,μ events removed (fb)
pp→gg,gq,qq (q=u,d,s):				
inclusive (m_t=40)	3.1 μb (*)	2.1 nb	304.3	130.7
inclusive (m_t=200)	2.7 μb (*)	2.0 nb	93.1	39.6
indirect heavy flavour:				
g→c$\bar{c}$,b$\bar{b}$ (no top)	3.0 μb (*)		20.5	
g→t$\bar{t}$ (m_t=40)	43.0 nb (*)		283.7	
g→t$\bar{t}$ (m_t=200)	25.5 pb (*)		75.3	
direct heavy flavour:				
pp→c+X,b+X	127.4 nb (*)	75.1 pb	11.9	7.7
pp→t+X (m_t=40)	476.2 μb	19.8 pb	179.3	98.8
pp→t+X (m_t=200)	743.3 pb	9.6 pb	170.3	68.6
W→eν	21.4 nb	450.0 fb	107.9	26.1
W→$\mu\nu$	21.4 nb	450.0 fb	107.9	17.1
W→$\tau\nu$	21.4 nb	450.0 fb	196.3	132.7
Z^0→$\nu\nu$	13.1 nb	380.0 fb	370.0	359.2
Z^0→$\tau^+\tau^-$	2.3 nb	65.7 fb	6.0	2.6
pp→WW	65.0 pb	33.5 fb	3.3	1.2
pp→WZ^0	13.1 pb	6.5 fb	2.5	1.8
pp→Z^0Z^0	19.0 pb	14.3 fb	4.2	3.6
TOTAL (m_t=40)			1293.6	781.5

(*) $p_T^{jet} > 100$ GeV/c

The first two rows give cross sections for inclusive production of light quarks and gluons. The next three rows give cross sections for "indirect" heavy flavour flavour production, where a heavy quark pair is produced in light quark or gluon fragmentation. Rows 5-7 give the cross sections for "direct" heavy flavour production where the c,b or t quark is produced in the original 2→2 hard scatter.

Given separately for each process are (1) the total cross section, (2) the cross section with the p_T of the hard scatter (i.e. the p_T of the outgoing jet or W or Z^0 boson) greater than 500 GeV/c, (3) the cross section for missing transverse energies greater than 500 GeV/c, and (4) the same as (3) but after events containing an identified electron or muon have been removed.

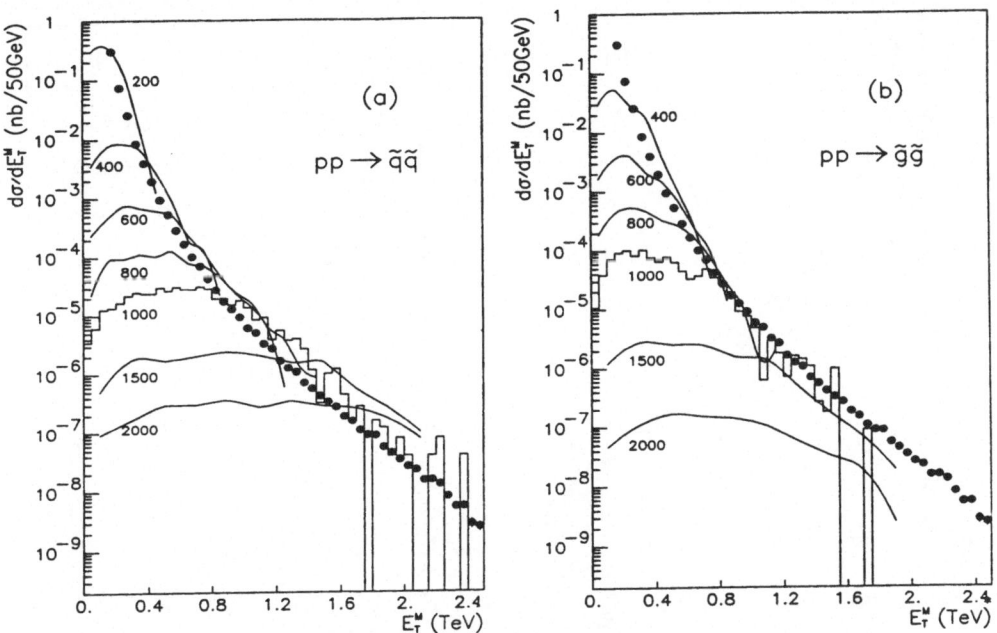

Figure 3. Distribution of true missing transverse energy for (a) squark and (b) gluino pair production (solid curves and histograms), for several different values of squark or gluino mass. The points with error bars show the total standard model background from heavy flavour and weak vector boson production, assuming a top quark mass of 40 GeV/c^2.

separate contributions from QCD heavy flavour production (for a top quark mass of 40 GeV/c^2), from $Z^0 \rightarrow \nu\nu$ and $W \rightarrow l\nu$ decays, and from W and Z^0 pair production. For values of missing E_T below ~ 500 GeV/c^2, the dominant contribution is QCD heavy quark production. For higher values, it is $W^\pm$ and Z^0 production (mainly $Z^0 \rightarrow \nu\nu$ decays). The $Z^0 \rightarrow \nu\nu$ curve in figure 4 is essentially just the Z^0 production cross section $d\sigma/dp_T^Z$. The contribution from pair production of W's and Z's is small, as expected.

Figure 4(b) shows the total QCD (heavy flavour) background for the two different values (40 and 200 GeV/c^2) chosen for the top quark mass. The "non-top" component of the QCD background (from direct and indirect c and b quark production) is shown separately and is seen to be small in comparison to the background from top quark production. Although c and b quarks are produced with much larger cross sections, the harder top quark fragmentation function (essentially a δ-function at $z=1$) produces a much harder missing transverse energy distribution for top.

The backgrounds computed with $m_t = 40$ GeV/c^2 and $m_t = 200$ GeV/c^2 differ appreciably only for values of missing E_T below ~ 500 GeV. As can be seen from Table 2, although the *total* cross section for direct top quark production (pp$\rightarrow t\bar{t}$+X) drops considerably as the top quark mass increases, the cross section at high top quark p_T (and therefore at large missing E_T) is relatively insensitive to the top quark mass. Qualitatively, the same is true of "indirect" top quark production.

For $m_t = 40$ GeV/c^2, indirect top quark production accounts for approximately two-thirds of the total QCD background, but for $m_t = 200$ GeV/c^2, *direct* top quark production has become the dominant contribution. Production of a $c\bar{c}$ or $b\bar{b}$ pair in the initial or final state evolution in ISAJET (i.e. $g \rightarrow c\bar{c}$ or $g \rightarrow b\bar{b}$) occurs in almost every event (see the first column of Table 2). For a top quark mass $m_t = 40$ GeV/c^2, a $t\bar{t}$ pair is produced in a few percent of events. For $m_t = 200$ GeV/c^2, the probability of finding a $t\bar{t}$ pair is some three orders of magnitude smaller than this.

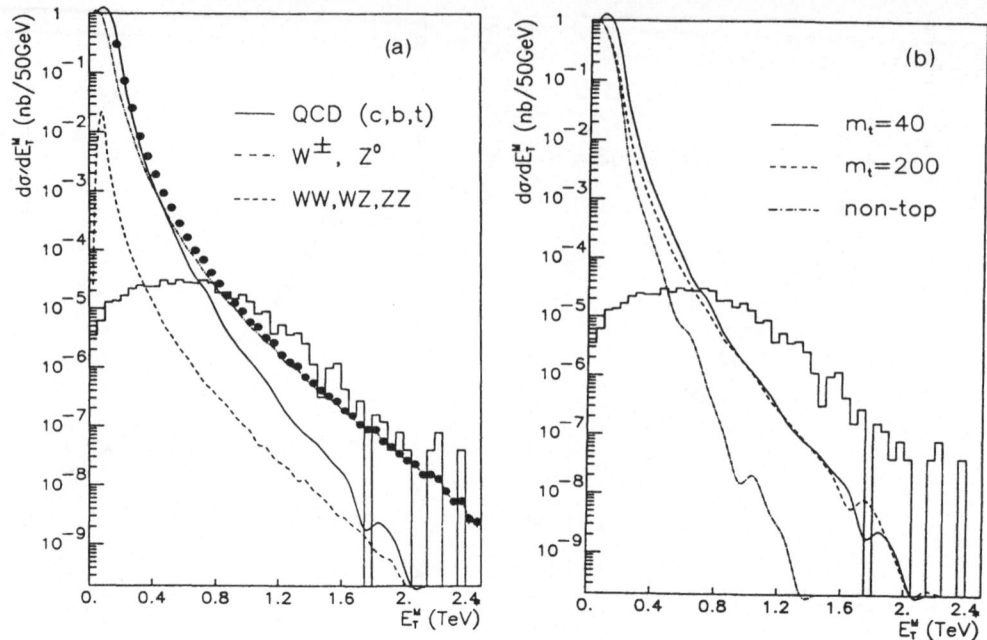

Figure 4. (a) Inclusive missing transverse energy distributions from standard model backgrounds, namely QCD heavy flavour production (solid curve), pp→W,Z^0+X (dot-dash curve), and W,Z pair production (dashed curve). The total standard model background is given by the points with error bars. (b) Comparison of total QCD background for top quark masses of 40 GeV/c^2 (solid curve) and 200 GeV/c^2 (dashed curve). The dot-dash curve shows the background from events with no top quark production, i.e. from direct and indirect c and b quark production alone.

Integrated cross sections for missing E_T greater than 500 GeV for each background process and for direct and indirect heavy flavour production are summarised in the third column of Table 2.

3.1.2 Background Rejection

At this point, a crude calorimeter simulation and calorimeter jet finding were carried out. The detector smearing assumed a calorimeter granularity of $\Delta\eta$=0.05 and $\Delta\phi$=0.05 with complete coverage out to rapidities $|y|$ <5. Each generated particle was assumed to deposit all its energy in a single cell. Calorimeter resolutions of $\sigma(E)/E$=10%/$\sqrt{E}$+1% were assumed for electrons and photons and $\sigma(E)/E$=50%/$\sqrt{E}$+5% for hadrons. The detector smearing in fact had only a small effect on the missing E_T distributions of figure 4.

Jet finding was carried out on the calorimeter cells using a jet algorithm similar to that used by UA1. The highest remaining calorimeter cell with E_T>10 GeV was used to initiate a new jet, and all cells within ΔR<1 of this initiator cell were associated to the jet, where $(\Delta R)^2$=$(\Delta\eta)^2$+$(\Delta\phi)^2$. Only cells with E_T>1 GeV were considered.

The cuts used to reduce the background for the La Thuile study [17] were based on those used in the recent UA1 analysis[8] of SPS Collider data:

1. *Reject events containing a muon or an isolated electron.* This largely removes the background from W→eν and W→$\mu\nu$ decays, and also reduces somewhat the heavy flavour background. It was assumed that all muons with transverse momentum greater than 15 GeV/c could be recognised as such, even if the muon was inside a jet. Similarly, events containing an electron with p_T^e>15 GeV/c were removed provided that the electron was *isolated*, namely that $\Sigma p_T/p_T^e$ <0.1, where Σp_T is the total transverse momentum of all other particles (excluding neutrinos and muons) in a cone of half angle ΔR<0.4 around the electron.

2. *jet multiplicity*: select events containing two or more reconstructed calorimeter jets. This would be expected to reduce the background from $Z^0 \to \nu\nu$ decays for example where low jet multiplicities should dominate;

3. *event topology*: select events in which the missing E_T is not aligned with one of the jets in the event, or in which the jet activity is not coplanar in azimuth. In background events due to heavy quark decay, we expect the missing transverse energy direction often to lie close to one of the jets in the event. Further, we would expect that in general the jet activity would be correlated back-to-back in the transverse plane. For squark production on the other hand, we expect two energetic jets which will not in general be back-to-back. For gluino production, we expect the jets to be produced rather isotropically.

The effect of the charged lepton cuts can be seen by comparing the last two columns of Table 2. As expected, the backgrounds from $W \to e\nu$ and $W \to \mu\nu$ decays are significantly reduced. In addition, the heavy flavour background is reduced by more than a factor of two. The reduction factor is greater for larger quark masses (i.e. greater for $m_t = 200$ than for $m_t = 40$ than for c and b), reflecting the greater isolation of the charged lepton from the semi-leptonic decay as the heavy quark mass increases. Overall, the background is reduced only by a factor ~ 2 (for missing $E_T > 500$ GeV) because of the large contribution from $Z^0 \to \nu\nu$ decays.

Removing events with identified electrons and muons improves the signal to background ratio by approximately a factor of two for squarks, but only marginally for gluinos. This can be understood by considering the heavy flavour content of the squark and gluino events themselves. Squark pair production is dominated by the production of u and d squarks, with c,b and t squarks accounting for only $\sim 20\%$ of the total. Gluino decays however have over 50% total branching ratio into channels containing heavy quarks ($\tilde{g} \to c\bar{c}\tilde{\gamma}$, $\tilde{g} \to b\bar{b}\tilde{\gamma}$, $\tilde{g} \to t\bar{t}\tilde{\gamma}$). A larger fraction of gluino events are therefore rejected by the cuts used to remove electron and muon events.

A comparison of the number of reconstructed calorimeter jets found per event for the squark and gluino signal (for a squark or gluino mass of 1 TeV/c^2) and for the standard model background is shown in figure 5. The measured missing transverse energy has been required to be greater than 300 GeV/c, and a threshold of 250 GeV has been placed on the transverse energy of the jets. For squark and gluino production, most events contain two or more jets. Standard model background processes however are dominated by monojet events (for this particular choice of jet transverse energy threshold).

In figure 6, the measured missing E_T distribution for the signal and background is compared for different requirements on the total number of calorimeter jets in the event. Monojet production (figure 6(a)) is completely dominated by $Z^0 \to \nu\nu$ decays for all values of the measured missing transverse energy. For multijet events (figures 6(b) and (c)), and for values of missing E_T below ~ 1 TeV/c^2, QCD heavy flavour production dominates. For higher missing E_T values, $Z^0 \to \nu\nu$ decays again become the largest contribution. For monojet topologies, the squark signal never rises significantly above the background. For multijet topologies however, a signal to background ratio of ~ 5-10 is seen for larger values of the missing transverse energy.

To study the topology of signal and background events, the following quantities were defined:

1. $\Delta\phi$, the difference in azimuthal angle between the missing transverse energy direction and the direction of the highest transverse energy jet;

2. $\Delta\phi_{12}$, the azimuthal angle between the two highest transverse energy jets;

3. Circularity C, computed from the transverse projection of the calorimeter cells:

$$C = \tfrac{1}{2} \min (\Sigma E_T . \hat{n})^2 / (\Sigma E_T^2)$$

where the minimisation is carried out over all choices of $\hat{n}$, a unit vector in the transverse plane. Thus, C=0 for a pencil-like event and C=1 for an isotropic event (in the transverse plane). The vector $\hat{n}$ for which C is minimised defines the circularity axis.

4. x_{out} [19] , the component of the missing transverse energy perpendicular to the circularity axis, normalised to the total scalar transverse energy ΣE_T measured in the calorimetry:

$$x_{out} = (E_T^{miss}.\sin(\Delta\phi_c))/\Sigma E_T$$

where $\Delta\phi_c$ is the difference in azimuthal angle between the missing E_T direction and the circularity axis.

We note that circularity C and x_{out} are independent of the algorithm used to reconstruct calorimeter jets.

A comparison of $\Delta\phi$, $\Delta\phi_{12}$, circularity and x_{out} distributions for squark production and for background are shown in figure 7, for events with measured missing $E_T > 300$ GeV. The differences in event topology anticipated above are clearly seen in each case.

3.1.3   Results

A comparison of the measured missing E_T distributions for supersymmetry and for background is shown in figure 8 for some representative selections based on circularity and $\Delta\phi$: (a) C>0.25, $N_{jet}>3$ and (b) $\Delta\phi_{12}<130^0$, $N_{jet}=2$. The first of these is aimed at finding isotropic multijet events expected from gluino pairs, while the second is aimed at events with two non-coplanar jets expected from squark pairs. Signal to background ratios of over 10 to 1 are achieved at large missing E_T. Qualitatively similar results are obtained from selections based on the variables $\Delta\phi_{12}$ and x_{out} [17].

In Table 3, the number of signal and background events expected for an integrated luminosity of 10^{40}cm^{-2} (10fb^{-1}) is shown with these selection cuts applied. The errors quoted are the statistical errors associated with the Monte Carlo generation. For each squark or gluino mass

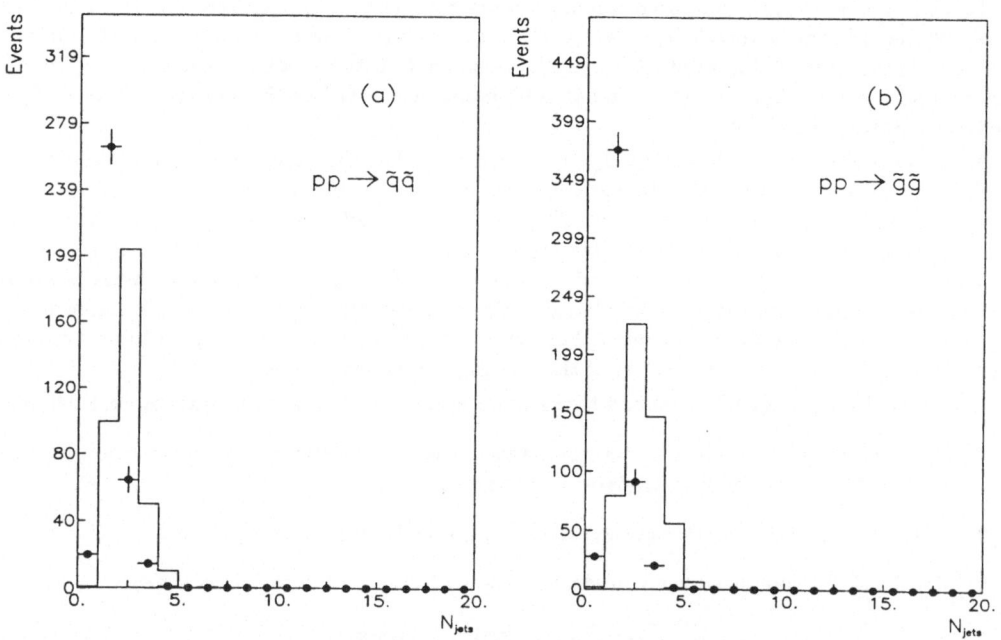

Figure 5. Distribution of the number of calorimeter jets per event for (a) squark and (b) gluino production (histograms) and for background (points with error bars). The signal and background have been normalised to the same total area. The plots are for measured missing $E_T > 300$ GeV and for jet transverse energies >250 GeV. The squark or gluino mass is 1 TeV/c^2.

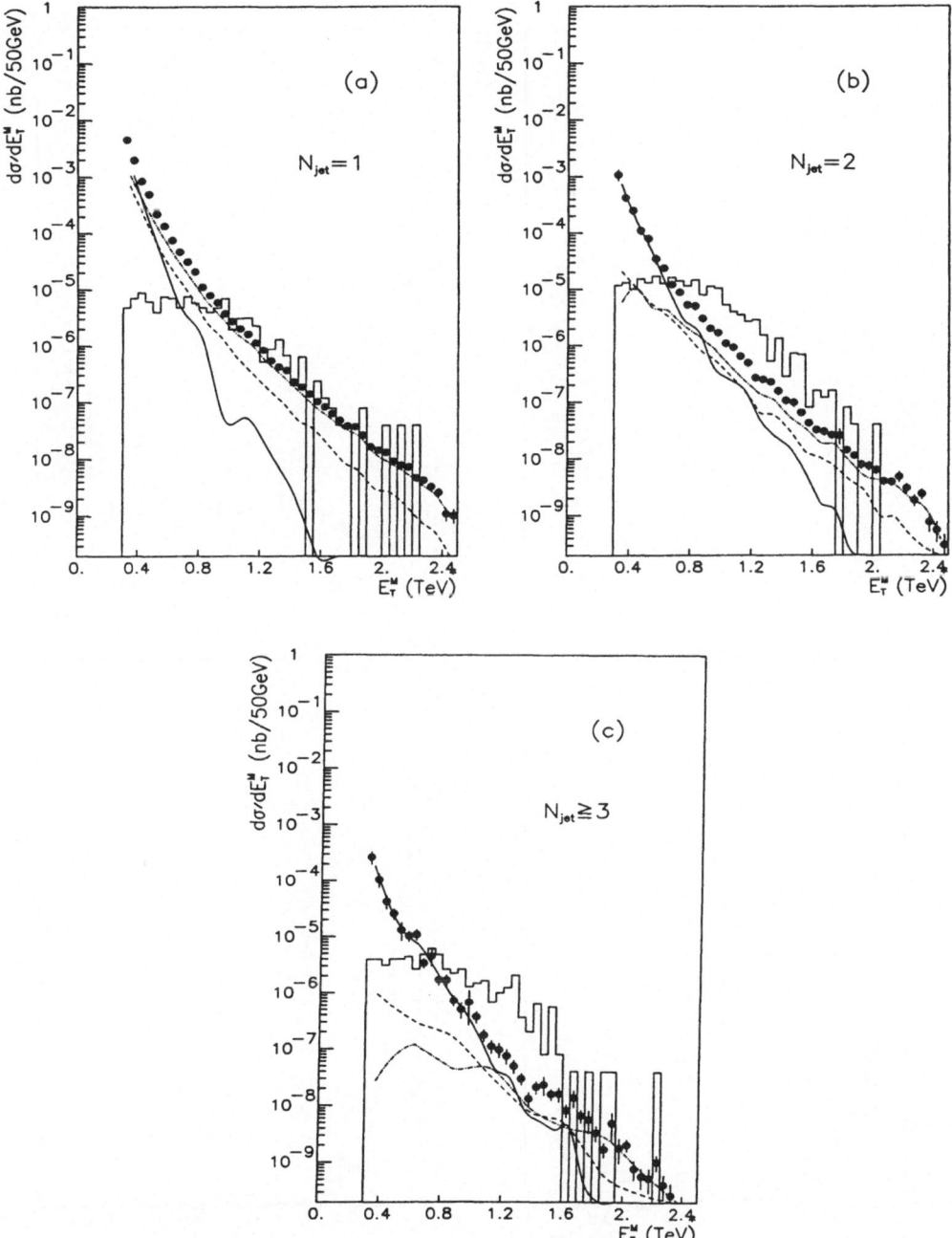

Figure 6. Distribution of measured missing transverse energy for a squark mass of 1 TeV/c² and for background, (a) for events with one and only one jet ("monojets") (b) for events with exactly two jets and (c) for events with three or more jets, for a jet transverse energy threshold of $E_T > 250$ GeV. The points with error bars represent the total standard model background, while the curves show the separate contributions from QCD heavy flavour production (solid curve), from $Z^0 \to \nu\nu$ decays (dot-dash curve), and from other $W^\pm$ and Z^0 processes (dashed curve).

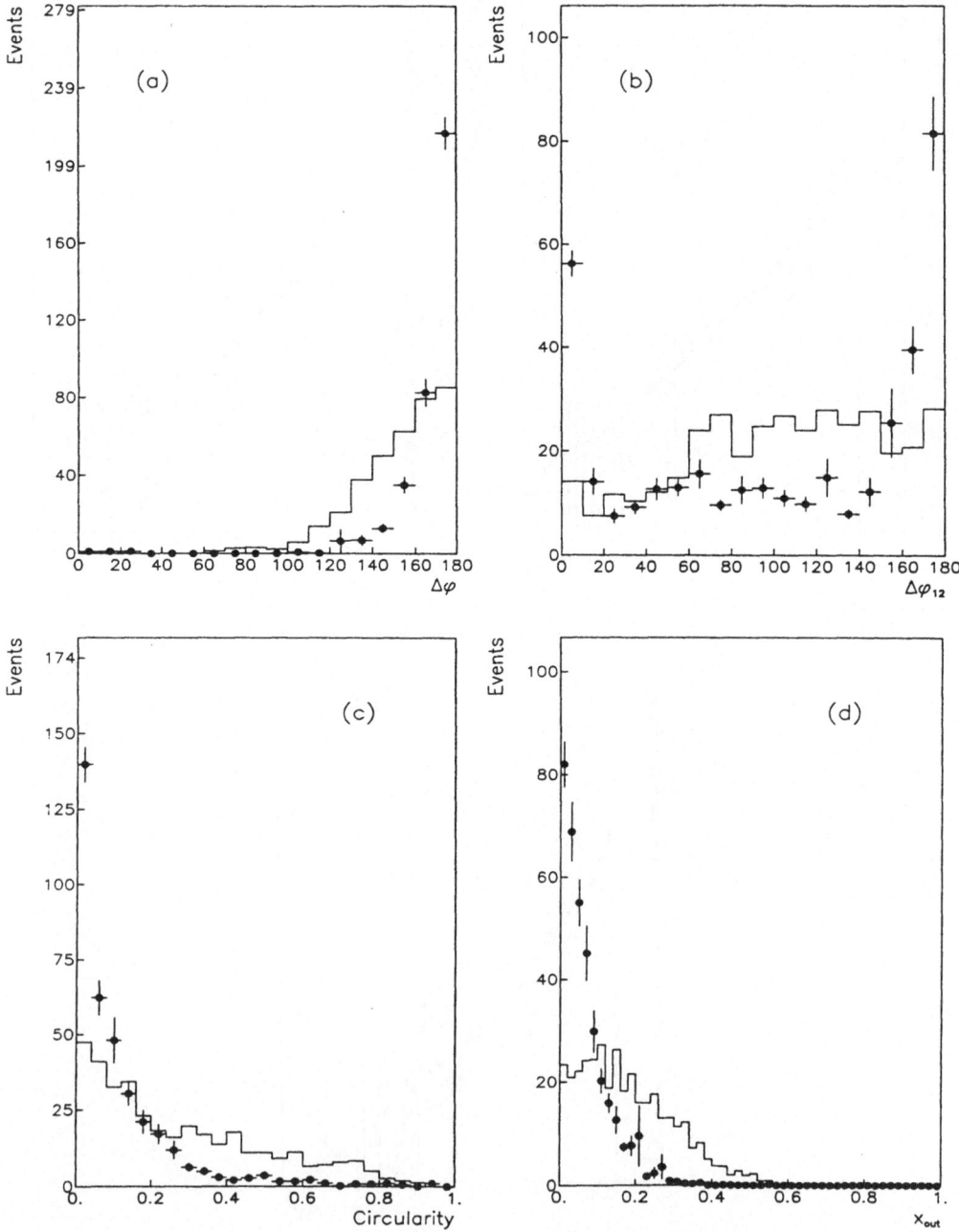

Figure 7. Distribution of (a) $\Delta\phi$ (b) $\Delta\phi_{12}$ (c) circularity C and (d) x_{out} for squark production, for a squark mass of 1 TeV/c^2 (histogram), compared with background (points with error bars). The background contribution has been normalised to the signal. The measured missing E_T is required to be greater than 300 GeV.

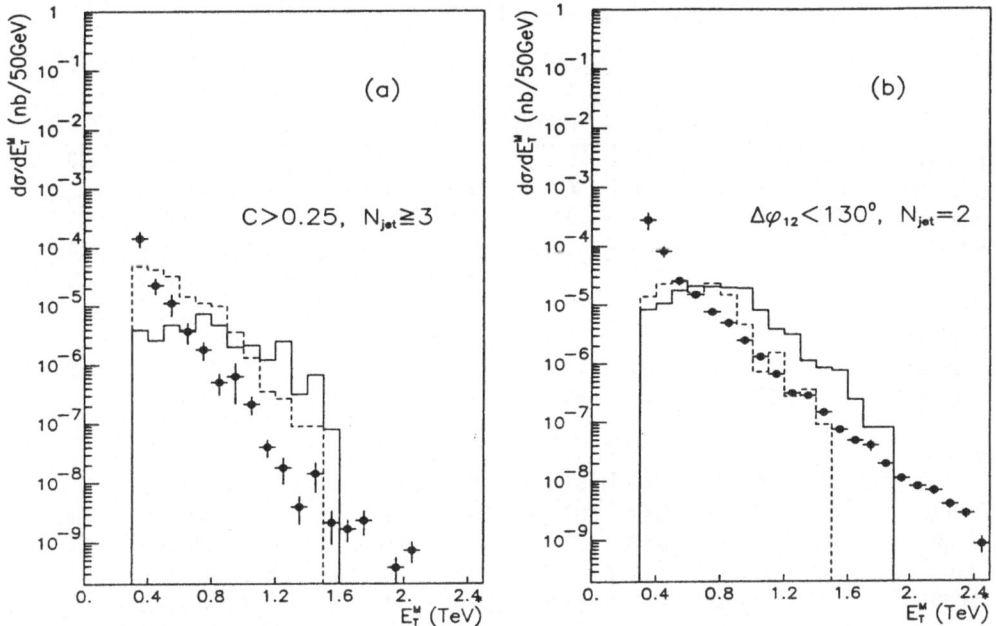

Figure 8. Distribution of measured missing transverse energy for squark (solid histogram) and gluino (dashed histogram) production and for background (points with error bars) after final selection cuts. A squark or gluino mass of 1 TeV/c^2 is assumed. The selection cuts are (a) C>0.25, N$_{jet}$≥3 and (b) Δϕ_{12} <130^0, N$_{jet}$=2.

considered, a missing E$_T$ cut at or above the squark or gluino mass gives a signal to background ratio of typically 10:1. For masses above 1 TeV/c^2 however, the number of squark or gluino events surviving the cuts is typically ~100 or less. This probably defines 1 TeV/c^2 as a reasonable "discovery limit" for the LHC.

It should be emphasised that, in the time available, it was not possible to optimise fully these event selections. In particular, it is probable that the heavy flavour (top quark) background could be reduced substantially by cutting harder on the missing transverse energy isolation. The residual jet from the heavy quark fragmentation would be expected often to lie close to the missing transverse energy direction. Consider for example the quantity

- Δϕ_n, the difference in azimuthal angle between the missing transverse energy direction and the direction of the jet closest in azimuth to the missing E$_T$.

Figure 9 shows the Δϕ_n distribution for events with missing E$_T$ greater than 500 GeV passing the two selections presented in Table 3. Only jets with a transverse energy greater than 50 GeV were considered in computing Δϕ_n. The background distribution peaks near Δϕ_n=0, while the distribution for squark or gluino pair production is essentially flat. For the selection in the first column of table 3 for example (C>0.25, N$_{jet}$≥3, E$_T^{miss}$>500 GeV), applying the additional cut Δϕ_n>30^0 reduces the total background from 183.5 events to 11.4±6.2 events, but reduces the signal by less than a factor two (e.g. from 746.0 events to 446.4±60.8 events for a gluino of mass 1 TeV/c^2). This indicates that the QCD background could be reduced to a very small level, while keeping a large fraction of the signal.

There are many uncertainties involved in computing the background event rates, the more important of which are uncertainties in cross sections for heavy flavour production, lack of knowledge of the top mass, and the uncertainty in the cross section and the associated event topology for Z^0 production at large transverse momentum. These and other uncertainties combine to give an overall uncertainty in the predicted background rate of probably at least a factor 3-4. Also, the

Table 3. Event rates after selection cuts

Process	C>0.25, $N_{jet} \geq 3$		$\Delta\phi_{12} < 130^0$, $N_{jet}=2$	
	$E_T^{miss}>500$	$E_T^{miss}>800$	$E_T^{miss}>500$	$E_T^{miss}>800$
QCD (m_t=40)	167.3 ± 47.6	10.4 ± 4.5	167.0 ± 38.0	14.6 ± 5.9
QCD (m_t=200)	63.6 ± 17.0	6.3 ± 2.0	118.9 ± 39.6	8.9 ± 3.5
$Z^0 \to \nu\nu$	6.6 ± 1.8	2.1 ± 0.6	250.7 ± 22.0	59.6 ± 5.7
$W^\pm \to \tau\nu$	7.1 ± 1.8	1.9 ± 0.6	105.0 ± 13.5	22.4 ± 3.1
other	2.5 ± 1.1	0.3 ± 0.1	62.0 ± 18.1	6.8 ± 2.0
Total bgd.				
m_t=40	183.5 ± 47.7	14.6 ± 4.6	584.7 ± 49.4	102.6 ± 9.0
m_t=200	79.8 ± 17.2	10.6 ± 2.2	536.6 ± 50.2	97.7 ± 7.7
$pp \to \tilde{q}\bar{\tilde{q}}$:				
600	968.7 ± 184.6	151.4 ± 70.4	3948.0 ± 396.0	302.7 ± 99.5
800	519.0 ± 74.4	180.5 ± 31.9	2494.0 ± 176.0	639.7 ± 75.5
1000	302.5 ± 34.0	139.6 ± 21.0	1150.0 ± 67.1	564.8 ± 44.0
1500	56.5 ± 5.2	43.2 ± 4.5	177.0 ± 9.1	148.7 ± 8.2
2000	10.0 ± 1.0	8.5 ± 1.0	37.3 ± 2.0	34.9 ± 2.0
$pp \to \tilde{g}\tilde{g}$:				
600	1892.0 ± 533.0	0.0 ± 0.0	3247.0 ± 613.0	116.8 ± 52.2
800	1553.0 ± 214.0	86.5 ± 18.9	3175.1 ± 308.5	119.4 ± 22.2
1000	746.0 ± 78.1	160.2 ± 29.2	862.2 ± 81.0	222.3 ± 35.6
1500	95.0 ± 6.3	45.3 ± 4.1	51.5 ± 4.4	35.1 ± 3.4
2000	9.0 ± 0.5	5.4 ± 0.4	2.5 ± 0.3	2.1 ± 0.3

Expected event rates for signal and background after final selection cuts, for a total integrated luminosity of 10^{40} cm^{-2} (10 fb^{-1}). The quoted errors are the statistical errors arising from the Monte Carlo generation. For each selection, the event rates are given for two different cuts on the measured missing transverse energy (500 GeV and 800 GeV). Jet transverse energies are required to be greater than 250 GeV.

particular combinations of squark and gluino masses considered here (i.e. either $m(\tilde{q}) \gg m(\tilde{g})$ or $m(\tilde{g}) \gg m(\tilde{q})$) are a somewhat pessimistic choice. Production cross sections are typically a factor 2-3 higher for a more equal choice of masses, i.e. $m(\tilde{q}) \simeq m(\tilde{g})$.

It therefore seems that squark and gluino masses of up to ~1 TeV/c^2 are accessible at the LHC, though the uncertainties mentioned above must be born in mind. The SSC (for the same integrated luminosity) will clearly be able to explore somewhat beyond this. The much larger signal cross sections for squark and gluino masses above 1 TeV combined with a relatively small increase in the background cross sections probably give a comparable limit for the SSC of ~1.5-2 TeV/c^2 [19]. This is confirmed by an analysis of squark and gluino production where similar analysis procedures were used at the two different energies [18].

3.2 LHC High Luminosity Studies: $\mathcal{L}=5 \times 10^{34}$ cm^{-2}s^{-1}

The design luminosity for both the SSC and the LHC is $\mathcal{L}=10^{33}$ cm^{-2}s^{-1}. It is natural to ask whether the LHC's disadvantage in energy can be offset by running the LHC at higher luminosity. This requires a very small time seperation between bunch crossings, and leads to multiple pp interactions on each bunch crossing. At a luminosity of $\mathcal{L}=5 \times 10^{34}$ cm^{-2}s^{-1} for example, assuming

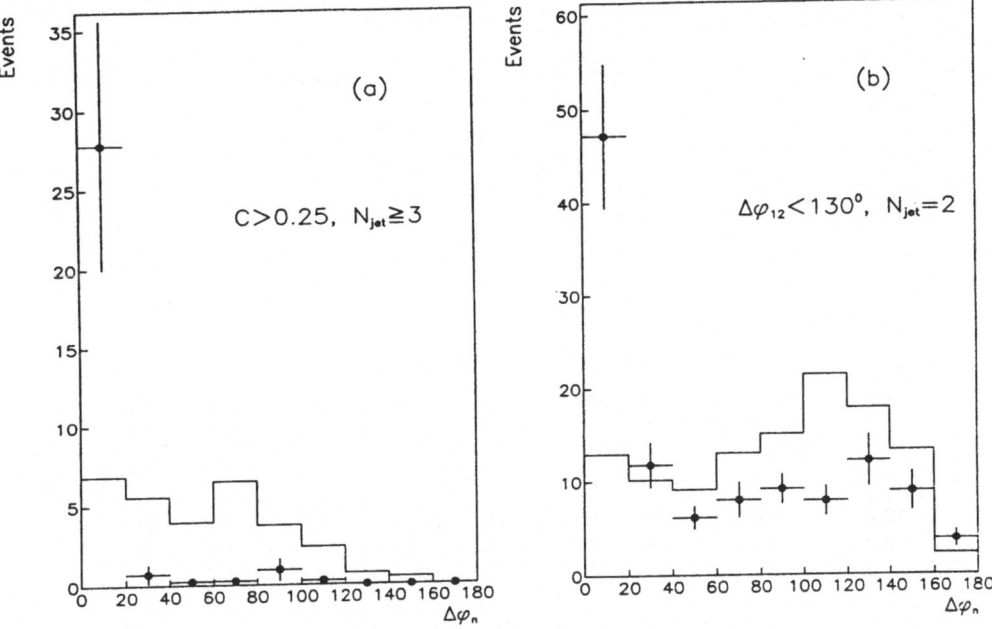

Figure 9: Distribution of $\Delta\phi_n$ for a squark mass of 1 TeV/c^2 (histogram) and for background (points with error bars) for events with measured missing $E_T > 500$ GeV after final selection cuts. The selection cuts are (a) circularity C>0.25, $N_{jet} \geq 3$ and (b) $\Delta\phi_{12} < 130^0$, $N_{jet} = 2$.

5ns seperation between bunch crossings and a total pp cross section of 100mb, on average there will be a total of 25 pp interactions per bunch crossing. Each squark or gluino event will therefore come superimposed with ~25 low p_T events ("pile-up").

The studies described in the previous section have recently been extended to consider the feasibility of carrying out squark and gluino searches with the LHC running at a luminosity $\mathcal{L} = 5 \times 10^{34}$ cm^{-2}s^{-1} [23]. Squark and gluino masses of 1 TeV/c^2 and 2 TeV/c^2 were considered. The low luminosity analysis [17] (section 3.1) was modified as follows [23]:

1. *simulation of event pile-up.* Pile-up was simulated by mixing together several low p_T jet events with each (ISAJET) gluino event. Since, at LHC energies, the jet cross section computed in perturbative QCD saturates the total pp cross section for jet p_T's of order 10 GeV, these additional events were represented by low p_T QCD jet events ($10 < p_T^{jet} < 100$ GeV/c) generated with ISAJET. The number of low p_T events to be added was generated according to a Poisson distribution with a mean of 25.

2. *take account of reduced electron and muon identification.* Although charged track information will probably not be available, it may be possible to identify isolated high p_T ($p_T^e > 50$ GeV/c) electrons using TRD's[22]. The isolation requirement ($\Sigma E_T / E_T^e < 0.1$ in a cone of half angle $\Delta R < 0.4$ around the electron) was left unchanged. Similarly, it was assumed that all muons with transverse momentum greater than 50 GeV/c (instead of 15 GeV/c) could still be recognised.

3. *optimised jet-finding algorithm.* The cone size used for the jet finding algorithm was reduced from $\Delta R < 1$ to $\Delta R < 0.5$ in order to reduce the amount of energy drawn into the jet from the beam fragmentation and from the additional low p_T events.

Distributions of E_T^{miss}, N_{jet}, $\Delta\phi_{12}$ and circularity C for $\tilde{g}\tilde{g}$ events are shown in figure 10, with and without pile-up. The missing E_T distribution is almost completely unaffected by the presence

Table 4. Event rates at $\mathcal{L}=5\times10^{34}$ cm^{-2}s^{-1} after selection cuts

Process	C>0.25 $N_{jet}\geq3$		$\Delta\phi_{12}<130^0$ $N_{jet}=2$	
	$E_T^{miss}>800$	$E_T^{miss}>1200$	$E_T^{miss}>800$	$E_T^{miss}>1200$
QCD	655 ± 410	11 ± 6	1235 ± 550	20 ± 12
$Z^0\to\nu\nu$	80 ± 32	9 ± 2	2830 ± 290	340 ± 30
$W^\pm\to\tau\nu$	70 ± 18	5 ± 2	1195 ± 210	100 ± 13
other	5 ± 3	1 ± 1	515 ± 260	60 ± 35
Total bgd.	810 ± 410	25 ± 8	5775 ± 750	520 ± 52
$pp\to\tilde{q}\tilde{q}$:				
1000	8300 ± 1200	1580 ± 650	25960 ± 1950	2600 ± 580
2000	495 ± 50	355 ± 50	1850 ± 92	1310 ± 85
$pp\to\tilde{g}\tilde{g}$:				
1000	6660 ± 710	270 ± 150	12150 ± 2060	1350 ± 600
2000	335 ± 35	110 ± 15	105 ± 11	50 ± 9

Expected event rates for signal and background after final selection cuts, for a total integrated luminosity of 5.10^{41} cm^{-2} (500 fb^{-1}). The quoted errors are the statistical errors arising from the Monte Carlo generation. For each selection, the event rates are given for two different cuts on the measured missing transverse energy ($E_T^{miss}>800$ GeV and $E_T^{miss}>1200$ GeV). Jet transverse energies are required to be greater than 250 GeV.

of the additional pile-up events. The same is true for the jet multiplicity and the jet topology (with the re-tuned jet algorithm). The circularity distribution however shows clearly that the gluino events are more isotropic when pile-up is added.

A comparison of the measured missing E_T distributions for gluino production and for background is shown in figure 11 for the two selections (a) C>0.25, $N_{jet}\geq3$ and (b) $\Delta\phi_{12}<130^0$, $N_{jet}=2$. Events containing identified electrons or muons have been removed. Jets were required to have transverse energies $E_T^{jet}>250$ GeV. A comparison of figures 8(a) and (b) (from the 10^{33} cm^{-2}s^{-1} analysis) with figures 11(a) and (b) shows rather little change at high luminosity. Signal to background ratios of ~10:1 are again achieved at high missing E_T.

In Table 4, the number of signal and background events expected for an integrated luminosity of 500fb^{-1} (1-2 years running at high luminosity) is shown. The errors quoted are the statistical errors associated with the Monte Carlo generation. Even for squark or gluino masses of 2 TeV/c^2, $\gtrsim$100 events survive the final selection cuts.

In summary, it seems that pile-up has only a small effect on the selection efficiency for high mass (>1 TeV) squark and gluino events, though the data quality must presumably be poorer at high luminosity. Signal to background ratios similar to those at lower luminosity can be obtained. Thus, if the detector problems associated with experimentation at high luminosity can be overcome, there seems to be no fundamental reason why the additional luminosity should not extend the discovery limit for squarks and gluinos to masses of ~2 TeV/c^2, comparable to that for the SSC.

3.3   Squarks and gluinos: other decay modes

To end this somewhat lengthy chapter on squark and gluino searches, we return to consider the more complex signatures that might arise for high mass squarks and gluinos (see section 2.3). Cascade decays via heavy winos, zinos or higgs bosons will produce final states which often contain

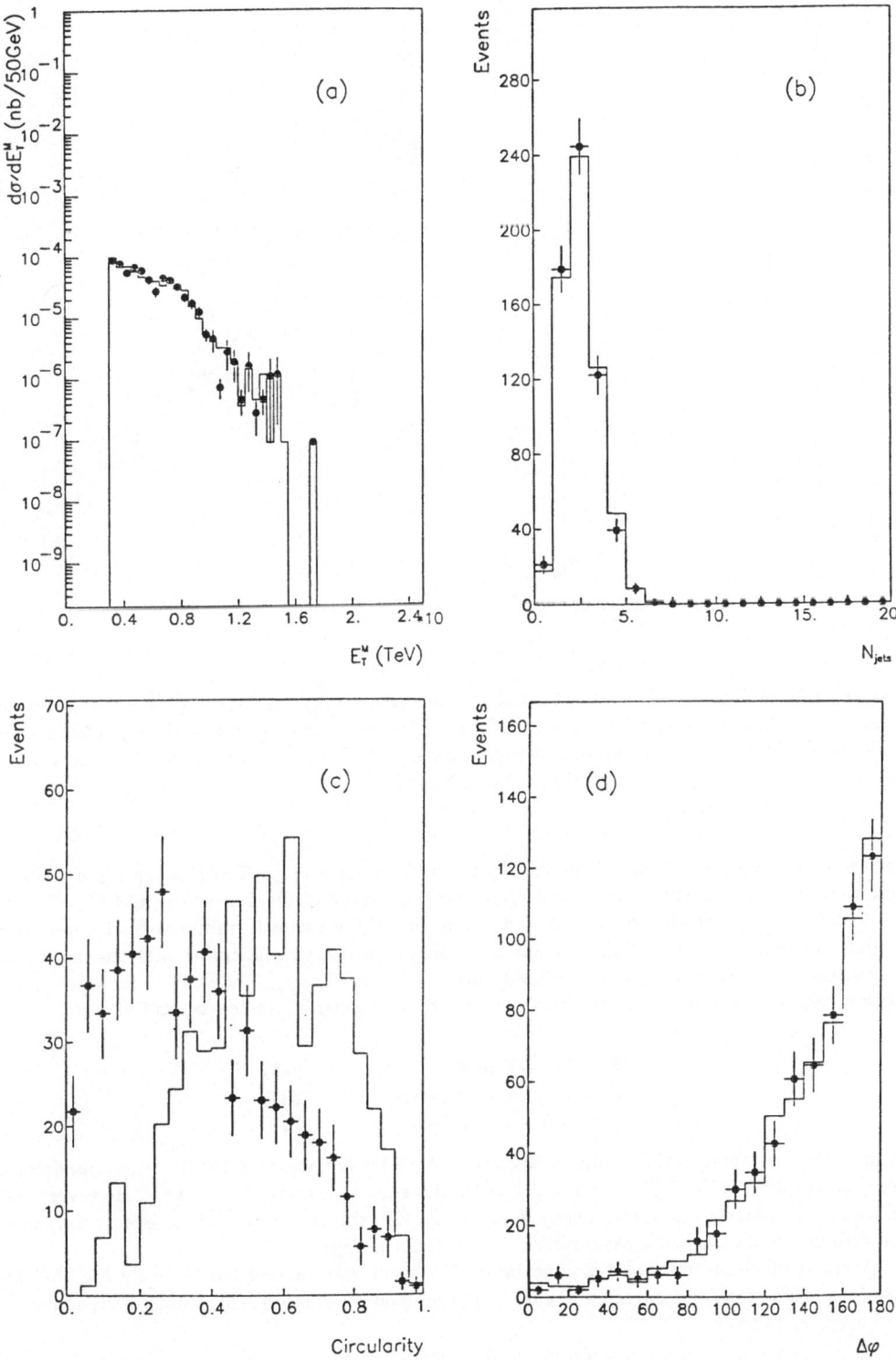

Figure 10. Distribution of (a) E_T^{miss} (b) N_{jet} (c) $\Delta\phi_{12}$ and (d) circularity C for gluino production, with (histogram) and without (solid points) pile-up, for a gluino mass of 1 TeV/c².

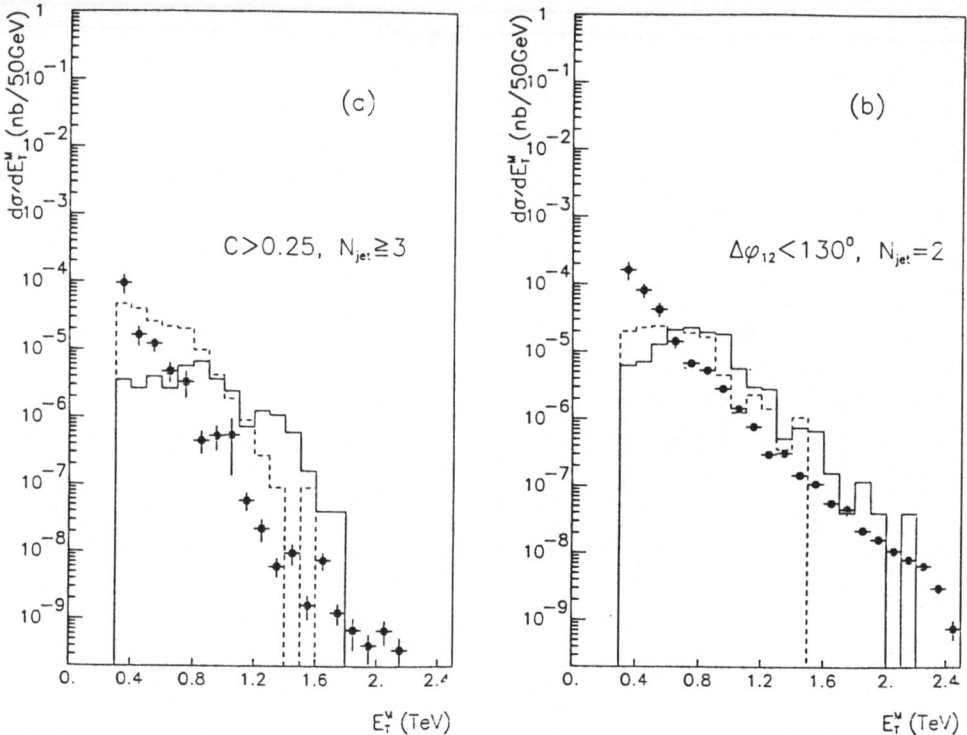

Figure 11. Distribution of measured missing transverse energy for squark (solid histogram) and gluino (dashed histogram) production and for background (points with error bars) after final high luminosity selection cuts. A squark or gluino mass of 1 TeV/c^2 is assumed. The selection cuts are (a) $C>0.25$, $N_{jet} \geq 3$ and (b) $\Delta\phi_{12}<130^0$, $N_{jet}=2$.

$W^\pm$ or Z^0 bosons, as well as two or more jets and missing energy. The missing E_T distribution will however be softer than that found above for the case of direct decays to the LSP.

A detailed presentation of squark and gluino branching ratios in minimal N=1 supergravity models can be found in refs [14,15]. These branching ratios vary significantly with the input model parameters and with the squark or gluino mass. However, for $m_{\tilde{g}} \gtrsim 750$ GeV/c^2 for example, and for a wide range of input parameter values, the gluino branching ratios are as follows:

$$\tilde{g} \to W^\pm + X \text{ or } Z^0 + X: \qquad 58\%$$
$$\tilde{g} \to H^\pm + X \text{ or } H^0 + X: \qquad 28\%$$
$$\tilde{g} \to \text{LSP (direct decay)}: \qquad 14\%$$

In fact, the branching ratio for direct decays to the LSP is always $\leq 14\%$ (for these models), i.e. only for at most $\sim 2\%$ of $\tilde{g}\tilde{g}$ events will both gluinos decay directly to the LSP. However, $\sim 25\%$ of events will contain one direct decay $\tilde{g}\to$LSP+X, and the topology of these events may not be too disimilar from the usual assumption of two direct decays.

A study of gluino final states containing Z^0 bosons was carried out recently for SSC energies [15]. This study considered the case of gluino pair production where both gluinos decay to Z^0 bosons as follows:

$$\tilde{g} \to q\bar{q}\widetilde{Z_i^0}, \qquad \widetilde{Z_i^0} \to \widetilde{Z_j^0}Z^0, \qquad Z^0 \to l^+l^-.$$

The final state therefore contains at least four hard jets (from the primary decay of the two produced gluinos), plus four charged leptons from the leptonic decay of two Z^0 bosons. The largest backgrounds are probably

1. pp → Z⁰Z⁰+jets, where the jets occur as initial state radiation;

2. high p_T Z⁰ production pp → qZ⁰+jets where the quark radiates an additional Z⁰ boson;

3. QCD jet production, pp → gq$\bar{q}$ for example, where each quark radiates a Z⁰ boson.

Detector smearing and jet finding was carried out along similar lines to those described above for the La Thuile studies. Cuts were made on the transverse energies of the three highest E_T jets, on the total scalar E_T, and on $\Delta\phi_{12}$. The background from pp → Z⁰Z⁰+jets (the first background in the list above) was evaluated explicitly using ISAJET. After all selection cuts, and for a gluino mass of 750 GeV/c², the signal (for an integrated luminosity of 10^{40} cm^{-2}) was 56 events compared to a background from Z⁰Z⁰ production of 0.5 ± 0.3 events. The other backgrounds were not evaluated explicitly, though arguments can be given that they will also turn out to be small [15]. Thus, the first look at these more complex decay modes gives reasonably promising results, though it would clearly be interesting to pursue this analysis with more detailed background studies.

4 Electroweak Sparticles

Relatively little consideration has been given to searches for electroweak sparticles such as sleptons and winos. The review by EHLQ [6] contains a systematic survey of production mechanisms and possible signatures, some of which have been pursued in more detail for SSC [19,25] or LHC [24] energies.

Searches for selectrons and winos at the LHC were considered by Mansoulié [24] at the La Thuile workshop. Two processes were studied, namely selectron pair production and squark-wino production:

$$(1) \quad pp \to \tilde{e}^+\tilde{e}^- + X$$
$$\llcorner \to e^- \tilde{\gamma}$$
$$\llcorner \to e^+ \tilde{\gamma}$$

$$(2) \quad pp \to \tilde{q}\widetilde{W} + X$$
$$\llcorner \to e\nu\tilde{\gamma}$$
$$\llcorner \to q\tilde{\gamma}$$

In each case, production cross sections were computed using the matrix elements from ref. [7], and a simple detector smearing was applied using the resolutions quoted in section 3.1.2.

4.1 Searches for selectrons

In the case of selectron pair production, the signature is a high mass electron pair accompanied by large missing transverse energy. The total cross section is rather small: $\sigma(pp \to \tilde{e}^+\tilde{e}^-)$=6x10^{-6} nb for a selectron mass of 300 GeV/c² for example. Two background processes considered by Mansoulié were:

1. Drell-Yan: pp→$(\gamma, Z^0 \to e^+e^-)$ + X

2. WW pair production: pp→$(W^- \to e^-\nu)$ + $(W^+ \to e^+\nu)$ + X

To reduce these backgrounds, the selection cuts m(e⁺,e⁻)>200 GeV and E_T^{miss}>100 GeV were applied. Assuming a "standard model like" ordering of sparticle masses ($m_{\widetilde{W}} \gg m_{\tilde{e}} \gg m_{\tilde{\nu}}$), Mansoulié finds a signal after cuts of 30 events for $m_{\tilde{e}}$ =300 GeV/c² for a total integrated luminosity of 10^{40} cm^{-2}, to be compared with a background from Drell-Yan and WW production of $\simeq$15 events. He therefore concludes that a discovery limit $m_{\tilde{e}} \sim$300 GeV/c² is reasonable.

However, this conclusion depends critically on the assumed sparticle mass spectrum. In minimal models of the supergravity type for example, supersymmetric particle masses are given in terms of two bare mass parameters $(m_0, m_{\frac{1}{2}})$ by [26]:

$$m_{\tilde{q}}^2 \simeq m_0^2 + 7m_{\frac{1}{2}}^2 \quad m_{\tilde{g}} \simeq 3m_{\frac{1}{2}} \quad m_{\tilde{\gamma}} \simeq 0.47m_{\frac{1}{2}} \quad m_{\widetilde{W}} \simeq 0.84m_{\frac{1}{2}}$$

$$m_{\tilde{l}_R}^2 \simeq m_0^2 + 0.15m_{\frac{1}{2}}^2 \quad m_{\tilde{l}_L}^2 \simeq m_0^2 + 0.5m_{\frac{1}{2}}^2$$

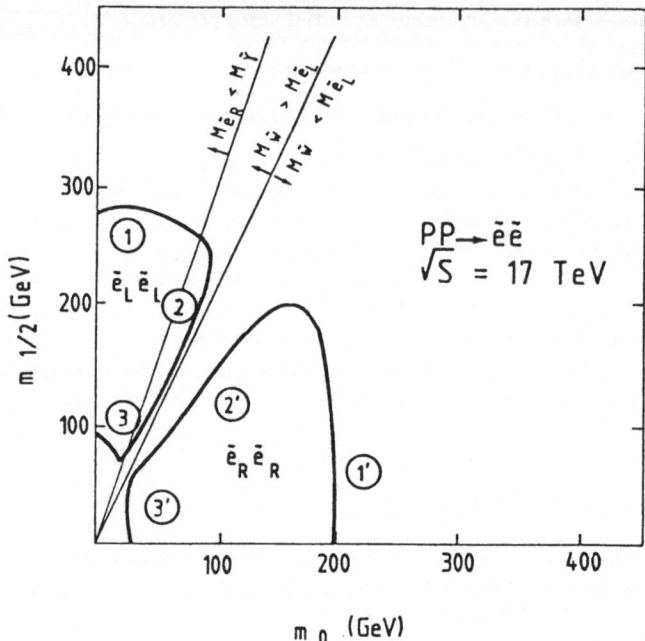

Figure 12. Discovery limits in the $(m_0, m_{\frac{1}{2}})$ plane obtained from selectron pair production in pp collisions at $\sqrt{s}=17$ TeV.

Production cross sections, decay modes and signatures can vary considerably throughout the $(m_0, m_{\frac{1}{2}})$ plane. In the case of selectron pair production being considered here for example, the decay mode $\tilde{e} \to e\tilde{\gamma}$ will sometimes be forbidden or may have a branching ratio much smaller than 100%. For this particular model, and for this particular signature, the discovery limits in the parameters m_0 and $m_{\frac{1}{2}}$ are shown in figure 12. These limits were based on observing at least 30 events passing the cuts above, and correspond to selectron mass limits somewhat below the limit of 300 GeV/c^2 found above (only ~ 200 GeV/c^2 if $m_{\frac{1}{2}}=0$ for example).

4.2 Searches for winos

Turning now to the case of wino production, pp $\to \bar{q}\widetilde{W}$, here the signal is an electron and a jet plus large missing E_T. The most important background process is probably single high p_T $W^{\pm}$ production: pp $\to$ (W$\to$eν) + jets + X.

Figure 13 shows the electron-E_T^{miss} transverse mass for $\widetilde{W}\bar{q}$ production and for the $W^{\pm}$ background. The background shows a prominent Jacobean peak around the W mass, but there is a long slowly falling tail extending up to transverse masses of several hundred GeV. The dashed and dot-dashed curves in figure 13 were calculated assuming that the wino decays into a light sneutrino ($\widetilde{W} \to e\tilde{\nu}$) and that $m_{\widetilde{W}} \simeq m_{\tilde{q}}$. With these assumptions, the $\bar{q}\widetilde{W}$ signal rises clearly above the background for transverse masses close to the wino mass. The discovery limit is determined by the pp$\to \bar{q}\widetilde{W}$ cross section and lies in the region of $m_{\widetilde{W}} \sim 800$ GeV/c^2.

Again however, this conclusion depends on the assumed sparticle mass spectrum. For the minimal supergravity model considered above, the signal is always at least two orders of magnitude below the background. The transverse mass distribution for one of the more favourable combinations of m_0 and $m_{\frac{1}{2}}$ ($m_0=140$ GeV, $m_{\frac{1}{2}}=370$ GeV) is shown in figure 13, with the $\widetilde{W}$ decaying as $\widetilde{W} \to \tilde{e}\nu$, $\tilde{e} \to e\tilde{\gamma}$. There are two main reasons for the drastically reduced signal: (1) the squark mass is always much greater than the wino mass, thus reducing the pp$\to \bar{q}\widetilde{W}$ cross

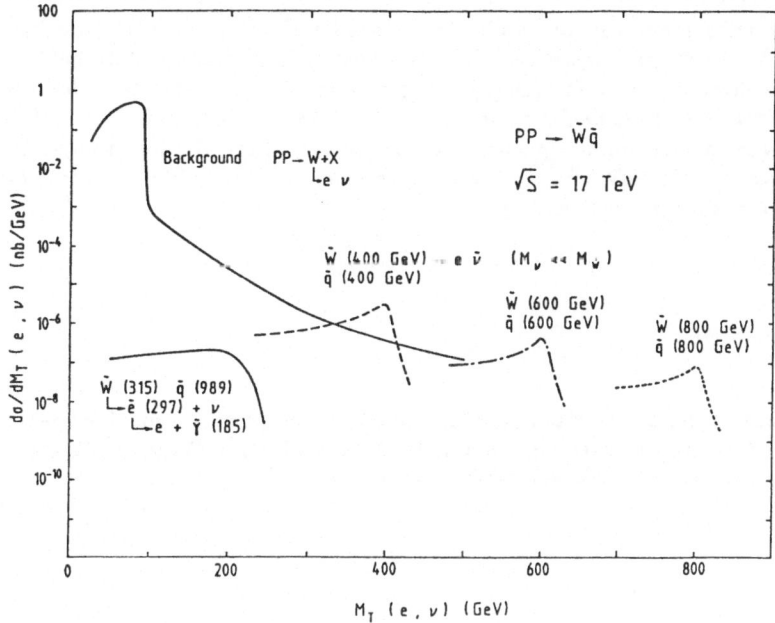

Figure 13. electron- missing E_T transverse mass for $\widetilde{W}\tilde{q}$ production and for $W^\pm$ background in pp collisions at $\sqrt{s}=17$ TeV.

section, and (2), the higher masses of the $\widetilde{W}$ decay products shifts the transverse mass to lower values where the background is more important. Mansoulié therefore concludes that wino masses of up to $\sim$800 GeV/c^2 are attainable at the LHC *provided* the wino behaves roughly like a heavy W. For more general models, the wino signal may be swamped by the background.

5   Summary and Conclusions

Studies of supersymmetric particle detection at future high energy pp colliders (LHC and SSC) have concentrated on searches for the strongly interacting squarks and gluinos. The usual assumption is that squarks and gluinos decay directly to the lightest supersymmetric particle (usually assumed to be a light photino): $\tilde{q}\rightarrow q\tilde{\gamma}$ or $\tilde{g}\rightarrow q\bar{q}\tilde{\gamma}$, giving a signature of large missing E_T plus two or more high E_T jets. The main backgrounds in this case are found to be from top quark production and from $Z^0\rightarrow\nu\nu$ decays. With relatively simple cuts on missing E_T, on jet multiplicity and on event topology, it seems that the standard model background can be reduced sufficiently to achieve signal to background ratios of 10:1 or better.

Because of the theoretical and practical uncertainties involved in these studies, it is difficult to quote accurate "discovery limits" for supersymmetric particles. With simple selection cuts applied, acceptable squark and gluino event rates ($\gtrsim$100 events in 1–2 years running) are achieved for squark and gluino masses up to $\sim$1 TeV/c^2 at the LHC and up to $\sim$1.5–2 TeV/c^2 at the SSC. By running the LHC at high luminosity ($\mathcal{L}$=5x10^{34} cm^{-2}s^{-1}), it seems possible to compensate somewhat for the SSC's energy advantage and achieve sensitivity up to 2 TeV/c^2 also, though there are formidable experimental problems to be overcome in this case.

For high mass squarks and gluinos, direct decays to the LSP may account for only a small fraction of squark and gluino decays. Decays involving heavier winos and zinos (e.g. $\tilde{q}\rightarrow q'\widetilde{W}$ or $\tilde{g}\rightarrow q\bar{q}\widetilde{Z}$) tend to dominate instead. This will soften the missing E_T distribution and will produce rather complex final states with several jets, and often containing $W^\pm$ and Z^0 bosons. Some studies of final states containing Z^0 pairs have begun recently for SSC energies. The initial results are promising, but a more detailed background evaluation would be welcome.

Searches for electroweak sparticles such as selectrons and winos will probably have to rely on signatures involving isolated high p_T charged leptons. Whether signals for selectrons and winos can be extracted from backgrounds such as Drell-Yan production or $W \rightarrow l\nu$ decays depends critically on the supersymmetric particle mass spectrum. The LHC for example is sensitive to wino masses up to $\sim$800 GeV/c^2 and selectron masses up to $\sim$300 GeV/c^2, but only for favourable assumptions for other sparticle masses.

Acknowledgments

I would like to thank Felicitas Pauss for numerous discussions on missing energy and super-symmetry, and for help in carrying out this work. I would also like to thank the organisers of this conference for an enjoyable and interesting meeting.

References

[1] M.Jacob (ed.), Proc. ECFA-CERN Workshop on a Large Hadron Collider in the LEP tunnel, Lausanne and CERN, 1984, ECFA 84/85 and CERN 84-10 (1984)

[2] J.Mulvey (ed.), Proc. Workshop on Physics at Future Accelerators, La Thuile and CERN (1987), CERN 87-07 Vols I and II (1987)

[3] Proc. DPF Summer Study on the Design and Utilization of the Superconducting Super Collider, Snowmass, 1984 (AIP, New York, 1985)

[4] Proc. DPF Summer Study on Physics at the SSC, Snowmass, 1986

[5] For recent reviews of supersymmetry, see for example:
H.Haber and G.Kane, Phys.Reports 117 (1984) 279
H.P.Nilles, Phys. Rep. 110 (1984) 1
J.Ellis, CERN-TH.4255/85, Lectures presented at the 28th Scottish Universities Summer School in Physics, Edinburgh, Scotland (1985)

[6] E.Eichten, I.Hinchcliffe, K.Lane and C.Quigg, Rev.Mod.Phys. 56 (1984) 579

[7] S.Dawson, E.Eichten and C.Quigg, Phys.Rev. D31 (1985) 1581

[8] UA1 Collaboration, C.Albajar et al., CERN-EP/87-148 (1987)

[9] UA2 Collaboration, R.Ansari et al., CERN-EP/87-117 (1987)

[10] ASP Collab., G.Bartha et al., Phys. Rev. Lett. 56 (1986) 685
ASP Collab., C.Hearty et al., Phys. Rev. Lett. 58 (1987) 1711

[11] H.Baer and E.Berger, Phys.Rev. D34 (1984) 1461; erratum, Phys.Rev. D35 (1987) 406
E.Reya and D.Roy, Z.Phys. C32 (1986) 615

[12] A.Böhm and W.Hoogland (ed.), Proc. ECFA Workshop on LEP 200, Aachen, 1986 (CERN 87-08 and ECFA 87/108)
M.Chen, C.Dionisi, M.Martinez and X.Tata, CERN-EP/87-143 (1987)

[13] R.J.Cashmore et al., Phys. Rep. **122** (1985) 275
G.Wolf, DESY 86-089 (1986)

[14] H.Baer et al., University of Wisconsin preprints MAD/PH/316 (1986), MAD/PH/357 (1987) and MAD/PH/362 (1987)

[15] R.Barnett et al., LBL-24150 (1987), Proc. Workshop on Experiments, Detectors and Experimental Areas for the Supercollider, Berkeley, July 1987.

[16] T.Åkesson et al., ref [1], p.165.

[17] R.Batley, ref [2], Vol II, p.109

[18] A.Savoy-Navarro and N.Zaganidis, ref [2], Vol II, p.82

[19] S.Dawson and A.Savoy-Navarro, ref [3], p.263
S.Dawson, Proceedings of the Oregon Workshop on Super High Energy Physics (1985), p.171

[20] R.M.Barnett and H.E.Haber, LBL-21947 (1986)
R.M.Barnett, ref [4] p.262.

[21] F.Paige and S.Protopopescu, BNL 38034 (1986)

[22] LHC High Luminosity Study Group, T.Åkesson et al., CERN report, in preparation.
J.Ellis, CERN-TH.4888/87

[23] F.Pauss, R.Batley, M.Marquina and A.Nandi, ref [22].

[24] B.Mansoulié, ref [2], Vol II, p.126

[25] F.Ukegawa, Y.Takaiwa and K.Kondo, ref [4], p.276.

[26] J.Ellis, CERN-TH.4277/85 and Proc. Int. Symposium on Lepton and Photon Interactions at High Energies, Kyoto (1985)

SIGNATURES OF COMPOSITENESS IN VERY HIGH ENERGY pp COLLISIONS

Barbara Schrempp

Sektion Physik, Theoretische Physik
Universität München
D-8000 München, Fed. Rep. of Germany

INTRODUCTION

Compositeness has become the headline of a vast field of activities, supplying many, mostly model dependent, predictions for future colliders. Fortunately, the energy range addressed at this meeting on future pp collisions is comfortably high, ranging from $\sqrt{s}$ = 17 TeV at the LHC at CERN over 20 to possibly 40 TeV at SSC to 100 TeV at the further reaching ELOISATRON project. So, I can afford to focus on scenarios of compositeness which are distinguished by reasonably model independent predictions.

Of course, all predictions crucially depend on the size of the compositeness scale Λ which is of the order of the inverse quark-lepton radius R

$$\Lambda \sim R^{-1} \tag{1}$$

Given $\sqrt{s}$, there are two generic scenarios, $\Lambda \gtrsim \sqrt{s}$ and $\Lambda \ll \sqrt{s}$, which are characterized by distinctly different signatures of compositeness. Following this distinction, the first two topics to be discussed are

(i) $\Lambda \gtrsim \sqrt{\bar{s}}$: Contact interactions, in particular those in the four-fermion subprocesses $q\bar{q}' \rightarrow q\bar{q}'$ and $q\bar{q} \rightarrow l^+l^-$. These are the cleanest signatures of quark-lepton substructure, if Λ is much larger than the centre of mass energy $\sqrt{\hat{s}}$ available in the four-fermion subprocess

$$\sqrt{\hat{s}} \ll \Lambda. \tag{2}$$

(ii) $\Lambda \ll \sqrt{s}$: pp collisions as a unique <u>source of new colored composites</u> with mass m ~ O(Λ). The least model dependent predictions can be made for <u>pair production</u> of colored composites via color forces.

A third topic addresses very high energies,

$$\sqrt{s} \gtrsim 40 \text{ TeV}. \tag{3}$$

It is concerned with the issue of

(iii) <u>strong $W_L W_L \rightarrow W_L W_L$ interactions</u>, measured in the WW fusion subprocess $qq \rightarrow qq$ WW of pp collisions, as a crucial probe of the nature of the electroweak symmetry breaking sector.

Four-fermion contact interactions in $q\bar{q}^{(-)} \rightarrow q\bar{q}^{(-)}{}'$ and $q\bar{q} \rightarrow l^+l^-$ are discussed in Sect. 2. The "reach" or sensitivity of future pp colliders for such contact interactions, as function of $\sqrt{s}$ and $\int Ldt$, is reviewed. A comparison with the reach attainable at future e^+e^- and ep collider projects is performed. A qualitative but illuminating <u>analytical insight</u> into the numerical results is provided. A rough estimate of the reach for gauge invariant contact interactions in processes involving gluons, $gg \rightarrow q\bar{q}$ and $gg \rightarrow l^+l^-$ is also given.

In Sect. 3 a brief exposé of heavy colored composites, as typically predicted in composite models, is given. Their pair production via color gauge forces in high energy pp collisions is discussed. Expected event rate limits and some discovery limits for their mass, the only free parameter, are presented. Single production of colored composites is also briefly discussed.

In Sect. 4 a general framework for strongly interacting $W_L W_L \rightarrow W_L W_L$ scattering of (elementary or composite) longitudinally polarized W bosons is reviewed. Generalized low-energy theorems are quoted and supplemented by reasonable extrapolations to higher energies. Rates for the WW fusion subprocess $qq \rightarrow qqW_L W_L$ for strongly interacting $W_L W_L \rightarrow W_L W_L$ vis-à-vis the background annihilation process $q\bar{q} \rightarrow WW$ in very high energy pp collisions are given.

CONTACT INTERACTIONS

Four-Fermion Contact Interactions

If quarks and leptons are composite, the binding forces among their constituents will effect a short-range residual interaction among the quarks and leptons (much in the same way as the QCD forces among quarks are the origin of the strong interactions among composite hadrons). Whichever the dynamics of these residual interactions among four fermions f, f = q,l, at "low energies", $\sqrt{\hat{s}} \ll \Lambda$, they reduce to four fermion contact interactions[1,2]

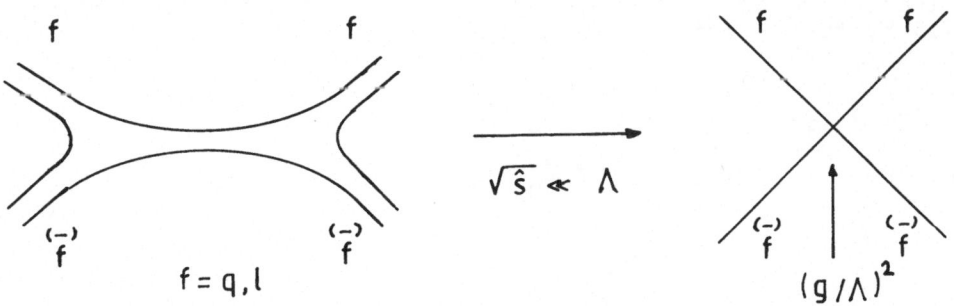

with some unknown effective coupling g^2/Λ^2 and unknown helicity structure. Such contact interactions would show up through characteristic interference effects with the standard model amplitudes in the relevant observables.

Thus, four-fermion contact interactions are a clean, fairly model independent "low-energy" signal of quark-lepton substructure. Let me emphasize, however, that they are more generally the classical "low-energy" signature of any theory involving new heavy bosons, for instance new heavy gauge bosons, with mass $m \gg \sqrt{\hat{s}}$ and a coupling to $f\bar{f}$ or ff pairs. Prominent examples are the Z' boson or the leptoquarks appearing in superstring inspired models. This clearly enhances the importance of a search for four-fermion contact interactions at future colliders also beyond the scope of compositeness.

In the framework of compositeness, the safest four-fermion contact interactions to look for are in flavor diagonal processes

$$\begin{array}{ll} \text{in pp:} & uu \to uu, \quad dd \to dd, \\ & u\bar{u} \to u\bar{u}, \quad d\bar{d} \to d\bar{d}, \ \ldots \\ \text{in } e^+e^-: & e^+e^- \to e^+e^-. \end{array} \qquad (4)$$

While e^+e^- collisions test lepton substructure, pp collisions independently test quark substructure; this distinction becomes relevant, if quarks and leptons have different radii which is not inconceivable. Further four fermion processes

$$\begin{array}{ll} \text{in pp:} & q\bar{q} \to l^+l^- \\ \text{in } e^+e^-: & e^+e^- \to q\bar{q} \\ \text{in ep:} & eq \to eq \end{array} \qquad (5)$$

will have contact interactions only, if the participating fermions have constituents in common.

Following Eichten, Lane and Peskin[1] and Eichten, Hinchliffe, Lane and Quigg[2] I concentrate on the search for the most likely, i.e. helicity conserving four fermion contact interactions mediated by color singlet isoscalar exchanges. They are described by the effective Lagrangian

$$
\begin{aligned}
L_{eff} = k_{if}\, \frac{g^2}{\Lambda^2} \, (\,& \eta_{LL}\bar{f}_L\gamma^\mu f_L \bar{f}_L\gamma_\mu f_L \\
+\ & \eta_{RR}\bar{f}_R\gamma^\mu f_R \bar{f}_R\gamma_\mu f_R \\
+\ & \eta_{LR}\bar{f}_L\gamma^\mu f_L \bar{f}_R\gamma_\mu f_R \\
+\ & \eta_{RL}\bar{f}_R\gamma^\mu f_R \bar{f}_L\gamma_\mu f_L\,)\,.
\end{aligned}
\tag{6}
$$

k_{if} is 1/2 if the initial state is equal to the final state and 1 else, summation over color indices is implied if f stands for a quark, Λ is the unknown compositeness scale and g the unknown dimensionless coupling. Only the ratio Λ/g is measurable. Instead of quoting bounds for Λ, while assuming $g^2 = 4\pi$ as is conventionally done[1,2] , I prefer to quote bounds for

$$
\Lambda^* = \Lambda\,/\,\sqrt{g^2/4\pi}
\tag{7}
$$

which coincides with Λ for $g^2 = 4\pi$. The ignorance about the helicity structure in the ansatz (6) is reflected in the unknown weights η_{ij} with $0 \leqslant |\eta_{ij}| \leqslant 1$. It has become customary to consider the eight standard helicity configurations summarized in Table 1.

A contact interaction is measured via its interference with the standard model amplitude (QCD in $q\bar{q}' \to q\bar{q}'$, electroweak in $q\bar{q} \to l^+l^-$, $eq \to eq$, $e^+e^- \to q\bar{q}, l^+l^-$). Figs. 1a),b), taken from Ref. 2, show the effect of a contact interaction in pp → jet + X and pp → l^+l^- + X, respectively, for $\sqrt{s}$ = 20 TeV and 100 TeV and various values of Λ^* and the two LL helicity combinations with η_{LL}= -1 (———) and η_{LL}= +1 (---).

Table 1. Standard helicity configurations

	η_{LL}	η_{RR}	$\eta_{LR}= \eta_{RL}$
LL	± 1	0	0
RR	0	± 1	0
VV	± 1	± 1	± 1
AA	± 1	± 1	∓ 1

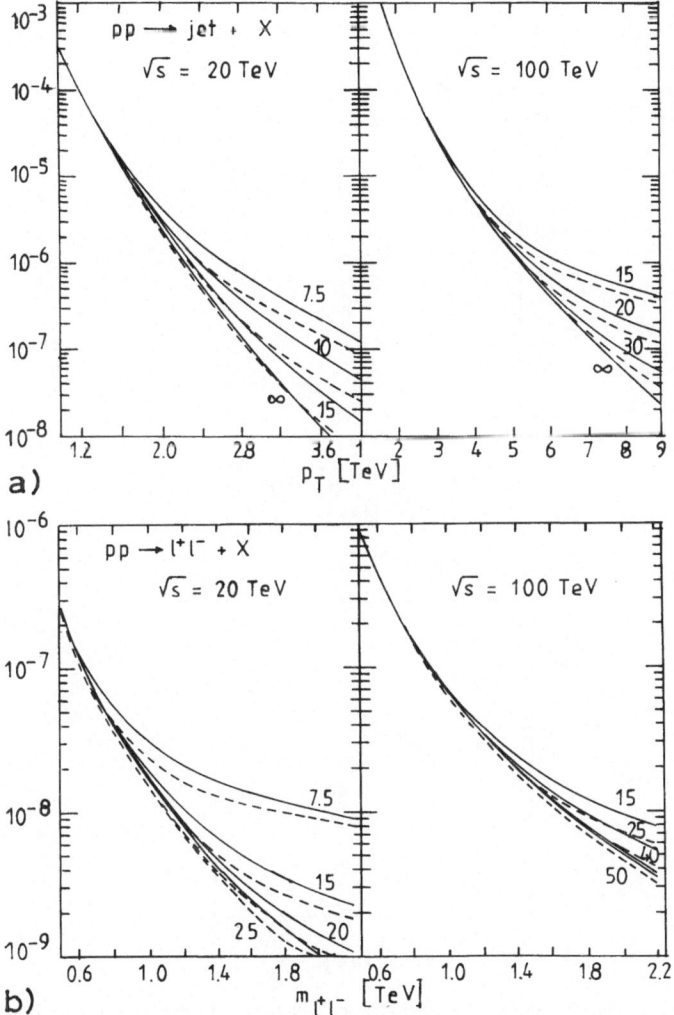

Figure 1. four fermion contact interactions for various Λ^*
a) $pp \rightarrow \text{jet} + X$, $d\sigma/dp_T dy \big|_{y=0}$ [nb/GeV]
b) $pp \rightarrow l^+ l^- + X$, $d\sigma/dmdy \big|_{y=0}$ [nb/GeV], from Ref. 2.

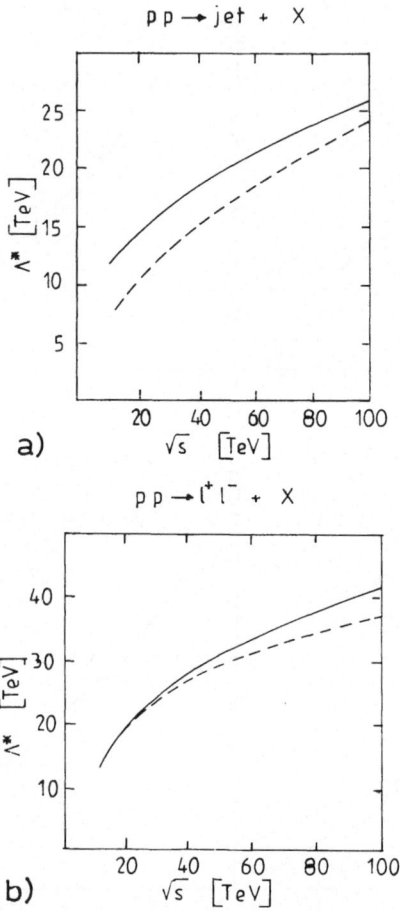

Figure 2. reach for four fermion contact interactions of a pp
collider with ∫Ldt= 10^4 pb^{-1} in terms of a bound on
Λ^*, a) pp → jet + X,b) pp → l^+l^- + X, from Ref. 2.

Table 2. The reach for four-fermion contact interactions at future colliders in terms of a bound on Λ^*

	LHC $\quad$ 17 TeV, 10^4 pb^{-1}	SSC $\quad$ 20/40 TeV, 10^4 pb^{-1}	ELOISATRON $\quad$ 100 TeV, 10^4 pb^{-1}	$(s_{eff} \int L\,dt)^{1/4}$
pp	pp → jet + X $\quad$ 9.5 – 13.8 TeV	10.4 – 14.6/ 15.1 – 18.5 TeV $\quad$ pp → l$^+$l$^-$X	24 – 26 TeV	
LL only	17.4 TeV Ref. 2	19/ 26 – 28 TeV Ref. 2	37.2 – 41.4 TeV Ref. 2	

		HERA($e^{\pm}$p) $\quad$ 314 GeV, 100 pb^{-1}each	LHC, ep – Option $\quad$ 1.4 TeV, 10^3 pb^{-1}	$E_q = E_p/6$, i.e.$s_{eff}=s$/6
ep		2.9 – 6.9 TeV (pol.: 3.3 – 8.3 TeV) Refs. 9	7 – 12 TeV (pol.: 8 – 14 TeV) Ref. 10	HERA: $\quad$ 5 $\quad$ TeV LHC-ep: $\quad$ 19 $\quad$ TeV
LL,RR,LR,RL, VV,AA				

	PETRA $\quad$ ≤ 46.8 GeV, 103 pb^{-1}	LEP II $\quad$ 190 GeV, 500 pb^{-1}	CLIC $\quad$ 2 TeV, 2 · 10^4 pb^{-1}	$s_{eff}=$ s
e^+e^-	1.4 – 3.1 TeV	6 – 13.5 TeV (pol.: 7.6 – 13.5 TeV)	53 – 105 TeV	PETRA : $\quad$ 2.8 TeV ($\sqrt{s} \approx$ 40 GeV) LEP II: $\quad$ 9 TeV CLIC: $\quad$ 75 TeV
LL,RR,AA,VV	Refs. 11	Refs. 4,5	Refs. 4,6	

The "reach" or detection sensitivity for contact interactions of a given collider is expressed by the attainable

$$95 \text{ \% CL bound on } \Lambda^* = \Lambda^*(\sqrt{s}, \int Ldt) \tag{8}$$

as a function of $\sqrt{s}$ and the integrated luminosity $\int Ldt$. This reach depends to some extent on the helicity configuration of Table 1. (It is quite instructive to add a reaction dependent "worst case" helicity configuration to this list, which is defined by minimal interference with the respective standard model amplitude. Such a worst case study was proposed and performed for e^+e^- reactions in Refs. 3-6).

Fig. 2, also taken from Ref. 2, summarizes the reach of pp colliders with $\int Ldt = 10^4$ pb^{-1} as a function of $\sqrt{s}$. The bounds on Λ^* are obtained from the requirement that the nominal ratio between the interference term and the standard model term is at least 1/2 at a value of p_T above which there are 50 events for pp → jet + X (at least 1 at a value of the lepton pair mass above which there are 75 events). Again only the LL helicity combinations, as specified in Figs. 1 a,b) , are considered.

In Table 2 the reach for four-fermion contact interactions of the future pp collider projects LHC, SSC (for $\sqrt{s}$ = 20 and 40 TeV) and ELOISATRON, as read off from Fig.2, is summarized. An analysis[7] of the jet-jet angular distribution in pp → jet + jet + X leads to a very similar reach, Λ^* = 12 TeV, at LHC. For comparison, Table 2 also summarizes the reach of future ep and e^+e^- collider projects. Besides HERA and LEP II also an ep option of LHC with $\sqrt{s}$ = 1.4 TeV and $\int Ldt$= 10^3 pb^{-1} and a proposed CERN linear e^+e^- collider CLIC with $\sqrt{s}$ = 2 TeV and $\int Ldt$= $2 \cdot 10^4$ pb^{-1} are included. The latter two, together with LHC, make up the set of pp, ep and e^+e^- colliders, the physics potential of which has been extensively studied in Ref. 8. The results for the 95 % CL bounds on Λ^*quoted in Table 2 for LEP II[4,5], CLIC[4,6], HERA[9] and the ep option of LHC[10] also take into account systematic errors and are specified for unpolarized electron or positron beams as well as, in brackets, for polarized beams. In all cases the variation of the reach over the helicity configurations specified in the left-most column is quoted. A discussion of Table 2 is deferred to the end of the next subsection.

Analytical insight

A qualitative analytical insight[12] into the results of Table 2 may be obtained by generalizing considerations[3-6] for e^+e^- reactions to any four-fermion process. The differential cross section of a $f\bar{f} \to f'\bar{f}'$ subprocess with cm energy $\sqrt{\hat{s}}$ may be approximated by

$$\frac{d\sigma}{d\cos\theta} \simeq \frac{d\sigma}{d\cos\theta}\bigg|_{SM} + 2 \cdot f(\theta)\eta \frac{\sqrt{\hat{s}}}{\Lambda^{*2}} \sqrt{\frac{d\sigma}{d\cos\theta}\bigg|_{SM}} + O\left(\frac{\hat{s}}{\Lambda^{*4}}\right) \tag{9}$$

the standard model (SM) cross section and the interference term.

Typically

$$\int_{-1}^{+1} d\cos\theta \; f(\theta)^2 \sim O(1) \tag{10}$$

holds. The parameter η, confined to

$$0 \leqslant |\eta| \leqslant 1, \tag{11}$$

is supposed to measure the degree of interference of the contact with the standard model term (depending on the helicity configuration and possible phases of the amplitudes involved). The square of the relative deviation from the standard model

$$\Delta_{th}^2 = \left[\frac{\delta \frac{d\sigma}{d\cos\theta}}{\frac{d\sigma}{d\cos\theta}} \right]^2 \simeq 4 \; f^2(\theta)\eta^2 \; \frac{\hat{s}}{\Lambda^{*4}} \; \frac{1}{\frac{d\sigma}{d\cos\theta}} \tag{12}$$

has to be compared to the square of the statistical error

$$\Delta_{exp}^2 = \frac{1}{dN(\cos\theta)} = \frac{1}{\int Ldt \cdot \frac{d\sigma}{d\cos\theta} \cdot d\cos\theta} \tag{13}$$

where $N(\cos\theta)$ is the number of events into angles between $\cos\theta$ and $\cos\theta + d\cos\theta$. Both quantities are inversely proportional to $d\sigma/d\cos\theta$. The 95 % CL bound on Λ^* is obtained from

$$\Delta\chi^2 = 4 = \int_{\cos\theta} \frac{\Delta_{th}^2}{\Delta_{exp}^2} = 4 \; \eta^2 \; \frac{\hat{s}}{\Lambda^{*4}} \int Ldt \int_{-1}^{+1} d\cos\theta \; f^2(\theta) \tag{14}$$

Using the estimate (10) we obtain

$$\text{95 \% CL bound on } \Lambda^* \sim O(1) \; \sqrt{\eta} \; (\hat{s} \int Ldt)^{1/4}. \tag{15}$$

Following Refs. 3-6, we conclude that Λ^* increases only with the fourth root of $\int Ldt$, i.e. that "patience does not really pay". A further important observation is that, apart from the size of η, the bound on Λ^* does not depend on the size and the details of the standard model amplitude.

Clearly, this exercise for the $\overset{(-)}{f}\overset{(-)}{f'} \to \overset{(-)}{f}\overset{(-)}{f}$ subprocess can only be applied directly to $e^+e^- \to f\bar{f}$ reactions with $s = \hat{s}$. In ep reactions the rule of thumb that effectively each of the three valence quarks in the proton carries 1/6 of its energy (while the gluons carry the other half) may be applied. As displayed in the last column of Table 2, the naive estimates

$$\Lambda^* \sim (s \int Ldt)^{1/4} \text{ for } e^+e^- \text{ and}$$
$$\Lambda^* \sim (\tfrac{1}{6} s \int Ldt)^{1/4} \text{ for ep} \tag{16}$$

compare quite well with the bounds determined professionally.

The comparatively small bounds quoted for pp initiated reactions reflect the inherent handicaps: i) in pp → 1^+1^-X the effective energy available in qq → 1^+1^- is strongly reduced, since antiquarks carry only a very small fraction of the proton energy, ii) pp → jet + X and pp → jet jet + X are dominated by gg and gq initiated subprocesses; the $q\bar{q}'$ → $q\bar{q}'$ subprocesses only win at the high energy end of phase space, where the $q\bar{q}$ luminosities are very small. The pp bounds in Table 2 were calculated assuming a LL helicity configuration for the four-fermion contact interaction only. It may be strongly argued that other helicity configurations lead to higher bounds on Λ^*. Firstly, a recent determination[6] of bounds on Λ^* in e^+e^- → $q\bar{q}$ reactions shows that the AA configuration is the most favorable one, leading to

$$\Lambda^*_{AA} \approx 1.7\ \Lambda^*_{LL} \quad \begin{cases} \text{in } e^+e^- \to q\bar{q}, \\ \text{in } q\bar{q} \to 1^+1^- \\ \text{in } pp \to 1^+1^- + X \end{cases} \tag{17}$$

which applies equally well to pp → 1^+1^- + X, since the relation (17) is energy independent. Secondly, in pp → jet + X the QCD amplitude for $q\bar{q}'$ → $q\bar{q}'$ involves a vectorlike coupling; therefore it seems quite likely that maximal interference is obtained for the VV helicity configuration and not for LL.

To summarize, in comparison to e^+e^- and ep colliders pp colliders are not particularly suitable tools to detect <u>first signatures</u> of quark-lepton compositeness in terms of <u>four-fermion contact interactions</u>. Nevertheless, it should be kept in mind that they become our only probe of quark substructure, if i) quarks and leptons have no constituent in common or ii) the lepton radius is much smaller than the quark radius. The reach of pp colliders quoted in Table 2 may become almost twice as large for more favorable helicity configurations. A professional analysis of the RR,VV and AA combinations is strongly advocated.

Contact interactions involving gluons

Since pp → jet + X and pp → jet jet + X are dominated by gg initiated reactions, it is worthwhile to investigate their potential to detect quark-lepton substructure through gauge invariant contact interactions in gg → $q\bar{q}$ subprocesses. Notice, however, that the appearance of such contact terms is more model dependent than that of four-fermion contact terms. They are the low-energy limit ($\sqrt{\hat{s}} \ll \Lambda$) of possible new colored heavy fermion exchanges in the crossed channel

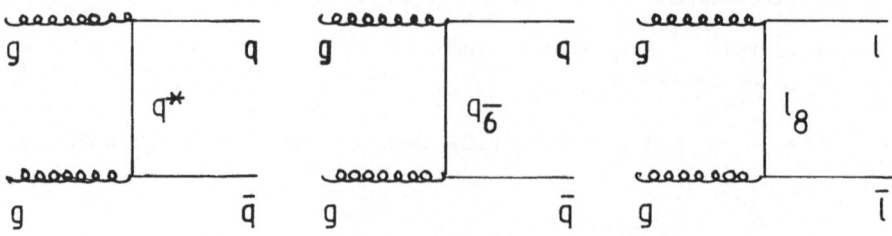

with masses of the order of Λ. The most straightforward exchange is an excited quark q^*. Another example is the exchange of a composite color antisextet quark $q_{\bar{6}}$ which automatically accompanies a composite quark, if the quark consists of at least two colored constituents. This is illustrated by the simple exercise

$$\underline{\bar{3}} \times \underline{\bar{3}} = \underline{3} + \underline{\bar{6}} . \qquad (18)$$
$$\phantom{\underline{\bar{3}} \times \underline{\bar{3}} = }\underset{q}{\uparrow}\underset{q_{\bar{6}}}{\uparrow}$$

The couplings of spin $1/2$ q^* and $q_{\bar{6}}$ to $g\bar{q}$ is given in terms of dimension five operators. The appropriate helicity conserving, CP conserving gauge invariant gqq^* coupling is described by

$$L = g_c(\Lambda) \frac{g}{2\Lambda} \bar{q}^*(x) \sigma^{\mu\nu}(1{\overset{+}{\underset{-}{}}}\gamma_5) \frac{\lambda^a}{2} q(x) G^a_{\mu\nu}(x) + \text{h.c.} \qquad (19)$$

where $g_c(\Lambda)$ is the QCD coupling constant at momenta of the order of Λ, g a coupling conventionally chosen of the order of $\sqrt{4\pi}$, $G^a_{\mu\nu}$ is the gluon field strength tensor, λ^a the Gell-Mann matrices and summation over $a = 1, \ldots, 8$ is implied. The $gqq_{\bar{6}}$ coupling is constructed analogously.

The contribution of the contact interaction in the $gg \rightarrow f\bar{f}$ subchannel with cm energy $\sqrt{\hat{s}}$ by interference with the standard model QCD amplitude may then be estimated in complete analogy to eq. (9) to be

$$\frac{d\sigma}{d\cos\theta} \simeq \frac{d\sigma}{d\cos\theta}\Big|_{SM} + 2g_c^2(\Lambda)\, f(\theta) \cdot \eta\, \frac{\sqrt{\hat{s}}^{\,3}}{\Lambda^{*4}} \sqrt{\frac{d\sigma}{d\cos\theta}\Big|_{SM}} \qquad (20)$$

where $g^2 = 4\pi$, $m_f = \Lambda = \Lambda^*$ and $\sqrt{\hat{s}} \ll \Lambda^*$ have been assumed. Repeating the arguments following Eq. (9) we find the

$$95\ \%\ \text{CL bounds on } \Lambda^*_{gg\rightarrow q\bar{q}} = O(1)\ (g_c^2(\Lambda) \cdot \eta)^{1/4}\ (\hat{s}^3 \textstyle\int Ldt)^{1/8}$$
$$(21)$$

With $g_c^2(\Lambda) = 4\pi\alpha_c \lesssim O(1)$, $\eta = 1$, and, very tentatively, $s_{eff} = s/10$ (motivated by the fact that $x^3 G(x, Q^2)$ has a maximum at $x \sim 0.3 \sim \sqrt{10}$, where $xG(x, Q^2)$ is the gluon distribution) I obtain

$$\Lambda^* \simeq ((\tfrac{s}{10})^3\ \textstyle\int Ldt)^{1/8} \simeq \begin{cases} 23 & \text{for LHC} \\ 26\ /\ 44 & \text{for SSC} \\ 88 & \text{for ELOISATRON} \end{cases} \qquad (22)$$

These optimistic "guestimates" are sufficiently interesting to advocate a professional determination of the reach of pp colliders for contact interactions in $gg \rightarrow q\bar{q}$ due to quark substructure.

Finally, let me mention a gauge invariant contact term in $gg \rightarrow l^+l^-$, due to the exchange of a composite color octet lepton[13] l_8 with mass of order Λ in the crossed channel. Composite l_8 automatically accompany composite leptons, if the leptons consist of at least two colored constituents, as is illustrated by

$$\underline{3} \times \underline{\bar{3}} = \underline{1} + \underline{8} .$$

$$\begin{array}{cc} \uparrow & \uparrow \\ 1 & 1_8 \end{array}$$

$$(23)$$

Since there is no standard model amplitude for $gg \to 1^+1^-$, the contact term adds <u>incoherently</u> to the Drell-Yan process in $pp \to 1^+1^- + X$. Correspondingly one expects a substantially smaller reach than in Eq. (22). Indeed, the bounds on Λ^*, quoted in Ref. 14 and transposed to my convention ($g^2 = 4\pi$), are

$$\Lambda^*_{gg \to 1\bar{1}} \quad \begin{cases} 7.5 \text{ TeV for } \sqrt{s} = 20 \text{ TeV} \\ 10.4 \text{ TeV for } \sqrt{s} = 40 \text{ TeV.} \end{cases} \quad (24)$$

pp COLLISIONS AS SOURCE OF COLORED COMPOSITES

The real potential of very high energy pp colliders to unravel substructure shows up in scenarios with a compositeness scale Λ well below $\sqrt{s}$, for 17 TeV < $\sqrt{s}$ < 100 TeV:

$$\Lambda \sim O \ (0.5 - 5 \text{ TeV}). \quad (25)$$

This discussion includes, as an extreme case, also "nearby compositeness" with Λ equal to the Fermi scale, $\Lambda \sim (\sqrt{2}G_F)^{-1/2}$ $\sim$ 250 GeV, where besides the quarks and leptons also the $W^\pm, Z$ bosons are composites.

Any composite model predicts, besides composite quarks and leptons (and possibly $W^\pm, Z$), <u>new heavy composites</u> with

$$\text{mass} \sim O(\Lambda), \quad (26)$$

among them also <u>colored composites</u>, like e.g. excited quarks q^*, color antisextet quarks $q_{\bar{6}}$, color octet leptons 1_8, leptoquarks, color octet bosons etc. New particles, directly produced, are clean signatures of compositeness. Clearly, very high energy pp colliders are an ideal source for new colored particles. They are even <u>unique</u>, in comparison to all presently discussed e^+e^- and ep collider projects, for <u>pair production</u> of heavy colored composites. First of all, there is sufficient phase space available for pair production of particles with masses of the order of a few TeV. Secondly, colored particles can be pair produced via QCD forces with reasonable rates. Finally, the predictions are to a high degree independent of the details of the underlying composite model.

Examples of Colored Composites

Among composite <u>colored fermions</u>, besides excited quarks q^*, most frequently color antisextet quarks $q_{\bar{6}}$ and color octet leptons 1_8 are predicted. As already mentioned earlier, they naturally appear in any composite model in which quarks and leptons harbour at least two colored constituents (see e.g. Refs. 15,16). They decay almost to 100 % into

$$q_{\bar{6}} \rightarrow q + \text{gluon}, \quad l_8 \rightarrow l + \text{gluon} \qquad (27)$$

A rich spectrum of exotic composite <u>colored bosons</u> is e.g. predicted by the strongly coupled standard model[17,18], a prominent model for nearby compositeness, which is distinguished by its intimate relationship to the standard electroweak model. It is based on a confining SU(2) gauge theory on the constituent level; the constituents are fermionic SU(2) doublets Q_i, L_i and a complex scalar doublet ϕ, where i= 1,2,3 counts the generation, Q_i are color triplets, L_i and ϕ color singlets. The SU(2)-singlet bound states involving the constituent ϕ are identified with the quark doublets q_i and lepton doublets l_i, the vector bosons $W^{\pm}$, Z and a scalar boson φ

$$q_i = \begin{pmatrix} Q_i\phi \\ Q_i\phi^* \end{pmatrix} \quad , \quad l_i = \begin{pmatrix} L_i\phi \\ L_i\phi^* \end{pmatrix} \quad , \quad W^{\pm},\ Z = \begin{pmatrix} \phi\ \phi \\ \phi\ \phi^* \\ \phi^*\phi^* \end{pmatrix} \quad , \varphi = \phi\phi^*.$$
$$(28)$$

New composite bosons appear as bosonic SU(2) singlet bound states of the fermions Q_i and L_j

Spin 1
i,j= 1,2,3
$$\begin{cases} L_i\bar{L}_j & \text{color } \underline{1} \quad \text{bosons} \\ Q_i\bar{L}_j & \text{color } \underline{3} \quad \text{leptoquarks with Q= 2/3} \qquad (29) \\ Q_i\bar{Q}_j & \text{color } \underline{1} + \underline{8} \quad \text{bosons} \end{cases}$$

Spin 0,1
i,j= 1,2,3
$$\begin{cases} L_iL_j & \text{dileptons with } Q = \underline{-1} \\ Q_iL_j & \text{color } \underline{3} \text{ leptoquarks with Q= -1/3} \qquad (30) \\ Q_iQ_j & \text{color } \underline{\bar{3}} + \underline{6} \text{ diquarks with Q= 1/3.} \end{cases}$$

Each boson carries two generation indices i,j and has two decay channels with branching ratio of 50 % each (unless they coincide). The decay channels of the colored composites are

$$Q_i\bar{Q}_j \diagup^{\displaystyle q^i_{2/3} \quad \bar{q}^j_{2/3}}_{\displaystyle q^i_{-1/3} \quad \bar{q}^j_{-1/3}} \qquad Q_iQ_j \diagup^{\displaystyle q^i_{2/3} \quad q^j_{-1/3}}_{\displaystyle q^i_{-1/3} \quad q^j_{2/3}} \qquad (31)$$

$$Q_i\bar{L}_j \diagup^{\displaystyle q^i_{2/3} \quad \bar{\nu}^j}_{\displaystyle q^i_{-1/3} \quad l^{+j}} \qquad Q_iL_j \diagup^{\displaystyle q^i_{2/3} \quad l^{-j}}_{\displaystyle q^i_{-1/3} \quad \nu^j}$$

This model has no problem with flavor changing neutral currents.

Pair Production of Colored Composites

Given the spin and color assignment ($\underline{3}, \underline{6}, \underline{8}, \ldots$) of a composite particle, its pair production rate in $\bar{p}p$ collisions via color forces is calculable as a function of its mass, <u>the only</u> <u>free parameter.</u> In order to determine discovery limits for the

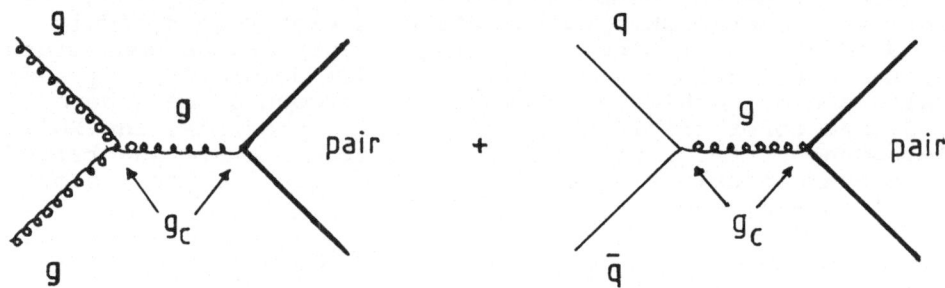

mass, some more model dependence enters, since the main decay channels and their branching ratios have to be specified. Finally background calculations have to be performed.

As examples let me present in Fig. 3 the cross sections for pair production in pp collisions of i) spin 0, color $\underline{3}$ leptoquarks at LHC from Ref. 19, ii) spin 1/2, charged color octet leptons at SSC from Ref. 20 and iii) Spin 1/2, color $\underline{3}$ fermions like q^* at 20,40 and 100 TeV from Ref. 2. For a given energy they are, not surprisingly, of comparable size. Fig. 3 shows that for pair production of leptoquarks at LHC the event rate limit is

$$m_{leptoquark} \simeq 2 \text{ TeV at } \sqrt{s} = 17 \text{ TeV}. \tag{32}$$

The pair of leptoquark $\rightarrow$ q + l decays would give $l^+ + l^-$ + jet + jet final states. The dominant background from pair production of heavy quarks (e.g. $t\bar{t}$), followed by their semileptonic decays, was estimated in Ref. 19. The authors conclude that the discovery limit is equal to the event rate limit (32). Since the pair of color octet lepton $\rightarrow$ l + gluon decays gives the same final state, $l^+ + l^-$ + jet + jet, the same conclusions may be drawn for color octet leptons. For ELOISATRON, with $\sqrt{s}$ = 100 TeV and $\int Ldt = 10^4$ pb^{-1}, the event rate limit for new colored composites is roughly

$$m_{colored\ composites} \sim 5 \text{ TeV for } \sqrt{s} = 100 \text{ TeV}. \tag{33}$$

Single production of colored composites

All the exotic colored composites may also be singly produced in pp collisions

$$gg, q\bar{q} \rightarrow 1_8 \bar{1}, \quad gg, q\bar{q} \rightarrow q^* \bar{q}, q_{\underline{6}} \bar{q},$$

$$gq \rightarrow \text{leptoquark } \bar{l}, \quad gq \rightarrow \text{diquark} + \bar{q}, \quad qq \rightarrow \text{diquark}. \tag{34}$$

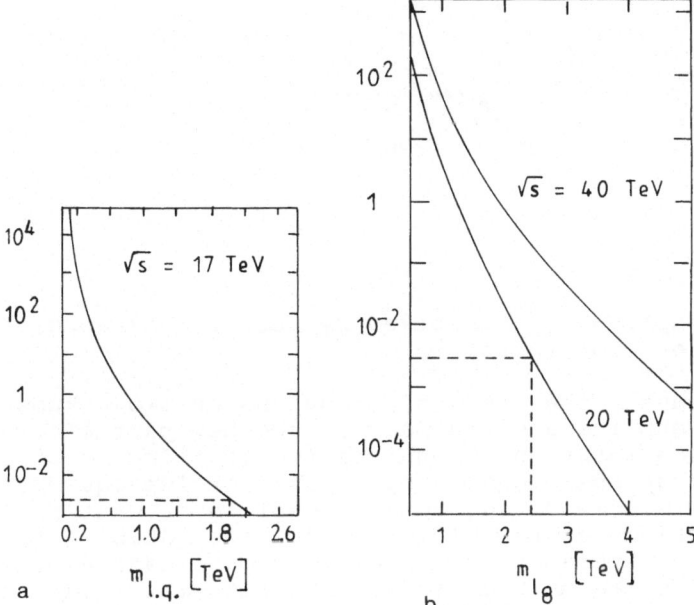

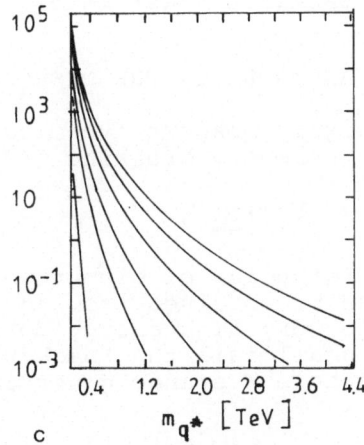

Figure 3. σ [pb] for pair production of colored composites in pp → pair + X for a) leptoquarks, from Ref. 19, b) color octet leptons from Ref. 20 and c) excited quarks from Ref. 2.

A further interesting mechanism is $gg, q\bar{q} \to$ color $\underline{8}$ vector boson V_8

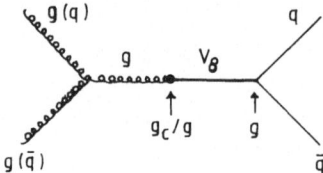

where the coupling g_C/g is fixed by vector dominance, generalized[21] to the gluon sector.

In all cases, the predictions depend on two unknown parameters, the coupling and the mass of the new particle. For $1_8, q^*, q_{\bar{6}}$ gauge invariance requires the coupling to be $g_C(\Lambda) \cdot g/\Lambda$, as spelt out in Eq. (19), for leptoquarks and diquarks it is some dimensionless coupling g. What is needed is a calculation of the single production rates in terms of the two unknown parameters and a background estimate. The detection limit for the mass of a new particle will result as a function of the coupling g. Such a calculation for $qq \to$ diquarks $\to$ jet + jet is mentioned in Ref. 19, with the conclusion that the signal is drowned in the QCD background for any plausible magnitude of the coupling g. A few calculations of single production rates are available (see e.g. Ref. 22), but clearly more work has to be done on background estimates.

$W_L W_L \to W_L W_L$ SCATTERING AS PROBE OF THE HIGGS SECTOR

The issue of the largely unknown dynamics responsible for spontaneous electroweak symmetry breaking (SSB)

$$SU(2)_L \times U(1)_Y \xrightarrow{SSB} U(1)_{em} \qquad (35)$$

is central for our understanding of electroweak interactions. Valuable information can be obtained from an effort to isolate the WW fusion process in very high energy pp collision, $qq \to WWqq$, from what I shall call the "background" process, $q\bar{q} \to WW$ annihilation, (where W stands generically for $W^\pm$ or Z)

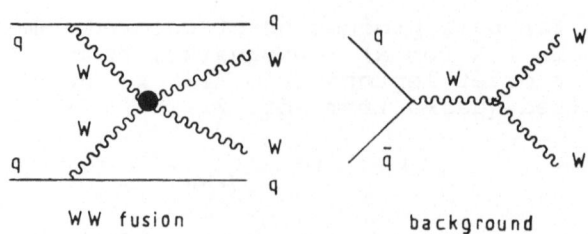

WW fusion   background

The attainable information from such an effort has been anticipated [23-25] in form of a "no-lose corollary":

i) <u>Either</u> the SSB sector provides one or more light particles (m << 1 TeV) which perform the functions of the standard model Higgs; these are <u>directly</u> observable signatures of the SSB sector. In that case the WW fusion process, of order $O(g^2 m^2/16\pi v^2) << O(g^2)$, becomes unobservably small as compared to the annihilation background which is of $O(g^2)$. Here g is the $SU(2)_L$ gauge coupling and v the electroweak symmetry breaking scale $v = (\sqrt{2}G_F)^{-1/2} \simeq 246$ GeV.

ii) <u>Or</u> there is no light particle with mass $\lesssim$ 1 TeV. In this case <u>indirect</u> information about the SSB sector comes from $W_L W_L \rightarrow W_L W_L$ which becomes strongly interacting and will eventually dominate the background process for sufficiently high energies. Unfortunately, at $\sqrt{s} \simeq 40$ TeV the effect turns out [23-25] to be still marginal. Since it increases with increasing energy, the interest in the WW fusion subprocess is a strong argument for a 100 TeV pp project like ELOISATRON.

Strongly interacting $W_L W_L \rightarrow W_L W_L$ scattering

Rather strong statements can be made within a general framework for strongly interacting $W_L W_L \rightarrow W_L W_L$ scattering. The framework is defined by

i) a local $SU(2)_L \times U(1)_Y$ gauge theory;

ii) the gauge bosons $W^{\pm}$, Z aquire masses due to the spontaneous symmetry breakdown (35);

iii) there are no light particles, in particular no light Higgs boson (masses $\gtrsim$ 1 TeV).

Otherwise the SSB is unspecified. For instance, it could be effected

i) by a standard model Higgs sector with a heavy Higgs boson with mass $m_H \gtrsim$ 1 TeV,

ii) by a dynamical composite Higgs à la technicolor or by a heavy composite (pseudo-) Goldstone boson Higgs[26], both schemes leading to composite longitudinal components $\vec{W}_L$ of the $\vec{W}$ triplet or

iii) by a framework, in which $\vec{W}_L$ and quarks and leptons are composite etc.

The SSB sector is permitted to have any global symmetry with symmetry group $G \supseteq SU(2)_L \times U(1)_Y$, spontaneously broken to some subgroup H

$$G \xrightarrow{\text{SSB}} H \tag{36}$$

with $H \supseteq U(1)_{em}$ (provided that no pseudo-Goldstone bosons become lighter than 1 TeV).

For illustration purposes, let me briefly recall the standard model Higgs sector for <u>small</u> Higgs mass, $m_H <<$ 1 TeV. It has a global $G = SU(2)_L \times \overline{SU(2)}_R$ symmetry which is spontaneously broken to $H = SU(2)_{L+R}$. In terms of the physical Higgs field H and the triplet $\vec{\Pi}$ of Goldstone boson fields associated with the SSB the Higgs potential reads

$$V(\vec{\Pi}, H) = \frac{\lambda}{4}(\vec{\Pi}^2 + H^2)^2 + \lambda v(\vec{\Pi}^2 + H^2)H + \frac{m_H^2}{2}H^2 \tag{37}$$

The coupling

$$\frac{\lambda}{8\pi} = \frac{m_H^2}{16\pi v^2}$$ (38)

is a measure of the $\Pi\Pi$ interaction strength. It is small for $m_H \ll 1$ TeV. Due to the Higgs mechanism in the presence of $SU(2)_L \times U(1)_Y$ gauge interactions the $\vec{\Pi}$ degrees of freedom furnish the longitudinal modes $\vec{W}_L$ of the $\vec{W}$ gauge bosons; their tree amplitudes become equal for $\sqrt{s} \gg m_W$

$$M(W_L W_L \rightarrow W_L W_L) = M(\Pi\Pi \rightarrow \Pi\Pi) + O(\frac{m_W}{\sqrt{s}}).$$ (39)

The tree contribution of the $\frac{\lambda}{4}(\vec{\Pi}^2)^2$ interaction term in Eq. (37) for $\sqrt{s} \gg m_W$

$$M(W_L^+ W_L^- \rightarrow Z_L Z_L) \simeq \frac{s}{v^2}$$

$$M(W_L^+ W_L^- \rightarrow W_L^+ W_L^-) \simeq -\frac{u}{v^2}$$ (40)

$$M(Z_L Z_L \rightarrow Z_L Z_L) \simeq 0$$

is prevented from violating the unitarity bound by the Higgs exchange contribution

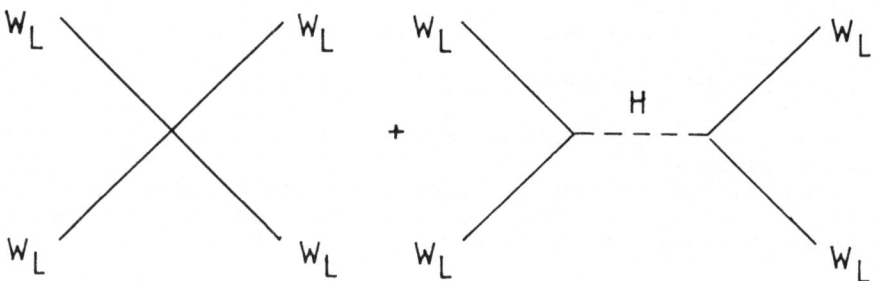

If m_H increases beyond 1 TeV, λ increases

$$\frac{\lambda}{8\pi} > O(1),$$ (41)

leading to strong $\Pi\Pi$, resp. strong $W_L W_L$ interactions.

The important point is now that as well the tree level equivalence relation (39) as a generalized version of the tree level "low-energy" amplitudes (40) have been shown [23,27] to hold to all orders in the strong coupling (λ) within the general framework of strongly interacting $W_L W_L$ as specified above. (Cf. also the talk by Z. Kunszt at this workshop). At the basis of this generalizability is the fact that strong $W_L W_L$ interactions are traced back to strong interactions of Goldstone bosons, which nonlinearly realize the full $SU(2)_L \times U(1)_Y$ symmetry in a well-prescribed way.

The generalized low energy theorems for strongly interacting $W_L W_L$ with spontaneous $SU(2)_L \times U(1)_Y \to U(1)_{em}$ symmetry breakdown are

$$
\begin{aligned}
M(W_L^+ W_L^- \to Z_L Z_L) &\simeq \frac{s}{v^2} \frac{1}{\rho} \\[6pt]
M(W_L^+ W_L^- \to W_L^+ W_L^-) &\simeq -\frac{u}{v^2} (4 - \frac{3}{\rho}) \\[6pt]
M(Z_L Z_L \to Z_L Z_L) &\simeq 0
\end{aligned}
\tag{42}
$$

valid in the interval

$$
m_W \ll \sqrt{s} \ll \Lambda_{SB} = \min (4\pi v, M_{SB}),
\tag{43}
$$

where M_{SB} is the characteristic scale of the spectrum in the SSB sector, $v = 2m_W/g \simeq 246$ GeV, $s + t + u \simeq 0$ and ρ the rho parameter

$$
\rho = \left(\frac{m_W}{m_Z \cdot \cos\theta_W} \right)^2.
\tag{44}
$$

All other WW amplitudes may be obtained from eq. (42) by crossing. Clearly, in standard model like theories with

$$
G = SU(2)_L \times SU(2)_R \to H = SU(2)_{L+R} : \rho = 1,
\tag{45}
$$

as also required by experiment to within a few percent. A three-fold proof for the low energy theorems (42) was given in Refs. 27

i)   by means of a power counting analysis carried out in a perturbative framework (with respect to the strong coupling). The corrections to the low energy $W_L W_L$ amplitudes (42) turn out to be either screened by α_w/π (with $\alpha_w = g^2/4\pi$) or to be suppressed by powers of s/Λ_{SB}^2;

ii) by exploiting $SU(2)_L$ current algebra and the principle of partially conserved left-handed current. This derivation is in complete analogy to Weinberg's derivation of the low-energy theorems for strongly interacting π mesons based on $SU(2)_L \times SU(2)_R$ current algebra and PCAC (the rôle of the pion decay constant f_π being replaced by v).

iii) by setting up an effective Lagrangian which realizes the G symmetry nonlinearly in terms of the G/H Goldstone boson fields. To my opinion this is the most transparent approach.

Rates in pp → WW + X; WW fusion versus $q\bar{q}$ → WW

The issue is to estimate the contribution of the WW fusion subprocess to pp → WW + X for <u>strongly interacting $W_L W_L$</u> in comparison with the background annihilation process as a function of the collider energy $\sqrt{s}$. For this purpose the low energy theorems (42) for strongly $W_L W_L$ interactions are an important lead, however, they are not sufficient. A simple extrapolation of the behaviour (42) to higher energies would violate the unitarity bound for $s \simeq 16 \pi v^2$. Assumptions about the dynamics responsible for unitarity restauration have to be made. Following Refs. 23-25 two generic constellations are discussed.

i) <u>Heavy Higgs</u>. A fairly heavy standard model Higgs boson with mass $m_H = 1$ TeV is assumed which (marginally) allows to satisfy the unitarity bound. In my opinion the main interest in this choice comes from non-perturbative approaches to the standard model for very large (bare) Higgs mass ($m_H \gg 1$ TeV), resp. for very strong coupling λ. Firstly, in Ref. 28 the dynamics of the $SU(2)_L \times SU(2)_R \simeq O(4)$ symmetric standard model Higgs sector is approximated by an $O(2N)$ symmetric theory to leading order in $1/N$; this approach implies a non-perturbative treatment to all orders in the strong coupling λ. The result is a broad Higgs like $W_L W_L$ bound state with (renormalized) mass $m_H \sim 800$ GeV. (See also Ref. 29). A quantitative confirmation comes from lattice calculations[30] for strong coupling ($\lambda \to \infty$) which lead to $m_H/m_W \simeq 8.5$.

ii) <u>S wave unitarity saturation</u>. A straightforward assumption is, to extrapolate[23-25] the low-energy behaviour (42) in the S wave, $a_{J=0}(s)$, until the unitarity bound is hit at s_0, followed by a saturation of the bound, $a_{J=0}(s) = 1$ for $s \geqslant s_0$. In comparison with scattering of π mesons, this assumption is quite adequate for the isospin I=0 amplitude, see[31] Fig.4, underestimating the I=1 amplitude, which has a strong P wave enhancement due to the ρ meson resonance, and overestimating the I=2 amplitude.

Following Refs. 23-25 the WW fusion contributes to $pp \to WW + X$ for $WW = ZZ, W^+W^-, W^{\pm}Z, W^{\pm}W^{\pm}$

$$(46)$$

$$\sigma(pp \to WW + X) = \int_\tau \frac{\partial L}{\partial \tau}\Big|_{qq/pp} \sum_{VL} \int_x \frac{\partial L}{\partial x}\Big|_{V_L V_L/qq} \sigma(V_L V_L \to W_L W_L)$$

with

$$\tau = s_{qq}/s_{pp}, \quad x = s_{VV}/s_{qq} \tag{47}$$

as a convolution of the luminosity distribution to find q beams in the incident protons with the luminosity distribution to find longitudinally polarized gauge bosons V_L in the incident q's and finally with the $2 \to 2$ scattering of the longitudinally polarized gauge bosons. In the effective W approximation the luminosity function for $W_L^+ W_L^-$ pairs is

$$\frac{\partial L}{\partial x}\Big|_{W_L W_L/qq} = \frac{\alpha^2}{16\pi \sin^4\theta} \frac{1}{x} [(1+x)\ln\frac{1}{x} + 2x - 2]. \tag{48}$$

A conspicuous feature of this behaviour it that doubling the qq energy increases the WW luminosity by a huge factor of order 20. This <u>very rapid increase of rates for increasing collider energy $\sqrt{s}$</u> reflects the fact that the four-body final state in

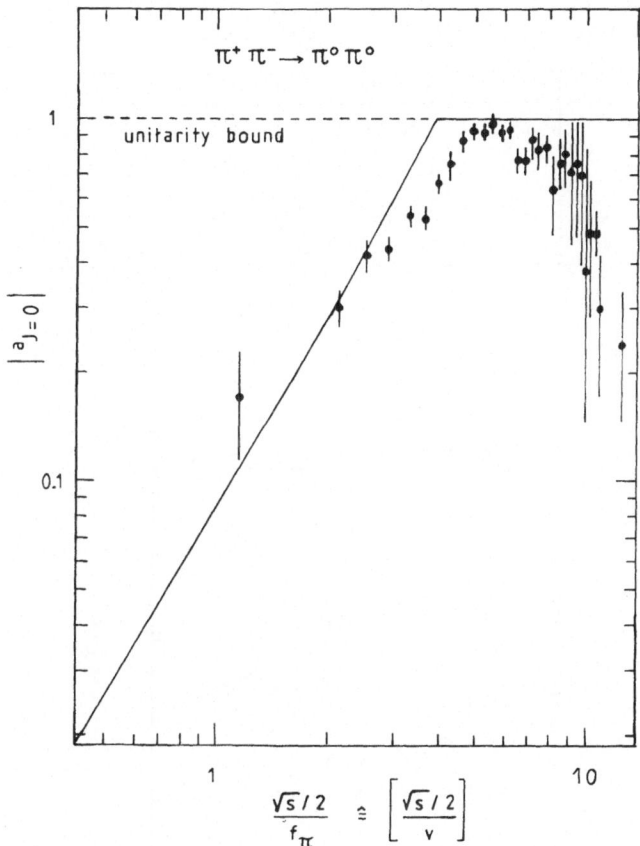

Figure 4. Taken from Ref. 31. Data for the $\pi^+\pi^- \to \pi^0\pi^0$ isospin $I = 0$ S wave amplitude $a_{J=0}(s)$ plotted versus $\sqrt{s}/(2f_\pi)$, which plays the same role as $\sqrt{s}/(2v)$ in $W_L^+W_L^- \to ZZ$. The solid curve indicates the "S wave unitarity saturation" approximation, used in $W_LW_L \to W_LW_L$.

Table 3. Event rates per 10^4 pb^{-1} for 20 and 40 TeV colliders, from Refs. 23-25. The triplets of numbers denote the rates for the annihilation background / WW fusion with S wave unitarity saturation / WW fusion with a 1 TeV standard model Higgs, respectively.

	20 TeV	40 TeV
ZZ	150 / 90 / 200	370 / 470 / 890
W^+W^-	660 / 12o / 410	1600 / 630 / 1790
$W^{\pm}Z$	290 / 120 / 0	670 / 670 / 0
$W^+W^+ + W^-W^-$	0 / 200 / 0	0 / 1200 / 0

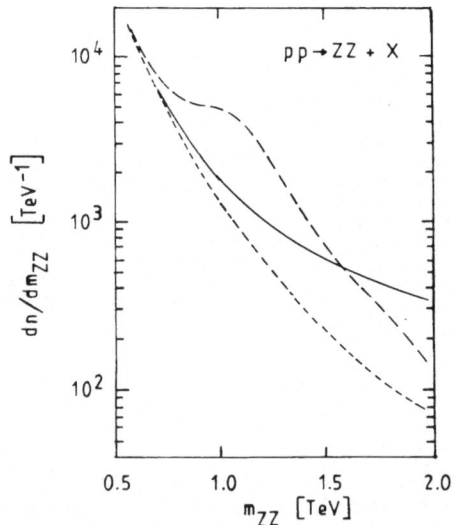

Figure 5. The signal for strongly interacting WW → ZZ scattering in qq → qqZZ i) for S wave unitarity saturation (———), ii) for a 1 TeV standard model Higgs boson (— — —), superimposed incrementally on the background from qq̄ → ZZ (-------) for a √s = 40 TeV ∫Ldt = 10^4pb^{-1} pp collider (requiring $|y_Z| < 1.5$). From Refs. 23-25.

Table 4. Expected number of events for pp → WW + X in the various decay channels of WW (W=W$^\pm$,Z) at $\sqrt{s}$ = 40 TeV, ∫Ldt = 10^4 pb^{-1}. The three numbers correspond to background from pp → WW / WW fusion, S-wave unitarity saturation / WW fusion, standard model Higgs with m$_H$ = 1 TeV, respectively.

final state and experimental cuts	number of events
ZZ → e$^+$e$^-$/μ$^+$μ$^-$ + e$^+$e$^-$/μ$^+$μ$^-$; $\|y_Z\|$ < 1.5, m$_{ZZ}$ > 1 TeV	1 ; 2 / 4
ZZ → e$^+$e$^-$/μ$^+$μ$^-$ + νν̄ ; $\|y_Z\|$ < 1.5, p$_T$(l$^+$l$^-$) > 0.45 TeV, large missing p$_T$; no jet activity to veto background from W + jet	8 /15 /27
W$^\pm$Z → eν/μν + e$^+$e$^-$/μ$^+$μ$^-$; $\|y_{W,Z}\|$ < 1.5, m$_{WZ}$ > 1 TeV	3 / 7.5/0 + 1 +12 including technirho
W$^\pm$Z → eν/μν/τν + e$^+$e$^-$/μ$^+$μ$^-$; $\|y_{W,Z}\|$ < 1.5, m$_{WZ}$ > 1 TeV	4.5/11.3/0 +1.5 +18 including technirho
W$^+$W$^+$/W$^-$W$^-$ → l$^+$l$^+$/l$^-$l$^-$ + νν; experimentally meaningful cuts? ($\|y_W\|$ < 4, m$_{WW}$ > 1 TeV)	0 /33 / 0

the qq subprocess, qq WW, mainly needs phase space to compete
with the two-body final state in q̄q annihilation. This is a
strong argument to go for as high collider energies as possible
(100 TeV!).

Fig. 5, taken from Refs. 23,25, shows the yield computed
with Eqs. (46) and (48) for a pp collider (SSC) with $\sqrt{s}$ = 40 TeV
and ∫Ldt = 10^4 pb^{-1} for the two alternatives i) Higgs with
m_H = 1 TeV, ii) S wave unitarity saturation, each superimposed
on the qq annihilation background. A rapidity cut $|y_z|$ < 1.5 has
been applied to reduce the qq → ZZ background which is strongly
forward, while the fusion signal is relatively isotropic. The
expected numbers of events[23,25] at $\sqrt{s}$ = 20 and 40 TeV are displayed in Table 3 for an invariant mass cut m_{VV} > 1 TeV (as
suggestive from Fig. 5 and for $|y_{W,Z}|$ < 1.5 (except for the
W$^\pm$W$^\pm$ channels, where $|y_W|$ < 4 suffices, owing to the absence of
the annihilation background).

A rough estimate[25] of the corresponding rates at 40 TeV for
the various feasible decay channels of the WW final states is
collected in Table 4, along with the appropriate cuts. In the
W$^\pm$Z channel also the rates for a tentative P-wave enhancement
due to some technirho type spin 1 resonance has been included.

This table shows that a prospective top energy of 40 TeV
for the SSC is at best marginal for a detection of the WW fusion
process (see also Ref. 32).

CONCLUSIONS

The reach of a pp collider with ∫Ldt = 10^4 pb^{-1} for new
four fermion contact interactions as first signatures of quark-
lepton compositeness is given in terms of an attainable bound
on the compositeness scale

Λ^* = 10-15, 15-19, 24-26 TeV from pp → jet + X,

Λ^* =   19, 26-28, 37-41 TeV from pp → l$^+$l$^-$ + X at   (49)

$\sqrt{s}$ =   20,   40,   100 TeV, respectively,   (50)

for LL helicity configurations. Arguments are given which strongly suggest a considerably higher sensitivity for other helicity
combinations, VV in pp → jet + X and AA in pp → l$^+$l$^-$ + X. A
crude estimate for the reach for gauge invariant ggqq̄ contact
interactions (g = gluon) leads to bounds

Λ^* = 26, 44, 88 TeV   (51)

at the respective energies (50). An analytical insight into the
dependence of Λ^* = Λ^* ($\sqrt{s}$ ∫Ldt) for various contact terms and
the various colliders, pp, e$^+$e$^-$ and ep, is given. In comparison
to e$^+$e$^-$ and ep machines, pp colliders appear to be not the most
suited for the detection of first"far-away" signals of quark-
lepton compositeness.

On the other hand, very high energy pp colliders are
unbeatably   suited for the detection of new colored composite
states which are typically predicted in composite models. Among

these composites are fermions like color $\bar{6}$ quarks, color $\underline{8}$ leptons, excited quarks and bosons like color $\underline{8}$ bosons, color $\underline{3}$ leptoquarks and color $\bar{3} + \bar{6}$ diquarks. Of particular interest is their pair production via color gauge forces with a reasonable production rate, which is completely determined as a function of the mass of the pair produced particle, once its spin and color assignment are given. For a pp collider with $\int Ldt = 10^4$ pb^{-1} the event rate limit is roughly reached for a mass

$$m \simeq 2 \text{ TeV for } \sqrt{s} = 20 \text{ TeV and}$$

$$m \simeq 5 \text{ TeV for } \sqrt{s} = 100 \text{ TeV.} \tag{52}$$

For $\sqrt{s} \simeq 20$ TeV and pair production of leptoquarks and of color $\underline{8}$ leptons the event rate limit was shown to be also the discovery limit. Single production of colored composites is also briefly discussed.

Finally, the improtance of very high energy pp colliders ($\sqrt{s} \gtrsim 40$ TeV) for an investigation of the scattering of longitudinally polarized W_W, $W_L W_L \to W_L W_L$, as a crucial probe of the electroweak symmetry breaking sector has been reviewed in detail. Whichever the nature of the physics responsible for the spontaneous $SU(2)_L \times U(1)_Y \to U(1)_{em}$ symmetry breakdown, if it does not provide light particles ($m \lesssim 1$ TeV) besides the (elementary or composite) longitudinal degrees of freedom $\vec{W}_L$ of the gauge boson triplet $\vec{W}$, $W_L W_L \to W_L W_L$ scattering becomes strongly interacting. For $\sqrt{s} = 40$ TeV, $\int Ldt = 10^4$ pb^{-1}, the sensitivity for strongly interacting $W_L W_L$ contributing to the qq $\to$ qq WW fusion process vis-à-vis the qq $\to$ WW annihilation background is marginal. Higher energies are needed to allow for more phase space for the four-body final state of the fusion process, leading to a dramatic increase of the signal. A 100 TeV collider project (ELOISATRON) would be ideal to settle this issue of crucial importance for our understanding of the dynamics of electroweak symmetry breaking.

REFERENCES

1. E. J. Eichten, K.D. Lane and M.E. Peskin, New Tests of Quark and Lepton Substructure, Phys. Rev. Letters, 50 : 811 (1983).
2. E. Eichten, I. Hinchliffe, K. Lane and C. Quigg, Supercollider Physics, Reviews of Mod. Phys. 56 : 579 (1984).
3. F. Schrempp, Physics at Future e^+e^- Colliders, Proc. XXIII Int. Conf. on High Energy Physics, Berkeley/USA, Vol. II: 1243 (1986).
4. B. Schrempp, F. Schrempp, N. Wermes and D. Zeppenfeld, Bounds on New Contact Interactions from Future e^+e^- Colliders, Nucl. Phys. B296: 1 (1988).
5. U. Baur, A. Blondel, D. Bloch, D. Dominici, H. Fesefeldt, K. Hamacher, L. Levinson, M. Lindner, L. Lyons, P. Mättig, P. Méry, E. Milotti, M. Perrottet, F. Renard, D. Schildknecht, B. Schrempp, F. Schrempp, K.H. Schwarzer, D. Treille, N. Wermes and D. Zeppenfeld, Compositeness at LEP II, Proc. ECFA Workshop on LEP 200, Aachen 1986, CERN 87-08 and ECFA 87-108, Vol. II: 414 (1987).

6. N. Wermes,(B. Schrempp, F. Schrempp and D. Zeppenfeld), Probing Quark and Lepton Substructure at a 2 TeV e^+e^- Collider, Proc. of the Workshop on Physics at Future Accelerators, La Thuile, CERN 87-07, Vol. II: 305 (1987).

7. A.K. Nandi, an Achievable Limit on the Energy Scale of Compositeness for Quarks from pp Collisions at $\sqrt{s}$ = 17 TeV, Proc. of the Workshop on Physics at Future Accelerators, La Thuile, CERN 87-07, Vol. II: 270 (1987).

8. Proc. of the Workshop on Physics at Future Accelerators, La Thuile, CERN 87-07, Vol. I and II (1987).

9. F. Cornet, H.-U. Martyn, R. Rückl, contribution to the DESY Workshop on Physics at HERA, Hamburg (1987). F. Cornet, Some Exotic Physics at HERA, DESY 87-131 (1987). H.-U. Martyn, Contact Terms and Substructure at HERA, PITHA 87-40 (1987).

10. F. Cornet and R. Rückl, Compositeness in ep Collisions at LEP-LHC, Proc. of the Workshop on Physics at Future Accelerators, La Thuile, CERN 87-07, Vol. II: 287 (1987).

11. TASSO Collaboration, M. Althoff et al., An Improved Measurement of electroweak Couplings from $e^+e^- \rightarrow e^+e^-$ and $e^+e^- \rightarrow \mu^+\mu^-$, Z. Phys. C22: 13 (1984). JADE Collaboration, W. Bartel et al., Tests of the Standard Model in Leptonic Reactions at PETRA Energies, Z. Phys. C30: 371 (1986).

12. B. Schrempp and F. Schrempp, to be published.

13. H. Harari, Colored Leptons, Phys. Lett. 156B:250 (1985).

14. F.M. Renard, Tests of Colored Substructure of Leptons, Phys. Lett. 166B: 229 (1986).

15. H. Harari and N. Seiberg, A Dynamical Theory for the Rishon Model, Phys. Lett. 98B: 269 (1981); The Rishon Model, Nucl. Phys. B204: 141 (1982).

16. H. Fritzsch and G. Mandelbaum, Weak Interactions as Manifestations of the Substructure of Leptons and Quarks, Phys. Lett. 102B: 319 (1981).

17. M. Claudson, E. Farhi and R.L. Jaffe, Strongly Coupled Standard Model, Phys. Rev. D34: 873 (1986).

18. B. Schrempp, Composite Vector Bosons, Proc. XXIII Int. Conf. on High Energy Physics, Berkeley, Vol. I:454 (1986) and references quoted therein.

19. J. Ellis and H. Kowalski in V. Angelopoulos et al., Beyond the Standard Model, Proc. of the Workshop on Physics at Future Accelerators, La Thuile, CERN 87-07, Vol. I: 80 (1987).

20) Y. Nir, Colored Lepton Production in Hadron Colliders, Phys. Lett. 164B: 395 (1985).

21. W. Buchmüller, Heavy Composite Vector Bosons and Low-Energy Weak Interactions, Phys. Lett 145B: 151 (1984). B. Schrempp and F. Schrempp, Massive Yang-Mills - An Effective Lagrangian for Composite W,Z and New Colored Vector Bosons, DESY 84-055 (1984). U. Baur and K.-H. Streng, Phenomenology of Composite Colored Weak Bosons, Z. Phys. C30: 325 (1986).

22. U. Baur and K.H.G. Schwarzer, Signatures for Isoscalar Weak Vector Bosons at pp and $p\bar{p}$ Colliders, Phys. Lett. 180B: 163 (1986). U. Baur and K.-H. Streng, Colored Lepton Mass Bounds from $p\bar{p}$ Collider Data, Phys. Lett. 162B: 387 (1985).

23. M.S. Chanowitz and M. K. Gaillard, The TeV Physics of Strongly Interacting W's and Z's, Nucl. Phys. B261: 379 (1985).

24. M.S. Chanowitz, Universal W,Z Scattering Theorems and No-Lose Corollary for the SSC, Proc. of the XXIII Int. Conf. on High Energy Physics, Berkeley, Vol. I: 445 (1986).

25. M.S. Chanowitz, Probing Electroweak Symmetry Breaking at a Multi-TeV Collider, Proc. of the Rencontres de Physique de la Vallee d'Aoste, La Thuile (1987).

26. M.J. Dugan, H. Georgi and D.B. Kaplan, Anatomy of a Composite Higgs Model, Nucl. Phys. B254: 299 (1985).

27. M.S. Chanowitz, M. Golden and H. Georgi, Universal Scattering Theorems for Strongly Interacting W's and Z's, Phys. Rev. Lett. 57: 2344 (1986), Low Energy Theorems for Strongly Interacting W's and Z's, Harvard Univ. preprint HUTP-87/AO26 (1987).

28. M.B. Einhorn, Speculations on a Strongly Interacting Higgs Sector, Nucl. Phys. B246: 75 (1984).

29. R. Casalbuoni, D. Domonici and R. Gatto, Effective Lagrangian Description of the Possible Strong Sector of the Standard Model, Phys. Lett. 147B: 419 (1984).

30. A. Hasenfratz, K. Jensen, C.B. Lang, T. Neuhaus and H. Yoneyama, The Triviality Bound of the 4-Component ϕ^4 Model, Florida State Univ. preprint FSU-SCRI-87-52 (1987).

31. F. Schrempp, from a Talk on "Physics of the Higgs Sector"

32. C.H. Llewellyn Smith, Physics at Future High Energy Colliders, Proc. of the XXIII Int. Conf. on High Energy Physics, Berkeley, Vol. I: 255 (1986).

LEPTOQUARKS AT 100 TeV: PROTON-ANTIPROTON COLLIDER

D. Haidt

DESY

1. Introduction

Leptoquarks are hypothetical particles with lepton and quark properties [1]. Such difermions have sofar not been observed experimentally. If they exist, their production will proceede in a $\bar{p}p$ collision via the subprocess

$$q\bar{q} \rightarrow LQ + \overline{LQ}$$
$$gg \rightarrow LQ + \overline{LQ} \tag{1}$$

provided the leptoquark mass is not too large. To be specific the following assumtions are made:

- Spin 0 leptoquarks using matrix elements of Altarelli and Rückl [2]

- Decay: $LQ \rightarrow \tau t$
  mass of t-quark 40 GeV

- Centre of mass energy of $p\bar{p}$ collider: 100 TeV

- Mass of leptoquark 1 TeV

Leptoquarks with the above properties have been incorporated in the PYTHIA program [2] which is the parton-cascade type with LUND strings. It was ensured that the *Leptoquark* signal and the QCD background were treated on the same footing concerning event generation, fragmentation and decay.

The production cross section for leptoquarks is 0.13 pb. Fig. 1 shows the sum of transverse energy of the detectable final state particles. For reconstructing jets the detector is considered to be a calorimeter with a cellular structure $\delta\phi \times \delta y = 0.1 \times 0.1$ extending over the full azimuthal angular range and over rapidities from -4 upto $+4$. This granularity is adequate for the present study. A detailed discussion of how to build such calorimeters is given in ref. [4].

It is convenient to express the 4-momentum p of a particle with mass m in terms of its rapidity y, its azimuthal angle ϕ and its transverse momentum p_T, resp. transeverse energy E_T defined with respect to the laboratory frame as follows:

$$p = (E_T \cosh y, E_T \sinh y, p_T \cos\phi, p_T \sin\phi)$$

The representation of an event on the (y, ϕ)-plane is shown in Fig. 2.

When reconstructing jets the cell size $\Delta\phi \times \Delta y$ enters. Particles falling into the same cell are no longer distinguished. To each cell with transeverse energy E_T^c and angles (y_c, ϕ_c) a 4-momentum vector is assigned:

$$p_{cell} = E_T(\cosh y_c, \sinh y_c, \cos \phi_c, sin\phi_c)$$

The distance between two cells with coordinates $(y, \phi)_1$ and $(y, \phi)_2$ is given by:

$$d(\Delta y, \Delta\phi) = 2\sqrt{(\sinh^2 \Delta y/2 + \sin^2 \delta\phi/2)}$$

with $\Delta\phi = \phi_1 - \phi_2$ ($|\Delta\phi| \leq \pi$) and $\delta y = y_1 - y_2$. For closeby cells the above formula approaches the usual expression: $\sqrt{\Delta y^2 + \Delta\phi^2}$.

The cluster algorithm in the PYTHIA program package [5] is used for the reconstruction of jets. Jets are *defined* as local E_T clusters within a cone of radius d_J on the (y, ϕ) plane having summed transverse energy bigger than E_J. The *jet mass* is defined as the invariant mass of all cell 4-vectors forming the cluster. For a study of jet characteristics in TeV range see ref. [6].

2. Detection of Leptoquarks

The final state of a leptoquark event consists of

- $\tau + t\text{-jet}$

- $\bar{\tau} + \bar{t}\text{-jet}$

- $g\text{-jets}$

leading to characteristic multi-jet configurations. Fig. 2 shows a typical example. Since the mass of the decaying leptoquark is 1 TeV the τ decay products have a typical distance of $d_{\tau t}$ of 1 ± 0.4 from the t-jet, whereas τ and $\bar{\tau}$ occur at an average $d_{\tau\bar{\tau}}$ of 2. Due to negligbly small mass of the τ its decay products fall well within one cell. A τ can therefore be classified as an *isolated jet occupying a single cell*. The τ-vector is well determined regarding y_τ and ϕ_τ, whereas $E_{T\tau}$ is systematically underestimated due to the missing ν_τ (in leptonic decays even two neutrinos). This suggests a fit procedure which may be investigated in a further study.

Selection criteria:

- $\Sigma E_T \geq 3.5$ TeV
  The sum runs over all detectable final state particles

- At least 1 jet with $d_J = 1.0$ and $E_J = 1.75$ TeV
  The invariant jet mass distribution show already some resonance structure near and below 1 TeV (Fig. 3)

- The structure of each jet is further investigated by reconstructing subjets with $d_{SJ} = 0.15$ and $E_{SJ} = 50$ GeV. Jets are retained, if they have at least two subjets. The invariant mass distribution of the subjets (Fig. 4) shows a characteristic peak at $m = 0$, which is mainly due to τ-decay products.
  A τ-candidate is defined as a subjet with $E_T \geq 300$ GeV and $m \leq 0.1$ GeV and no other subjet within $d = 0.5$. Jets which do not have a τ-candidate are rejected.

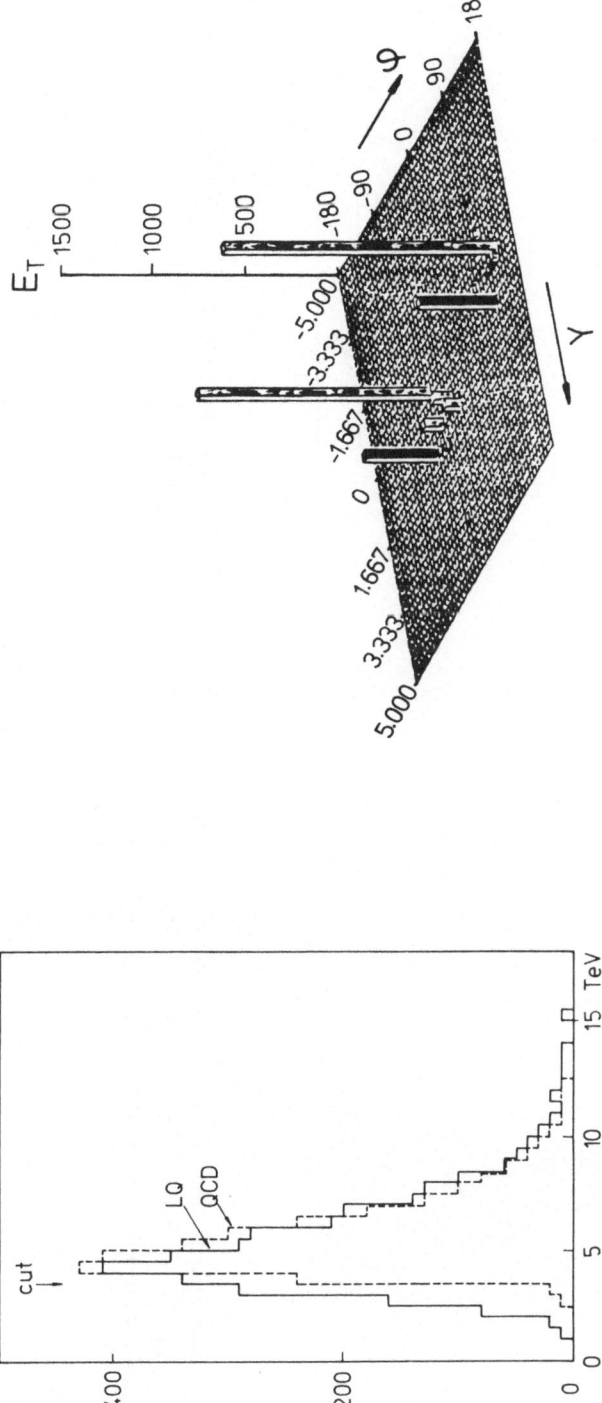

Figure 2. Display of a typical Leptoquark event on the rapidity-azimuth plane. The four distinct transeverse energy clusters correspond to the two τ leptons and to the two *top* quark jets. The insert lists the relevant kinematical quantities.

Figure 1. Distribution of the total transeverse energy for Leptoquark (solid line) and QCD events (dotted line). The QCD event distribution has been generated with transverse momentum threshold, otherwise it would steeply rise towards low transverse energies.

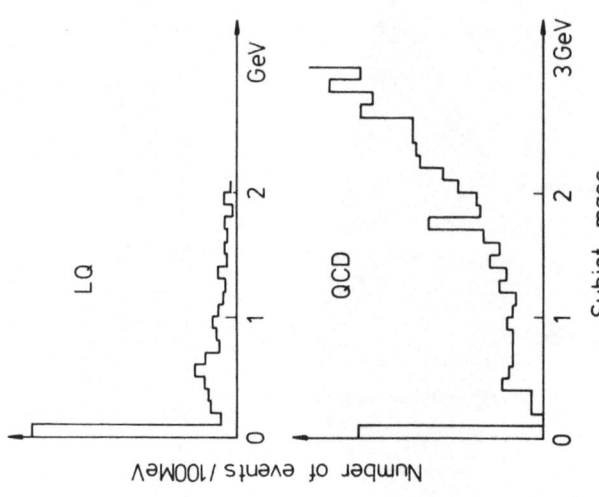

Figure 4. Subjet mass distribution of Leptoquark (a) and QCD (b) events for jets with at least two subjets.

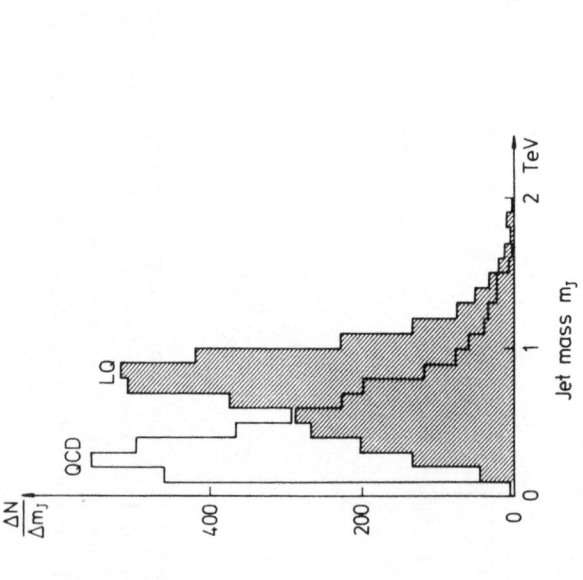

Figure 3. Jet mass distribution of Leptoquark (shaded) and QCD background events for jets with $d_J = 1$ and jet energies larger than 1.75 TeV for all events with total transverse energy larger than 3.5 TeV.

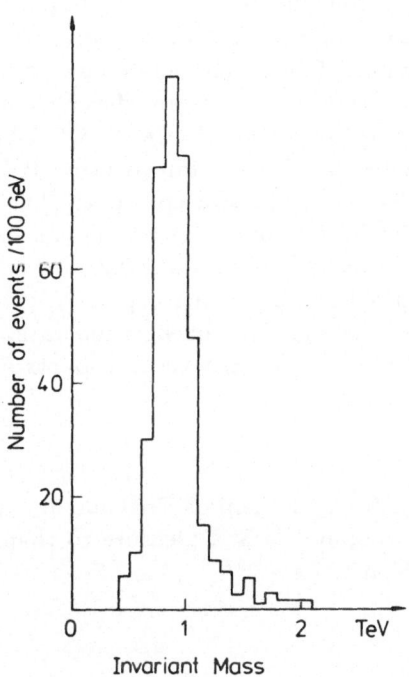

Figure 5. Invariant mass distribution of leptoquark candidates. The peak occurs at 850 GeV (instead of 1000 GeV) due to missing neutrinos.

The invariant mass distribution of leptoquarks events selected by the above criteria is shown in Fig. 5.

3. QCD-Background

The only background considered in this study are the conventional QCD processes. Even with the cutoff $\Sigma E_T \geq 3.5$ TeV this background has a cross section which is more than three orders of magnitude larger than the signal. The contribution of the gg-subprocess is about 5 times stronger than the $q\bar{q}$-subprocess at production. Since signal has a threshold the QCD-processes have been generated with an appropriate threshold such that the actual ΣE_T distribution looks similar as shown in Fig. 1. The invariant mass distribution of all reconstructed jets after cutting off events with $E_T \leq 3.5$ TeV differs in shape from the one for the leptoquark signal. Next events are selected, which have at least one jet with two or more subjets. Fig. 4 shows the invariant subjet mass distribution of QCD and leptoquark events. The characterstic behaviour near zero invariant masses allowes for an efficient reduction of background. After applying all above mentined selection criteria the QCD background is reduced to $(1.2 \pm 0.3)10^{-4}$. This number has been obtained using PYTHIA with full fragmentation requiring a considerable amount of cpu-time. The corresponding figure for leptoquarks is 0.1. Taking into account the production cross sections and the acceptance factors the signal to background ratio is 0.2.

Since there should be no problems to produce copiously leptoquarks, if they exist with the assumed properties, the signal to background ratio can be enhanced to about 100 by applying the above selection criteria to two jets per event.

Acknowledgement

This study profitted much from the help of Dr. Gunnar Ingelman in making efficient use of the PYTHIA program. It is a pleasure to thank Dr. Ahmed Ali for creating a fruitful atmosphere at Erice.

References

[1] B. Schrempp, Contribution to this Workshop.

[2] G. Altarelli and R. Rückl, Phys. Let. 114B (1984) 126.

[3] H.-U. Bengtsson and T.B. Sjörstrand, Comp. Phys. Comm. 46 (1987) 43.

[4] G. Wolf and J. Kirkby, Contributions to this Workshop.

[5] Subroutine CALJET is a modification of LUCELL in JETSET 6.2 by G. Ingelman.

[6] P.N. Burrows and G. Ingelman, Jet characteristics at TeV enrgies, Z. Phys. C C34 (1987) 91 and Proc. Workshop on Physics at Future Accelerators, La Thuile, Italy, Vol. 2 p. 369, CERN 87-07; L. Dilella, contribution to this Workshop.

PHYSICS OF SUPERHEAVY ONIA

Ken-ichi Hikasa[*]

Deutsches Elektronen-Synchrotron
2000 Hamburg 52
Federal Republic of Germany

The properties of quark-antiquark bound states made up with a superheavy quark (mass $\gtrsim m_W$) are discussed along with their production and decay at super-high-energy proton-proton colliders. Various aspects of superheavy onia go beyond the naïve extrapolation from known light and heavy onium systems. The empirical "constant" law for the leptonic width no longer holds; The pseudoscalar is the predominantly produced state in pp collisions.

Three categories of onia are described: (1) Nonexistent quarkonia, which literally do not exist as bound states; An important example is the superheavy toponium. (2) Heteronia, made of a quark with a mass generated by the $SU(2) \times U(1)$ breaking. A typical example is the fourth-generation quarkonium with a small KM mixing. (3) Homonia, whose mass is unrelated to the weak scale. An example is the state made of an extra charge-$\frac{1}{3}$ "quark" in the E_6 model.

The second category is phenomenologically the most interesting: Dominant decay modes include the final states with weak gauge bosons and/or Higgs bosons, specific mode being different for each onium state. At super pp colliders this kind of onia can be a rich source of Higgs bosons, which can be tagged by weak bosons.

Related topics of the extra Z boson as a source of the Higgs and the production of gluinonium and technipionium are also discussed.

INTRODUCTION

A proton-proton collider is a difficult machine for the physics of the future. Our recent experience tells us that it is hard to discover or exclude a new particle at hadron machines, even if it would be copiously produced. On the other hand, it is much easier to give a definite statement on the existence of a new particle in the clean environment of electron-positron colliders, something which we can see already from the long list of negative results of new particle searches.

However, if we look back to the recent history of new particle discoveries in the 1970's, we observe that the two new quark flavors (charm and bottom) were found in hadron reactions no later than at e^+e^- storage rings. For charm, the discovery was made simultaneously with

---

* On leave from National Laboratory for High Energy Physics (KEK), Oho-machi, Tsukuba, Ibaraki 305, Japan

Table I. Quarks

Category	Definition	Example
Light	$m_q \lesssim \Lambda_{\rm QCD}$	u, d, s
Heavy	$\Lambda_{\rm QCD} \lesssim m_q \lesssim m_W$	$c, b, t(?)$
Superheavy	$m_Q \gtrsim m_W$	$t(?), \ldots (?)$

πp and e^+e^- reactions (hence the name J/ψ), and for bottom, the discovery in pp interactions preceded its confirmation at an e^+e^- machine by a year. The moral is that hadron machines can do well if there is a distinctive signature of the new phenomena (lepton pairs in the two examples). This was of course the case for the recent discovery of the W and Z bosons at the CERN $p\bar{p}$ collider.

One more point which emerges from the above example is that hidden-flavor states were found before the discovery of open-flavor states. In other words, quarkonia preceded flavored mesons. This is also related to the fact that a quarkonium can provide a clearer trace (decay modes) of its existence.

So, in considering the physics possibility at future super-high-energy proton colliders, it seems worthwhile to study* the production and decay of a new onium, the bound state of a quark and its antiquark (or the bound state of more exotic particles).

Actually, it turns out that the properties of onia changes drastically when the mass of its constituent quark exceeds the W mass. In contrast to light quarks (u, d, s) and heavy quarks (c, b), we call such a heavy quark ($m_Q \gtrsim m_W$) a *superheavy* quark. It is on these aspects which I will concentrate in this talk.

HETEROQUARK *vs* HOMOQUARK

The first qualitative novelty we encounter is that one should really distinguish two types of superheavy quarks. Although this distinction applies to lighter quarks also, there occurs significant phenomenological difference in the onia properties *only* if the quark is superheavy. We shall call these two classes of quarks *heteroquarks* and *homoquarks*. More precisely we define them as follows:

(1) A heteroquark is massless if the weak SU(2)×U(1) symmetry is not broken, and acquires mass through the weak symmetry breaking. In a formal language, the left-handed and right-handed counterparts are in different SU(2)×U(1) representations. All the quarks we know belong to this category. A fourth-generation quark is by definition a heteroquark.

(2) A homoquark, on the other hand, can be massive before the weak symmetry breaking. The left- and right-handed quarks are in the same SU(2)×U(1) representation and can have an invariant mass term at the Lagrangian level. Some examples include the exotic "d quark"

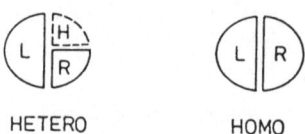

HETERO   HOMO

Fig. 1. Heteroquark and homoquark

---

* A large part of this talk is based on the collaboration with V. Barger, E.W.N. Glover, W.-Y. Keung, M. G. Olsson, C. J. Suchyta III, and X. Tata. For details I refer to Ref. 1. See also Ref. 2 for a related work.

present in the superstring-inspired E_6 model. Broadly speaking, the gluino, the supersymmetric partner of the gluon, belong to this category, although it is not a color triplet but an octet and it is a Majorana fermion.

UPPER LIMITS ON THE QUARK MASS

As we are considering very heavy quarks, it is natural to ask how heavy a quark can one tolerate. For a homoquark the answer is simple: There is no limit on its mass. This is because the origin of the quark mass is totally unrelated to the weak interaction. Of course, at the aesthetic level, one is faced with the heralded problem of hierarchy and naturalness. In a renormalizable theory, however, one may choose any value of the quark mass, without referring to "bare" quantities which are the source of the problem. The resulting theory is healthy after proper renormalization is done.

On the other hand, the situation is totally different for a heteroquark. The mass comes from the Higgs mechanism, so the Yukawa coupling of a heteroquark with the Higgs is proportional to the mass. If we want to make the quark mass larger, we are forced to go to a strong-coupling regime. Thus the limit of large quark masses is singular here and many strange phenomena can happen.

There are some theoretical "limits" on heteroquark masses:

ρ *parameter:* The neutral-to-charged current coupling ratio ρ is shifted from the canonical value $\rho = 1$ if there is a split weak multiplet.[3] A generous bound $|\rho - 1| < 0.02$ would give a limit $|m_{t'} - m_{b'}| \lesssim 200\,\text{GeV}$ for any pair of quarks.[†] There is no limit for a degenerate doublet, however, even if the mass is very large.

Breakdown of perturbation theory: The Yukawa coupling exceeds unity for a very heavy quark. To be quantitative, partial-wave unitarity for tree scattering amplitudes is violated if $m_Q \gtrsim 500\,\text{GeV}$.[5] This is not a *real* limit; it tells us only that above this value we are dealing with a strongly interacting theory, in which higher order contributions are no longer negligible and we cannot (as yet) make any reliable prediction.

Vacuum stability: From the stability of the broken vacuum, derived from the one-loop effective potential, one gets $m_Q \lesssim 100\,\text{GeV}$ if $m_H \sim 0$.[6] However the limit goes away if there exist heavy scalar particles (heavy Higgs, supersymmetry, *etc.*).

Renormalization group limits: Because the Yukawa coupling is not asymptotically free, one may deduce a limit on the coupling at the weak scale by demanding that the coupling does not diverge (or become bigger than unity) up to some scale, which can be the GUT scale,[7] the Planck mass,[8] *etc.*[*] This kind of limit is always better than the simple perturbation limit because one can take advantage of the renormalization-group evolution to push down the allowed Yukawa strength *at the weak scale*. These limits are, however, rather prejudice-dependent, since one has to assume a desert between the weak scale and the high scale.

One may conclude that there is no rigorous upper limit on the heteroquark mass. I will, however, restict myself to the mass range $m_Q < 500\,\text{GeV}$, because otherwise one cannot rely on predictions based on perturbation theory at all.

SPECTRUM OF ONIA

I will discuss only the lowest-lying states of the quarkonium system, the S- and P-wave

---

[†] In a recent global analysis of the "neutral current" data, Amaldi *et al.*[4] claim that $m_t < 180\,\text{GeV}$ (at 90% CL and if $m_H < 100\,\text{GeV}$).

[*] The so-called "triviality" bound should be in this category. In my opinion this bound is more than questionable because it is obtained by requiring the validity of the gauge-Higgs theory itself far above the Planck mass, where gravity should certainly play an important role.

Table II. Spectrum of onia. There are, of course, radial
and higher orbital excitations.

J^{PC}	$S = 0$	$S = 1$	Names	$S = 0$	$S = 1$
$L = 0$	0^{-+}	1^{--}	$L = 0$	η	ψ
$L = 1$	1^{+-}	$0^{++}, 1^{++}, 2^{++}$	$L = 1$	h	χ_0, χ_1, χ_2

ground states. The spectrum is predicted by a simple quark model as is well known (see Table II).

In this talk I will use the simplified names for these states, as given in Table II. If one has to be more specific, one can add a subscript to indicate the flavor content, e.g., η_c, ψ_b ($= \Upsilon$), etc. Please remember that "η" is not *the* η meson, connected to the U(1) problem! Note that for the gluinonium, there are no $C = -$ states (ψ, h), due to the Majorana nature of gluinos.

NOVEL FEATURES OF SUPERHEAVY ONIA

I summarize now some of the new features of superheavy onia, which may lie beyond a simple extrapolation from known quarkonia. Each of these properties will be explained below in more detail.

(1) The empirical "constant law" ($\Gamma_{ee}/e_q^2 \sim$ const.) for the leptonic width no longer holds. Eventually one expects that this quantity scales linearly with the quark mass (up to a logarithmic correction).

(2) Quarkonium production cross section in hadronic interactions is only a tiny fraction of the quark-pair production cross section of the same flavor, in contrast to the charmonium and bottomonium cases.

(3) There is a strong hierarchy in the production cross section of states with different quantum numbers. The pseudoscalar η accounts for most of the onium cross section. The vector ψ production is much suppressed, and P-wave χ states are even more suppressed.

(4) For the heteronia (onia made of heteroquark), "weak" decay modes are enhanced and they actually become the dominant modes of annihilation decays. It is important here to discriminate heteronia from homonia, for there is no enhancement at all for the latter.

ANNIHILATION STRENGTH

To give a detailed discussion of the points mentioned above, it is convenient to introduce the concept of "annihilation strength," which has the dimension of mass and is defined as follows: For an S-wave state

$$F_S = \frac{|R_S(0)|^2}{M^2} ,$$

and for a P-wave state

$$F_P = \frac{|R'_P(0)|^2}{M^4} .$$

Here M is the mass of the onium (also used throughout the paper), $R_S(0)$ is the (nonrelativistic) radial wave function of the S state at the origin, and $R'_P(0)$ is the derivative of the radial wave function of the P state at the origin. The merit of introducing the quantity F is that the annihilation decay width Γ of a state can be estimated simply as the annihilation strength multiplied by some appropriate coupling factor

$$\Gamma \sim (\text{coupling})^2 \cdot F ,$$

where "coupling" is, for instance, g_s^2 for the two-gluon decay, and e^2 for leptonic decays.

An important relation for superheavy onia is

$$F_S \gg F_P ,$$

which should be contrasted with $F_S \sim F_P$ for light onia. The latter can be inferred from $f_\rho \sim f_{A_1}$ for up/down quarks, and $\Gamma(\eta_c) \sim \Gamma(\chi_0)$ for charm. (In the charmonium case the P-wave annihilation is actually beginning to be suppressed.)

It is easy to understand the above inequality by remembering that for a very heavy onium the one-gluon exchange approximation for the potential becomes appropriate (for the ground states at least; the confining interaction is still important for higher excitations). In this approximation, the annihilation strengths can be read off any elementary quantum mechanics textbook, with appropriate insertion of color factors and substitutions:

$$F_S = \frac{4}{27}\alpha_s^3 M ,$$

$$F_P = \frac{1}{5832}\alpha_s^5 M ,$$

$$F_P/F_S = \frac{1}{864}\alpha_s^2 \sim 10^{-5} .$$

The reason for the large ratio above is that for a Coulomb potential the shape (wave function) of a bound state is essentially determined by the Bohr radius $\sim (\alpha_s M)^{-1}$, and therefore F_P is smaller than F_S by a factor of α_s^2. The additional small numerical factor $(\frac{1}{864})$ is in part due to the somewhat inadequate definition of F_S and F_P (the onium mass is four times the reduced mass).

For a realistic potential, which agrees with the expected asymptotic behaviors and reproduces the charmonium and bottomonium spectra, such as the Richardson[9] and the Wisconsin[10] potentials which I will use to estimate the rates in this talk, F_P is somewhat larger than the Coulomb result but the overall tendency remains correct.

In Fig. 2 the S-wave annihilation strength F_S of heavy quarkonia is plotted. The region sandwiched by the Richardson and Wisconsin curves should give a rather realistic estimate. Included in the same figure are the experimental data of the "reduced" leptonic widths of the ground-state vector mesons. The plotted quantity $\Gamma/(4\alpha^2 e_q^2)$ is equal to F_S in the quarkonium calculation in the lowest order. One should remember that the QCD correction factor to the rate $(1 - \frac{16}{3\pi}\alpha_s)$ is not at all small; Actually it amounts to $\sim 50\%$ for charmonium and $\sim 30\%$ for bottomonium.

A conclusion one can draw from the figure is that the empirical "constant law," which is quite good for the five known states, cannot continue to hold for heavier quarkonium (in particular for toponium). The rate should eventually scale as $\Gamma \sim M$, with a logarithmic correction due to asymptotic freedom.

The P-wave annihilation strength F_P is shown in Fig. 3. Realistic values are larger than the Coulombic estimate, because the P-wave state is more extended than the S state and is thus more sensitive to the confining part of the potential. Still, they are much smaller than the corresponding F_S values.

There are two important implications of the fact that $F_S \gg F_P$.

(1) The annihilation decay of the P states is much slower than that of the S states. This makes the P-wave annihilation decay less competitive to the other decay mechanisms, notably the constituent decay, even when the S state annihilation is dominating (see later).

(2) The production cross section of the P states will be much less than that of the S states, as we discuss in some detail below.

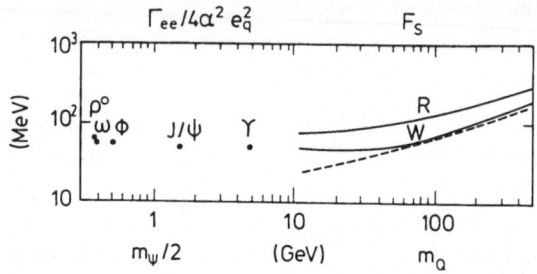

Fig. 2. Annihilation strength F_S for the ground S state of a quarkonium. Solid curves: Richardson (R) and Wisconsin (W) potentials. Dashed curves: Coulomb potential approximation. Also shown are the corresponding experimental data extracted from the leptonic width of known vector mesons, which displays the empirical "constant" law.

ONIA PRODUCTION IN HADRON COLLISIONS

I will now turn to the discussion of the production of onia in pp collisions. Before going to the specific production processes, I note that the cross section of superheavy onia is much smaller than the corresponding cross section of the open-flavored quark. This may be explained with a duality argument for the dominant production process $gg \rightarrow Q\bar{Q}$, for which the part below the flavor threshold is supposed to turn into the quarkonia states. Although this kind of calculation cannot be trusted quantitatively (there is an ambiguity in the selection of the quark mass value, and also a problem of the division into each onium state), it can provide a qualitative understanding of the relative suppression of the onia production.

To be specific, let me contrast a superheavy onium with charmonium. For charmonium, the resonance region extends from 2.98 GeV to 3.73 GeV, which is not a small interval compared

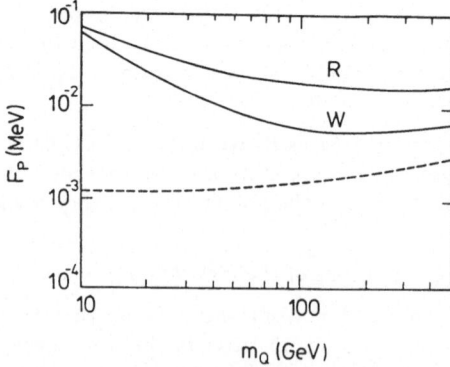

Fig. 3. Annihilation strength F_P for the ground P state of a quarkonium. Curves are the same as in Fig. 2. Notice the difference in the scale.

Table III. Production of onia in pp collisions

State	Mechanism	Cross section
η	$gg \to \eta$	1(Normalization)
ψ	$gg \to \psi g$	10^{-2}
	$gg \to \chi_{0,2} \to \psi\gamma$	$< 10^{-3}$
	$q\bar{q} \to \psi$	10^{-4}
	$WW \to \psi$	$< 10^{-5}$
h	$gg \to hg$	10^{-4}–10^{-5}
	$gg \to \eta' \to h\gamma$	10^{-3}
χ_0, χ_2	$gg \to \chi_0, \chi_2$	10^{-3}
χ_1	$gg \to \chi_1 g$	10^{-4}–10^{-5}
	$gq \to \chi_1 q$	10^{-4}–10^{-5}
	$q\bar{q} \to \chi_1 g$	$< 10^{-6}$
	$q\bar{q} \to \chi_1$	10^{-7}

to the charm mass itself. For superheavy onia with a mass of a couple of hundred GeV, on the other hand, the typical mass scale is orders-of-magnitude larger than that for charm, but the extent of the resonance region does not change much (at most a few GeV). Thus the importance of the resonance region in the cross section is much diminished.

There are various mechanisms for onia production in proton interactions. These are summarized in Table III.

The most important is two-gluon fusion into the pseudoscalar η. The cross section is proportional to the decay width $\Gamma(\eta \to gg)$, which gives the inverse reaction rate. The η cross section decreases by many orders of magnitude when the onium mass changes from 100 to 1000 GeV. Fig. 4 shows the η cross section in pp interaction at $\sqrt{s} = 40$ TeV. At this energy the cross section is ~ 100 pb for $M = 250$ GeV, and ~ 1 pb for $M = 750$ GeV. Note, however, that the expected number of η per year at the SSC is quite large; For the nominal luminosity of 10^{33} cm^{-2} sec^{-1}, a cross section of 1 pb corresponds to 10^4 events per year.

The vector (ψ) cross section is $\sim 1/100$ of the η cross section for the mass range under consideration. The suppression is because there is no gg coupling to the ψ, so that the higher-order process $gg \to \psi g$ must be invoked.* The cross section of χ_0 production is further suppressed and is only $\sim 1/1000$ of the η production. This is due to the previously stressed relation $F_P \ll F_S$, in spite of the fact that the two-gluon fusion channel is possible. The χ_2 cross section is similar ($\frac{4}{3}$ times that for χ_0). Other states h and χ_1 have even smaller cross sections.

These production properties are the same for heteronia and homonia, since they have identical color interactions. The production cross section of gluinonium is larger since the larger color charge of the gluino leads to a larger annihilation strength.

DECAYS OF SUPERHEAVY ONIA

I now turn to the decay of superheavy quarkonia. According to their decay properties one

---

* The feeddown from $\chi \to \psi\gamma$, which is important for charmonium, is here much smaller than the direct production due to the smaller χ production cross section.

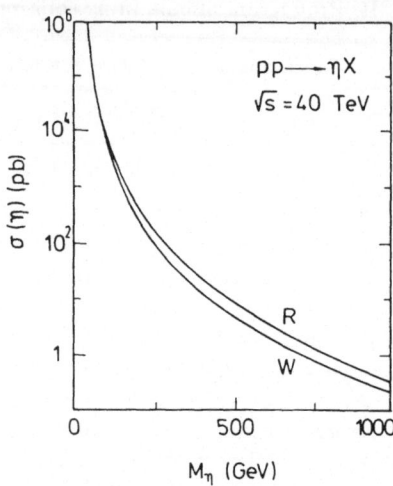

Fig. 4. η cross section at the SSC. Potential dependence enters here because the ηgg coupling is proportional to the quarkonium wave function.

can classify three types of superheavy onia.

(1) *Nonexistent quarkonium*
(2) *Heteronium*
(3) *Homonium*

The first one, although sounding nonsensical, is an important class, because a superheavy toponium would belong to this category. An example of the second class (heteronia) is the lighter of the fourth-generation quarkonia (probably $(b'\bar{b}')$), if the Kobayashi-Maskawa (KM) mixing to lighter quarks is sufficiently small. The heavier of the fourth-generation quarkonia $(t'\bar{t}')$ is in either (1) or (2) depending on the mass splitting. The last class (homonia) includes the E_6 exotic onia.

Nonexistent quarkonia

If the decay width of a bound state exceeds the binding energy, it follows from the uncertainty principle that it is not possible to distinguish the bound state from the continuum (see Fig. 5).

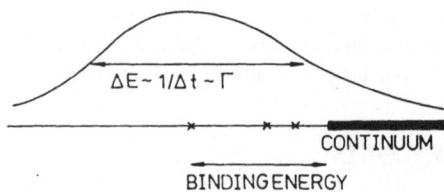

Fig. 5. Bound states with large widths are indistinguishable from the continuum.

This situation actually occurs for a heavy toponium[11] when $m_t \gtrsim 150\,\text{GeV}$, as the decay rate $\Gamma((t\bar{t}) \to t\bar{b}W)$ is proportional to m_t^3 while the binding energy $2m_t - M$ is roughly proportional to m_t:

$$\Gamma \sim \frac{\alpha m_t^3}{m_W^2}, \qquad \text{binding energy} \sim \alpha_s m_t .$$

Existent quarkonia: three classes of onium decay

Just as there are three classes of onia, one also finds three types of decay modes.

1. *Annihilation decay:* Here constituent quark and antiquark annihilate to produce the final state (Examples: $\eta \to gg, \gamma\gamma, \psi \to ggg, \ell^+\ell^-$).

2. *Transition:* Transition inside the onium family can occur by emitting light particles. There are two subtypes: hadronic transitions (Example: $\psi' \to \psi\pi\pi$) and radiative transitions (Examples: $J/\psi \to \eta_c\gamma, \chi_J \to \psi\gamma$).

3. *Constituent decay:* The constituent quark or antiquark decays singly, while the other remaining as a spectator. No known example exist yet, but it will be important for heavier onia (Example: $\psi_t \to t\bar{b}\ell\bar{\nu}$ or $t\bar{b}W$).

Constituent decay

The weak decay rate of a (hetero) quark is proportional to m_Q^5 if $m_Q < m_W$ or m_Q^3 if $m_Q > m_W$. As the quark mass increases, the weak decay starts competing with the strong and electromagnetic decays. The decay of a superheavy quarkonium can be dominated by the weak decay (constituent decay). The decay rate is shown in Fig. 6. If there is no mixing suppression, the large width $\Gamma((Q\bar{Q}) \to Q\bar{q}W$ or $\bar{Q}qW)$ totally masks other (interesting!) decay modes. However, if the mixing is less than a few percent, the constituent decay is much suppressed and the S-wave states (η, ψ) mainly decay by annihilation. (To get a feeling, the annihilation rate for $\eta \to gg$ is between 5 and 10 MeV for the relevant mass range.) For the P-wave states a smaller mixing $(\lesssim 10^{-3})$ is required for this to occur. The constituent decay is also suppressed by the available phase space, if the quark mass difference $m_Q - m_q$ is not so large $(\lesssim 50\,\text{GeV})$, which could happen for $t' \to b'$. In this case a real W cannot be produced and the decay rate $\Gamma((t'\bar{t}') \to t'\bar{b}'\ell\bar{\nu})$ *etc.* receives a suppression factor of $(m_{t'} - m_{b'})^5$. The constituent decay of a superheavy toponium is never suppressed, since we know that the mixing U_{tb} should be essentially unity. On the other hand, there are models which predicts a small mixing for the fourth generation quarks.

For a homoquark, the constituent decay is not important since it usually occurs only via a small mixing with a heteroquark.

Transition to lower-lying states

The M1 transition $\psi \to \eta\gamma$ is completely negligible $(\Gamma \ll 1\,\text{eV})$. The E1 transitions $\chi_J \to \psi\gamma$ and $h \to \eta\gamma$ have comparable rates to annihilation decays, typically $\Gamma \sim 5\,\text{keV}$. The hadronic transitions such as $\chi_J \to \eta\pi\pi, \psi\omega$, and $h \to \psi\pi\pi$ may be estimated by a QCD multipole expansion and are expected to be negligible. Thus the only important transitions are the E1 radiative decays from the P states to the S states.

Annihilation decays

The annihilation decay rate is related to the annihilation strength discussed earlier. For the strong decay to two gluons, we have $\Gamma \sim \alpha_s^2 F$ and for the electromagnetic decay modes to $\gamma\gamma, e^+e^-$ the rate is $\Gamma \sim \alpha^2 F$, which is smaller than the strong decay by a factor of $(\alpha/\alpha_s)^2$. For these two kinds of decay there is no difference between heteronia and homonia.

The weak (annihilation) decay modes to ZZ, W^+W^-, ZH, ... are more involved. Here heteronia show totally different behavior than homonia. The decay rates of homonia are of the normal order $\Gamma \sim \alpha^2 F$. However, the rates for heteronia are sometimes enhanced, depending on the final state and on the quantum number of the initial state. In some cases the rate is doubly enhanced ($\Gamma \sim (M^4/m_W^4)\alpha^2 F$), rather than singly enhanced ($\Gamma \sim (M^2/m_W^2)\alpha^2 F$). It can also happen however that there is no enhancement at all ($\Gamma \sim \alpha^2 F$).

The annihilation decay rates of the six states are summarized in Table IV for heteronia (explicit formulas for these rates can be found in Ref. 1). This table should be compared to the corresponding table (Table V) or homonia.

The dominant decay mode of heteronia differs for each onium state; They are

$$\eta \rightarrow ZH \ ,$$
$$\psi \rightarrow W^+W^- \ ,$$
$$h \rightarrow \eta\gamma \ ,$$
$$\chi_0 \rightarrow HH \ ,$$
$$\chi_1 \rightarrow ZH \ ,$$
$$\chi_2 \rightarrow W^+W^-, ZZ \ .$$

The branching fraction for these decays can reach nearly 100% provided that the mode is not kinematically suppressed.

The dominant decay of homonia, on the other hand, is always the strong decay mode to gg, ggg, or $gq\bar{q}$ (and possibly the radiative transition for the P states).

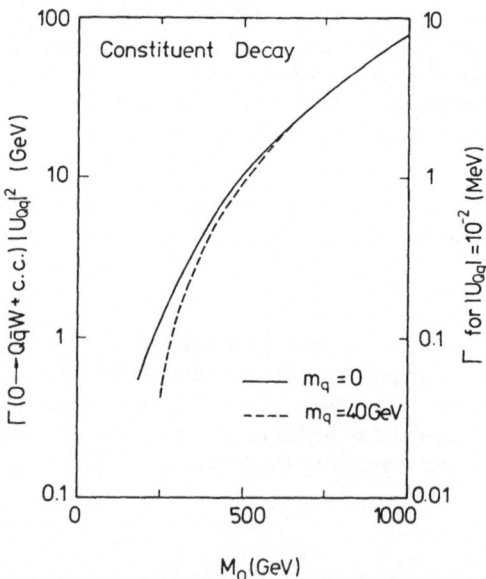

Fig. 6. Single quark decay rate of a heteronium

Table IV. Heteronia annihilation decay. The symbols are: — : the decay does not occur (at least in lowest order); o: the decay is of normal strength; ↑: singly enhanced; ↑↑: doubly enhanced; ↓: singly suppressed; ↑↗ shows that the decay is doubly enhanced but there is an additional helicity suppression factor, resulting in the net factor of $M^2 m_L^2 / m_W^4$.

Mode	η	ψ	h	χ_0	χ_1	χ_2
gg	o	—	—	o	—	o
$ggg/gq\bar{q}$	o	o	o	o	o	o
$\gamma\gamma$	o	—	—	o	—	o
$\ell^+\ell^-, q\bar{q}$	—	o	—	—	o	—
$L^+L^-, t\bar{t}$	↑↗	o	—	↑↗	o	—
$Z\gamma$	o	↑	↑	o	↓	o
ZZ	o	↑	↑	↑↑	↑	↑↑
W^+W^-	o	↑↑	↑	↑↑	↑	↑↑
γH	—	↑	↑	—	—	—
ZH	↑↑	↑	↑	—	↑↑	↑
HH	—	—	—	↑↑	—	↑↑

Table V. Annihilation decays of a homonium. Symbols are the same as in Table IV. The parenthesis means that the decay occurs only if the constituent quark is a weak non-singlet.

Mode	η	ψ	h	χ_0	χ_1	χ_2
gg	o	—	—	o	—	o
$ggg/gq\bar{q}$	o	o	o	o	o	o
$\gamma\gamma$	o	—	—	o	—	o
$\ell^+\ell^-, q\bar{q}$	—	o	—	—	o	—
$L^+L^-, t\bar{t}$	—	o	—	—	o	—
$Z\gamma$	o	—	—	o	↓	o
ZZ	o	—	—	o	↓	o
W^+W^-	(o)	o	(o)	(o)	(↓)	(o)
γH	—	—	—	—	—	—
ZH	—	o	—	—	—	—
HH	—	—	—	—	—	—

The mechanism causing the enhancement of the weak-boson modes for heteronia is the following: The spin vector of a longitudinally polarized massive spin-1 particle (say W) is proportional to E_W/m_W at high energies (where $E_W \sim \frac{1}{2}M$ for onium decay), giving the possibility of a single enhancement factor per W. Actually, the spin vector is approximately proportional to the momentum $\epsilon^\mu(\text{longitudinal}) = (1/m_W) k^\mu + O(m_W/E_W)$. If the current which couples to the W is conserved, this possible enhancement does not appear in the physical amplitude, since the remaining part of the spin vector is suppressed as the energy increases. For an axial vector coupling of a heavy quark, the current is not conserved and the divergence will be proportional to the quark mass, resulting in an enhancement factor of M/m_W in the amplitude. When there are two longitudinally polarized W's in the final state, one can thus get a double enhancement of M^2/m_W^2 in the amplitude.

In addition, if there is a Higgs boson in the decay product, the Yukawa coupling (proportional to the quark mass) enhances the amplitude, giving a factor of M/m_W relative to the gauge coupling.

On the contrary, there is no enhancement at all for homonia. This is because a homoquark

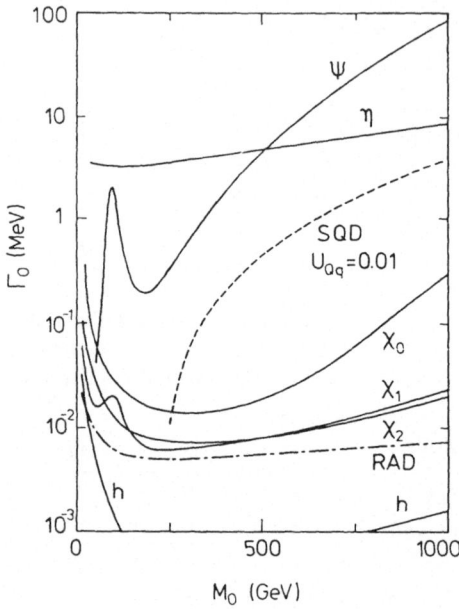

Fig. 7. Heteronium total annihilation width (for a charge-$\frac{1}{3}$ quark) compared to the constituent decay and radiative (E1) transition widths. Wisconsin potential is used.

couples to the weak bosons only through a vector coupling. For instance, the vector and axial couplings of a fermion to the Z^0 are

$$v = I_{3L} + I_{3R} - 2Q\sin^2\theta_W \ ,$$
$$a = I_{3L} - I_{3R} \ .$$

For a vectorlike (homo) fermion the axial coupling vanishes. If there is only a vector coupling, the possible enhancement for longitudinal W's is killed by current conservation. To complete the argument one should note that there is no (enhanced) Yukawa coupling to the Higgs boson.

The total width of the heteronium ($b'\bar{b}'$) is shown in Fig. 7. The Higgs boson is assumed here to be very massive. The η width is dominated by the two-gluon mode and increases only moderately with the mass. On the contrary, the ψ width at higher masses is controlled by the WW mode and rapidly increases with the mass. These two annihilation widths are larger than the constituent decay (SQD) with a 1% mixing.

188

For the modes with the Higgs boson, the $\psi_{b'} \to H\gamma$ branching ratio can reach more than 10% for some mass range (even four times bigger for the $\psi_{t'}$), the $\eta \to ZH$ and $\chi_1 \to ZH$ can be almost 100% for higher masses, and the $\chi_0 \to HH$ mode can be close to 100%.

DETECTION OF ONIA AT SUPERCOLLIDERS

At proton-proton colliders the most important backgrounds for quarkonium signals come from the generalized "Drell-Yan" processes, which produce ($q\bar{q}$ originated) electroweak final states such as e^+e^-, $\gamma\gamma$, ZZ, W^+W^-, etc. These backgrounds are more important (relative to the signal) for higher onium mass, because the gluon-gluon luminosity responsible for onia production is steeper than the quark-antiquark luminosity. For a fixed mass, the background is less important at higher pp c.m. energies, because of the same reason.

At SSC energies, the backgrounds are in general severe and the expected onia signal tends to be buried in the background, especially when one takes into account the fact that one cannot observe the W and Z directly and that only the leptonic decays are likely to be observable. The best prospects lie in the decays $\psi \to e^+e^-$, $\mu^+\mu^-$, for masses under the W pair threshold ($M \lesssim 200\,\text{GeV}$), and $\eta_{t'} \to \gamma\gamma$ [12] ($\eta_{b'} \to \gamma\gamma$ is sixteen times smaller).

The most intriguing signal is the decay $\eta \to ZH$ which can have almost 100% branching ratio. The η is the state with the largest production cross section. Furthermore, the signature is rather clean. One can tag the event by the Z^0, which has a Jacobian peak in p_T, and the whole system is highly constrained by the kinematics, $M(ZH) = M_\eta$, etc.

For $m_H > 2m_W$, one has three-boson events with $M(ZZ, WW) = m_H$ and $M(ZZZ, WWZ) = M_\eta$. For $m_H < 2m_t$, one can look for the decay $h \to \tau^+\tau^-$ with $\sim 4\%$ branching ratio. (One can reconstruct two τ four-momenta from the observed directions of the τ decay products and the missing transverse momentum, provided that the two taus are not parallel.)

For the notorious intermediate mass range [13] $2m_t < m_H < 2m_W$,[*] the dominant Higgs decay mode is $H \to t\bar{t}$. The most serious background comes from the process $gg \to Zt\bar{t}$, if the top quark can be identified through its decay. This background has been examined and it is concluded that one can achieve a signal-to-background ratio of better than 1:1 under generous cuts on the $t\bar{t}$ and $Zt\bar{t}$ masses.[14] Thus one may have the possibility of finding the fourth generation quarkonium and the Higgs boson at the same time!

MORE EXOTIC STATES

Gluinonium

Due to Fermi statistics, we have only the η and χ_J states of gluinonium. The production cross section at pp colliders is ~ 77 times that for a quarkonium, due to the larger color factor. There are no electroweak decay modes. The main decay mode is to two gluons, and possibly to $q\bar{q}$ via squark exchange. It is very difficult to detect these states, because one has to look for a peak in the jet-jet invariant mass where there is a huge QCD background. Constituent decay is also possible via $\tilde{g} \to \tilde{q}\bar{q}$ or $\tilde{g} \to q\bar{q}\tilde{\gamma}$. These would give a missing momentum signal, but the continuum $\tilde{g}\tilde{g}$ production is much larger and essentially indistinguishable from the onium signal.

Technipionium

In extended technicolor models many pseudo-Goldstone bosons (called technipions in this section) are predicted. Among these there are color-octet technipions, which are also weak

---

[*] If $m_t > m_W$ (in which case no "intermediate" region), look for τ pair for $m_H < 2m_W$, W^+W^-/ZZ for $m_H > 2m_W$. The branching ratio for $H \to W^+W^-/ZZ$ is large unless $m_W \ll m_t \lesssim \frac{1}{2}m_H$.

triplets (π_8^+, π_8^0, π_8^-). These states can form color-singlet bound states below threshold. Assuming that their gauge couplings are not suppressed by form factors, one can predict the properties of these states.[15] Because the constituents are color octet, the production cross section is similar to gluinonium production. Unlike the gluino case, the constituents have electroweak charges, leading to distinctive electroweak decay modes such as W^+W^-, $\gamma\gamma$, and $Z\gamma$. The branching ratios are in the range of 0.1–0.3% for these modes. There are no enhancement factors. However due to the large cross section the signal should stand out over the background.

Extra Z

Although it is not a bound state but an elementary gauge boson, I will touch upon this subject, since some of the overall characteristics resemble that of the onia. In particular, the extra U(1) gauge boson in the superstring-inspired E_6 model may have a substantial branching fraction to W^+W^- and to ZH at the few percent level.[16] There is no enhancement in these decays: For the WW mode, the double enhancement coming from the two longitudinal W's is cancelled by the small Z'–Z mixing. In fact, the $Z'WW$ vertex can come only from the SU(2) non-Abelian vertex and the SU(2) impurity in the extra U(1) boson is suppressed by $(m_Z/m_{Z'})^2$. The ZH mode is singly enhanced by the longitudinal Z, but counter-suppressed by the dimensional $Z'ZH$ coupling proportional to m_Z. The interesting fact that there is a nonvanishing off-diagonal $Z'ZH$ coupling arises because there is more than one Higgs field in the theory.

Though the branching ratios are not terribly large, the production rate is much larger, because the Z' has an elementary gauge coupling strength to the initial quarks.

ONIA PRODUCTION AT HIGHER ENERGIES

Since this Workshop is aimed at the ELOISATRON, it is appropriate to ask what the gain would be if one goes from 40 TeV to 200 TeV. The production rate of onia from gluon fusion increases by an order of magnitude (although there are more forward events). The $q\bar{q}$-induced background increases more slowly. One can achieve an improvement in the signal-to-background ratio of up to a factor 2–3. For the elementary gauge bosons produced by $q\bar{q}$, one has an increase of the cross section, but there is no change in the S/B ratio.

CONCLUSION

To summarize:
(1) At super proton-proton colliders the pseudoscalar onium (η) is substantially produced for a wide mass range. The vector (ψ) production cross section is 10^{-2} times that of the η, and the scalar and tensor ($\chi_{0,2}$) are another order of magnitude smaller.
(2) Superheavy toponium does not exist.
(3) The decay of homonium (*e.g.* E_6 exotic onium) is mostly strong (to gluons) and only a small fraction goes to electroweak final states.
(4) Heteronia (*e.g.* the fourth-generation onium with small KM mixing) have enhanced electroweak decay modes. $\eta \to ZH$ and $\psi \to W^+W^-$ are dominant.
(5) The decay η(hetero) $\to ZH$ provides a good opportunity for Higgs detection, with manageable background.
(6) An extra U(1) gauge boson, with the decay $Z' \to ZH$, is an alternative to a heteronium for a Higgs search.

ACKNOWLEDGMENT

I would like to thank the authors of Ref. 1 for a enjoyable collaboration. I am very grateful to Ahmed Ali for inviting me to this very stimulating workshop, and to Roberto Peccei for his hospitality at DESY and for improving my English.

REFERENCES

1. V. Barger, E.W.N. Glover, K. Hikasa, W.-Y. Keung, M. G. Olsson, C. J. Suchyta III, and X. R. Tata, Phys. Rev. Lett. **57**, 1672 (1986); Phys. Rev. D **35**, 3366 (1987).
2. I. Bigi, Y. Dokshitzer, V. Khoze, J. Kühn, and P. Zerwas, Phys. Lett. B **181**, 157 (1986).
3. M. Veltman, Nucl. Phys. **B123**, 89 (1977);
   M. B. Einhorn, D.R.T. Jones, and M. Veltman, Nucl. Phys. **B191**, 146 (1981);
   A. Cohen, H. Georgi, and B. Grinstein, Nucl. Phys. **B232**, 61 (1984).
4. U. Amaldi *et al.*, Phys. Rev. D **36**, 1385 (1987).
5. M. S. Chanowitz, M. A. Furman, and I. Hinchliffe, Phys. Lett. **78B**, 285 (1978); Nucl. Phys. **B153**, 402 (1979).
6. P. Q. Hung, Phys. Rev. Lett. **42**, 873 (1979);
   H. D. Politzer and S. Wolfram, Phys. Lett. **82B**, 242 (1979); **83B**, 421(E) (1979);
   R. A. Flores and M. Sher, Phys. Rev. D **27**, 1679 (1983);
   M. J. Duncan, R. Philippe, and M. Sher, Phys. Lett. **153B**, 165 (1985).
7. N. Cabibbo, L. Maiani, G. Parisi, and R. Petronzio, Nucl. Phys. **B158**, 295 (1979).
8. L. Maiani, G. Parisi, and R. Petronzio, Nucl. Phys. **B136**, 115 (1978).
9. J. L. Richardson, Phys. Lett. **82B**, 272 (1979).
10. K. Hagiwara, S. Jacobs, M. G. Olsson, and K. J. Miller, Phys. Lett. **130B**, 209 (1983).
11. K. Fujikawa, Prog. Theor. Phys. **61**, 1186 (1979);
    T. G. Rizzo, Phys. Rev. D **23**, 1987 (1981).
12. E.W.N. Glover, Cavendish Lab. preprint HEP/87/6 (1987).
13. J. F. Gunion, P. Kalyniak, M. Soldate, and P. Galison, Phys. Rev. D **34**, 101 (1986).
14. J. F. Gunion and Z. Kunszt, University of California–Davis preprint UCD-86-21 (1986), in *Proceedings of the Snowmass Workshop*, 1986;
    H. Baer, D. Dicus, M. Drees, and X. Tata, Phys. Rev. D **36**, 1363 (1987).
15. V. Barger and W.-Y. Keung, Phys. Lett. B **185**, 431 (1987).
16. R. Kleiss and W. J. Stirling, Phys. Lett. B **180**, 171 (1986);
    F. Del Aguila, M. Quirós, and F. Zwirner, Nucl. Phys. **B284**, 530 (1986);
    R. Najima and S. Wakaizumi, Phys. Lett. B **184**, 410 (1987);
    R. Najima, Prog. Theor. Phys. **77**, 926 (1987);
    T. G. Rizzo, Phys. Rev. D **34**, 1438 (1986);
    S. Nandi, Phys. Lett. B **181**, 375 (1986);
    H. Baer *et al.*, Ref. 14;
    C. Dib and F. J. Gilman, Phys. Rev. D **36**, 1337 (1987);
    V. Barger and K. Whisnant, University of Wisconsin preprint MAD/PH/351 (1987).

NEW TRIGGER SYSTEMS FOR HIGH LUMINOSITY COLLISIONS

Andrea   Contin

Dipartimento di Fisica
Università di Bologna
Bologna, Italy

1. INTRODUCTION

In order to exploit at best the possibilities of new physics discoveries opened up by a multi-TeV hadron collider like ELOISATRON, the trigger and data acquisition system of any detector, either general purpose or specialized, is of the outmost importance.

In this talk, the requirements arising from the special environment of multi-TeV hadron collisions are reviewed, and indicative solutions are presented.

Various parameters have a strong influence in the design of the trigger and data acquisition system for experiments in a multi-TeV hadron collider. The first and most important is the *luminosity*, which determines the total interaction rate. Then, the *hermeticity* of the detector with respect to energy flow out of the interaction and the need for a *reasonable cost* impose that the electronics used for digitization and first level triggering must be extremely compact. The *granularity* of the detector determines the event size, i.e. the throughput capabilities of the data-acquisition system. Finally, the *physics* to which the detector must be sensitive to, calls for a certain degree of flexibility in the triggering system.

In section 2, limits on the luminosity are set based on present views and technical limitations on detectors and electronics.

In section 3, various trigger levels are shown to be necessary to reduce the trigger rate to 1Hz, to match with the mass storage capability and with a reasonable amount of off-line computing. The electronics which might be used to implement this trigger structure (Application Specific Integrated Circuits, Look-up Tables, Gate Arrays, Microprocessor farms, etc.) is also briefly described.

Finally, section 4 presents some recommendations on the R&D necessary in the electronics design and construction in order to be ready when real detectors will come on stage.

2. LUMINOSITY AND TRIGGER RATES

The limiting factors on luminosity of a multi-TeV hadron collider are:
  i) the bunch spacing;
 ii) the average number of events per crossing;
iii) the radiation dose on detectors and electronics.

The *bunch spacing* is a relatively flexible parameter, which is mostly dominated by the constraints on the detector (pile-up effects, drift times, etc.) and on the electronics (clocking speeds) performances. The bunch spacing will be quantized by the choice of the Radio-Frequency for the injector (once this is decided, multiples of it can be used as bunch spacing). A lower limit for the bunch spacing is, probably, 25 ns.

The *average number of events per crossing* strongly depends on the vertex and tracking chambers capability to clearly separate multiple vertices in the same bunch crossing. It is also limited by the pile-up in the detectors and by the on-line data processing throughput. The ideal situation will be to have only one event, on average, per bunch crossing. However, this means that 37% of the crossing will have no event, 37% will have one event, 18% two events, 8% three events, etc.,i.e. the trigger system of a future detector will be confronted with a not negligible rate of multiple events, and it must therefore be able to cope with it.

Due to the high number of channels, the electronics should be mounted on the detectors where it will be irradiated by the particles produced in the interactions and by the particles accompanying the beams. This implies that the *radiation dose on detectors and electronics* will be very high. Present limits on the maximum dose which electronics can support before failing (radiation hardness) are 10^{4+5}Rad.
The radiation dose on detectors and electronics, at a distance R from the interaction point, can be computed from the mean charged particle multiplicity $\langle n \rangle$ produced in the interaction (supposed uniformly distributed), the hadronic cross section σ, and the luminosity L:

$$\text{Dose [mips cm}^2] = \frac{N_e \langle n \rangle}{4 \pi R^2} \, ,$$

with:

$$N_e = \sigma \int L dt = \text{number of events} \, .$$

Taking into account that a minimum ionizing particle, i.e. a pion of 500 MeV/c, releases, by passing through matter, an energy:

$$\frac{dE}{dx} = 1.75 \text{ MeV cm}^2 \text{gm}^{-1}$$

so that:

$$1 \text{ Rad} = 100 \text{ erg cm}^{-1} = 3.5 \times 10^{7} \text{ mips cm}^{2},$$

and assuming the expected values[1]:

$$\langle n \rangle = 60 \text{ and } \sigma = 100 \text{ mb },$$

the total dose in one year at R=10 cm from the interaction region is:

$$\text{Dose} = 1.7 \times 10^{-27} \times L \text{ Rad year}^{-1}.$$

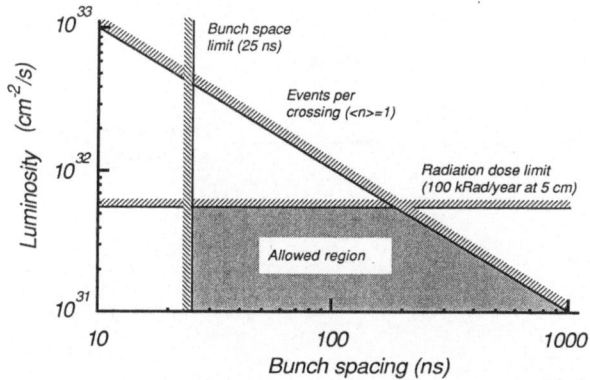

Fig.1 – Luminosity as a function of bunch spacing. The limits correspond to the three extreme values: bunch spacing = 25 ns, average events per crossing = 1, radiation dose = 100 kRad/year. The gray region is allowed.

Present electronics is resistant to radiation doses as high as 10^{4+5} Rads in one year running, which correspond to a luminosity of $L=0.6 \times 10^{31+32}$ cm^{-2}s^{-1}.

Figure 1 summarizes the limits on luminosity from bunch spacing (25 ns), average events per crossing (1) and radiation dose (100 kRad/year). The higher limit on luminosity, $L=0.6 \times 10^{31}$ cm^{-2} s^{-1}, is determined by radiation hardness. A big effort must therefore be made to improve it beyond the present limits.

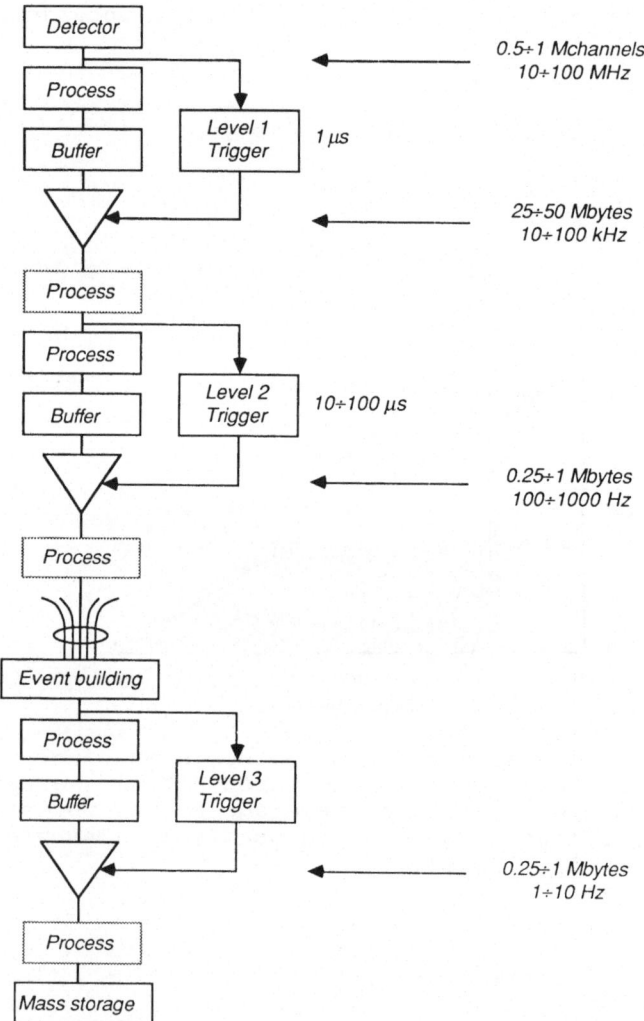

Fig.2 - Principle design of a multi-level trigger[2]

Luminosities above $L = 0.6 \times 10^{31}$ cm^{-2}s^{-1} requires either lower bunch spacing or a higher number of events per crossing. In the following a reference luminosity of $10^{31 \div 32}$ cm^{-2}s^{-1} is assumed, which corresponds to a total event rate of $10^{7 \div 8}$/s.

3. TRIGGER SCHEMES

The rate of accepted events cannot be higher than the recording and off-line analysis capacity, which is estimated at about $1 \div 10$ event/s. A reduction in rate of about $10^{7 \div 8}$ can be achieved with high efficiency only through different trigger levels, each working with more complex algorithms and more refined data.

The following sections will present the requirements for a 3-levels scheme, based on the general principles outlined in Fig.2. In each level the trigger is pipelined in parallel with some processing and/or buffering. The first two levels process signals from single detector elements. Only the third level uses data from the full detector for a more complex analysis. The reduction in rate and the event size foreseen at each level are also shown in Fig.2.

3.1 First Level Trigger

The first level trigger should be deadtimeless, thus each detector signal must be pipelined. Up to some μs decision time may be allowed. The output rate must be of the order of $10^{4 \div 5}$Hz.

A first level trigger providing this kind of reduction in rate can be done using the signatures from the calorimeter (isolated high-energy electromagnetic or hadronic energy, total transverse energy, total missing energy), the vertex chamber (high p_T charged particles), and the muon detector (high p_T penetrating particles).

Fast comparators (50ns response time) should be used to discriminate the energy level in calorimeters and to define "hits" in position detectors. In the framework of the LAA Project, a fast comparator is foreseen to be developed at CERN[3]. It will have an auto-setting of the threshold (just above the noise level), a threshold checking via test inputs to the front end electronics, and a mask to suppress noisy channels.

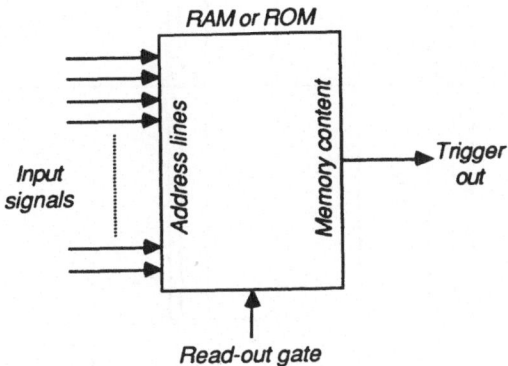

Fig.3 – Look-up Table: principle of operations

Programmable combinatorial electronics - Look-up Tables or
Gate Arrays - can be used for track or energy cluster finding at
the first level trigger.

A Look-up Table, shown in Fig.3, consists of a memory (RAM or
ROM) where the detector signals act as address lines. The memory
content is 1 only if the input configuration is a good trigger.

Since the memory size increases as 2n (with n=number of input
signals) this is a good solution only for a limited number of
input signals (up to 16).

More complicate trigger schemes require dedicated circuits
which can be implemented in Gate Array technology.

An example is the Forward Muon Detector trigger in the ZEUS
experiment at HERA[4]. The principle of the trigger is shown in
Fig.4. The Forward Muon Detector covers an angle between 5° and
35° with respect to the outgoing proton direction. Muons produced
in the interactions are deflected by magnetized iron toroids
(FBAC and FMUT) and detected in two planes of Limited Streamer
Tubes (FMUI and FMUO) with polar coordinates read-out. The radial
LST pads give directly the coordinate in the deflection plane, so
that a trigger with a fixed p_T cut can be implemented as shown in
Fig.5.

Given a radial coordinate in the FMUI plane, only a fixed interval in the FMUO plane is allowed. In order to reduce acciden-
tals and to achieve the desired spatial resolution, each plane is
divided into 8 octants, each with 128 radial pads. The principle
design of the trigger electronics is shown in Fig.6. A matrix
(M_{ij}; i,j=1,128) of 1-bit memory cells contains the allowed combinations of input and output coordinates.

The input plane configuration (I_j; j=1,128) is transformed by
the matrix into the vector

$$O_i^{'} = \sum_{j=1}^{128} M_{ij} I_j$$

of the allowed output configurations. A bit-to-bit AND with

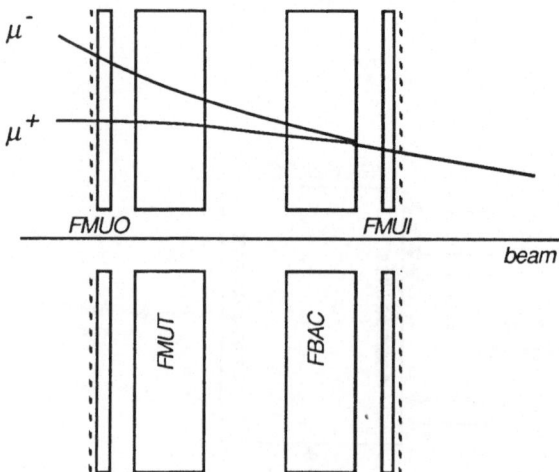

Fig.4 - The Forward Muon Detector of the ZEUS experiment

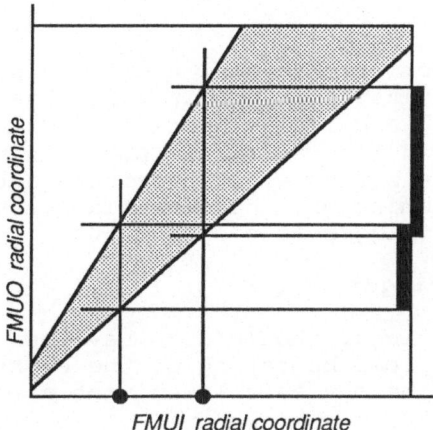

Fig.5 – Correlation between input and output planes for a
          fixed p_T cut in the Forward Muon Detector of the
          ZEUS experiment.

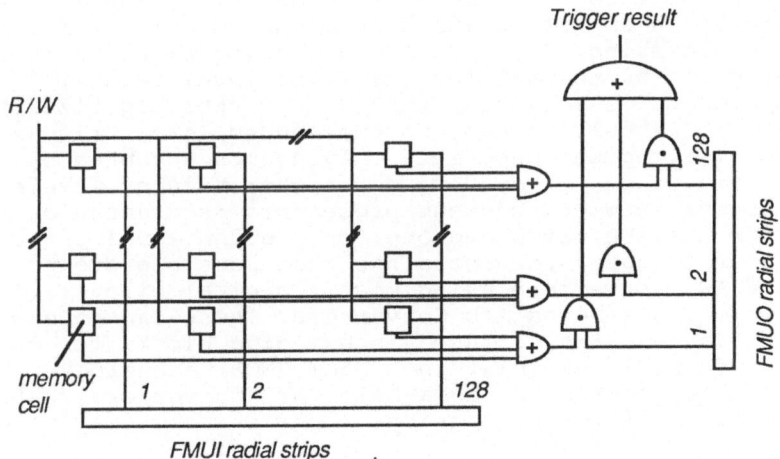

Fig.6 – Block diagram of the electronics needed to imple-
          ment the 1st level trigger in the Forward Muon
          Detector of the ZEUS experiment.

the actual output signals (O_i; i=1,128) followed by a final OR
determines the trigger:

$$T = \sum_{i=1}^{128} O_i^{'} O_i \ .$$

Given the big number of channels, the Look-up Table solution
is highly impractical. Therefore, the trigger has been implemented using Gate Arrays. The building block is a 32×16 matrix which
is being developed by an Italian firm, LABEN, in Milan[5]. A complete matrix of 32 gate arrays will constitute a trigger module.
The control and programming will be done via the VMEbus. The
response time of this trigger is of about 50ns.

3.2 Second Level Trigger

The second level trigger should also have as low a dead-time
as possible, therefore some buffering of the events is necessary.
Up to 100μs decision time may be allowed with a reduction in rate
of a factor 100.

The second level trigger should repeat the same analysis of
the single detector components as the first level trigger, but
using more refined data digitized by very fast Analog-to-Digital
(A/D) converters. A second level trigger processor for a multi-
TeV hadron collider will certainly make extensive use of ASIC
(Application Specific Integrated Circuits) which are presently
developed in many laboratories around the world.

An example of ASIC is the HARP – Hierarchical Analog Read-out
Processor – project, under development at CERN in the framework
of the LAA Project for detector development[3].

The block diagram of the processor is shown in Fig.7. In a
single chip the complete processing of 32 detector signals can be
done. Each signal is amplified and discriminated for the first
level trigger processing. In parallel, an analog pipeline is fed
with the analog signal to wait for the first level trigger
result. If the event is accepted, an A/D converter digitizes the
analog signal. The result is sent to the second level trigger
processor and to a digital pipeline. Only events accepted by the
second level trigger can proceed to the higher trigger levels.
The communications between adjacent processors is granted by two
busses, one for the 1st level and one for the 2nd level trigger.

This kind of trigger processors has been possible due to the
development of A/D converters using mostly digital circuitry,
like for example Sigma-Delta A/D Converters. These can be easily
integrated in single chips, or used as building blocks for more
complicate chips. As an example, in 1.25μm CMOS technology, 30mm[2]
of chip area could contain 32 Sigma-Delta A/D Converters, 256k of
RAM and 10000 gates of logic functions.

3.3 Third Level Trigger

The third level trigger will perform the analysis of the full
events, using digitized data corrected for calibration, in order
to reconstruct physical quantities. It will connect different detector components to cut on complex event topologies.

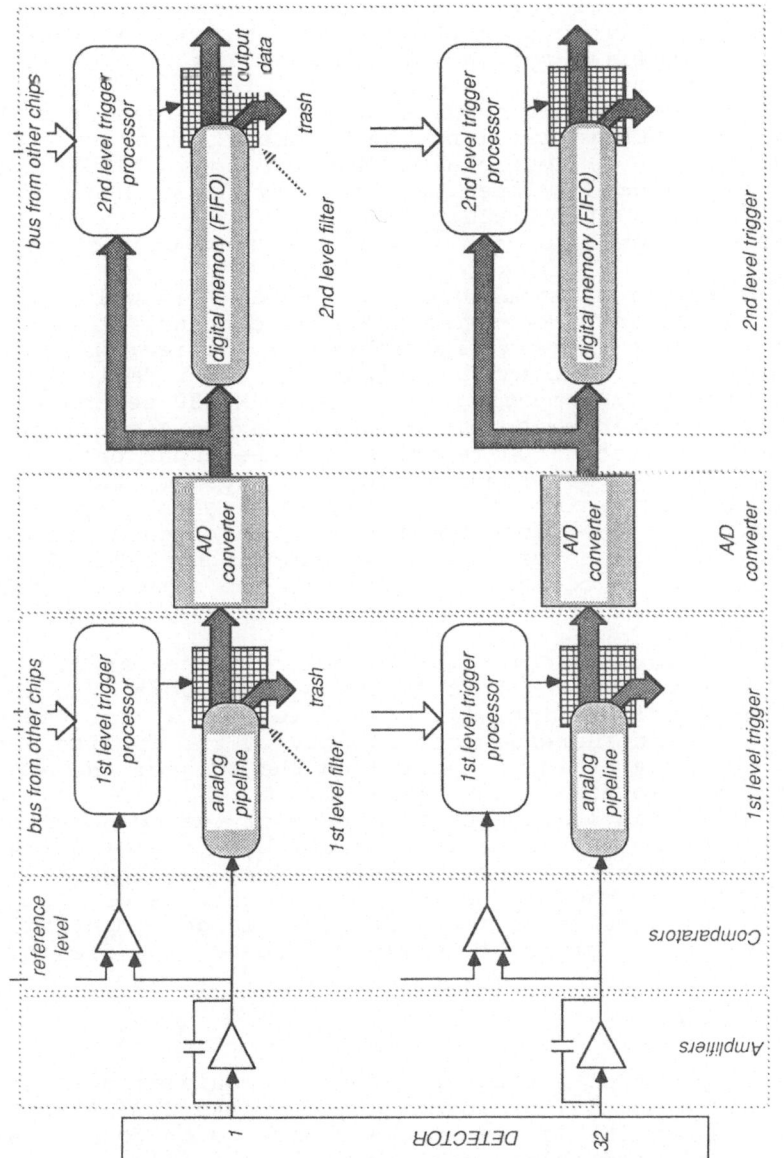

Fig.7 - The HARP processor

The dead-time will be kept at a minimum by making use of multi-processor farms. Up to 100ms per event may be allowed and a reduction in rate of a factor 100 is required in order to have a final rate of events on mass storage not greater than 10Hz.

The principle design of the third level trigger is shown in Fig.8. Data from each detector element are collected into a matrix of "detector" memories (one row per event, one column per detector element). Each row, i.e. each complete event, is then read-out into one "micro-processor" memory through a high-speed bus on a first-in first-out basis, and starting from the first processor in the row.

Different types of micro-processors are candidate for the implementation of a third level trigger. Some of them are commercially available, e.g. Motorola MC60020 or Inmos RISC Transputers, other are being developed specifically for high-energy physics, like the 3081/E Emulator.

Two different philosophies in the treatment of data may apply:

i) A hierarchical structure (multi-stack) in which different levels of processing are foreseen and, at each level, more complex algorithms are applied, with a consequent increase in the computing time[6]. Table I gives the needs in computing power (as VAX/780 seconds), with some reasonable assumption on the event size (1Mbyte), the input rate (100events/s) and the input bandwidth (1280 Mbytes/s), and with an hypothesis on the reduction factor (10) which can be obtained at each step.

ii) A single micro-processor stack in which each processor analyzes events with a common algorithm. Table II gives the needs in computing power with the same assumptions as above and assuming a computing time of 10s/event.

Even if the assumed computing times are rather arbitrary, it is clear that hypothesis (i) is favoured. Therefore, the third level trigger configuration will most likely be a micro-processor multi-stack with an increasing complexity of the applied algorithms. In this way a higher control on the trigger and a greater flexibility is also possible.

The industry can provide us with many of the components required to implement such a configuration, but to engineer working and flexible systems from the components much development is needed on software. The main effort should be towards high level languages in a real-time environment and a large compatibility with the off-line programs for maximum feedback and flexibility.

4.   CONCLUSIONS

The development of a trigger and data acquisition system for experiments in a multi-TeV hadron collider like ELOISATRON is in its infancy. A completely new approach have to be found starting from few general principles which are outlined below.

The first requirement is compatibility and standardization between the various detector component. On hardware, a "pool" of ASIC components should be constituted, and a standard in micro-"pool" of ASIC components should be constituted, and a standard in micro-processor have to be chosen. On software, efficient programming languages and operating systems have to be found and to be made generally available. The rapid evolution of the digital industry prevents somehow an effective standardization. So a lot of effort must be put on the flexibility of the system to accept new solutions in all its components.

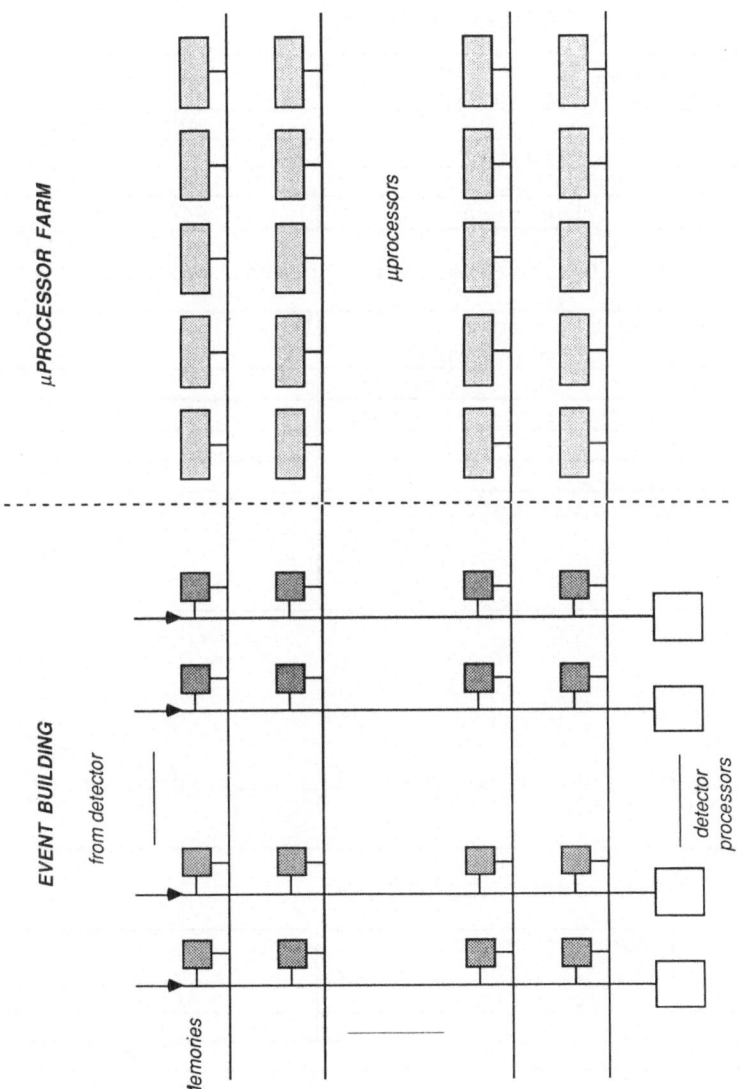

Fig.8 – The third level trigger Event Builder/micro-proces-
sor stack(s)

Table I - Computing power needed with a micro-processors multi-stack

INPUT PARAMETERS		
N	number of vertical busses in from the detector	32
B	bandwidth	40MBytes/s
D	duty factor	0.5
S	event size	1MByte
$R_0 = N \times B \times D/S$	max. input rate	640events/s
DATA FLOW (4-stack structure: a,b,c,d)		
f_a, f_b, f_c, f_d	fraction of events passed to the next stack	1/5
t_a	time per event - stack a	0.1s
t_b	time per event - stack b	1s
t_c	time per event - stack c	10s
t_d	time per event - stack d	100s
$T = R_0 \times (t_a + f_a t_b + f_a f_b t_c + f_a f_b f_c t_d)$	total time required	960s

Table II - Computing power needed with a micro-processors single stack

INPUT PARAMETERS		
N	number of vertical busses in from the detector	32
B	bandwidth	40MBytes/s
D	duty factor	0.5
S	event size	1MByte
$R_0 = N \times B \times D/S$	max. input rate	640events/s
DATA FLOW (1-stack structure)		
f	rejection factor	1/640
t	time per event	10s
$T = R_0 \times t$	total time required	6400s

The approach "home-built electronics with commercial on-line computer" is now antiquated. A much more differentiate approach must be taken, with full awareness of the market possibilities, expertise in connecting components, and joint efforts with industry to fill the remaining gap. In order for a multi-TeV detector to be affordable, the cost per channel should be to kept below 10÷50 US$, and this requires mass production of components.

The radiation hardness of all the front-end, and possibly also of the first and second level trigger electronics, must be improved much above the present limits, in order to take full advantage from the technical possibilities of the machine.

Acknowledgements

I would like to thank Professor A. Ali for the prefect organization of the Workshop. Under his continuing efforts, this series of Workshops on the Physics at ELOISATRON are bound to succeed in the following years.

I would also like to thank E. Heijne and R. Böck for very useful discussions during the preparation of the LAA Project for detector development, which have helped me very much in preparing this report.

References

1. See, for example, CERN 84-10.
2. A.J. Lankford and G.P. Dubois, Proceedings of the Workshop on Triggering, Data Acquisition and Computing for High Energy/High Luminosity Hadron-Hadron Colliders, Fermilab, November 11-14, 1985, p. 185.
3. A. Zichichi, Report on the LAA Project, 25 June 1987, Volume 2.
4. Technical Proposal: "The ZEUS detector", March 1986.
5. LABEN, "Moduli Trigger per Forward Muon Detector basati su Gate Array", result of preliminary studies, September 1987.
6. T. Devlin, Proceedings of the Workshop on Triggering, Data Acquisition and Computing for High Energy/High Luminosity Hadron-Hadron Colliders, Fermilab, November 11-14, 1985, p. 244.

PARTICLE DETECTORS BASED ON THE NEW MULTICHANNEL PHOTOMULTIPLIERS

R. Meunier

SACLAY*

* on secondment at CERN, Geneva, Switzerland

It is hard to conceive an extrapolation of our present collider detectors at around 1 TeV to the next energy range $\sim$ 40 TeV. Therefore a discussion of an alternative approach may be appropriate.

It is pointed out in this note why and how the detection of the Cerenkov light emitted by the particles coming from the interaction point can form the basis of a collider detector.

BASIC CONSIDERATIONS

The Cerenkov pattern of light emission contains the directional parameter (angle in space) of the particle.

The Cerenkov light is emitted in a very small solid angle (like a laser beam), it can be transported and focused at any distance with 100% efficiency.

The Cerenkov energy is in a very thin conical pancake about 10^{-12} sec thick attached to the particle.

The Cerenkov radiation is produced in transparent media. We suggest Helium at atmospheric or subatmospheric pressure to precisely track in angle the particles, and water for a less precise tracking and an analysis of the development of the cascades - electromagnetic, hadronic and muonic.

The light emission is not affected directly by temperature or purity. The radiators are radiation hard. The calibration is stable in time.

Cerenkov light absorption is negligible in gas and in pure water. Absorption length $\sim$ 20 m.

Light of different sources do not interfere, the light signals of many particles are not mixed up. If we point a telescope to the sky the star images are separated, although all the light has been processed by a single instrument.

It has been shown that for particles emitted from a small interaction point, all the Cerenkov light can be focused to a point. An optical system designed for conical wave (instead of the usual spherical one) can perform as a telescope, but with a fundamental difference. The telescope sees the direction of the stars from the position of the telescope, the Cerenkov detector the direction from the I.P.

With such an optical system (focusing detector), double track resolution is limited by multiple scattering and light diffraction. 210^{-6} radian seems the practical limit, but far exceeds the probable needs. Mirrors and optics made of fused silica are radiation hard.

Clearly the tracking of a particle cannot be simpler.

Some measurement of the γ of the particle is possible. With gas radiation up to a practical maximum of 2000 for which $\Delta p/p \sim 10\%$.

The essential differences of Cerenkov light as a signal to detect particles, compared with ionisation, are the following :

Isotropic light emission (scintillation) cannot be focused nor collected efficiently (unless one covers with photo cathode the scintillator).

Therefore unlike for charges produced by ionisation which are difficult to transport and separate for a multitrack event, there is no necessity to place radiation sensitive amplifiers and electronics close to the I.P.

HOW TO IMPLEMENT THESE CHARACTERISTICS IN A DETECTOR

The idea of an imaging Cerenkov detector is very old. Hutchison in 1954 had already presented it. The idea could not be implemented at that time as it was not possible to cover the surface of the radiator with segmented photocathodes. Only phototubes are sensitive enough, and they cost $\sim$ \$100 for only a pixel.

The proton detectors IMB-KAMIOKANDE were built 30 years later on this principle. Special tubes were developed in Japan. The success of these detectors is remarkable. They have pushed the limit of the proton lifetime, opened the neutrino astronomy, and have searched for underground muons from cosmic source. They are used to study solar neutrino.

The particles are detected over a span in energy from a few MeV (neutrino), to GeV (p decay) and TeV (muon bunches).

These detectors have shown tracking, identification and calorimetric properties.

The problem of these large detectors and collider detectors are essentially different on one point :

- In proton detectors, most of the volume is the fiducial volume.
- In colliders we know the interaction point.

Therefore focusing is of no help in proton detectors, it would violate the 2nd principle of thermodynamics but not in a collider detector.

At CERN we have developed optics along these lines for many years, but we could only prove the principles because <u>multianode photomultipliers</u> were not commercially available at that time.

After years of contact with P.M. manufacturers, these tubes just exist now and we have tested that they can fulfil the job.

We propose to use these tubes in a collider detector according to the figure (1). This detector has no magnetic field. For future collider experiments, it is not known if any useful magnetic deflection system may be designed into a detector.

The first Cerenkov ball is essentially Helium. It is segmented similarly to a multiple mirror telescope MMT, there are no gaps in efficiency. This ball is a tracking device with some measurement of γ, the relativistic factor.

The second ball is a spherical shell of water. The light from particles of interest are collected by lenses (of the Fresnel type) oriented at the necessary angle to collect light emitted at 42° along the particle trajectory. The light is detected on an array of phototubes, or a large MPM (multianode PM).

This part of the detector is reminiscent of Cerenkov detectors for Ultra High energy γ ray astronomy, or Fly's Eye detector. The surface of photocathode to collect the useful light is much reduced compared to the photocathode coverage of KAMIOKANDE.

SPECIAL CHARACTERISTICS OF THE DETECTOR

It is very fast. The light from all the particles from the I.P. fall on the photocathode of an MPM almost synchronously in a time interval set by the difference in flight time between a cone and a sphere of around 10 ns. As each row of pixels in the MPM is swept in time, the timing of each pixel is much narrower, ~ 0.5 ns.

Multianode phototube pulses are fast, ~ 5 ns, full width of the pulse. They will be the limiting factor.

There is no memory effect in the detector.

One would observe that the PM anode would draw current in bursts similar to the bursts of the machine. An event may produce particle decaying in the detector = (secondary vertices).

The light pulses from the particles coming from the secondary vertices will not be as sharp but will be spread in time around the burst time by an amount equal approximately to the impact parameter of the particle. Therefore these events have a different electronic signature, and may be used in a fast trigger.

The light yield is copious (~ 40 photoelectrons per cm of track). Obviously the water ball, a "diffuse calorimeter", does not need 100% sampling.

Ultra fast scanning phototubes can conceivable be used at some places to improve the secondary vertex reconstruction (only about 20 cm with the MPMs).

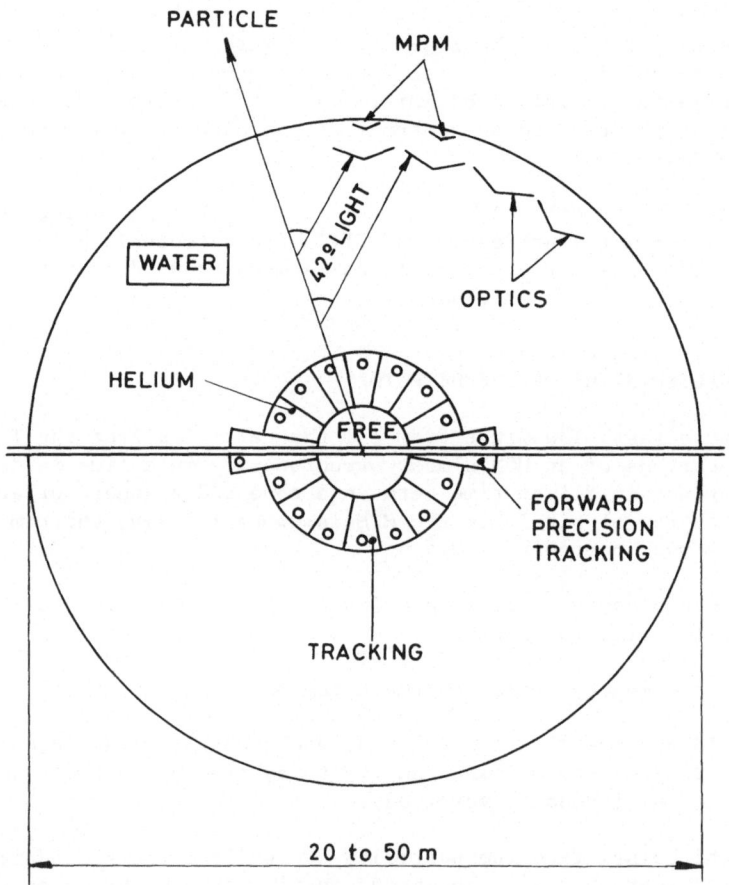

Figure 1.

This detector does not contain large amounts of expensive or rare material. It is mostly a civil engineering project. It can therefore be built in a short time ($\sim$ 2 years) when the R & D is finished.

Most of the cost is in the phototubes and electronics. R & D in parallel-access electronics is needed (Pixel electronics).

Magnetic deflection, even of very few microradians, can be measured with focusing Cerenkov detectors. This well-known property has been used in high resolution spectrometers, but we do not know how to include it in a full solid angle geometry. This is a shortcoming of this proposed detector.

This detector is not new in its principle, it is only an adaptation of detector techniques of non-accelerator physics and of optical studies which have been well and mostly exclusively developed at CERN. Although it needs lots of R & D, in the present version it does not contain speculative or untested ideas.

The data for an event consist of a set of hits, i.e., only space coordinates, which are exactly synchronous for particles from the I.P. The time coordinates are relevant only for the secondary vertices. The data flow is peaked and follows the bursts of the machine over a short time interval. This fact may have favourable consequences for the choice of the bunch spacing and the luminosity of the collider. The central tracking gas sphere gives the angles of the particles without any computation, it has been done by the optics. The hits in the water sphere, as for a proton decay detector, must be analysed, but the particle track is known from the gas sphere.

Extensive simulation of events is necessary to finalize the design of this detector.

Comparison of this detector with conventional ones calls for the same remarks as for proton decay detectors :

Three of them are based on Cerenkov or scintillation detection with poor granularity and P.M. readout. The other three are of the fine grain calorimeter type, carefully designed to detect nucleon decays. Only the former could cope with unexpected phenomena and have reported supernova neutrinos, because they are much faster and have a much lower energy threshold than required for nuclear decays studies.

It is expected that the proposed detector would be able to follow the same philosophy.

BEAUTY PHYSICS AT THE ULTRAHIGH ENERGIES
OF THE ELOISATRON

Brad Cox
Fermilab
Batavia, IL

ABSTRACT

The potential for experimentally studying B physics at the proposed INFN 100 TeV ELOISATRON (Euroasiatic Long Intersecting Superconducting Accelerator Synchrotron) is compared with possibilities at 40 TeV at the Superconducting Super Collider. The effect of the increase in center of mass energy on the production and decay of B mesons has been investigated, particularly with respect to the accumulation of large samples of B hadron decays necessary for the detection of CP violating effects.

INTRODUCTION

Ambitious plans have been advanced for the construction of a Euroasiatic machine, the ELOISATRON, which would have considerably higher energy (100 TeV) than the 40 TeV presently planned for the Superconducting Super Collider. This makes it interesting to evaluate the effect that such an increase in collider energy would have on the production and decay of beauty quark states. In particular, the possibilities for producing large samples of various B meson states with the objective of detecting and measuring CP violation in their exclusive decay modes may be significantly enhanced by the higher energy. While the CP violating effects in various exclusive decay modes are expected to be large[1,2,3], the small cross sections for b quark production at present accelerator energies make accumulation of a large statistical sample of any exclusive decay mode difficult.

The possibility of experimentally studying B physics at the SSC has been investigated in the Snowmass 84 and 86 meetings[4,5,6] and other forums[7]. In addition, the possibilities for doing such physics at Fermilab in the TeV II fixed target program[8] and at the TeV I collider[9] are just beginning to be evaluated. The experimental problems and the potential of producing, detecting and completely reconstructing large samples of exclusive B decays at TeV II, TeV I, the SSC, and at the ELOISATRON are quite different and should be compared in detail. The issue of collider verses fixed target experimental configuration will not be addressed in detail in this paper although, as will be mentioned below, there are advantages and liabilities in both approaches. It is interesting, however, to note that the most energetic B's for any of the present or proposed future experimental configurations are being produced now in TeV II fixed target hadroproduction experiments (even in comparison to B's produced the 100 TeV ELOISATRON interactions!). Accordingly, the potential for doing this sort of physics in a fixed target environment with external beams at the energy of the ELOISATRON beams is an interesting subject for further study at another time. The purpose of this document is to investigate whether the ultrahigh energy of the ELOISATRON collider gives special advantages in comparison to the SSC.

There are obvious enhancements in the capability of doing this sort of physics that come with higher energy. These enhancements occur in several areas:

CROSS SECTIONS

The ratio of the gluon fusion dominated $b\bar{b}$ cross section to the total cross section grows rapidly with $\sqrt{s}$ since the $b\bar{b}$ cross section grows rapidly with energy (see Fig. 1) while the total cross section remains relatively constant. The rate of increase of the cross section with energy shown in Fig. 1 is that predicted by PYTHIA[10] but the level of the cross sections have been adjusted to take into account the meager experimental data that is available on the hadroproduction of beauty. The absolute level of the TeV II cross section has been determined from the WA78 measurement[11] of the B production cross section in $\pi^{-}U$ interactions at 320 GeV/c. This measurement results in a $\pi^{-}N$ cross section of $4.5 \pm 1.5 \pm 1.5$ nb when an A^1 dependence of b production on atomic number is assumed. Since this π^{-} nucleon cross section agrees well with the first order calculations of E. Berger[12], we have used the ratio of $\pi^{-}N \to b\bar{b}$ at 320 GeV/c to $pN \to b\bar{b}$ cross sections 900 GeV/c as calculated by Berger to estimate a pN cross section of approximately 14 nb per nucleon at $\sqrt{s} \approx 40$ GeV. This is approximately one quarter of the cross section for $pN \to b\bar{b}$ predicted by PYTHIA.

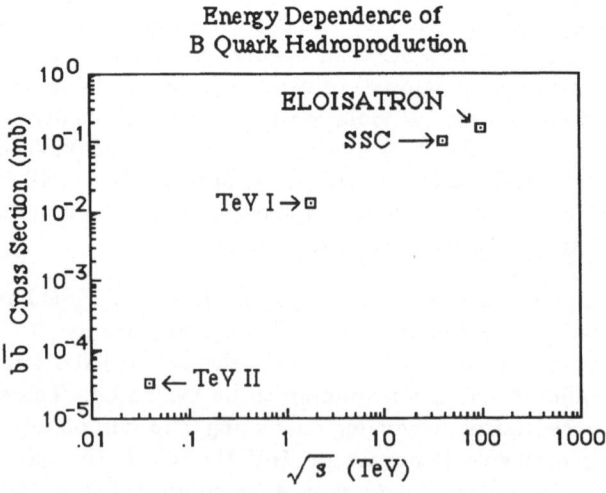

Figure 1. Growth of $b\bar{b}$ cross section in pp interactions as a function of $\sqrt{s}$

The second piece of experimental evidence on $b\bar{b}$ production used to determine the absolute level of the cross sections of Fig. 1 is the $\sqrt{s} = 630$ GeV $p\bar{p} \to$ high p_t dimuon data[13] of UA1. This data has been used to infer a $\bar{p}p \to \bar{b}b$ cross section of $1.2 \pm 0.1 \pm 0.2$ μb for the portion of the cross section at $p_t > 5$ GeV/c. Extrapolating to 1.8 TeV pp interactions by use of ratios of $p\bar{p}$ and pp cross sections obtained from PYTHIA, we obtain a $pp \to b\bar{b}$ total cross section of approximately 13 μb integrating over all p_t. This estimate is, once again, approximately one quarter of the 50 μb cross section estimated by PYTHIA. Therefore, we have decreased all cross sections calculated by PYTHIA by a factor of four but have retained the energy dependence for purposes of Fig. 1. If this energy dependence is valid we expect only 60% more cross section at the ELOISATRON that at the SSC.

Table I gives these cross sections and their ratios to total cross sections and indicates the relative richness of the higher energy collider interactions in $b\bar{b}$ production. The TeV II cross sections for Fig. 1 and Table I are calculated for an intermediate A target such as silicon.

Table I*

	TeV II	TeV II	SSC	ELOISATRON
$\sqrt{s}$ (TeV)	.040	1.8	40	100
$\sigma(b\bar{b})$ (cm^2)	$\approx 3 \times 10^{-32}$	$\approx 1.3 \times 10^{-29}$	$\approx 1.0 \times 10^{-28}$	$\approx 1.6 \times 10^{-28}$
$\sigma(b\bar{b})/\sigma_T(pN)$	$\approx 10^{-6}$	$\approx 10^{-4}$	$\approx 10^{-3}$	$\approx 1.6 \times 10^{-3}$
$\#b\bar{b}/10^7$ sec	$\approx 10^8$	$\approx 10^9$	$\approx 10^{11}$	$\approx 1.6 \times 10^{11}$

*The assumptions made in calculating the event rate for each machine is that the various detectors would be limited to 10^7 interactions per second, each experiment would run for 10^7 seconds of beam time and that the luminosity in the case of the TeV I collider would be 10^{31} cm^{-2} $sec{-1}$ (after upgrading).

The kinematic distributions of the b quarks produced in 100 TeV pp interactions (as predicted by PYTHIA) are shown in Fig. 2. The b quarks are produced at relatively low transverse momentum and with a relatively flat rapidity distribution.

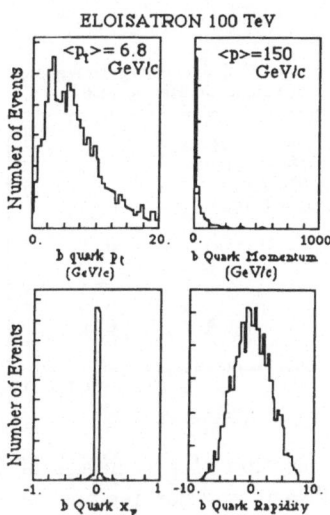

Figure 2. Production distributions of b quarks in p_t, momentum, x_F and rapidity in $\sqrt{s} = 100$ TeV interactions

PRODUCTION ANGULAR DISTRIBUTIONS OF THE BEAUTY QUARKS

The dominant gluon fusion production mechanism for $b\bar{b}$ production, at the higher energies of the SSC and the ELOISATRON, leads to highly collimated $b\bar{b}$ events with both b and b both traveling along one or the other beam directions (see Fig. 3a). As has been pointed out[5,6], at the SSC, this collimation and correlation of the b and the $\bar{b}$ quarks makes possible the design of smaller solid angle spectrometers which are more heavily instrumented for precision spectroscopy with particle identification but which can still capture the decay products of b's a significant percentage of the time. The forward peaking of the b quark production increases approximately linearly with log(s) between the TeV I and the energy regime of the SSC and the ELOISATRON. This is shown in Fig. 3b for the TeV I, the SSC and the ELOISATRON collider configurations. The required solid angle coverage of a B spectrometer decreases both because of the increase in collimation and correlation with increasing s and because of the increase of the momentum of the b's in the forward direction with s discussed below.

MOMENTUM OF THE b QUARKS AND B HADRONS

There is also a strong correlation of momentum of the b quark (and its corresponding B hadron) with laboratory angle (see Fig. 4a). The average momentum of b quarks in various angular regions is shown in Fig. 4b for TeV II, TeV I, the SSC and the ELOISATRON. Only in the most forward regions along either beam direction in the collider experiments does the momentum of the b's become appreciable while the central region on average contains only very low energy b's. The increase of s in the collider experiments causes an increase in the laboratory momentum of the b quark at all angles but the increase is most dramatic in the forward direction,

ELOISATRON 100 TeV

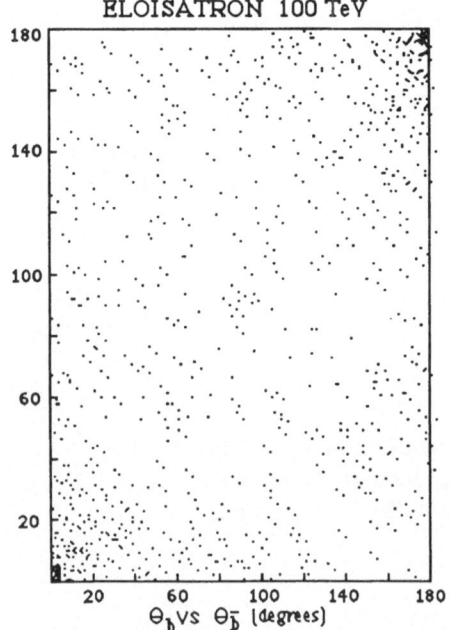

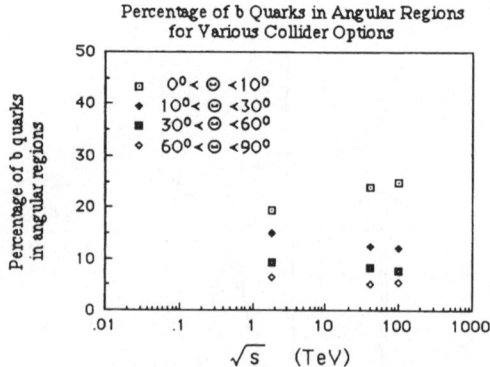

Figure 3a. Correlation of b and b quarks in laboratory angle at $\sqrt{s} = 100$ TeV

Figure 3b. Variation of percentage of b quarks in different angular regions as a function of $\sqrt{s}$

when the *b* quark and the resulting *B* hadron momenta are averaged over the entire solid angle, both averages appear to increase approximately linearly with $ln(\sqrt{s})$ for the collider configurations (see Fig. 4c). As noted above, the highest momentum *b* quarks and *B* hadrons for the four experimental configurations are already being produced in the Fermilab TeV II hadroproduction experiments. The increase in average momentum of *B* hadrons is considerably slower than increase in the average momentum of the *b* quarks due to the hadronization process leading to the *b* quarks. The *b* quarks are highly excited in the production process and must radiate gluons to return to the mass scale appropriate for hadronization into a physical *B* hadron.

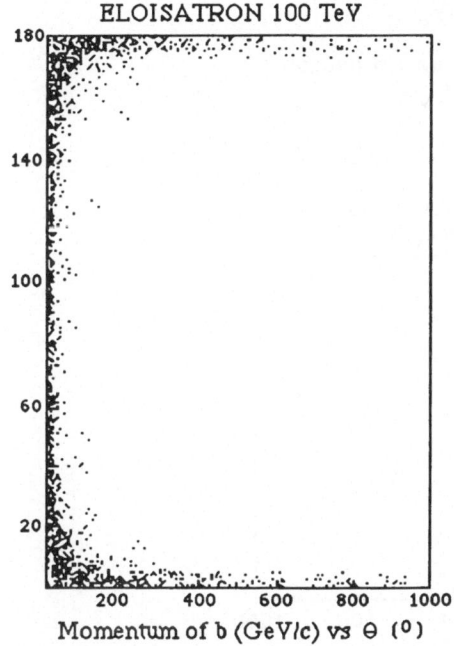

ELOISATRON 100 TeV

Momentum of b (GeV/c) vs Θ (°)

Figure 4a. Correlation of momentum with laboratory angle in 100 TeV interactions

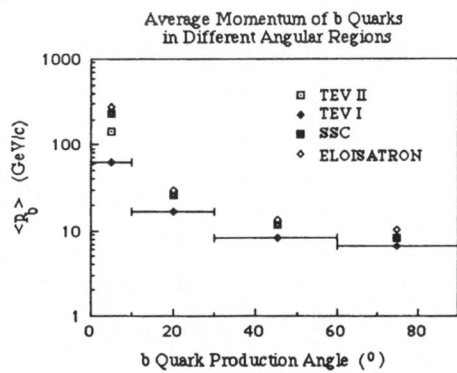

Average Momentum of b Quarks in Different Angular Regions

Figure 4b. Average *b* quark momentum as a function of angle of *b* quark production for TeV II, TeV I, the SSC, and the ELOISATRON

This gluon radiation, the level of which depends on the energy of the hard hadronic collisions, necessarily makes the resulting *B* hadrons considerably softer than the *b* quarks. Fig. 4c also indicates that the difference between the laboratory energy of the *b* quark and the *B* hadron is greatest for the TeV II experiments. While this is not fully understood, it may well to do with the combination of broadening of angular distributions in the hadronization process as well as the loss of energy in the hadronization of the *b* quark, both of which enter into the Lorentz transformation necessary to obtain the laboratory momentum of TeV II *B* hadrons.

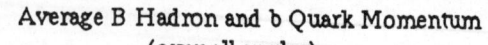

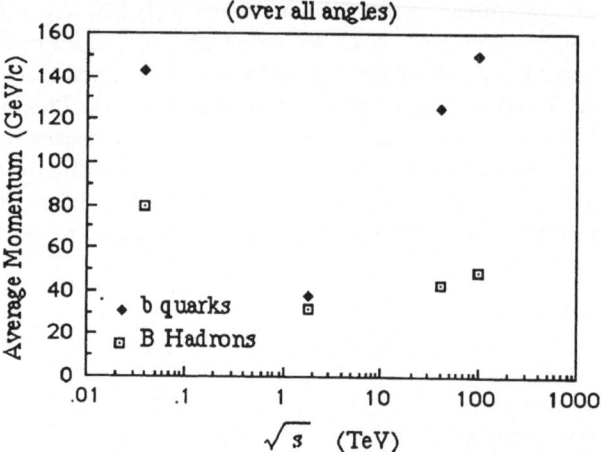

Figure 4c. Average b quark and B hadron momentum as a function of $\sqrt{s}$

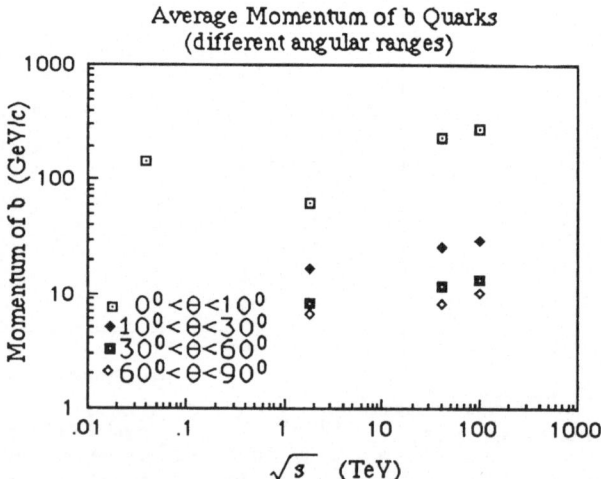

Figure 4d. Variation of Average b quark momentum with $\sqrt{s}$ for various laboratory angular regions

The momenta of the B hadrons and their decay products is critical for triggering on and reconstruction of the B's. While it is true that an impact parameter calculated for a B decay is invariant with B momentum since

Δ = impact parameter $\approx$ decay length $\times$ laboratory decay angle of B decay product

$$= \{\gamma\beta c\tau\}\,\{(1/\gamma)\,\tan\,(\theta*/2)\} \approx c\tau\,\tan\,(\theta*/2)$$

is independent of γ, the multiple scattering of the decay products of the B's inversely proportional to the momentum of the decay products. The resolution on the impact parameter is degraded for low momentum decay products passing through microvertex measurement stations.

In addition, the higher the momentum of the decay products of the B's the easier triggering on those b's becomes. In reference 6 the problems of triggering on the muon pair from the psi decay of the B's was discussed. There the challenge is to prepare a thick enough muon detector to range out hadrons and muons from decays of K's and π's produced in total cross section interaction while avoiding ranging muons from the B decay. The choice of thickness of the shield is a delicate balance between elimination of these trigger backgrounds and the retention of signal. The higher the momentum of the B's the more easily they are distinguished from the trigger backgrounds.

To illustrate the problem in achieving this separation the semileptonic decays of $B \rightarrow De\nu$ have been examined. In Fig. 5a the momentum distribution of the electron from this decay is plotted for the ELOISATRON. The average momenta of electrons from semileptonic decays in the four experimental configurations is shown in Fig. 5b. As indicated, the average momentum is 19.9 GeV/c at the ELOISATRON but this is deceptive. The distribution shows a long tail which skews the average much higher than it would be otherwise. More indicative is the

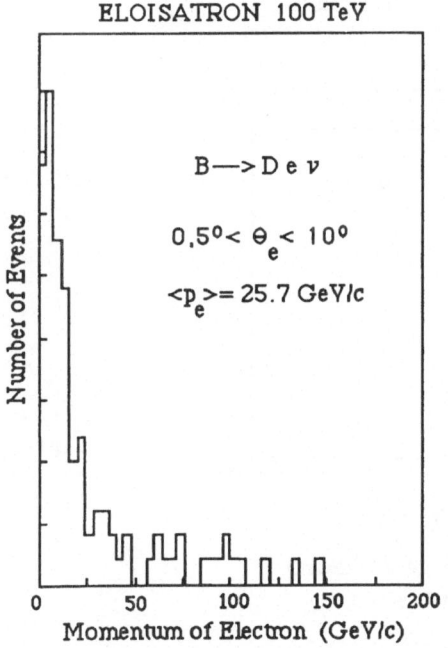

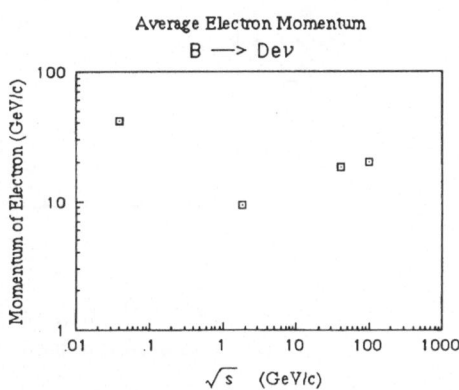

Figure 5a. Momentum of the electron from the semileptonic decay, $B \rightarrow De\nu$ in 100 TeV interactions

Figure 5b. Average electron momentum from the semileptonic decay, $B \rightarrow De\nu$ as a function of $\sqrt{s}$

percentage of the electrons from this decay that are less than 4 GeV/c. As shown in Fig. 5c, the fraction of the flux with momentum less than 4 GeV/c decreases slowly with $ln(\sqrt{s})$ from approximately 70% at TeV I to slightly greater than 30% at ELOISATRON energies. By contrast, the electrons from the semileptonic decays at TeV II are much more energetic with less than 10% having momenta less than 4 GeV/c.

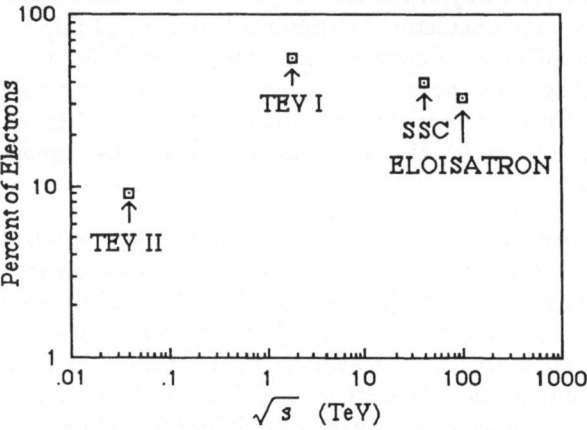

Figure 5c. Percentage of electrons from the semileptonic decay, $B \rightarrow De\nu$ with momentum less than 4 GeV/c as a function of $\sqrt{s}$

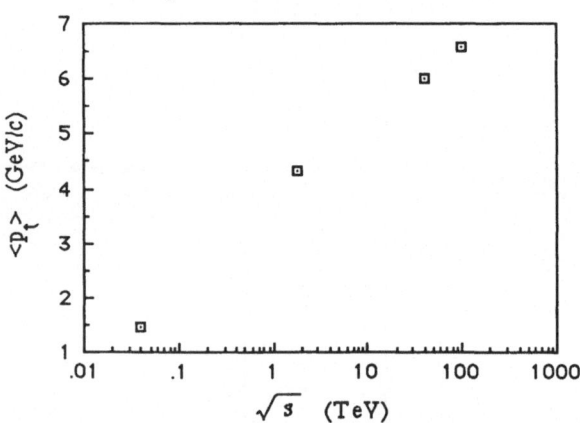

Figure 6. Growth of average B hadron transverse momentum with $\sqrt{s}$

On the other hand, if we examine the average transverse momenta of the B hadrons as a function of $\sqrt{s}$, an advantage of the higher energy collider interactions becomes obvious. As shown in Fig. 6 the average transverse momentum increases linearly with $\ln(\sqrt{s})$ from about 1.8 GeV/c at TeV II to 6.7 GeV/c at the ELOISATRON. This increase in the transverse momentum of the B's is mirrored in an increased separation of the primary and secondary vertices in the plane transverse to the beam direction. This is the plane in which the greatest resolution in the reconstruction of the secondary vertex position is obtained using the planar silicon detector configurations that the forward collimation of the B production makes possible. The separation of primary and secondary vertices at the ELOISATRON are shown in Fig. 7a assuming a B lifetime of 1.2×10^{-12} seconds and the given choice of mixing and CP violation parameters. The separation in the transverse plane shown in Fig. 7b averages approximately 178 microns, much larger than the expected resolution for reconstruction of the transverse separation of primary and secondary decay vertices. The primary decay vertex is constrained to be within the interaction region which will be of order 10 microns in radius for high energy machines like the SSC or the ELOISATRON. So the measurement error will come mainly from the determination of the secondary decay vertex and should be of order of a few tens of microns.

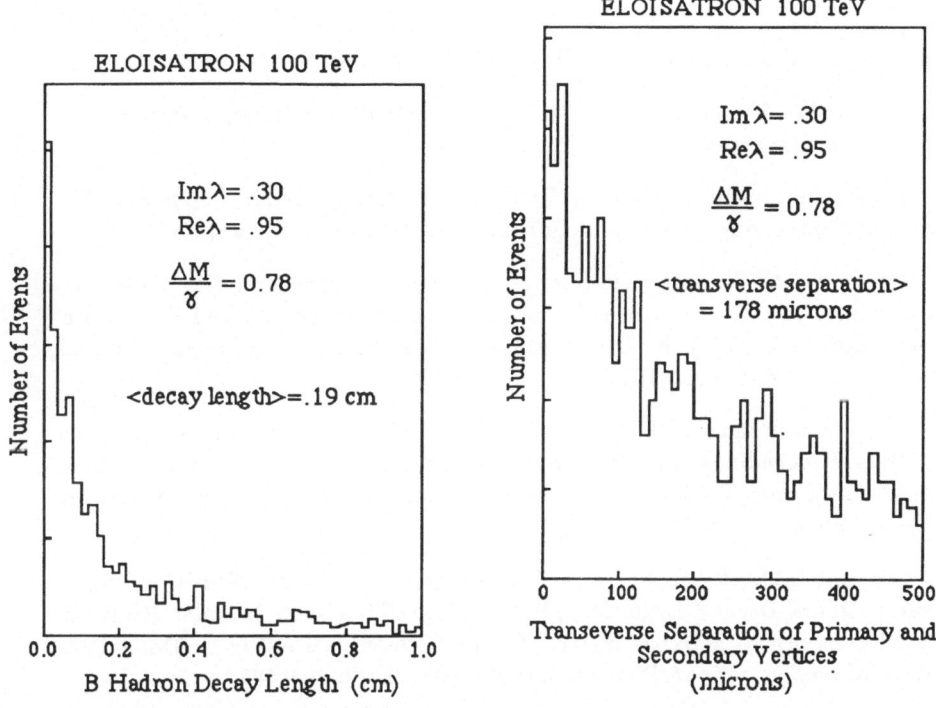

Figure 7a. Separation of primary and beauty hadron secondary vertices in 100 TeV pp interactions

Figure 7b. Separation of primary and beauty secondary vertices in the plane transverse to the beam direction

CONCLUSION

There are several advantages of the ELOISATRON relative to the SSC for collider type B experiments (larger cross sections, larger average momenta, and larger average transverse momenta plus more collimation and correlation of the $\bar{b}$ and b). However, these improvements are not large and would not seem to be crucial unless the final strategies for obtaining evidence for CP violation are marginal statistically. Since this may well be the case (see the Snowmass 86 references), more study of the ELOISATRON possibility is merited. Perhaps even more interesting may be the potential for B physics with fixed target beams at the ELOISATRON since the fixed target configuration has some experimental advantages. The possibilities of both the ELOISATRON collider and fixed target options for doing B physics should be investigated further as the machine parameters become better established.

ACKNOWLEDGEMENTS

I would like to acknowledge many conversations with many people concerning this topic. In particular Fred Gilman, J.D. Bjorken and Ed Berger have given useful theoretical guidance from time to time. I would also like to acknowledge the input of my experimental colleagues especially the members of the E-771 collaboration. I thank those associated with the ELOISATRON workshop and especially the director, Ahmed Ali for their hospitality and the opportunity to investigate the issues addressed in this paper.

References

1. I. Dunietz and J. Rosner, Phys. Rev. **D34** (1986)1404.

2. I.I. Bigi and A.I. Sanda, CP Violation in Heavy Flavor Decays: Predictions and Search Strategies, SLAC-PUB-3949 (May, 1986).

3. J.D. Bjorken, Fermilab Report, Estimates of Decay Branching Ratios for Hadrons Containing Charm and Bottom Quarks, (July, 86), unpublished.

4. J. Cronin *et.al.*, Summary Report on the Rare Decays and CP Violation, Proceedings of 1984 Snowmass Summer Study on the Design and Utilization of the SSC, edited by R. Donaldson and J.G. Morfin (American Physical Society, New York, 1984)161.

5. B. Cox, F.J. Gilman and T.D. Gottschalk, Summary Report on Heavy Flavors and Jets, Proceedings of the 1986 Summer Study on the Physics of the Superconducting Super Collider, edited by R. Donaldson and J. Marx, (American Physical Society, New York, 1986)33.

6. B. Cox and D.E. Wagoner, The J/ψ Tirgger/Tag for the Study of Weak Beauty Quark Decays at the SSC, Proceedings of the 1986 Summer Study on the Physics of the Superconducting Super Collider, edited by R. Donaldson and J. Marx, (American Physical Society, New York, 1986)83.

7. R.J. Morrison *et.al.*, A Beauty Spectrometer for the SSC, Berkeley SSC Detector Workshop, (July, 1988).

8. B. Cox *et.al.*, E-771 Collaboration, Fermilab Proposal E-771, (approved April, 1987).

9. N.W. Reay, Letter of Intent of a Tevatron Collider Beauty Factory (submitted, March, 1987); N.S. Lockyer *et.al.*, Proposal for a Bottom Collider Detector (submitted March, 1987).

10. H.U. Bengtsson and G. Ingelman, Computer Phys. Comm. **34**, (1985)351.

11. M.G. Catanesi *et.al.*, WA78 Collaboration, CERN report CERN-EP/86-177; P. Pistilli, Nucl. Phys. B, Proceedings Supplements: Proceedings of the Topical Seminar on Heavy Flavors, San Miniato, Italy, (May, 1987).

12. E.L. Berger, Argonne National Laboratory Reports, ANL-HEP-PR-87-87-53, (June,1987); ANL-HEP-PR-87-113, (Nov., 1987).

13. C. Albajar *et.al.*, UA1 Collaboration, Phys. Lett. **B186**, (1987)237.

LARGE TRANSVERSE MOMENTUM LEPTONS IN
(VERY) HIGH PROTON-(ANTI)PROTON COLLISIONS

B. van Eijk

CERN, Geneva, Switzerland

Abstract

We have extended an earlier study on open heavy flavour production in hadron-hadron collisions by including $O(\alpha_s^4)$ QCD tree level calculations. The effect on the inclusive muon and inclusive dimuon transverse momentum distributions and opening angle between the muons from the UA1 so-called non-isolated dimuon sample is estimated.

Using the CERN collider data as a reference, we have extrapolated open heavy flavour production cross-sections for future accelerators and we discuss the semileptonic decay of heavy flavours as a possible important background signal for neutral Higgs searches at LHC energies.

In addition, we have studied large transverse momentum J/ψ and Υ production at presently available collider energies and have compared our results with data recently obtained by the UA1 collaboration. Provided we include a K-factor of ~ 1.5 - 2 for direct production mechanisms, we find good agreement with theoretical predictions on both shape of the transverse momentum and rapidity distributions and absolute cross-section times branching ratio as measured by UA1: $\sigma.BR(J/\psi \rightarrow \mu^+ \mu^-) = 7.5 \pm 0.7 \pm 1.2$ nb. We estimate $\sigma(p\bar{p} \rightarrow \Upsilon X) = 0.8$ - 2 nb, which compares well with the UA1 result $\sigma(p\bar{p} \rightarrow \Upsilon X) = 0.98 \pm 0.21 \pm 0.19$ nb.

Since the bound states are mainly produced via gluon interactions we discuss the sensitivity of the production cross-sections on the parametrization of the gluon structure function in the colliding hadrons. Finally we investigate cross-section extrapolations for future colliders at various centre-of-mass energies.

1. Introduction

The analysis of 'prompt' muons signaled by the UA1 detector at the CERN Sp$\bar{p}$S collider has considerably added to our understanding of heavy flavour production mechanisms in hadron-hadron interactions. Despite the large background from decays in flight of pions and kaons, a rather 'clean' sample of prompt muon events was obtained. Especially the dimuon sample or, more general, a sample of events with more than one muon can be made relatively background free by applying reasonably large transverse momentum (with respect to the colliding beams) cuts on the muons . In addition to the continuum creation of heavy flavour pairs via strong interactions [1 - 4], convincing evidence has been found for sizable heavy flavour bound state production. The UA1 collaboration has reported clean J/ψ and Υ resonances via the observation of their muonic decay channels [1], [5].

The outline of this paper is as follows. In the next section we briefly review open heavy flavour production mechanisms at hadron colliders and describe the model and computer program we have used to calculate theoretical distributions. In section 3 we make a comparison of our theoretical model with data obtained by the UA1 collaboration. New results on $O(\alpha_s^4)$ QCD heavy flavour calculations are presented as well as cross-section extrapolations to FNAL-collider, LHC and SSC energies. Heavy flavour bound states are the subject of section 4, while comparisons with Sp$\bar{\text{p}}$S collider J/ψ and Υ data are presented in section 5. In the same section we show that provided sufficient statistics and good detector acceptance, onium signals carry information on the behaviour of the gluon structure function in the colliding (anti)proton at small values of x and may serve as calibration signals at future accelerators.

2. Open heavy flavour production

Because of the low centre-of-mass energies and the relatively small charm quark mass, early comparisons between FNAL-ISR data and QCD calculations have failed to give agreement. With about an order of magnitude more energy at hand at the Sp$\bar{\text{p}}$S collider, it is interesting to study heavy flavour production in detail in this new energy domain. It is particularly interesting to see whether we are able to use data taken at √s = 630 GeV to make reasonably reliable predictions for future collider projects as aimed at, at this workshop.

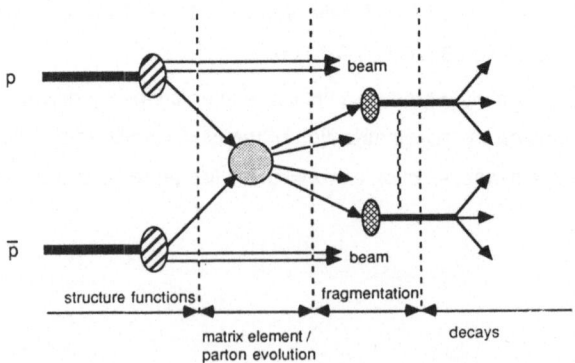

Fig. 1. Schematic description of inelastic hadron-hadron interactions.

In order to obtain numerical results on theoretical cross-sections, the conventional parton picture is used. That is to convolute the amplitudes squared of the matrix elements with the appropriate parton structure functions in the colliding hadrons. Since we are mainly interested in the properties of (prompt) leptons, the fragmentation of the bare partons and subsequent decay of (heavy) particles has to be treated in detail. They are the basic ingredients for either a analytical or Monte Carlo calculation of differential cross-section distributions. The procedure is schematized in figure 1. All theoretical distributions presented in this paper were obtained with the EUROJET Monte Carlo program for hard hadronic interactions [6]. At present, the EUROJET program contains five different sets of structure functions [7 - 9], which enables the user to compare different Q^2 evolutions for various choices of the QCD scale parameter Λ. We recall that Λ and the definition of Q^2 strongly effect the value of the strong coupling constant entering in the amplitude of the hard scattering process:

$$\alpha_s \sim \frac{1}{\log \frac{Q^2}{\Lambda^2}}$$

(1)

Since we will only consider tree level contributions in our Monte Carlo approach, diminution of Q will lead to an increase of the cross-section.

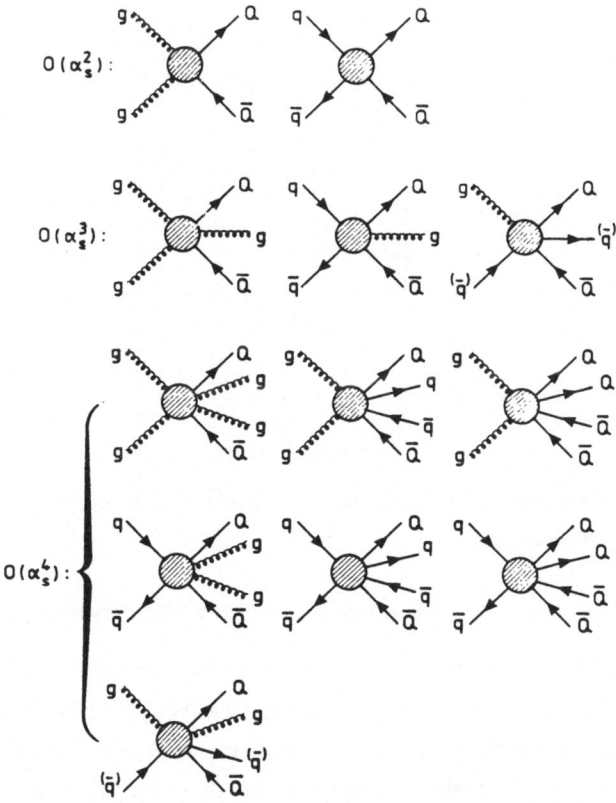

Fig. 2. Heavy flavour pair production mechanisms up to $O(\alpha_s^4)$. The $gg \to Q\bar{Q} X$ diagrams dominate at all orders of α_s. The heavy flavour contents of the interacting hadrons is assumed to be zero, avoiding double counting.

The $O(\alpha_s^2)$, $O(\alpha_s^3)$ and $O(\alpha_s^4)$ contributions to heavy flavour pair production are shown in fig. 2. The heavy flavour content of the colliding hadrons, as calculated in ref. [7] is neglected in order to avoid double counting via the gluon splitting diagrams.

The overall uncertainty in our calculations due to the variation of scale parameters is discussed in detail in ref. [3] and [6]. All distributions have been calculated with the structure functions of Eichten et al. [7] which were derived for Λ = 200 MeV. For the heavy flavour production processes depicted in fig. 2, the scale was fixed at $Q^2 = p_t^2 + M_Q^2$, motivated by the UA1 inclusive jet analysis [10], with M_Q the mass of the heavy quark. For the $O(\alpha_s^2)$ [11] and $O(\alpha_s^3)$ [12] matrix elements, the mass of the heavy quark was explicitly taken into account. As the lowest order cross-sections are finite, we only have to worry about the divergencies occurring at higher orders. To regulate the soft and collinear divergencies in the $O(\alpha_s^3)$ processes, we have introduced a cut-off on the transverse momentum of the additional jet to the heavy flavour pair. The cut-off value was fixed at p_t^{jet} > 5 GeV/c, giving a K-factor in agreement with leading logarithm calculations [13]. A comparison of these calculations with the UA1 measurement [14] of the number of D* inside jets is shown in fig. 3.

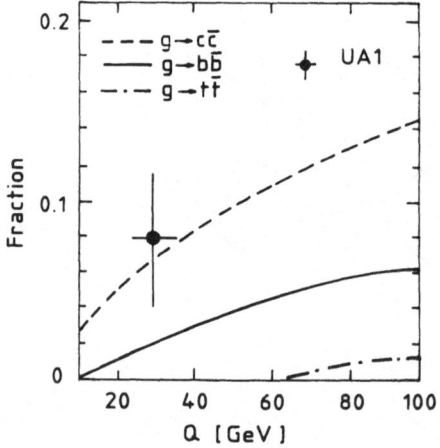

Fig. 3. *The UA1 inclusive D* measurement inside jets [14], is consistent with gluon splitting calculations for $g \to c\bar{c}$ [13].*

The $O(\alpha_s^4)$ expressions are only available in the massless limit [15]. However, we assume that for large enough heavy flavour transverse momenta, mass terms in the expressions for the matrix elements only affect the overall K-factor and do not significantly distort kinematical distributions. Note that we do simulate the correct phase-space properties of the event. The regularization of the divergencies at this order becomes rather complicated. We divert the discussion of this issue to a separate paper [16]. 'Reasonable' cuts are a cut-off on the transverse momentum of the (massless) outgoing parton and a minimum separation in rapidity-phi space of all incoming and outgoing partons:

$$p_t^{jet} > 7 \text{ GeV/c},$$

$$\Delta R = \sqrt{\Delta\eta^2 + \Delta\phi^2} > 0.5 \tag{2}$$

Quark fragmentation and particle decay properties have been carefully modeled. For details and comparisons with e^+e^- data we refer the reader to ref. [3] and [6].

3. Comparison with UA1 collider data and extrapolations to higher energies

In order to enhance the contribution from semileptonic heavy flavour decays to the dimuon sample, a minimum amount of hadronic activity around the muons is demanded. Motivated by the isolation studies of muons from the decay of weak charged bosons and the observation that fragmentation and decay of heavy flavours introduce additional hadronic activity, a highly efficient isolation variable has been defined to discriminate against W, Z decays and muons from genuine Drell-Yan pair production [1]:

$$S = \left(\sum_{\Delta R < .7} E_t \right)_{\mu_1}^2 + \left(\sum_{\Delta R < .7} E_t \right)_{\mu_2}^2 \tag{3}$$

The hadronic isolation of a single muon is visualized in fig. 4.

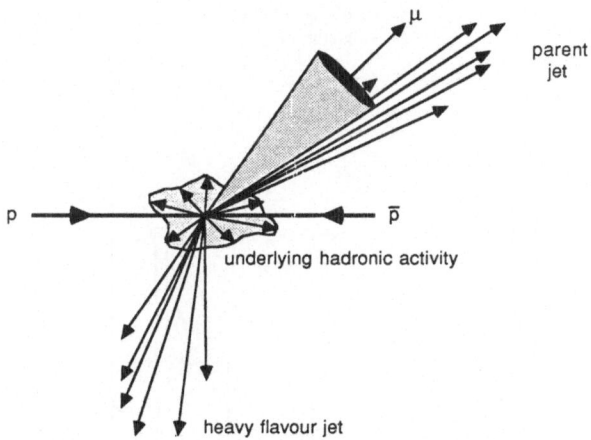

Fig. 4: Hadronic activity in a cone around the μ-vector determines the isolation of the muon.

All dimuon events from the non-isolated dimuon sample satisfy $S > 9$ GeV^2. We have studied two classes of events namely, one set of non-isolated dimuons with [1]:

$$p_t^{\mu_1}, p_t^{\mu_2} > 3 \text{ GeV/c},$$

$$M^{\mu\mu} > 6 \text{ GeV/c}^2 \tag{4}$$

and a second set with [2]:

$$p_t^{\mu_1} > 7 \text{ GeV/c}, \ p_t^{\mu_2} > 3 \text{ GeV/c},$$

at least one jet with $E_t^{jet} > 12 \text{ GeV}$ (5)

One of the distributions, that is highly sensitive to higher order contributions in α_s, is the opening angle between the muons, measured in the transverse plane with respect to the beams. In figure 5 a comparison is shown of our theoretical distributions with UA1 data. The muons satisfy the requirements of (4). In fig. 6a and 6b typical subsets of contributions from same sign and opposite sign dimuons are shown separately. The analysis is repeated for the cuts of (5) and is presented in figures 7 and 8. The comparison of our theoretical distributions with UA1 data shows for both dimuon samples a remarkably good agreement.

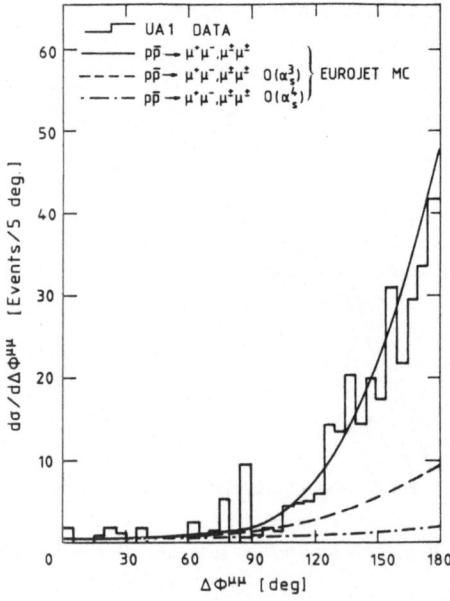

Fig. 5. $\Delta\phi^{\mu\mu}$ (opening angle in the azimuthal plane between the muons from the UA1 non-isolated dimuon sample) compared with fixed order QCD calculations.

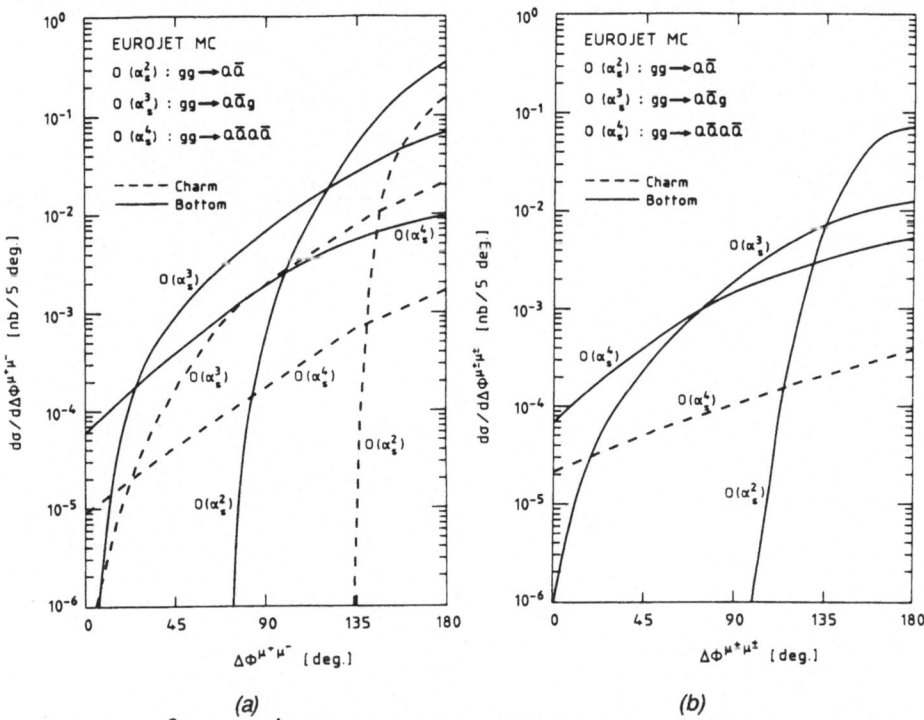

Fig. 6. $O(\alpha_s^3)$ and $O(\alpha_s^4)$ contributions to $d\sigma/d\phi^{\mu\mu}$ in fig. 5 for opposite sign dimuons (a) and same sign dimuons (b).

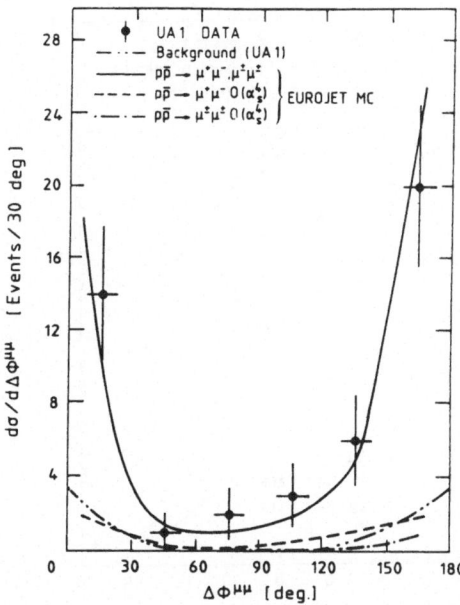

Fig. 7.   The $\Delta\phi^{\mu\mu}$ distribution for non-isolated dimuons, demanding $p_t^{\mu_1} > 3$ GeV/c, $p_t^{\mu_2} > 7$ GeV/c and at least one jet with $E_t > 12$ GeV, compared with tree level $O(\alpha_s^2)$, $O(\alpha_s^3)$ and $O(\alpha_s^4)$ QCD calculations.

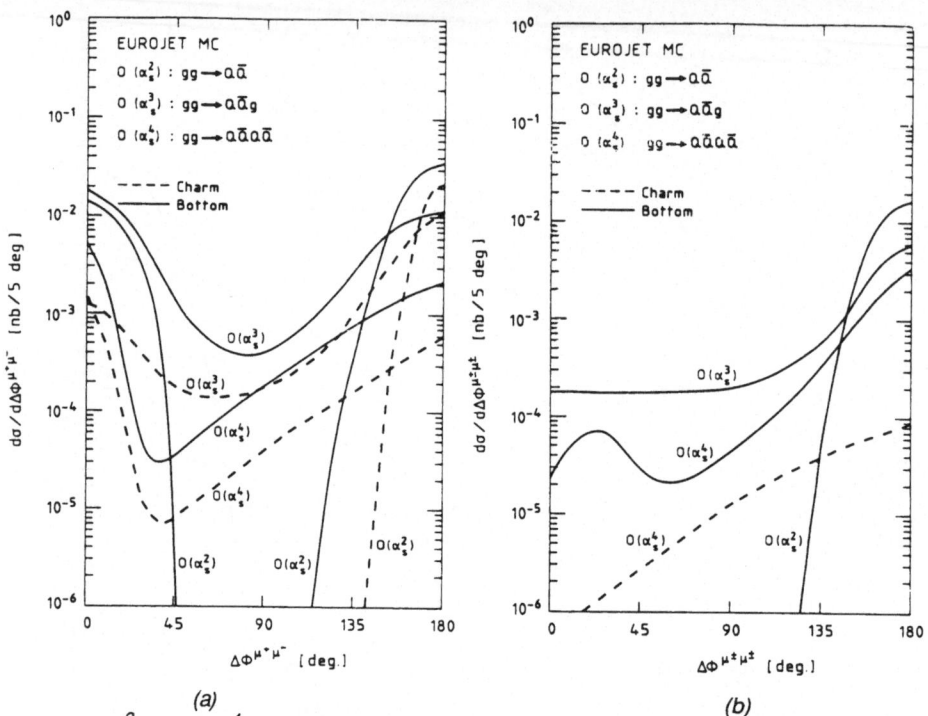

Fig. 8. $O(\alpha_s^3)$ and $O(\alpha_s^4)$ contributions to $d\sigma/d\phi^{\mu\mu}$ for opposite sign dimuons (a) and same sign dimuons (b). The curves satisfy the selection criteria applied to the data of fig. 7.

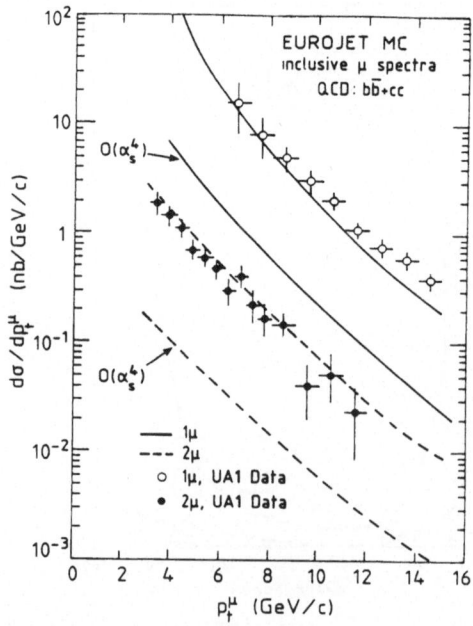

Fig. 9. Inclusive muon and inclusive (non-isolated) dimuon transverse momentum distributions [11] compared with QCD predictions from open charm and bottom production [10].

The O(α_s^4) contributions to the inclusive muon and inclusive dimuon signals are typically an order of magnitude smaller than the leading order contributions (fig. 9). A detailed topological event analysis and a detailed Monte Carlo event simulation for events with more than two well defined jets may give some more information on the absolute contribution of the O(α_s^4) processes to the heavy flavour cross-section at the CERN collider.

In fig. 10, the O(α_s^2) cross-sections for heavy flavour production at the FNAL-collider are shown. The cross-sections are based on extrapolations from collider data taken at $\sqrt{s}$ = 630 GeV. In fig. 10a the inclusive heavy quark transverse momentum distributions are presented after demanding $|\eta^Q| <$ 5. The inclusive rapidity distributions are given in fig. 10b, where each heavy satisfies a cut on the transverse momentum, $p_t^Q > 5$ GeV/c.

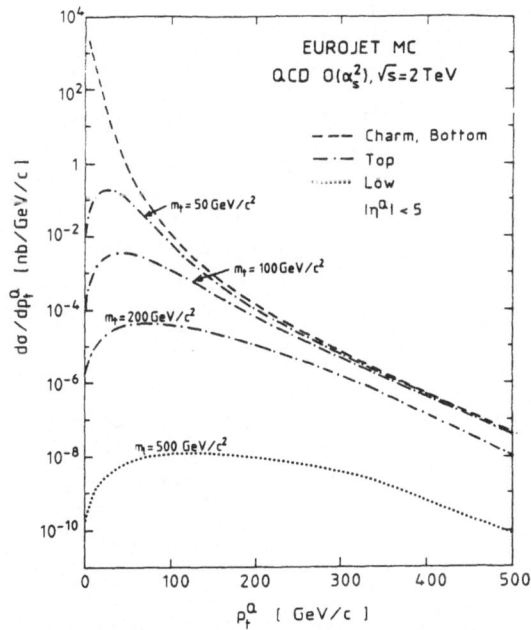

Fig. 10a. Inclusive transverse momentum distribution $d\sigma(p\bar{p} \rightarrow Q X)/dp_t^Q$ ($|y| < 5$), for Q = c, b, t and a heavy fourth generation 'down-like' quark at $\sqrt{s}$ = 2 TeV.

The search for (very) massive particles is one of the major motivations to increase beam energies and luminosity at future colliders. The accessible mass range will be determined by the type of colliding beams as well as both centre-of-mass energy and luminosity. In many searches for exotic particles, heavy flavours may play an important role. Their semileptonic decays for instance, can fake a direct or indirect leptonic decay of a 'new' particle when this particle is created in a complicated hadronic environment. In fig. 11a and 11b we present similar heavy flavour cross-section extrapolations as for the Fermilab collider, but at $\sqrt{s}$ = 16 TeV. An increase of one order of magnitude in $\sqrt{s}$, leads to roughly a two orders of magnitude increase in cross-section for top quarks.

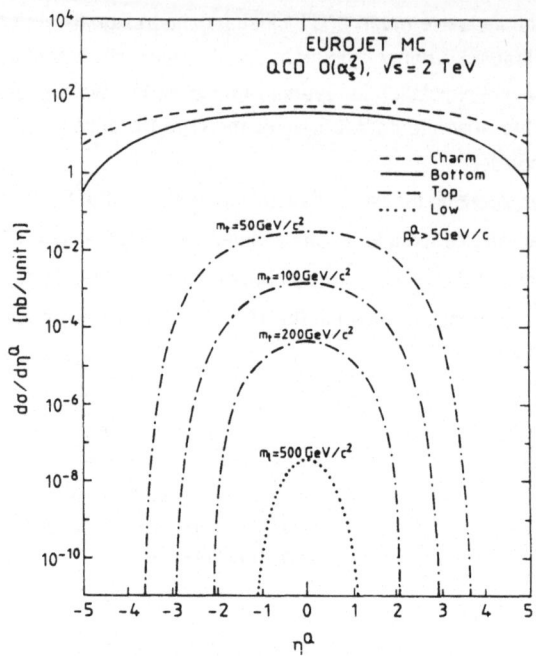

Fig. 10b. Inclusive rapidity distribution dσ(p̄p → Q X)/dy^Q (dp_t^Q > 5), for Q = c, b, t and a heavy fourth generation 'down-like' quark at √s = 2 TeV.

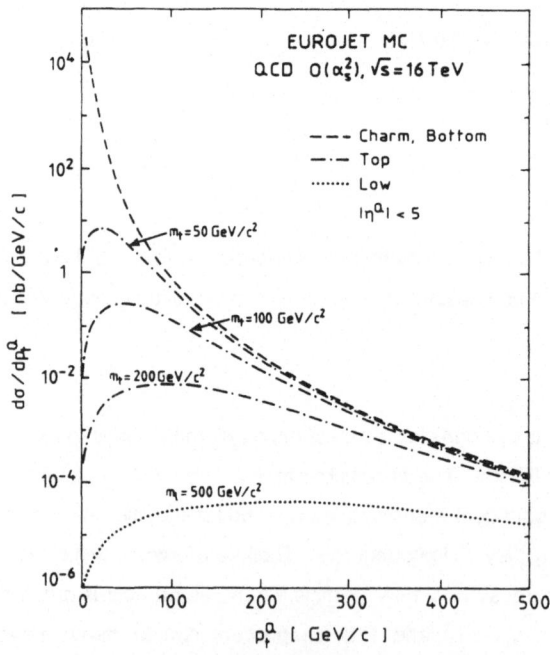

Fig. 11a. Inclusive transverse momentum distribution dσ(p̄p → Q X)/dp_t^Q (|y| < 5), for Q = c, b, t and a heavy fourth generation 'low' quark at √s = 16 TeV.

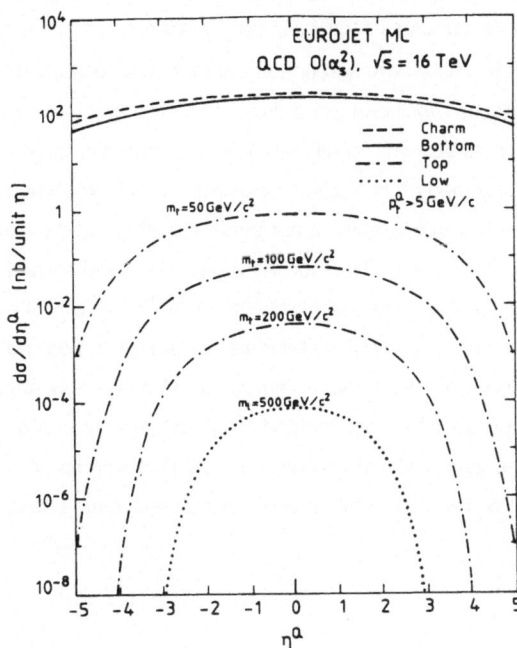

Fig. 11b. *Inclusive rapidity distribution $d\sigma(p\bar{p} \to Q\,X)/dy^Q$ ($dp_t^Q > 5$), for $Q = c, b, t$ and a low quark at $\sqrt{s} = 16$ TeV.*

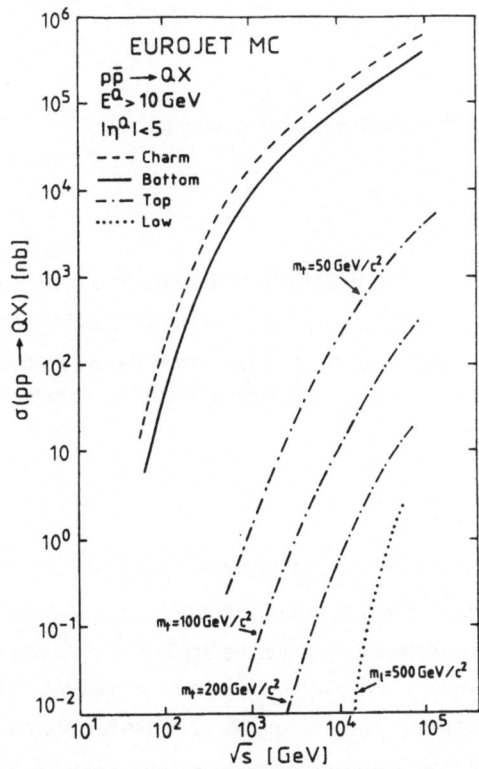

Fig. 12. *$O(\alpha_s^2)$ integrated inclusive heavy flavour cross-sections as a function of the centre-of-mass energy ($E^Q > 10$ GeV, $|y^Q| < 5$).*

Integrated $O(\alpha_s^2)$ inclusive heavy flavour cross-sections in $p\bar{p}$ interactions are presented in figure 12, where $E^Q > 10$ GeV and $|\eta^Q| < 5$. The charm and bottom quark cross-sections dominate over the full range of centre-of-mass energies. Included are a cross-section prediction for a heavy low quark of 500 GeV/c^2 and top quark cross-sections for different masses of the top quark.

High energy, high luminosity hadron-hadron interactions will cause huge experimental problems since the number of inelastic parton-parton interactions per beam crossing rapidly increases as a function of these parameters. At LHC ($\sqrt{s} = 16$ TeV) for instance, one could expect of order 15 inelastic interactions per bunch crossing at a design luminosity of $5 * 10^{34}$ cm^{-2}s^{-1}. It will become a complicated enterprise to trigger on very rare physics channels. However, a beam dump type of experiment, where muons emerging from the interaction vertex could be reliably separated from decay background, may still provide an important method to investigate new physics. As an example we discuss the search for a very massive neutral Higgs particle with its (in the Glashow-Weinberg-Salam scenario) dominant decay mode into a pair of neutral weak gauge bosons:

$$pp \rightarrow H^0 \rightarrow Z^0 Z^0 \rightarrow \mu\mu X \tag{6}$$

We assume that we can discriminate against decay backgrounds by imposing a relatively large transverse momentum cut on the muons. The process of (6) receives considerable background from large transverse momentum boson pair production [18],

$$pp \rightarrow Z^0 Z^0 \rightarrow \mu\mu X \tag{7}$$

and the 'standard' channel, where the Z^0 is recoiling against an energetic jet,

$$pp \rightarrow Z^0 X \rightarrow \mu\mu X' \tag{8}$$

Altarelli et al. [19] predict that 1.1 % of the Z^0 bosons will be produced with a transverse momentum larger than 140 GeV/c at $\sqrt{s} = 10$ TeV. For all processes (6) - (8) the invariant dimuon mass will have to match within the detector resolution with the Z^0 mass. Open heavy flavour production forms the third important background from 'known' physics. It is dominated by the $O(\alpha_s^3)$ contributions

$$pp \rightarrow b\bar{b} g \rightarrow \mu\mu X \tag{9}$$

Since we require muons at large p_t, the $b\bar{b}$ pair has to recoil against an energetic gluon jet. Therefore, we expect that tree level calculations are highly insensitive to the arbitrary soft gluon cut-off that we had to impose in our analysis of the muon data at the $Sp\bar{p}S$ collider. In figure 13a we have summarized the results for $m_H = 800$ GeV/c^2. We have assumed a dimuon invariant mass resolution to reconstruct the Z^0 mass of $M_Z \pm 5$ GeV/c^2. The total number of events is calculated assuming an integrated luminosity of $5 * 10^{41}$ cm^{-2}. The background from semileptonic bottom decays (9) drops rapidly as the transverse momentum of the dimuon pair increases. However, since the top quark mass is still a free, although restricted parameter, an even more important background may come from

$$pp \rightarrow t\bar{t} \, g \rightarrow \mu\mu \, X \tag{10}$$

in the rather peculiar situation that the invariant mass of the $t\bar{t}$ pair is close to or above the Z^0 mass, where the top quark decays into a hard muon and (soft) bottom jet. The background from this channel (for $m_t = 50$ GeV/c^2) is compared with the heavy flavour background from (9) and is shown in fig. 13b.

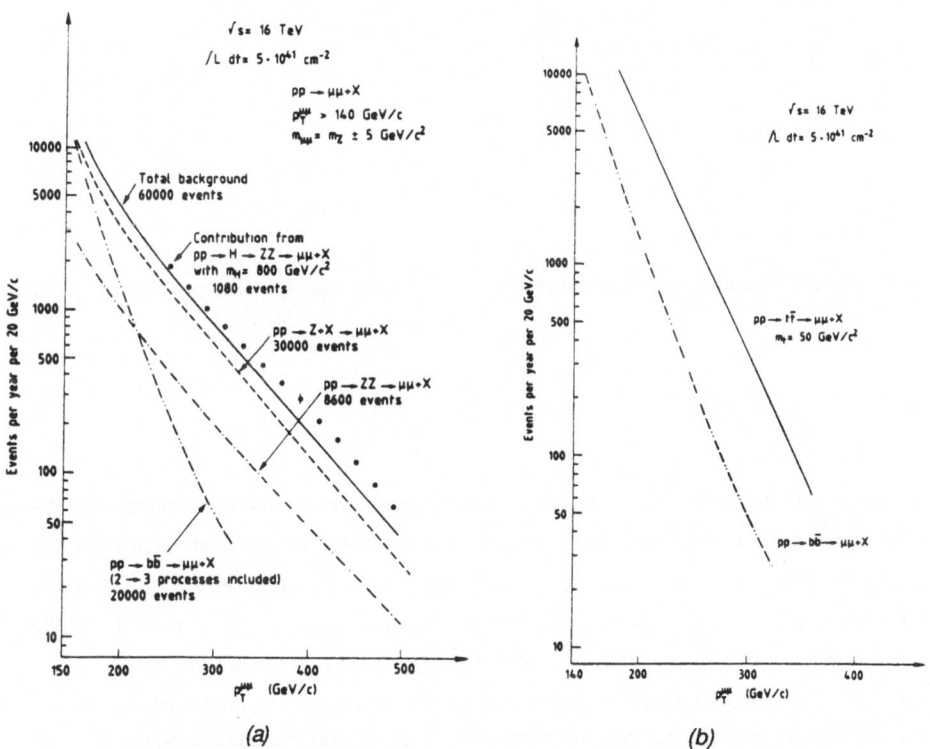

(a)   (b)

Fig. 13. *The background to* pp $\rightarrow$ H^0 $\rightarrow$ Z^0 Z^0 $\rightarrow$ $\mu\mu$ X (m$_H$ = 800 GeV/c^2) *from semileptonic decays of bottom quarks (a). The background from open top production exceeds the background from decaying bottom quarks considerably when the mass of the top quark is close to half the Z^0 mass (b) [20].*

Although the integrated inclusive top quark cross-section is much lower than the inclusive bottom quark cross-section (fig. 12), the window on the invariant dimuon mass enhances the background from (10) considerably. Details on the calculation can be found in ref. [20].

4. Heavy flavour bound state sources in hadron-hadron interactions

At the CERN $p\bar{p}$ collider, the UA1 collaboration has measured significant heavy flavour bound state cross-sections [1], [5]. The bound states are triggered via their muonic decay channels. The hardware of the UA1 muon trigger and additional software cuts impose a cut-off on the transverse momentum of the muons with respect to the colliding proton and antiproton beams. This leads to an effective transverse momentum cut-off on the charmonium state ($p_t^{J/\psi} > 4$ GeV/c). A close analysis of the dimuon invariant mass spectrum (where both muons are selected such that they appear to be relatively isolated) shows a distinct mass peak in agreement with expectations for $^1S_1$ bottom quark (Υ, Υ', Υ'') bound states [1]. Although the production of large transverse momentum vector bottomonium states is completely dominated by the so-called direct production mechanisms [21], [22],

$$
\begin{aligned}
q\bar{q} &\rightarrow {}^3P_j\, g \\
gq &\rightarrow {}^3P_j\, q \\
g\bar{q} &\rightarrow {}^3P_j\, \bar{q} \\
gg &\rightarrow {}^3P_j,\, {}^1S_1\, g
\end{aligned}
\tag{11}
$$

with $^3P_j \rightarrow {}^1S_1$, the charmonium cross-section receives a significant contribution from the decay of B hadrons,

$$
B \rightarrow {}^1S_1\, X
\tag{12}
$$

As we have seen in the previous sections, open heavy flavour cross-sections may be determined by correcting the inclusive single muon and inclusive dimuon cross-section by the appropriate charm hadron and bottom hadron semileptonic branching ratios, heavy quark fragmentation functions and experimental acceptance. The UA1 Υ trigger forms a subset of the inclusive dimuon trigger. Therefore, one expects many systematical errors for the heavy flavour sample to be similar to the ones for the Υ sample. In other words, taking cross-section ratios leads to cancellations of number of systematical uncertainties. In order to obtain a significant $^1S_1$ (J/ψ, ψ') signal the cut on the transverse momentum of one of the muons was largely reduced ($p_t^{\mu_1} > 3$ GeV/c, $p_t^{\mu_2} > 0.75$ GeV/c) whereas the mass window lies in the J/ψ mass range (2 - 4 GeV/c^2), consequently introducing different acceptance and trigger corrections. The obtained J/ψ sample turns out to be surprisingly clean and the invariant dimuon mass distribution may even indicate the presence of a ψ' signal [5]. Since the mass of the charmonium bound states are small compared to the total centre of mass energy at the collider, the interacting partons carry only a small fraction of the beam energy. The Υ and J/ψ signals may serve as a good probe for the (gluon) structure function behaviour at small values of x ($< O(10^{-2})$) [21], [22]. Detection of these signals at present (Sp$\bar{p}$S, Fermilab) and future (LHC, SSC) colliders may pin down the gluon structure function at these very small values of x [22].

The relevant Feynman diagrams contributing to the direct production of large transverse momentum bound states of eq. (11) are shown in figure 14. Note that unlike in e^+e^- annihilation, in $p\bar{p}$ collisions all possible quantum numbers of heavy quark bound states can be produced.

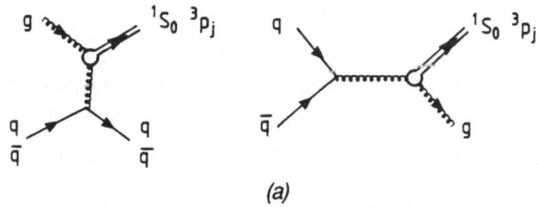

(a)

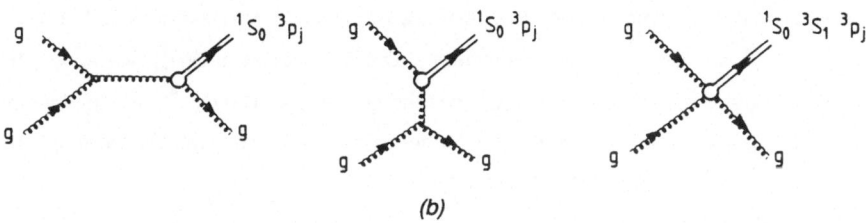

(b)

Fig. 14. *Feynman diagrams for large transverse momentum quarkonium production in hadronic interactions: (a) qg and $q\bar{q}$ scattering and (b) $gg \rightarrow {}^{2s+1}L_j\, g$, with L = S or P.*

The matrix elements squared have been calculated by several authors [23 - 24]. For the three gluon processes very compact formulas have been derived by Gastmans, Troost and Wu [24]. The differential cross-section for the various processes can be expressed in terms of a summation and integration over all parton species:

$$\frac{d\sigma}{dp_t^2\, dy} = \sum_{ij} \int dx_1\, dx_2\, F_i(x_1,Q^2)\, F_j(x_2,Q^2)\, \frac{d\sigma_{ij}}{dt} \tag{12}$$

Using the wave function formalism to calculate the absolute cross-section, $d\sigma_{ij}$ may be written as [26]:

$$\frac{d\sigma_{ij}}{dt} = |R(0)|^2\, \alpha_s^3(\Lambda^2,Q^2)\, |A|^2 \tag{13}$$

where R(0) is either the quarkonium wave function at the origin (S waves) or its derivative (P waves), α_s the strong running coupling constant and $|A|^2$ the sub-process matrix element squared. We have adopted the wave function values of ref. [26], [23], which were obtained from a QCD potential model analysis:

$$|R_s(0)|^2 = \frac{\Gamma_{ee}}{4\alpha^2 e_Q^2} M_{Q\bar{Q}}^2$$

$$|R_p(0)|^2 = 9.1 * 10^{-3} M_{\chi_c}^2 \qquad (14)$$

$$|R_p(0)|^2 = 1.5 * 10^{-2} M_{\chi_b}^2$$

The $O(\alpha_s^3)$ cross-sections are divergent in the forward direction, $t \rightarrow 0$, and contain an infrared divergence when the recoil jet to the onium becomes soft ($E^{jet} \rightarrow 0$). Although $O(\alpha_s^2)$ diagrams dominate the total integrated cross-section, their contribution to large p_t bound state production can savely be neglected [22]. All hadronic and leptonic decay branching ratios are summarized in ref. [22] and were mainly taken from [27].

The dominant open heavy flavour production mechanisms $gg \rightarrow Q\bar{Q}$ and $gg \rightarrow Q\bar{Q}g$ have been discussed in detail in the previous section. However, we would like to stress that since the bottom hadrons are produced at large transverse momentum at $O(\alpha_s^2)$ as well as at $O(\alpha_s^3)$, we expect that both will contribute roughly equally to the 'indirect' J/ψ (and ψ') signals. The $O(\alpha_s^4)$ contributions are again suppressed [16] and we will neglect them in further discussions. The spectator description of the weak decay of a bottom meson is depicted in figure 15.

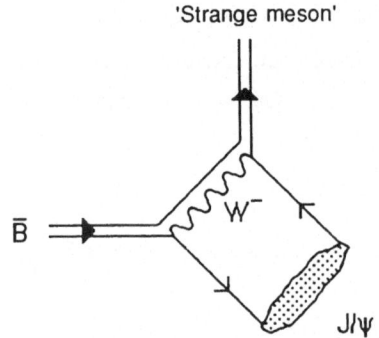

'Strange meson'

$\bar{B}$

W⁻

J/ψ

Fig. 15. Weak decay of the B meson: B → J/ψ X.

In our calculations we have used the decay branching ratio B → J/ψ ~ 1 % and leptonic decay ratio J/ψ → μ⁺μ⁻ ~ 7.4 % as measured by e⁺e⁻ experiments [27], [28]. Comparison of the J/ψ momentum spectrum (measured in the rest-frame of the B meson) with data obtained by the CLEO [29] and ARGUS [30] experiments (fig. 16) shows that our model describes the data rather well. QCD corrections to the spectator decay model hardly modify the momentum spectrum and are omitted.

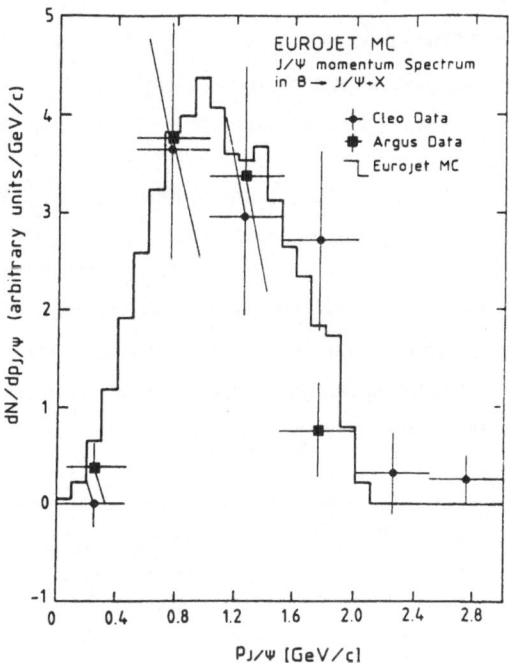

Fig. 16. *J/ψ momentum spectrum in the decay B → J/ψ X compared with CLEO [29] and ARGUS [30] data. The theoretical prediction is corrected for the Lorentz boost of the B meson produced at the Υ(4s) (p ~ 500 MeV/c).*

5. Comparison with data and extrapolations to higher beam energies

In comparing the absolute magnitude of theoretical cross-sections with the event rates observed by the UA1 collaboration, we are actually limited to a global shape analysis of various distributions. This is due to the large theoretical uncertainties in our calculations. Although the total cross-section for 'indirect' J/ψ is more or less determined by scaling the observed non-isolated dimuon rate, a number of uncertainties enter into the calculation of the 'direct' production processes. First, we have to regulate the divergencies for different kinematical regions which appear in the expressions for the matrix elements. Since we are still lacking the calculations of virtual corrections we impose a transverse momentum cut-off of 5 GeV/c on the bound state (hence, this is similar to the regularization of the open heavy flavour divergencies). By applying this cut we are satisfying both the experimental cuts and the requirement that we are well away from the divergencies. Secondly, although one expects that potential model analysis of bound states of two heavy quarks should be more reliable than similar analysis of single heavy quark mesons, the determination of the derivative of the wave function at the origin (and the wave function itself) is largely model dependent. Finally, the most important uncertainty is introduced by the unknown K-factor, Q^2 scale and QCD scale parameter Λ. For all presented calculations we have chosen the structure function parametrizations of Eichten et al. [7] with Λ = 200 MeV, whereas the scale was fixed at $Q^2 = p_t^2 + M_{Q\bar{Q}}^2$ for direct onium production and $Q^2 = p_t^2 + m_Q^2$ for the calculation of the open heavy flavour tree level amplitudes, unless stated otherwise. Although a K-factor of order 1.5 ~ 2 is by no means an unreasonable assumption, we have set K =1, obtaining a conservative cross-section estimate for the direct processes.

Tables 1a and 1b illustrate the contributions from the individual $O(\alpha_s^3)$ subprocesses to the charmonium and bottomonium cross-section, respectively, calculated at $\sqrt{s}$ = 630 GeV. Due to the large gluon flux at small x, the production of χ states is strongly dominated by the three gluon processes, whereas the $q\bar{q}$ contribution is negligible. The radiative χ_c decays form the main contribution to the J/ψ signal, whereas direct Υ production is typically of the same order of magnitude as Υ production via radiative χ_b decays. Because of their extremely small leptonic widths, η states do not contribute to the dimuon signal and are omitted.

	gg	gq	$q\bar{q}$
ψ	1.78	-	-
ψ'	0.91	-	-
ψ''	0.48	-	-
χ_0	29.6	6.81	0.002
χ_1	72.1	17.8	0.02
χ_2	43.2	11.7	0.007

Table 1a. The individual contributions from the subprocesses gg, gq and $q\bar{q}$ to $\sigma\,(p\bar{p} \to \psi,\, \chi_c X\,)$ [nb] at $\sqrt{s}$ = 630 GeV with $p_t^{onium} > 5$ GeV/c.

	gg	gq	$q\bar{q}$
Υ	1.29	-	-
Υ'	0.55	-	-
Υ''	0.34	-	-
χ_0	2.47	0.77	$4.0*10^{-5}$
χ_1	1.41	0.52	$2.0*10^{-3}$
χ_2	3.15	1.14	$1.0*10^{-3}$

Table 1b. The individual contributions from the subprocesses gg, gq and $q\bar{q}$ to $\sigma\,(p\bar{p} \to \Upsilon,\, \chi_b X\,)$ [nb] at $\sqrt{s}$ = 630 GeV with $p_t^{onium} > 5$ GeV/c.

In fig. 17a we show the p_t distributions of directly formed J/ψ, J/ψ from radiative ψ' and χ_c decays and from B meson decays at $\sqrt{s}$ = 630 GeV. The transverse momentum distribution for J/ψ from B meson decays is seen to fall off less steep than the other distributions. The rapidity distributions for the same processes are shown in fig. 17b.

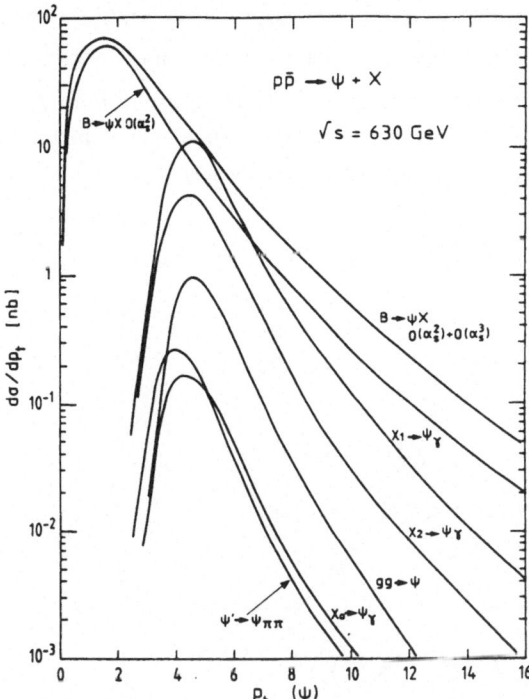

Fig. 17a. Transverse momentum distributions of the J/ψ for the so-called direct and indirect production mechanisms at √s = 630 GeV. Except for the p_t cut on the recoil jet (p_t^{jet} > 5 GeV/c) no additional cuts have been applied.

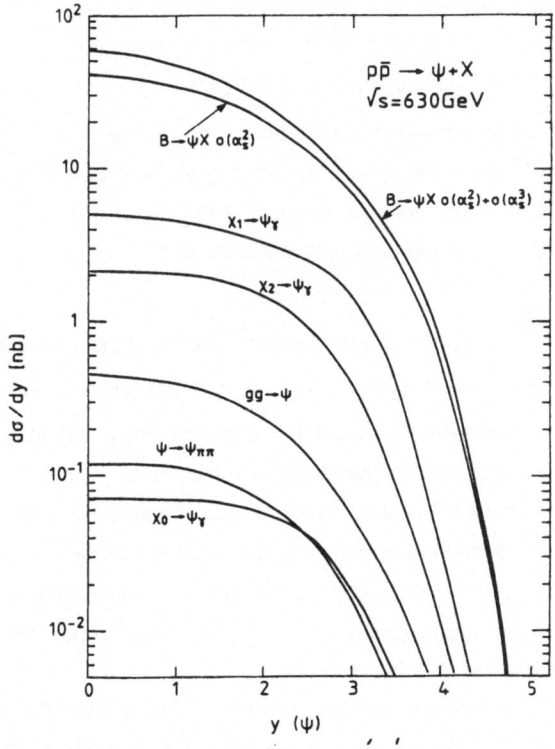

Fig. 17b. Rapidity distributions of the J/ψ for the various processes at √s = 630 GeV.

The J/ψ rapidity distribution from B decays tends to be slightly more peaked at small |y| values. Figure 18 illustrates the effect of the harder p_t spectrum of J/ψ from B hadron decays on the inclusive muon spectrum. The muon transverse momentum spectrum from direct processes becomes considerably softer. And indeed, a relaxed cut on the p_t of one of the muons in the UA1 J/ψ analysis considerably enhances the acceptance for the J/ψ signal.

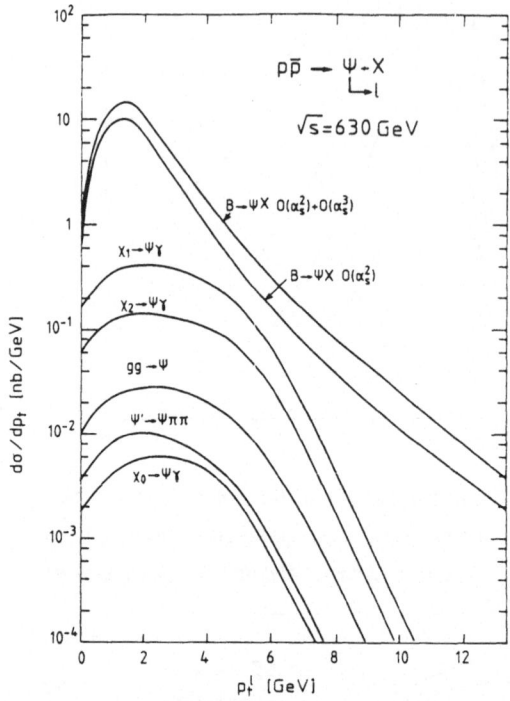

Fig. 18. *Inclusive differential muon cross-sections $d\sigma$ $(p\bar{p} \to {}^{2s+1}L_j X \to \mu^+ \mu^- X')/dp_t^\mu$ (L = S, P) and $d\sigma$ $(p\bar{p} \to b\bar{b} X \to J/\psi X' \to \mu^+ \mu^- X'')/dp_t^\mu$ at $\sqrt{s}$ = 630 GeV. Contributions from $\psi' \to \mu^+ \mu^-$ etc. are also included. The contributions from $O(\alpha_s^2)$ and $O(\alpha_s^3)$ open bottom production are shown separately.*

We expect that at relatively high p_t^μ (typically p_t^μ > 7 GeV/c), a large fraction of the muons from the J/ψ decay originate from J/ψ mesons produced in B decays. As long as the relative normalizations of the cross-sections for direct and indirect production stay within 'reasonable' limits, this may serve as a tool to separate the direct production mechanisms from the open $b\bar{b}$ channel.

To compare our predictions with the experimental results we have applied the same cuts on our simulated events as UA1 [5] has introduced in the search for the J/ψ and Υ signals via their muonic decay channels. The limited muon chamber geometry implies that only muons with |y| < 2 can be detected whereas at least one muon has to satisfy |y| < 1.3 in order to trigger the apparatus. Furthermore, a transverse momentum cut of 3 GeV/c on the muons is required. (recall that for the analysis presented in ref. [5], this cut was relaxed to allow one of the muons to have p_t > 0.75 GeV/c. In addition a cut p_t > 5 GeV/c is applied to the transverse momentum of the muon pair. Muon pairs in

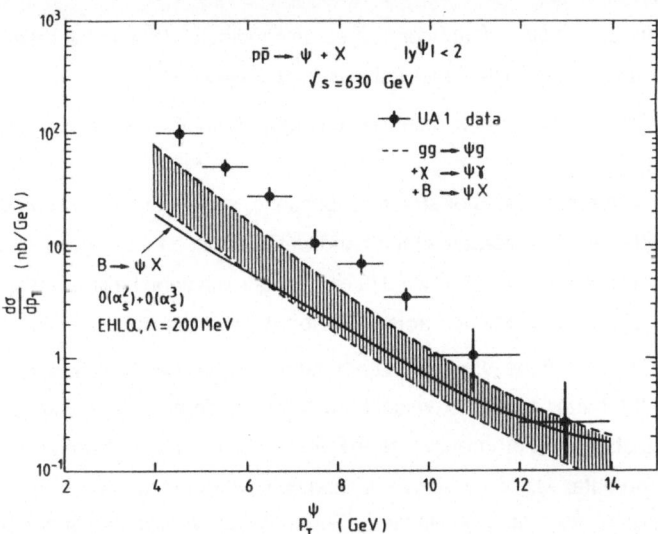

Fig. 19a. *Comparison of UA1 data with our theoretical predictions for dσ (p$\bar{p}$ → J/ψ + X)/dp$_t$. The band illustrates the uncertainties in the calculation of the direct processes due to different choices of scale and structure functions (K = 1). The drawn curve shows the O(α_s^2) + O(α_s^3) contribution from B-meson decays for a fixed set of parameters (see text).*

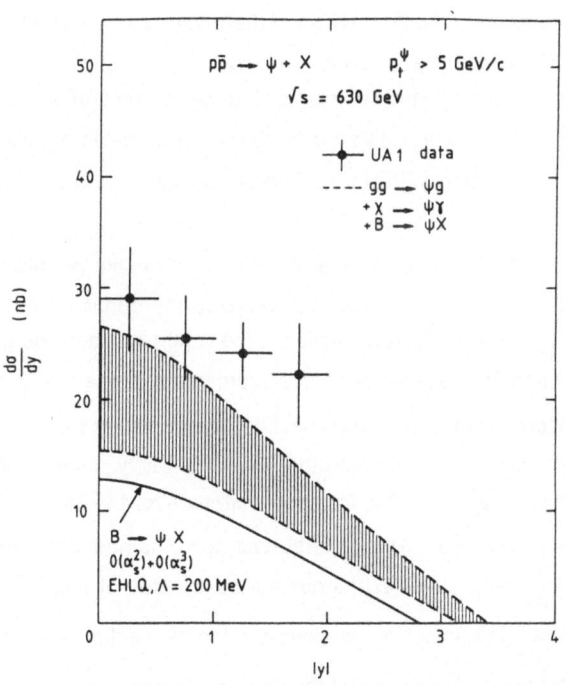

Fig. 19b. *Comparison of UA1 data and our theoretical predictions for dσ (p$\bar{p}$ → J/ψ + X)/dy. Note that p$_t^{J/\psi}$ > 5 GeV/c and we have assumed K = 1.*

243

the invariant dimuon mass range 2 - 4 GeV/c^2 are considered as candidates for the J/ψ resonance, while the mass range for Υ, Υ' and Υ'' candidates is limited to 8.7 - 10.5 GeV/c^2 .

The comparisons of the theoretical J/ψ transverse momentum and rapidity distributions with the UA1 data are presented in fig. 19.

The uncertainties due to the different choices of structure functions, scale etc. are indicated by a band between the two dashed lines. The borders of the band were obtained by using the structure function parametrizations of refs. [7], [8] and [9]. Two different scale definitions have been studied: $Q^2 = M_{Q\bar{Q}}^2$ and $Q^2 = p_t^2 + M_{Q\bar{Q}}^2$. The contributions from B-meson decays are calculated using our standard set of parameters. The sum of the $O(\alpha_s^2)$ and $O(\alpha_s^3)$ indirect processes is illustrated by the drawn curve. The borders of the bands show the typical form of the distributions if we sum over all subprocesses and hence the shape of both transverse momentum and rapidity distributions is in rather good agreement with the data. Using the above defined cuts we obtain a total cross-section of 25 - 65 nb for the production of J/ψ's at $\sqrt{s}$ = 630 GeV. The B-meson decays add about 20 - 30 nb, which is of the same order of magnitude as the contribution from direct production channels. The experimental cross-section times branching ratio: $\sigma.BR(\psi \rightarrow \mu^+ \mu^-)$ = 7.5 $\pm$ 0.7 $\pm$ 1.2 nb, quoted in ref. [5], leads to a cross-section of ~ 100 nb, which is a factor 1.5 - 2 larger than our predictions. A K-factor may account for the difference.

The decays are characterized by either the associated production of a strange particle (B $\rightarrow$ J/ψ 'S') or an associated photon (χ, $\psi' \rightarrow$ J/ψ γ). Therefore, correlations between the muons (J/ψ) and kaons may be an additional tool to separate direct and indirect mechanisms, although the associated kaons may be hard to disentangle from the strange particles originating from underlying beam fragments. A detailed discussions on hadronic isolation of the muons as well as whether a photon may be associated to the radiative decay of a bound state can be found in ref. [22].

A similar study was performed to obtain the contributions to the $^3S_1$ bottom quark bound states. The UA1 measurement: $\sigma(p\bar{p} \rightarrow \Upsilon X)$ = 0.98 $\pm$ 0.21 $\pm$ 0.19 nb [1] is, if we take into account all above discussed uncertainties, well in agreement with our theoretical calculations. We find $\sigma(p\bar{p} \rightarrow \Upsilon X)$ = 0.8 - 2 nb.

Based on the comparisons with the collider data we are able to make predictions for onium production at (future) colliders. In table 2 we have summarized the dominant direct production contributions to the J/ψ and Υ cross-sections as a function of the centre-of-mass energy. The cross-sections are obtained by integration over the transverse momentum of the resonance for p_t > 5 GeV/c. We have studied uncertainties in the cross-section determination by choosing different structure functions and scale parameters. We find that the uncertainty grows significantly with increasing beam energy. In figures 20a and 20b we show extremes for the $^3S_1$ $c\bar{c}$ and $b\bar{b}$ bound state cross-sections as a function of $\sqrt{s}$. The solid curves are obtained with the Glück et al. parametrizations [8] at $Q^2 = M_{Q\bar{Q}}^2$, while the dashed curves are a result of the Duke and Owens parametrizations with $Q^2 = p_t^2 + M_{Q\bar{Q}}^2$. Although we are aware of the momentum sum-rule violation by the D & O structure functions for Q > 10^3, we find that the Eichten et al. parametrizations lead to very similar results.

Table 2. $O(\alpha_s^3)$ contributions to the total cross-section σ ($p\bar{p} \rightarrow$ charmonium, bottomonium + X) [nb] at different centre-of-mass energies, with p_t^{onium} > 5 GeV/c.

	630 GeV.	2 TeV.	16 TeV.	40 TeV.
ψ	1.78	6.02	33.6	66.2
χ_c^0	36.4	146.	908.	$1.68*10^3$
χ_c^1	90.0	354.	$2.62*10^3$	$7.78*10^3$
χ_c^2	54.9	226.	$1.55*10^3$	$4.00*10^3$
Υ	1.29	5.69	45.1	96.7
χ_b^0	3.24	18.4	186.	419.
χ_b^1	1.93	11.2	122.	278.
χ_b^2	4.29	23.7	270.	682.

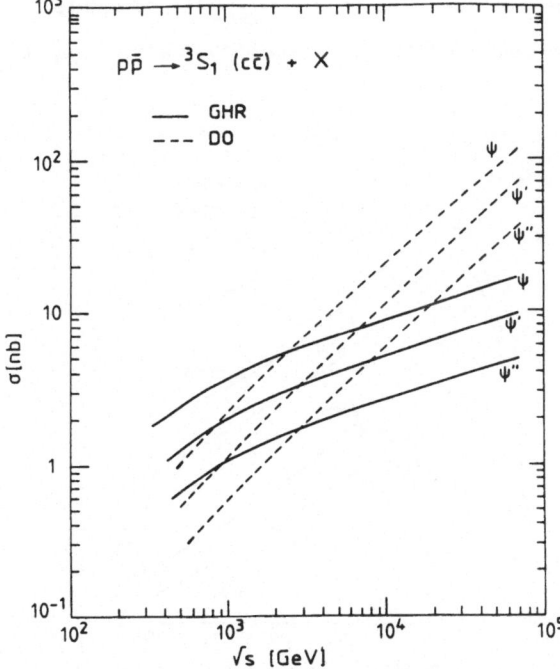

Fig. 20a. ψ cross-sections as a function of $\sqrt{s}$. The drawn curves correspond to the Glück et al. [9] structure functions with scale $Q^2 = M_{Q\bar{Q}}^2$. The dashed curves are calculated using the Duke and Owens parametrization [8] ($Q^2 = p_t^2 + M_{Q\bar{Q}}^2$).

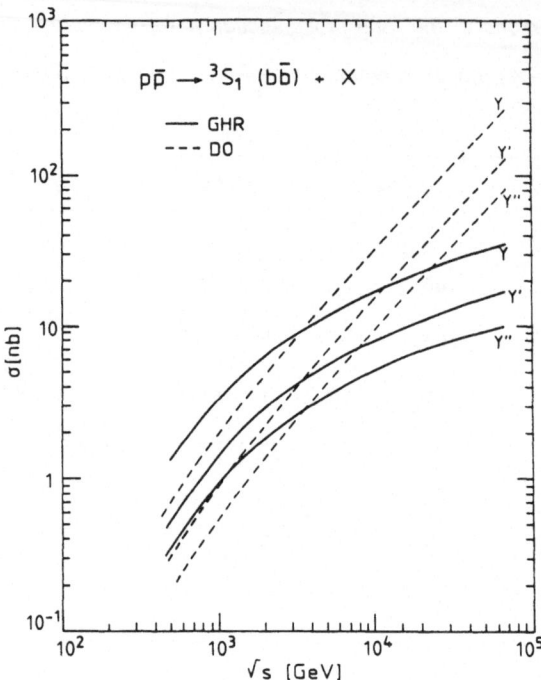

Fig. 20b. *Υ cross-sections as a function of √s. The drawn curves correspond to the Glück et al. [9] structure functions with $Q^2 = M_{Q\bar{Q}}^2$. The dashed curves are calculated using the Duke and Owens parametrization [8] with $Q^2 = p_t^2 + M_{Q\bar{Q}}^2$.*

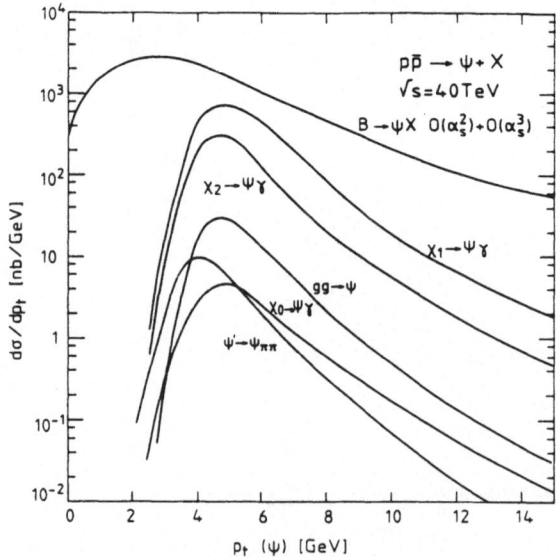

Fig. 21. *Transverse momentum distribution of the J/ψ for both direct and indirect production mechanisms at √s = 40 TeV. The various parameters entering into the calculation are defined in the text.*

At SSC energies there is almost an order of magnitude difference between the two extremes, reflecting the uncertainty on the poorly known gluon structure function at small values of x. Finally, we present in fig. 21 the J/ψ transverse momentum distribution extrapolated to SSC energies (√s = 40 TeV). Our standard choice of parameters, with K = 1, may be considered as very conservative. The p_t distributions are considerably flatter whereas the width of the rapidity distributions increase by about a factor two compared to the width of the distributions obtained at √s = 630 GeV. For the time being it remains an open question up to which centre-of-mass energies one will be able to disentangle bound state signals experimentally. However, at LHC and SSC energies, muons are believed to provide the most clean signals for existing and new physics phenomena. Therefore, known physics may provide important calibration signals.

Acknowledgements

It is a pleasure to thank Ahmed Ali and the organizing staff of the Ettore Majorana centre for inviting me to this very useful workshop. The prospects of having computing facilities at the Erice site available soon, will most certainly give an important boost to the studies proposed at this workshop. I would also like to acknowledge the help of and many stimulating discussions with Daniel Froidevaux, Ingrid ten Have, Ritva Kinnunen and my colleagues in the UA1 collaboration.

References

[1] C. Albajar et al. (UA1 Collaboration), Phys. Lett. 186B (1987) 237.

[2] C. Albajar et al. (UA1 Collaboration), Preprint CERN-EP/87-189 (1987), submitted for publication.

[3] A. Ali, B. van Eijk and I. ten Have, Nucl. Phys. B292 (1987) 1.

[4] C. Albajar et al. (UA1 Collaboration), Preprint CERN-EP/87-190 (1987).

[5] C. Albajar et al. (UA1 Collaboration), Preprint CERN-EP/87-175 (1987), submitted for publication,

M.J. Corden, private communication.

[6] A. Ali, B. van Eijk and E. Pietarinen, to be published,

B. van Eijk, PhD Thesis, University of Amsterdam, unpublished (1987).

[7] E. Eichten et al., Rev. Mod. Phys. 56 (1984) 579.

[8] D. Duke and J. Owens, Phys. Rev. D30 (1984) 49 and erratum.

[9] M. Glück, E. Hoffmann and E. Reya, Zeit. f. Phys. C13 (1982) 119.

[10] G. Arnison et al. (UA1 Collaboration), Phys. Lett. 172B (1986) 461.

[11] B.I. Combridge, Nucl. Phys. B1512 (1979) 429.

[12] Z. Kunszt and E. Pietarinen, Nucl. Phys. B164 (1980) 45.

[13] A.H. Mueller and P. Nason, Phys. Lett. 157B (1985) 226.

[14] R. Frey, Proc. of the APS Meeting of the Division of Particles and Fields, USA (1985) (World Scientific 1986).

[15]   Z. Kunszt and J.F. Gunion, Phys. Lett. 159B (1985) 167,
       Z. Kunszt and W.J. Stirling, CERN Report TH-4351/86 (1986).

[16]   B. van Eijk, NIKHEF Preprint (1988, inpreparation).

[17]   C. Albajar et al. (UA1 Collaboration), Phys. Lett. 186B (1987) 247.

[18]   D. Froidevaux, Proceedings of the Workshop on Physics at Future Accelerators, La Thuile
       and CERN, CERN 87-07 Vol. 1 (1987) 61.

[19]   G. Altarelli et al., Z. Phys. C27 (1985) 617.

[20]   B. van Eijk and D. Froidevaux, to be published in the report of the LHC study group.

[21]   E.W.N. Glover, F. Halzen, F. Herzog and A.D. Martin, Phys. Rev. D30 (1984) 700,
       E.W.N. Glover, F. Halzen and A.D. Martin, Phys. Lett. 185B (1987) 441,
       E.W.N. Glover, A.D. Martin and W.J. Stirling, Durham Preprint DTP/87/36 (1987).

[22]   B. van Eijk and R. Kinnunen, NIKHEF Preprint NIKHEF-H/87-15 (1987), submitted for
       publication.

[23]   R. Baier and R. Rückl, Phys. Lett. 102B (1981) 364, Z. Phys. C19 (1983) 251.

[24]   B. Humpert, CERN Report CERN-TH.4551/86.

[25]   R. Gastmans, W. Troost and T.T. Wu, Phys. Lett. 184B (1987) 257, Preprint-KUL-TF-87/4.

[26]   R. Baier and R. Rückl, Nucl. Phys. B208 (1982) 381.

[27]   M. Aguilar-Benitez et al. (Particle Data Group), Phys. Lett. 170B (1986).

[28]   M. Alam et al., Phys. Lett. 186B (1987) 237,
       H. Albrecht et al. (ARGUS Collaboration), Phys. Rev. D 34 (1986) 3279.

[29]   P. Haas et al. (CLEO Collaboration), Phys. Rev. Lett. 55 (1985) 1248.

[30]   H. Albrecht et al. (ARGUS Collaboration), Phys. Lett. 162B (1985) 395.

PROMPT PHOTONS IN VERY HIGH ENERGY COLLISIONS

Michel Fontannaz

Laboratoire de Physique Théorique et Hautes Energies
Université Paris-Sud, bâtiment 211
F-91405 Orsay, France

INTRODUCTION

For the last few years, hadronic collisions involving real photons turned out to be interesting reactions for testing perturbative QCD and for improving our knowledge of the proton [1,2,3,4]. The fact that the photon has a pointlike interaction with the quark makes it, indeed, a good probe of hadrons and allows precise QCD predictions. It turns out that, at present energies --13 GeV $\leqslant \sqrt{S} \leqslant$ 630 GeV--, the agreement between data and theory is spectacular, as well for photoproduction [5,6] than for prompt photon reactions [7,8]. The value of the QCD coupling constant and the gluon distribution in the proton will soon be extracted from high statistics experiments with a very good accuracy [9].

The issue I am going to discuss in this paper is the interest of Very High Energy (VHE) collisions for prompt photon physics. What shall we gain at these high energies, what will be the quality of the QCD tests, what shall we learn about the parton distributions in the proton ? It is clear that, in VHE reactions, we shall explore completely new kinematical regions, as well in p_T (the photon transverse momentum) than in $x_T = 2p_T/\sqrt{S}$. It is therefore worthwhile to examine what are the QCD predictions for this new energy domain and whether we can improve our knowledge of the proton.

Besides their standard QCD origins, prompt photons may also come from new reactions and be signals of some exotic phenomena. The QCD signal now becomes a background which must be carefully calculated in order to assess the possibility to observe the emergence of new physics.

Precise QCD tests or carefull background calculations clearly require the knowledge of the cross-sections beyond the Leading Logarithm Approximation (LLA). This point is discussed in the next section where the calculation of Higher Order (HO) corrections is briefly described. A short comparison between theory and present energy data is also presented. Section 3 deals with QCD predictions for collisions at TeV energies (2 TeV < $\sqrt{S}$ < 100 TeV). The possibility of exploring the very small x_T region (corresponding to very small x in the distribution functions) is emphasized. I discuss in section 4 the production of pairs of prompt photons. In section 5, new phenomena producing prompt photons are touched on and the QCD background considered. Section 6 contains the conclusions.

QCD PREDICTIONS FOR PROMPT PHOTON CROSS-SECTIONS*

At the lowest order in α_s, only two Born subprocesses contribute to the prompt photon cross-section, namely the Compton (Fig. 1a) and the annihilation (Fig. 1b) subprocesses. As we shall see, the Compton contribution is always the dominant one at VHE ; the prompt photon cross-section therefore gives us a direct access to the gluon distribution in the proton $G_{g/p}(x,Q^2)$. Let us also notice that the Born terms, being of order α_s, are sensitive to the value of Λ_{QCD}.

It is however well known that the LLA, in which the cross-section is given by the convolution of the Born terms with the scale breaking distributions of the partons, only leads to semi-quantitative results. The beyond LL corrections may be large and the arbitrariness in the choice of the scales appearing in the distribution functions and in α_s arises as an uncertainty in the value of the cross-section. Clearly the knowledge of beyond LL corrections is needed in order to get quantitative QCD predictions. HO diagrams (Fig. 2) were calculated during the last few years and their contributions to the prompt photon cross-sections are now available [8,10]. The single inclusive cross-section, containing the HO contributions, is written :

$$E \frac{d\sigma^{(2)}}{d\vec{p}} = \sum_{a,b} \int dx_a\, dx_b\, G_{a/p}(x_a,M) G_{b/p}(x_b,M) \{\frac{1}{\pi} \frac{d\hat{\sigma}}{d\hat{t}} (ab \rightarrow \gamma c)\delta(s+t+u)$$

$$+ \theta(s+t+u) \frac{1}{s^2} K_{ab}(\frac{t}{s}, \frac{u}{s}, \frac{M^2}{s}, \frac{\mu^2}{s})\} \qquad (1)$$

where $d\hat{\sigma}/d\hat{t}$ is the Born cross-section and K_{ab} the HO contribution coming from the subprocess $a+b \rightarrow \gamma+X$; s , t and u are the usual Mandelstam variables of the subprocess. We rewrite this expression in a more compact form

$$\sigma^{(2)}(M,\mu) = \alpha_s(\mu)\, \sigma^{BORN}(M) + \alpha_s^2(\mu)\, \sigma^{HO}(M,\mu) \qquad (2)$$

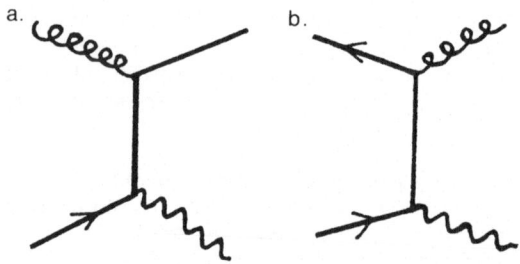

Fig. 1. The Compton (a) and annihilation (b) subprocesses

---

* The predictions presented in this paper are based on work done in collaboration with P. Aurenche, R. Baier and D. Schiff.

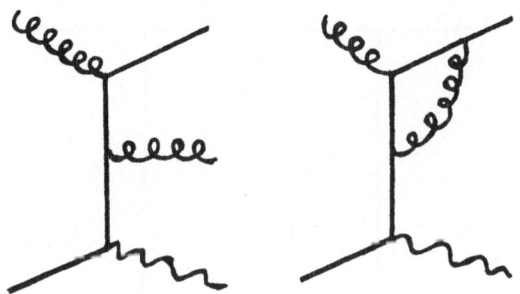

Fig. 2. Examples of HO diagrams

which makes more explicit the dependence on the strong coupling constant α_s and on the scales M and μ related to the factorization and renormalization procedures [11].

Renormalization group arguments tell us that M and μ are to be of the order of the large energy scale of the reaction, namely p_T, but they do not fix the exact relation between M, μ and p_T. Therefore there is still an ambiguity, in the second order QCD predictions, depending on the precise choice for M and μ. We are not going to discuss the various criteria allowing to determine an "optimal" definition of M and μ, and refer the interested reader to more detailed papers devoted to the subject [4,6,7, 8,11]. Here we content ourselves with the "standard" choice $M = \mu = c\, p_T$ where $1/2 \leq c \leq 2$.

Let us now consider numerically the HO corrections. Choosing the scales $M = \mu = p_T$, we show in Fig. 3 the ratio of the HO corrections to

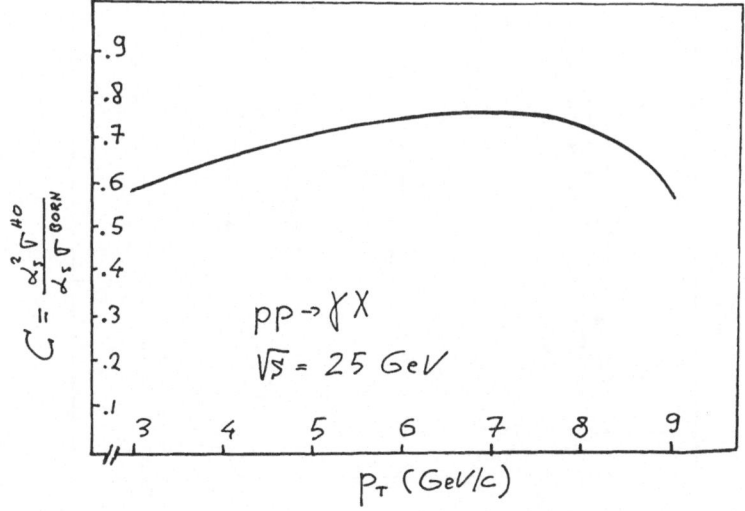

Fig. 3. The ratio of the HO terms to the Born cross-section

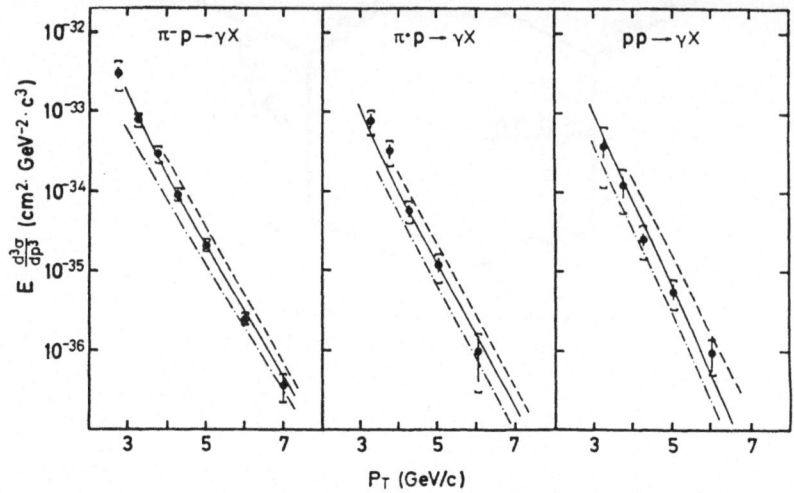

Fig. 4. Comparison between the NA24 data [12] and the theoretical
predictions (see text)

the Born cross-section $C = \alpha_s^2(\mu)\ \sigma^{HO}/\alpha_s(\mu)\ \sigma^{BORN}$ at the typical SPS
energy $\sqrt{S} = 25$ GeV. As we can see, the HO corrections are not negligible ;
clearly a quantitative comparison with experiment is only possible if such
HO terms are included in the calculations. Such a comparison is shown in
Fig. 4 for a CERN-SPS experiment [12]. The continuous curve is obtained
with the distribution functions of Duke and Owens, set 1 [13] (soft gluon
and $\Lambda_{QCD} = 200$ MeV), whereas the dashed curve is calculated with the set 2
of the DO distributions (hard gluon and $\Lambda_{QCD} = 400$ MeV). For these two
predictions, the Stevenson optimization procedure [14,11] is used to

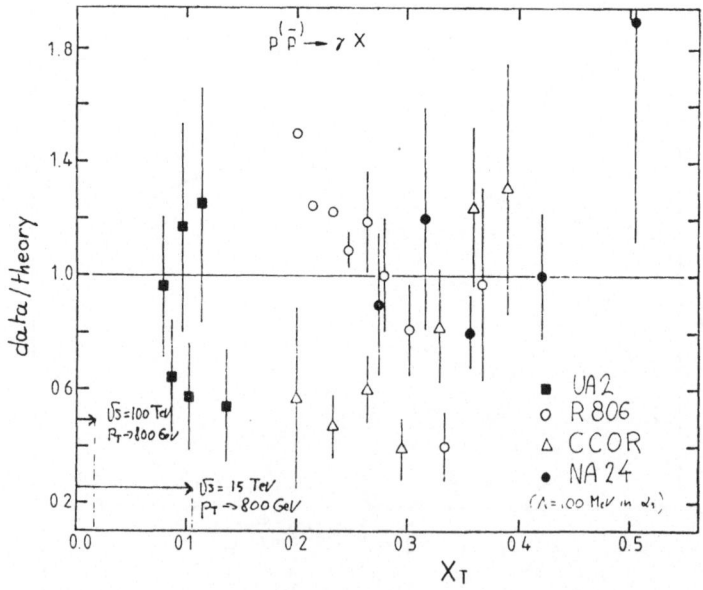

Fig. 5. Comparison between prompt photon data and theory

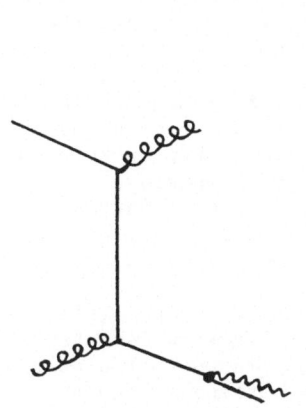

Fig. 6. Example of a bremsstrahlung graph
(anomalous component)

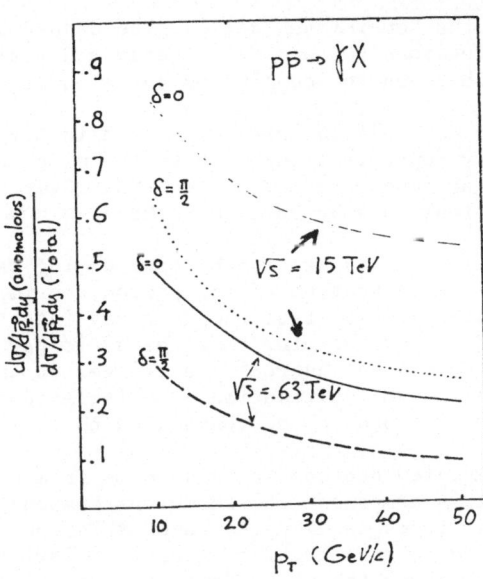

Fig. 7. Effect of the isolation
criterion

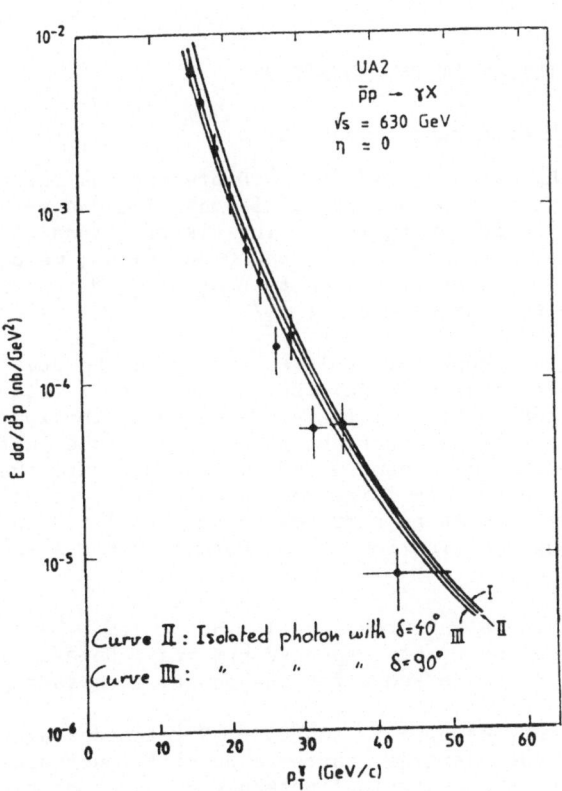

Fig. 8. UA2 data on prompt photon
production

determine the "optimized" scales. The dashed-dotted line corresponds to the distributions DO1 and to the standard scales $M = \mu = 2\ p_T$. The agreement between the continuous curve and the data is quite good.

Other data [15,16] are summarized in Fig. 5 which shows the ratio data/theory (optimized theory with the distribution DO1 and $\Lambda_{QCD} = 100$ MeV). Again the agreement is quite good [8]. (Over the explored x_T range, the cross-sections vary by more than three orders of magnitude).

Let us end this section by considering UA2 data [16] and a problem related to the detection of the photon. An important background to the prompt γ signal comes from the π^0 decay into two photons. This background can be reduced by noticing that the π^0 is part of a jet, therefore accompanied by other hadrons ; a trigger on an isolated photon partly suppresses the π^0 contribution. The isolated photon is defined by a cone centered on the photon and containing no other particle.

Accompanied photons may also come from bremsstrahlung graphs (Fig. 6). (We call this component the anomalous component). A trigger on isolated photons decreases the importance of this contribution. In our calculations, we took into account the requirement of isolation by putting a cone of aperture δ around the photon and excluding all configurations with a quark within the cone. The results for this trigger are shown in Fig. 7 ; the anomalous contribution which may be quite large at small x_T is considerably reduced. This fact is welcome, because the anomalous component is calculated within the LLA and a too large contribution would spoil the quantitative QCD predictions based on the Born and HO terms. The effect of the cone is also shown in Fig. 8 which displays UA2 data together with theoretical predictions [8].

PREDICTIONS FOR COLLISIONS AT TEV ENERGIES

Inclusive Spectrum in pp and p$\bar{\text{p}}$ Collisions

We discuss in this section [17] the QCD predictions for the single photon inclusive spectra in pp and p$\bar{\text{p}}$ collisions. The cross-sections (1) are calculated with the following input. The distributions of quark and gluon in the proton are taken from Duke and Owens [13] ; we use their set 1 corresponding to a soft gluon and to $\Lambda_{QCD} = 200$ MeV. For the scales M and μ, we stick to the choice $M = \mu = p_T/2$.

All predictions are done for isolated photons ; the aperture of the cone around the photon being $\delta = \pi/2$. The effect of this requirement at $\sqrt{S} = 15$ TeV may be seen in Fig. 7. The anomalous contribution becomes small enough to be treated as a correction as soon as the photon transverse momentum is larger than 30 GeV/c. Without cone, the anomalous contribution would be extremely large and the QCD predictions no more quantitative. For instance at $p_T = 30$ GeV/c and with $\delta = 0$, we get a ratio of the anomalous contribution to the Born + HO cross-section of about 1.5.

We shall often assume an integrated luminosity of $\mathcal{L} = 10^{40}/cm^2/year$ when quoting the number of events expected per year. Unless explicitly specified, the p_T spectrum is shown for the center of mass rapidity $y = 0$.

The p_T spectra for the pp $\to \gamma X$ and p$\bar{\text{p}} \to \gamma X$ reactions are shown in Figs. 9 and 10. At these energies, there is no difference between pp and p$\bar{\text{p}}$ collisions. The cross-section increases by a factor ten at $p_T = 30$ GeV/c when the CM energy varies from 0.63 TeV to 15 TeV. For this latter energy $3.5 \cdot 10^6$ events/GeV/year are expected at $p_T = 30$ GeV/c ; this number decreases to 1 event at $p_T = 1$ TeV/c.

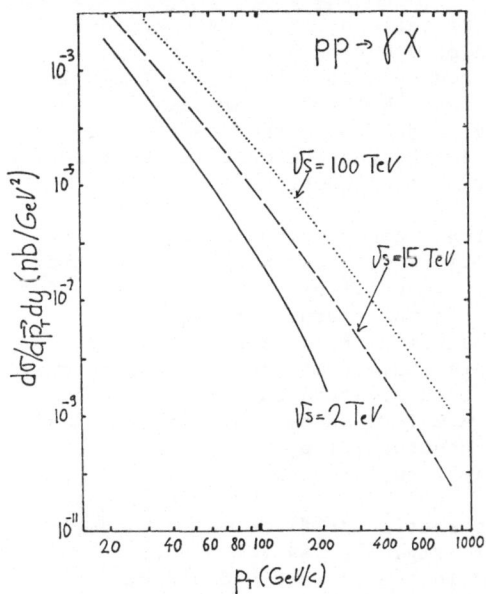

Fig. 9. The p$_T$-spectra of the
pp → γX reaction

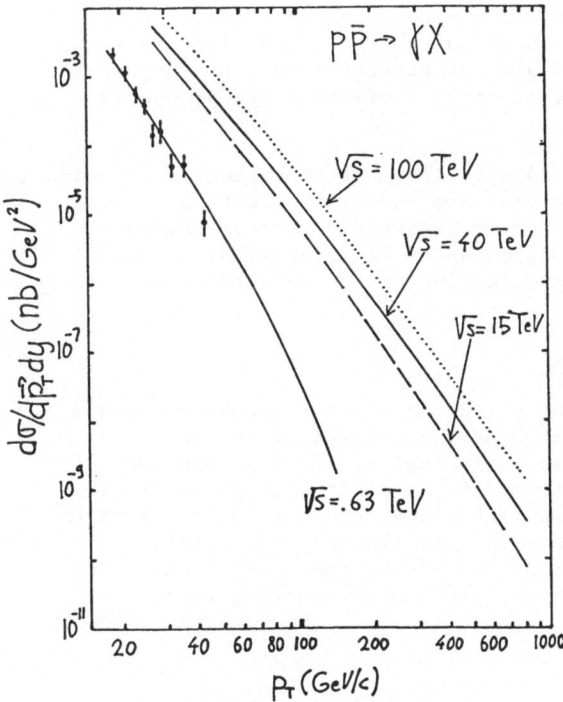

Fig. 10. The p$_T$-spectra of the
reaction pp̄ → γX. The
experimental points are from
the UA2 collaboration [16]

It is interesting to notice that the ratio of the cross-section $\sigma(pp \to \gamma X)$ to that for a hadronic jet $\sigma(pp \to$ hadronic jet+X) is smaller than 10^{-3}. For instance at $\sqrt{S} = 10$ TeV and $p_T = .6$ TeV/c, this ratio is equal to $0.5 \cdot 10^{-3}$. This comes from the fact that the subprocess initial states are quite different in jet and γ production ; the reaction gluon+gluon $\to$ a+b is very important in jet production whereas the reaction gg $\to q\bar{q}\gamma$ is just a correction in prompt photon reactions.

We verify in Fig. 11 that the dominant contribution comes from the Compton subprocess (Fig. 1a). In pp collisions, the annihilation contribution is negligible, even at very large value of p_T. We therefore have a direct access to the gluon distribution in the proton through the measurement of the prompt photon cross-section. Fig. 12 shows that the HO and anomalous ($\delta = \pi/2$) contributions are small compared to the Born one, with the exception of the phase space boundary where the cross-section becomes negligible (for instance at $\sqrt{S} = 2$ TeV and $p_T \simeq 0.8$ TeV/c). All contributions are therefore well under control for $p_T \geq 30$ GeV/c. The same pattern is observable at $\sqrt{S} = 40$ TeV and 100 TeV.

The width in rapidity of the cross-section, at fixed value of p_T, is large at $p_T \simeq 100$ GeV/c and $\sqrt{S} \geq 10$ TeV (Fig. 13). As y = 3 corresponds to 5° in the CM, many photons are expected in the forward and backward direction.

Associated Production

We are not going to quantitatively discuss the production of large p_T hadrons associated with a prompt γ, but let us evoke some qualitative features. First of all, due to the dominance of the Compton contribution (Fig. 1a), the recoiling jet is essentially a quark jet ; moreover the contribution of the quarks of charge 2/3 are enhanced by a factor 4 with respect to the contribution of the quarks of charge $- 1/3$. Therefore the contributions involving charm quarks are substantial ; for instance at $\sqrt{S} = 40$ TeV and $20 \leq p_T \leq 100$ GeV/c, about 25 % of the large p_T events contain charm quarks.

The subprocesses shown in Fig. 14 give the most important contributions to the cross-section and their signatures are one or two leptons, coming from the semi-leptonic decays of the charm quarks, in various region of the phase space. For instance the subprocess of Fig. 14c leads to one large p_T jet containing one photon and one lepton, associated to one recoiling jet containing one lepton.

Exploration of Very Small x_T Regions

In expression (1), at y = 0, the values of the fractional momenta x_1 and x_2 for which the integrand is large are controlled by that of $x_T = 2p_T/\sqrt{S}$ and we have $x_1 \simeq x_2 \simeq x_T$. This means that at small p_T and VHE, we essentially explore the small x-behavior of the distribution functions. This behavior may easily be estimated within the double leading logarithm approximation (DLLA) [18]. It consists in resumming all the leading contributions of the type $(\alpha_s \text{ Log } Q^2/\Lambda^2 \text{ Log } 1/x)^n$, coming from the gluon ladders of Fig. 15. The result*, for the gluon distribution, is :

---

* The fractional momentum $\bar{x}$ is that of the last gluon of the ladder with respect to that of the first one.

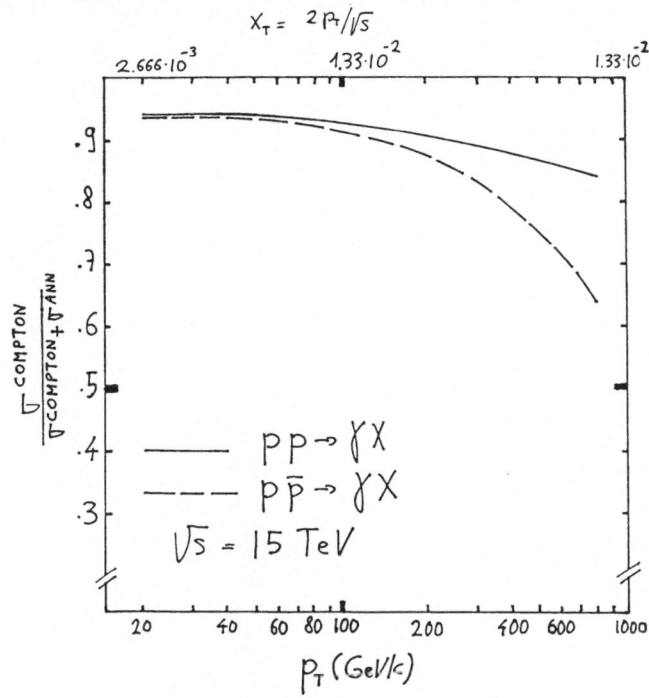

Fig. 11. Ratios of Born cross-sections

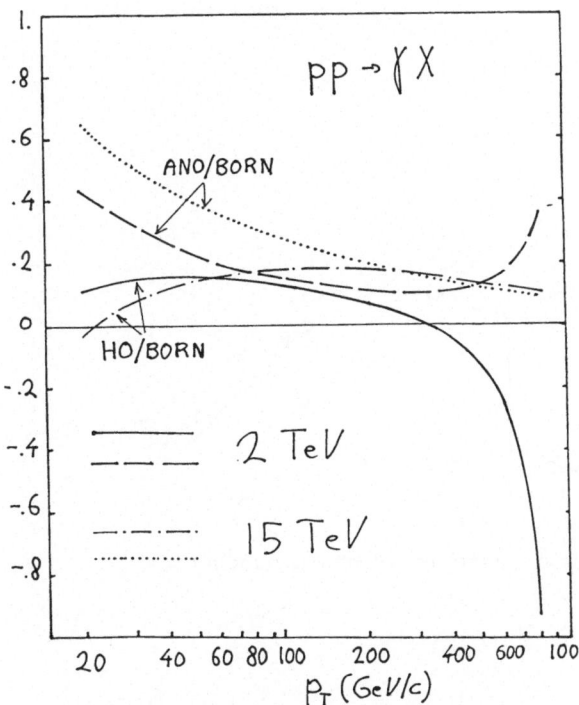

Fig. 12. Ratio of the HO and anomalous
contributions to the Born term

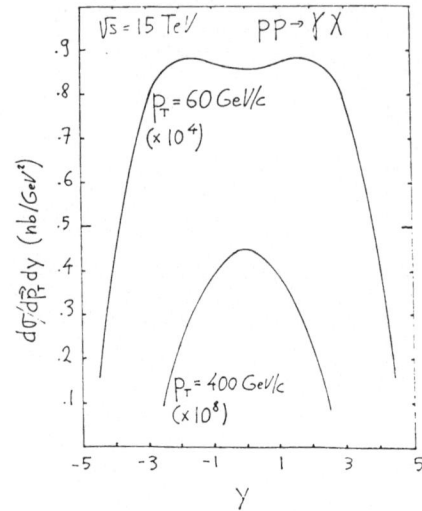

Fig. 13. Rapidity distributions of
the prompt photon cross-
sections

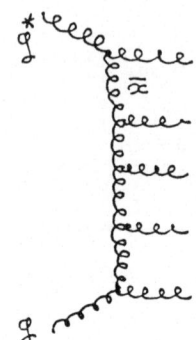

Fig. 15. Gluon ladder in
the g* (virtual
gluon)-g deep
inelastic collision

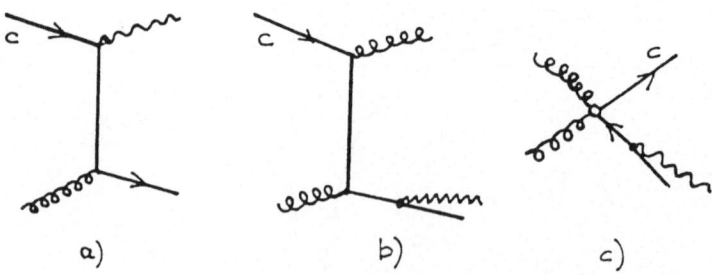

Fig. 14. Graphs leading to associated charm production

$$\bar{F}(\bar{x}, Q^2) = \bar{x} \; G_g(\bar{x}, Q^2) \sim e^{\sqrt{2\xi\bar{y}}}$$

$$\xi = \ell n \; \ell n \; Q^2/\Lambda^2$$

$$\bar{y} = \frac{8N_c}{b} \; \ell n \; \frac{1}{\bar{x}} \simeq 3 \; \ell n \; \frac{1}{\bar{x}} \quad , \quad b = \frac{11N_c - 2N_F}{3} \tag{3}$$

We immediately notice from (3) that $\bar{F}(\bar{x}, Q^2)$ violates the Froissart bound ; in a deep inelastic scattering experiment indeed, $\ell n \; 1/\bar{x} \sim \ell n \; S/Q^2$ and the cross-section increases with S faster than any power of ℓn S.

This phenomenon has been carefully discussed by Gribov, Levin and Ryskin [18]. Let us reproduce here one simple intuitive argument. The transverse deep inelastic scattering cross-section (we may imagine the deep inelastic scattering of a virtual gluon on a real gluon) is given in term of $F(\bar{x}, Q^2)$ by

$$\sigma_T \sim \frac{\alpha_s}{Q^2} \; \bar{F}(\bar{x}, Q^2) \tag{4}$$

The upper bound of this cross-section is determined by the radius R of the target and we must have

$$\sigma_T \leq R^2 \tag{5}$$

Relation (5) means that $F(\bar{x}, Q^2)$ cannot increase indefinitely and we get the upper bound

$$\bar{F}(\bar{x}, Q^2) \leq \frac{Q^2 R^2}{\alpha_s(Q^2)} \quad . \tag{6}$$

Using (3), we may estimate the validity domain of the DLLA

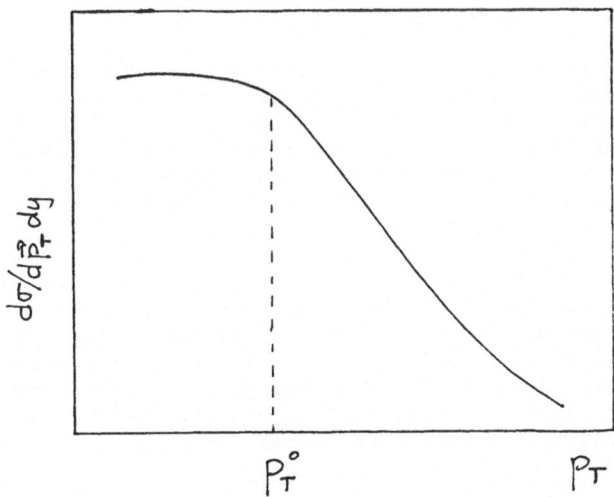

Fig. 16. Behavior of the prompt photon inclusive
cross-section at very small p_T

$$\ln \frac{1}{x} << \ln^2 Q^2 R^2 \tag{7}$$

Gribov, Levin and Ryskin have derived a more precise bound :

$$\ln \frac{1}{\bar{x}} < \frac{.21b}{8N_c} \ln^2 Q^2/\Lambda^2 \tag{8}$$

We therefore expect, for $\bar{x}$ smaller than the bound (8), the apparition of new phenomena, related to the unitarity constraints. Similar studies on the validity of the LLA and DLLA have also been done by Kwiecinski [19], and Mueller and Qiu [20].

Let us now come back to the prompt photon cross-section which may be written*

$$\frac{d\sigma}{d\vec{p}_T\,dy} \sim \frac{\alpha\,\alpha_s}{p_T^4} F_g(x,p_T^2)\,F_q(x,p_T^2) \tag{9}$$

At very small x, we expect from (6) $F(x,p_T^2) \sim p_T^2 R^2$ and therefore a complete change in the p_T behavior of the cross-section (9) for $p_T \leq p_T^0$, as illustrated in Fig. 16. We may roughly estimate the value of p_T^0 from (8), remembering that $\bar{x}$ is the fractional momentum of the last gluon in the ladder with respect to the first gluon. We therefore estimate x, the fractional momentum with respect to the proton, to be $x \sim \bar{x}/10$ $(x \simeq x_T)$ and find, at $\sqrt{S} = 10$ TeV, $p_T^0 \sim 5$ GeV/c.

Another point is that the input gluon distribution, at a small reference value Q_0^2, is even not known in the small x-region above the lower bound (8) where the unitarity constraint (5) is not yet important $(x \sim 10^{-3})$. The usual assumption, based on the Regge model, is that of a constant behavior of F_g at small x : $F_g(x,Q_0^2) \sim$ cte. But there are arguments [21] in favor of a more singular behavior in x :

$$F_g(x,Q_0^2) \sim 1/\sqrt{x} \tag{10}$$

It is clear that such a behavior would have strong consequences on the prompt photon cross-section at low p_T and very high $\sqrt{S}$.

For all the reasons given above, it is interesting to explore the small p_T regions. The corresponding QCD predictions are shown in Figs. 17 and 18. The anomalous and HO contributions are under control for $p_T \geq$ 10 GeV/c. These blow up at smaller p_T and more work is needed to understand this behavior. These predictions are based on the calculations described in section 2 and do not involve any of the very low x improvements discussed in ref. [18].

The problem of the observation of prompt photons at small p_T is a serious one, due to the very large background coming from π^0's. The ratio of isolated photon versus π^0 cross-sections is shown in Fig. 19. At $p_T = 25$ GeV/c, we have $\gamma_{ISO}/\pi^0 \simeq 0.5 \cdot 10^{-2}$; a small number compared to 0.1 observed in the UA2 experiment. At small values of p_T, a better trigger could be a low mass lepton pair originating from an almost real prompt photon ; the π^0 background should then become innocuous.

---

* F_g/x and F_q/x are the gluon and sea-quark distributions in the proton. We assume here that F_g and F_q have the same behavior at very small x.

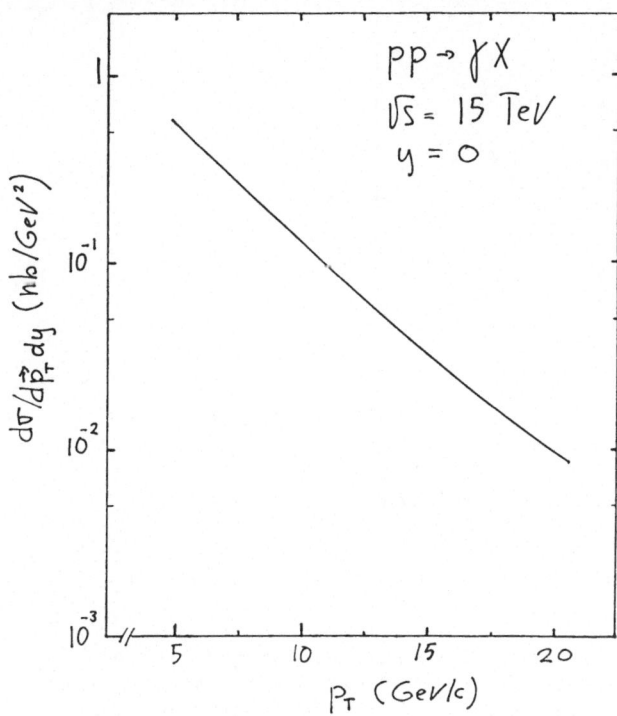

Fig.17. The pp → γX cross-section at very low
value of p_T

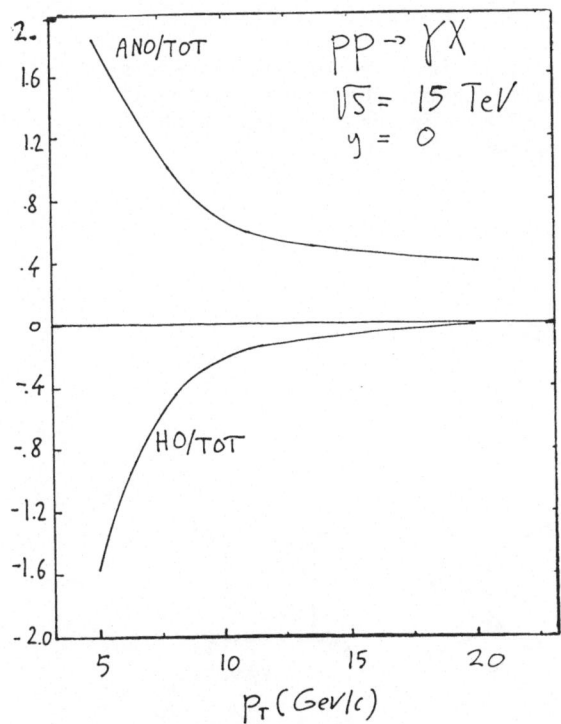

Fig. 18. The anomalous and HO
contributions at very low p_T

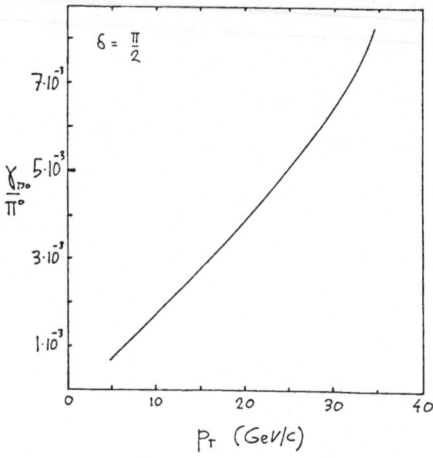

Fig. 19. The ratio of isolated
photons over π^0.

PRODUCTION OF TWO LARGE-P_T PHOTONS

Two photon reactions enable us to explore further the quark-gluon dynamics and the verification of perturbative QCD [22]. Among other things, the gluon-gluon → photon-photon cross-section should be observable. In hadronic collisions, two photons are produced through the Born reaction $q\bar{q} \rightarrow \gamma\gamma$ (Fig. 20a), the HO correction diagrams (Fig. 20b,c) and through the gg → $\gamma\gamma$ subprocess --the box contribution-- (Fig. 20d). When one photon is emitted collinearly by a quark, we get a bremsstrahlung contribution that we call anomalous contribution [22] (Fig. 21).

We describe the two photon reaction in terms of the transverse momentum p_{T_1} and rapidity y_1 of the trigger ; the second photon is described (Fig. 22) by the variable $Z = - \vec{p}_{T_1} \cdot \vec{p}_{T_2} / |\vec{p}_{T_1}|^2$. We shall consider either the integrated spectrum :

$$\frac{d\sigma}{d\vec{p}_{T_1} \, dy_1} = \int_{Z_{min}} dZ \, \frac{d\sigma}{d\vec{p}_{T_1} \, dy_1 \, dZ} \tag{11}$$

with $Z_{min} = p'_{T_{2}min}/p_{T_1}$, or the differential spectrum

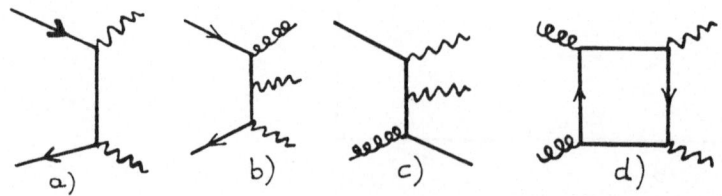

Fig. 20. Various contributions to the two-photon reaction
(see text)

262

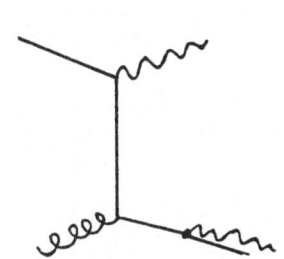

Fig. 21. The anomalous
contribution

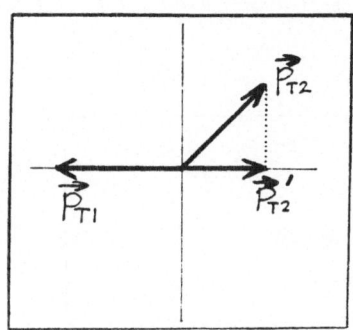

Fig. 22. The photon momenta
in the transverse
plane

$$\frac{d\sigma}{d\vec{p}_{T_1} \, dy_1 \, dZ} = \frac{1}{\Delta Z} \int_{Z - \frac{\Delta Z}{2}}^{Z + \frac{\Delta Z}{2}} dZ \ \frac{d\sigma}{d\vec{p}_{T_1} \, dy_1 \, dZ} \qquad (12)$$

This latter definition is used to regularize the differential cross-section which is a distribution in Z. Note that there is no cone around the photons.

At present energy, the agreement between theory and data is good [4]. It turns out that the full cross-section is twice as large as the Born cross-section. For higher energies, the integrated cross-sections are displayed in Fig. 23 (for p'_{T_2min} = 10 GeV/c). Here the Born cross-section is smaller by a factor ten than the fully corrected cross-section. The details of the various contributions are shown in Fig. 24. At small values of p_{T_1}, the box contribution is large ; the anomalous contribution is huge and increases with p_{T_1}.

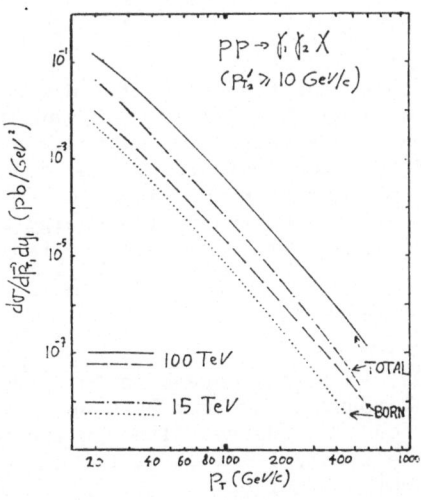

Fig. 23. The γ-γ cross-sections as
a function of p_{T_1}

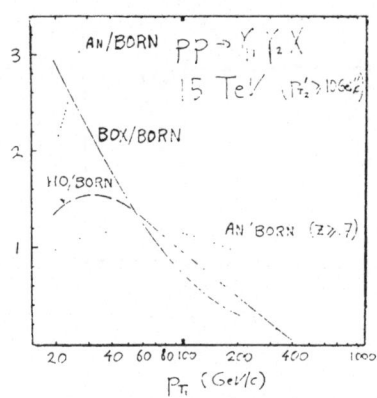

Fig. 24. The different contributions to the γ-γ
cross-section

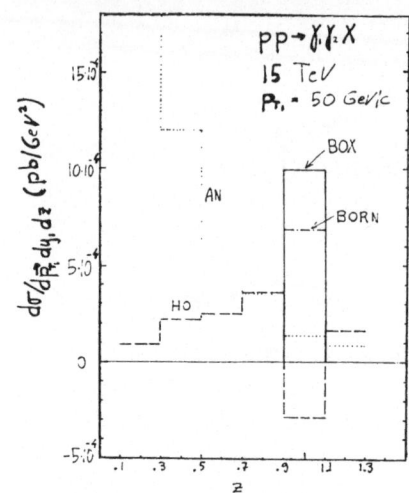

Fig. 25. The various contributions
to the differential
spectrum

The differential spectrum in Z is shown in Fig. 25. We see that the anomalous contribution is very large at small Z, which explains the behavior of this term in Fig. 24. In order to decrease this large contribution, a lower bound in Z_{min} independent of p_{T1} can be taken in the integral (11) ; the effect of such a lower bound can be seen in Fig. 24. The anomalous contribution is now well under control.

The cross-sections are not negligible. With p_{T1} = 50 GeV/c and p'_{T2min} = 10 GeV/c, 3000 events/GeV/year are expected. At p_{T1} = 300 GeV/c, this number decreases to 3 events. At small value of p_{T1} ($p_{T1} \leq$ 50 GeV/c), the box contribution becomes the most important one and could be observable. Notice however that this prediction strongly depends on the assumption on the gluon distribution in the proton at small x.

BACKGROUND TO NEW PHYSICS

Prompt photons may also be the signature of a new or exotic phenomenon [17]. I am going to study a few examples and to estimate the corresponding QCD background. This latter, as we shall see, is quite severe and it will be hard to disentangle the signal from the background by just looking at the inclusive photon spectrum. In few cases, the study of the event topology, of the missing transverse energy and of the associated production of leptons will be valuable to isolate the new physics event.

Radiative Decay of Toponia

The toponium state ${}^1S_0(\eta_T)$ mainly decays in two gluons, producing two large p_T jets. The QCD background is however huge and there is no hope to observe the η_T through its hadronic decay. Although the branching ratio for the decay in two photons is small ($\sim 10^{-3}$), the QCD background becomes comparable to the signal, which might be observed in this channel. Borrowing from Hikasa [23] the values of the η_T production cross-section --$\sigma(pp \to \eta_T x) \simeq$ 1 nb with $\sqrt{S}$ = 15 TeV and an η_T mass M = 100 GeV-- and the branching ratio $Br(\eta_T \rightarrow \gamma\gamma) \simeq 3.10^{-3}$, I consider the transverse mass differential cross-section

$$\frac{d\sigma^{\gamma\gamma}}{dM_T} = \sigma(pp \rightarrow \eta_T X).Br.\frac{M_T}{M}\frac{1}{\sqrt{M^2 - M_T^2}} \tag{13}$$

where $M_T = p_{T_1} + p_{T_2}$. This cross-section exhibits the usual Jacobian peak shown in Fig. 26. Assuming a resolution $\Delta M_T = 10$ GeV, I get

$$\sigma^{\gamma\gamma} = \int_{M - \Delta M_T}^{M} dM_T \; \frac{d\sigma^{\gamma\gamma}}{dM_T} \simeq \sigma.Br \sqrt{\frac{2\Delta M_T}{M}} \simeq 1 \text{ pb} \tag{14}$$

The QCD background may be directly estimated from Fig. 25 with the result ($\Delta y = 3$) :

$$\sigma_{QCD}^{\gamma\gamma} \simeq 1 \text{ pb} \tag{15}$$

We see that for this reaction, the signal is comparable to the background and the possibility to detect it essentially depends on the value of the resolution ΔM_T.

Associated Production of Wγ and Zγ

The rate for Wγ production is sensitive to the magnetic moment of the intermediate boson and that for the Zγ production may be a test of compositeness [24]. The measurement of these reactions may therefore give valuable informations on gauge boson physics. The calculation of the corresponding cross-sections is summarized in the paper by Eichten et al. [24], from which I take the result $\sigma(W^{\pm}\gamma) \simeq 1$ pb at $\sqrt{S} = 15$ TeV and $|y_{W,\gamma}| < 1.5$. This number is obtained in assuming that the Wγ invariant mass M is larger than 200 GeV. This condition may be translated in a bound on the photon transverse momentum. With the rough approximation $M_T \sim M$, I get $p_T\gamma_{min} \simeq 80$ GeV. From this value and Fig. 9, the one photon inclusive QCD cross-section may be calculated

$$\sigma_{QCD}^{\gamma} = \int_{80} d\vec{p}_T \; dy \; \frac{d\sigma}{d\vec{p}_T \; dy} \simeq 1 \text{ nb} \quad . \tag{16}$$

This number clearly shows that the signal cannot be seen in the one photon spectrum.

Missing transverse energy (for the $W \rightarrow \ell\nu$ channel) or photon-jet-jet topology (for the $W \rightarrow q\bar{q}$ channel) could enable a better extraction of the signal from the background. In the hadronic channel, a QCD background coming from the HO graphs of Fig. 2 may be important.

Excited Quarks

Excited quarks might be produced in the quark-gluon reaction $q+g \rightarrow q^*$ [25]. Here again the hadronic channel $q^* \rightarrow q+g$ is not suitable for the observation of the signal, due to the large QCD background. A better channel is $q^* \rightarrow \gamma+q$ where the signal is competing with the background [25].

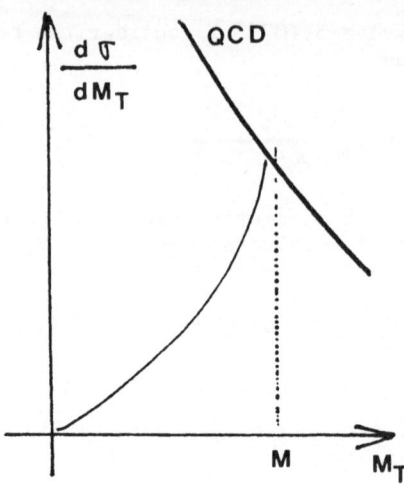

Fig. 26. The transverse mass $\gamma\gamma$ cross-
section

Down Quark of the Fourth Generation

When the mass of a heavy vector meson is larger than that of the Higgs boson, the radiative decay $V(Q_d\bar{Q}_d) \rightarrow H_o + \gamma$ may occur with a non negligible branching ratio [26]. Let us consider down quarks of the fourth generation Q_d and a vector meson mass $M_{Q\bar{Q}} = 300$ GeV. We find in ref. [23] the results $\sigma^V \sim 1$ pb for the production cross-section ($\sqrt{S} = 15$ TeV) and Br $\sim .1$ for the branching ratio $Br(V \rightarrow H_o + \gamma)$. The photon spectrum has a peak in the variable p_T ; assuming a resolution $\Delta p_T = 10$ GeV, we get for the cross-section (with $M_H = 100$ GeV)

$$\sigma^\gamma \simeq \sigma^V . Br. \sqrt{\frac{4p_T}{M^*}} \sim .04 \text{ pb} \tag{17}$$

where $M^* = (M_{QQ}^2 - M_H^2)/M_{QQ}$. For the corresponding QCD cross-section, we get from Fig. 9 $\sigma^\gamma_{QCD} \sim 50$ pb ($\Delta y = 3$).

Here again, the signal is completely hidden in the QCD one photon spectrum. If the Higgs boson can decay in a $t\bar{t}$ pair, the final state $\gamma t\bar{t}$ could be a better signature ; the QCD background would come from the HO graphs $gg \rightarrow t\bar{t}\gamma$.

Two Photon Decay of the Higgs Boson

The two photon decay of the Higgs boson is not completely negligible if the mass of this latter is below the WW on $t\bar{t}$ threshold. Assuming $\Delta M_T = 10$ GeV, we get for the two photon cross-section from expression (14) (with $M_H = 100$ GeV)

$$\sigma^{\gamma\gamma} \sim .05 \text{ pb} \quad ; \tag{18}$$

we have used $\sigma^H \sim .1$ nb [24] and $Br(H \rightarrow \gamma\gamma) \sim 10^{-3}$ [27]. The corresponding QCD background is

$$\sigma_{QCD}^{\gamma\gamma} \sim 1 \text{ pb} \qquad (\Delta y = 3) \quad . \tag{19}$$

The ratio signal/background is not very encouraging. It turns out however that it increases with the Higgs mass [28] and becomes more favorable.

CONCLUSION

In this paper, we studied the QCD predictions concerning the hadronic production of prompt photons at very high center of mass energy and at large transverse momentum. Our calculations include the anomalous (bremsstrahlung) contribution and the HO corrections. It turns out that, for p_T large enough ($p_T \geq 30$ GeV at $\sqrt{S} = 15$ TeV), the QCD predictions are well under control ; the ratios of the anomalous or HO terms versus the Born cross-section are small.

The one photon inclusive cross-section is dominated by the Compton subprocess $qg \rightarrow q\gamma$. The measurement of the prompt γ spectrum will therefore give valuable indications on the gluon distribution in the proton. The prompt photon yield is not negligible. With a luminosity $\mathcal{L} = 10^{40}$ cm^2/year, we expect $3.5.10^6$ events/GeV/year/Δy at $p_T = 30$ GeV/c. This yield decreases to 1 event for $p_T = 1$ TeV/c.

The exploration of the very small $x_T = 2p_T/\sqrt{S}$ region ($x_T \sim 10^{-3}$) should enable us to observe new behaviors, due to unitarity constraints, of the prompt photon p_T-spectrum. A better understanding of the gluon distribution at very small x is also expected from such measurements.

We also studied the production of a pair of photons at large p_T. Here again the QCD predictions are well under control. An interesting contribution to the cross-section is the $gg \rightarrow \gamma\gamma$ scattering (box diagram) which becomes observable at very high energy.

The last part of this paper is devoted to the study of new phenomena which could appear through the production of one or two prompt photons. An important example is that of the two photon decay of the Higgs boson. The QCD backgrounds are estimated and often found to be severe.

ACKNOWLEDGEMENTS

I would like to thank P. Aurenche, R. Baier and D. Schiff for valuable discussions.

REFERENCES

1. For recent reviews, see : G. Altarelli, Proceedings of the International EPS Conference on High Energy Physics, Bari 1985 ; P. Aurenche, VII International Workshop on Photon-Photon Collisions, Paris 1986 ; M. Fontannaz, in "The Quark Structure of Matter", World Sc. Publ. 1986, M. Jacob and K. Winter editors.

2. Prompt photon reactions are discussed in : T. Ferbel and W.R. Molzon, Rev. Mod. Phys. 56, 181 (1984) ; E.L. Berger, E. Braaten, R.D. Field, Nucl. Phys. B239, 52 (1984) ; J.F. Owens, preprint FSU-HEP 86/1021.

3. D. Treille, Proceedings of the International EPS Conference on High Energy Physics, Bari 1985.

4. M. Fontannaz, Proceedings of "Rencontres de Moriond" 1987, Tran Thanh Van editor.

5. NA14 collaboration, E. Augé et al., Phys. Lett. 168B, 163 (1986) and Phys. Lett. 174B, 458 (1986).

6. P. Aurenche, R. Baier, A. Douiri, M. Fontannaz and D. Schiff, Nucl. Phys. B286, 553 (1987).

7. R. Baier, XVII International Symposium on Multiparticle Dynamics, Seewinkel 1986. P. Aurenche, invited talk at the 1987 APS/DPF meeting, Salt Lake City, January 1987.

8. P. Aurenche, R. Baier, M. Fontannaz and D. Schiff, preprint LPTHE Orsay 87/30.

9. F. Richard, review talk at the 1987 International Symposium on Lepton and Photon Interactions, Hamburg, July 1987.

10. P. Aurenche et al., Phys. Lett. B140, 87 (1984).

11. P. Aurenche et al., Nucl. Phys. B286, 509 (1987).

12. NA24 collaboration, C. de Marzo et al., Phys. Rev. D36, 16 (1987).

13. D.W. Duke and J.F. Owens, Phys. Rev. D30, 49 (1984).

14. P.M. Stevenson, Phys. Rev. D23, 2916 (1981).

15. CCOR coll., A.L.S. Angelis et al., Phys. Lett. 94B, 106 (1980) and 98B, 115 (1981). R806 coll., E. Anasontzis et al., Z. Phys. C13, 277 (1982).

16. UA2 collaboration, Phys. Lett. B176, 239 (1986).

17. Prompt photon production at VHE has also been considered in the joint paper presented by A. Ali at the CERN-ECFA workshop, March 1984 ; CERN report 84-10, September 1984.

18. L.V. Gribov, E.M. Levin and M.G. Ryskin, Phys. Reports 100.

19. J. Kwiecinski, Zeitsch. für Phys. 29, 147, 561 (1986).

20. A.H. Mueller and J. Qiu, Nucl. Phys. B268, 427 (1986).

21. T. Jarosziewicz, Phys. Lett. 116B, 291 (1982). J.C. Collins, talk at the SSC workshop, UCLA, 1986.

22. P. Aurenche et al., Zeitsch. für Phys. C29, 459 (1985).

23. K. Hikasa, these proceedings. V. Barger et al., MAD/PH/300 preprint, January 1987.

24. E. Eichten et al., Rev. Mod. Phys. 56, 579 (1984), err. : Rev. Mod. Phys. 58, 1065 (1986).

25. A. De Rujula et al., CERN preprint TH 3779 (1983).

26. F. Wilczek, Phys. Rev. Lett. 39, 1304 (1977).

27. J. Ellis et al., Nucl. Phys. B106, 292 (1976).

28. R.K. Ellis, these proceedings.

DRELL-YAN PROCESSES IN HIGH ENERGY HADRON COLLISIONS

W. J. Stirling

Physics Department
Durham University
Durham, England

ABSTRACT

A short review of Drell-Yan processes at very high energy hadron colliders is presented. Standard W, Z cross-sections are discussed, with particular emphasis on the collider centre-of-mass energy dependence. New heavy gauge boson production is considered in a model independent way.

1. INTRODUCTION

The production of lepton pairs with large invariant mass in high energy hadron collisions provided one of the first and most significant tests of the parton model. Such lepton pairs are produced by quark-antiquark annihilation – the Drell-Yan mechanism. In the early years, the Drell-Yan process provided straightforward tests of the parton model i.e. that "hard scattering" cross-sections could be calculated quantitatively by convoluting parton cross-sections with universal structure functions parametrising the momentum distribution of partons in hadrons. With the evolution of the naive parton model into Quantum Chromodynamics, the Drell-Yan process continued to provide a variety of quantitative tests – higher order perturbative corrections ("K-Factors"), nuclear target dependence, Sudakov form factors for lepton pair production at small transverse momentum, scaling violations etc. Several important theorems concerning the separation of hard and soft components in hadronic cross-sections were established first for the Drell-Yan process.

In the original Drell-Yan mechanism the quark and antiquark annihilate to give a virtual photon. In present day proton-antiproton colliders, however, there is sufficient centre-of-mass energy for the production of on-shell W and Z bosons. It seems appropriate therefore when considering the next generation of hadron colliders to extend the definition of a Drell-Yan process to include the production of any massive colour-neutral state by quark-antiquark annihilation. (For simplicity, the production of heavy quarkonia states $q\bar{q} \to \psi_Q$ will not be discussed further here.) In practice, this means the Standard Model (SM) processes $q\bar{q} \to \gamma^*, W, Z$ at present energies with the possibility of producing, in addition, new heavy W- and Z-like gauge bosons in the future.

This, then, will be the subject of this short review. In the next section the basic cross-sections will be defined and the effective luminosity functions for quark-antiquark annihilation at future hadron colliders will be compared. Section 3 will review standard W and Z cross-sections at very high energies. There has been much discussion recently on physics beyond the Standard Model, in particular that which derives from superstring theories, in which additional $U(1)$ gauge symmetries survive at relatively low energies. The production of the corresponding Z' gauge bosons will be considered in Section 4 in a model independent way. Some conclusions are presented in Section 5.

2. CROSS-SECTIONS AND LUMINOSITY FUNCTIONS

In the QCD-improved parton model, the cross-section for the production process depicted in Fig. 1 is given by

$$\sigma^{AB \to X + \cdots} = \int dx_1 \, dx_2 \, [G_{q/A}(x_1, Q) \, G_{\bar{q}/B}(x_2, Q) \, \hat{\sigma}^{q\bar{q} \to X} \; + \; (q \leftrightarrow \bar{q})\,] \quad (1)$$

Here X is a large mass final state as defined in the previous section. For the case when X is a heavy, narrow resonance ($\Gamma_X \ll M_X$), or more generally a narrow mass window of a continuum mass distribution, the subprocess cross-section is simply

$$\hat{\sigma}^{q\bar{q} \to X} = C_q \, \delta(x_1 x_2 s - M_X^2) \quad (2)$$

Then eqn.(1) becomes, with $\tau = M_X^2/s$,

$$\sigma^{AB \to X + \cdots} = \frac{C_q}{s} \frac{\partial L_{q\bar{q}}}{\partial \tau}$$

$$\frac{\partial L_{q\bar{q}}}{\partial \tau} = \int dx_1 \, dx_2 \, \delta(x_1 x_2 - \tau) \, [G_{q/A}(x_1, Q) \, G_{\bar{q}/B}(x_2, Q) + \; (q \leftrightarrow \bar{q})\,] \quad (3)$$

The dimensionless function $\partial L_{q\bar{q}}/\partial \tau$ can be regarded as the effective quark-antiquark luminosity, representing the probability that $q\bar{q}$ annihilation will take place at centre-of-mass energy $\sqrt{\tau s}$. It is a very useful way to compare the cross-sections for producing states of different mass at different energies. More generally, the three most important luminosity functions for new physics at future colliders are

(i) $\partial L_{gg}/\partial \tau$ (for $gg \to H^0, \chi_Q, \eta_Q, \ldots$),

(ii) $\partial L_{q\bar{q}}/\partial \tau$ (for $q\bar{q} \to Z', W', \psi_Q, \ldots$)

(iii) $\partial L_{W_L W_L}/\partial \tau$ (for $W_L W_L \to H^0$).

The latter is obtained by convoluting quark distributions with the effective longitudinal W splitting function $q \to W_L q$. Fig. 2 shows these three luminosity functions (with the factor $1/s$ included for cross-section comparisons) as functions of $\sqrt{\hat{s}} = M_X$ for proton-proton collisions at the SSC energy (40 TeV). The luminosity functions are computed the using the EHLQ set 1 distributions [1]. Note that the $q\bar{q}$ luminosity function is defined as a straightforward sum over all quark flavours – this is appropriate for couplings C_q which are approximately flavour independent. For masses which are likely to be accessible, there is a very definite hierarchy $gg > q\bar{q} >$

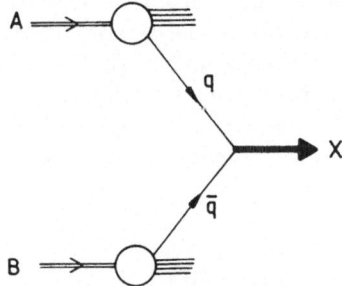

Fig. 1 . The Drell-Yan process for the production of a massive final state X in hadron-hadron collisions.

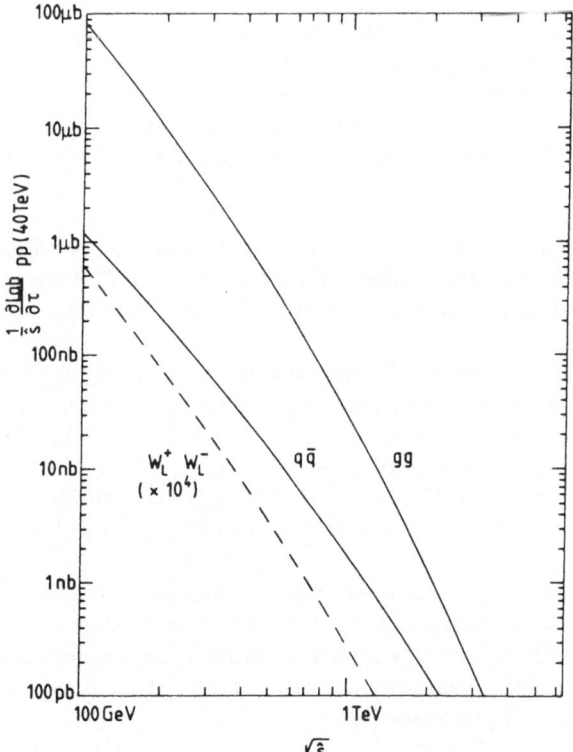

Fig. 2 . Luminosity functions as defined in the text for gg, $q\bar{q}$ and $W_L^+ W_L^-$ initial states in pp collisions at $\sqrt{s} = 40\ TeV$.

$W_L W_L$ at each of the three energies. It is also interesting to compare the *ratios* of the luminosity functions at different collision energies. These are shown in Fig. 3, where proton-proton collisions at the ELOISATRON (100 TeV), SSC (40 TeV) and LHC (17 TeV) are compared, with 40 TeV taken as the reference energy. It is noticeable that the advantage in going, for example, from 40 TeV to 100 TeV is least for $q\bar{q}$ induced Drell-Yan processes. The ratios of the gg and $W_L W_L$ luminosities rise much faster as the mass increases. Remember, however, that to take advantage of the rapid increase in these luminosity ratios there must be sufficient *absolute* rate for high mass production. Finally, note that for masses up to $O(1\ TeV)$, Drell-Yan cross-sections increase by a factor between 2 and 4 in going from LHC to SSC and from SSC to ELOISATRON energies.

3. W, Z PRODUCTION AT HIGH ENERGIES

W and Z production cross-sections as functions of collider energy are shown in Fig. 4. The solid (dashed) curves are for pp ($p\bar{p}$) colliders respectively. The next-to-leading order QCD corrections are included, but not the branching ratios. The data points are derived from UA1 and UA2 measurements at 630 GeV at the CERN $p\bar{p}$ collider [2,3]. Several comments can be made:

(i) At high energies ($\sqrt{s} > O(10\ TeV)$) there is no longer any advantage for a $p\bar{p}$ collider – valence-valence $q\bar{q}$ annihilation becomes negligible.

(ii) With a typical one year's integrated luminosity of $10^{40}\ cm^{-2} = 10^7\ nb^{-1}$ the next generation of pp colliders will produce of order 10^9 W and Z bosons per year.

(iii) The W/Z ratio varies very little with the collision energy. For example, the ratio is 3.1 for $p\bar{p}$ collisions at 630 GeV and 3.2 for pp collisions at 100 TeV with this particular set of parton distribution functions (EHLQ1).

(iv) The most effective detection will again use the leptonic decay modes. The only significant difference is that the decay leptons will have a much broader rapidity distribution at very high collision energy. This is illustrated in Fig. 5 where the electron rapidity distributions in $Z \rightarrow e^+ e^-$ production at 630 GeV ($p\bar{p}$) and 40 TeV (pp) are compared. A requirement of $|\eta_e| < 3$ for both final state leptons gives an acceptance of 97% for the former and 48% for the latter.

(v) The $W, Z \rightarrow 2\ jets$ decay mode will be even harder to detect at high energies than at present. The reason is that the 2 jet QCD background becomes more dominant as the collision energy increases and smaller x values are probed. This can be understood by considering the ratio of two jet production from W decay and from QCD $2 \rightarrow 2$ processes [4]:

$$\frac{\sigma^{W \rightarrow 2\ jets}}{\sigma^{jj}(|M_{jj} - M_W| < \Delta)} = \frac{6.1\ GeV}{\Delta} \cdot \left(\frac{\sigma^{u\bar{d} \rightarrow u\bar{d}}}{\sigma^{2 \rightarrow 2}}\right)_{\sqrt{\hat{s}} = M_W} \tag{4}$$

The second term on the right-hand-side is $O(10^{-2})$ for $p\bar{p}$ collisions at 630 GeV, but falls by an order of magnitude to $O(10^{-3})$ for pp collisions at LHC, SSC and ELOISATRON energies. the reason being the increasing dominance of two jet final states induced by gluon-gluon scattering as smaller x values are probed.

272

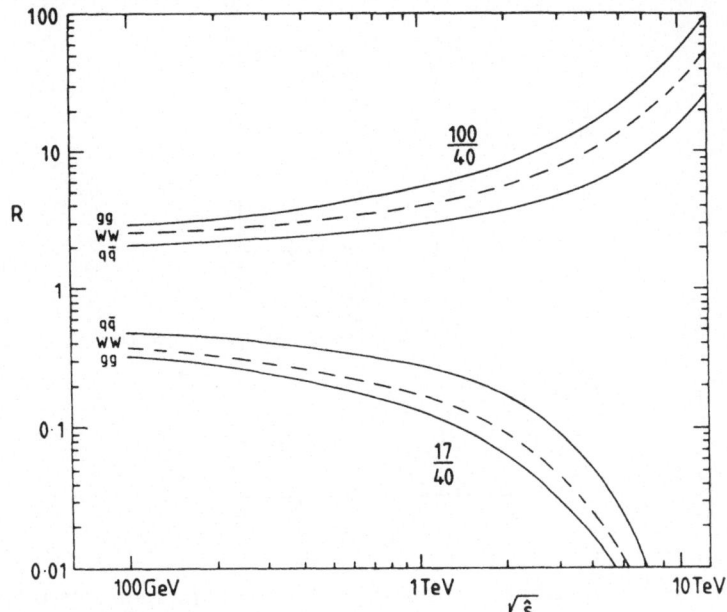

Fig. 3 . Ratios of the luminosity functions of Fig. 2 at three different pp collider energies, as a function of $\sqrt{\hat{s}}$.

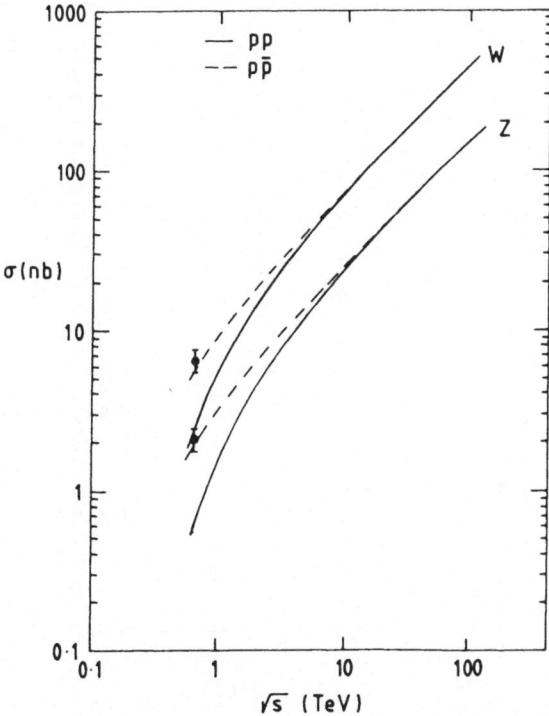

Fig. 4 . W and Z production cross-sections in pp (solid lines) and $p\bar{p}$ collisions (dashed lines) as a function of $\sqrt{s}$. The data points are from the UA1 [2] and UA2 [3] experiments at the CERN $p\bar{p}$ collider.

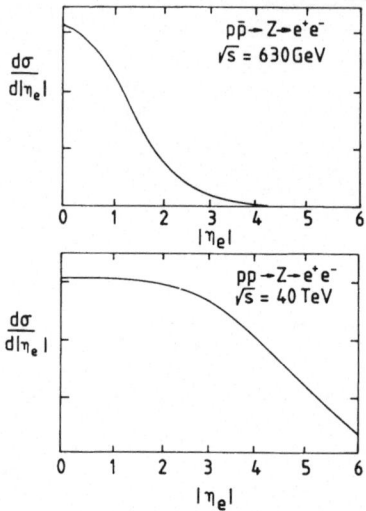

Fig. 5 . Electron rapidity distributions from Z production in $p\bar{p}$ collisions at 630 GeV (upper figure) and pp collisions at 40 TeV (lower figure).

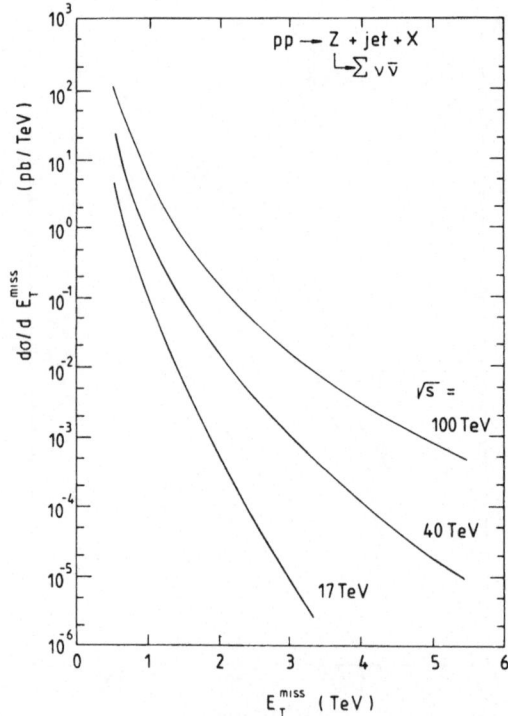

Fig. 6 . Missing transverse energy distribution from the process $pp \rightarrow Z(\rightarrow \Sigma\nu\bar{\nu}) + X$ at three different collider energies, computed in leading order QCD.

(vi) Such large numbers of W and Z events offers great potential for searching for "rare decays"which may signal new physics, e.g. $W \to \tilde{W}\tilde{\gamma}$.

It is already well-known that W and Z bosons produced with large transverse momentum are an important background for new physics searches. It is crucial, therefore, to be able to make as accurate predictions as possible for the production rates. As an example, Fig. 6 shows the "monojet"cross-section from the Standard Model process $pp \to Z(\to \Sigma\nu\bar{\nu}) + jet + ...$ as a function of the missing transverse energy at three collider energies. It is clear that the monojet rates are sizeable even at large E_T^{miss}. The curves are computed from the leading order parton subprocesses $qg \to qZ$ and $q\bar{q} \to gZ$, with the EHLQ set 1 parton distributions [1]. It is the former subprocess which dominates at these high collision energies, in contrast to the situation at present day $p\bar{p}$ colliders. Note also that at large values of the electroweak boson transverse momentum ($\sim O(100\ GeV)$) the rates for W, Z and photon production are comparable, and in an approximately constant ratio determined by the electroweak couplings:

$$
\begin{aligned}
\gamma \ : \ Z \ : \ W \ &= \ e_q^2 \ : \ \frac{v_q^2 + a_q^2}{16x_W(1-x_W)} \ : \ \frac{1}{4x_W} \\
&= \ 1 \ : \ 0.91 \ : \ 2.45 \quad (q = u) \\
&= \ 1 \ : \ 4.70 \ : \ 9.78 \quad (q = d)
\end{aligned}
\tag{5}
$$

for $x_W = \sin^2\theta_W = 0.23$.

4. NEW GAUGE BOSON PRODUCTION

In many extensions of the Standard Model (superstrings, left-right symmetric models, ...) new gauge bosons are predicted [5]. For example, in many superstring models the low energy gauge group is enlarged by an extra low energy $U(1)$ symmetry: $G \to G \times U(1)$, with an associated Z-like boson with mass $M_{Z'} > O(100\ GeV)$. Current lower mass limits on such bosons from the CERN $p\bar{p}$ collider are of order 200 GeV [6]. Rather than attempt a complete review of all possible models and predictions, we shall simply make some general model-independent statements which are appropriate for very high energy proton colliders. If we suppose that a new heavy weak Z' boson couples to light fermions in the usual way, then we may add an extra term to the standard model Lagrangian:

$$
\Delta L = \sum_\alpha \bar{\psi}_\alpha \gamma_\mu (v'_\alpha + a'_\alpha \gamma_5)\psi_\alpha Z'^\mu
\tag{6}
$$

The characteristic signature for detection at a hadron collider will again come from the process $q\bar{q} \to Z' \to l^+l^-$. The general form for the lepton centre-of-mass angular distribution, including the contributions from γ and Z exchange, is then

$$
\begin{aligned}
\frac{d\hat{\sigma}^{q\bar{q} \to l^+l^-}}{d\cos\theta} = \frac{\hat{s}}{384\pi} &\left[(1 + \cos\theta)^2 \left(\left| \sum_j \frac{(v_q^j + a_q^j)(v_l^j + a_l^j)}{D_j} \right|^2 \right. \right. \\
&\left. + \left| \sum_j \frac{(v_q^j - a_q^j)(v_l^j - a_l^j)}{D_j} \right|^2 \right) + (1 - \cos\theta)^2 \left(\left| \sum_j \frac{(v_q^j - a_q^j)(v_l^j + a_l^j)}{D_j} \right|^2 \right. \\
&\left. \left. + \left| \sum_j \frac{(v_q^j + a_q^j)(v_l^j - a_l^j)}{D_j} \right|^2 \right) \right]
\end{aligned}
\tag{7}
$$

where $j = 1 = \gamma$, $j = 2 = Z$, $j = 3 = Z'$, ... and the boson propagator is $D_j = \hat{s} - M_j^2 + iM_j\Gamma_j$.

If, for definiteness, we assume that $M_3 = M_{Z'} = 1\ TeV$ and that $v_f^3 = v_f^2, a_f^3 = a_f^2$, then we obtain the electron pair mass spectrum for $pp \to e^+e^- + X$ at $\sqrt{s} = 40\ TeV$ shown in Fig. 7. As expected, the Z and Z' peaks are clearly visible over the Drell-Yan background. The precise shapes of the tails of the resonance peaks are determined by the interference between the amplitudes for the different vector bosons and are therefore sensitive to the signs and magnitudes of the vector and axial couplings of the Z'. It is straightforward to rescale the Z' contribution for different fermionic couplings. In general the total Z' cross-section is obtained, in the narrow width approximation, from the simple formula given in eqn.(1) with $C_q = (v_q'^2 + a_q'^2)\pi/3$. Fig. 8 shows the production cross-sections for a Z' boson with the same couplings to light fermions as the standard Z at three different pp collider energies as a function of the mass $M_{Z'}$.

Turning to possible Z' decay modes, the leptonic channels will once again provide the cleanest signature. Fig. 9 compares the total Z' cross-section at 17 TeV with the QCD 2-jet background, integrated over a mass window ($M_{Z'} \pm 5\ GeV/c^2$). Although the background falls off slightly faster than the signal with increasing mass, there is little hope of detecting a new heavy boson in the hadronic decay channel. In addition to more exotic final states which may be accessible (e.g. slepton pairs etc.), an important decay mode (for $M_{Z'} > 2M_W$) may be $Z' \to W^+W^-$, given that such a coupling exists for the standard Z. In fact if there is a $Z'WW$ vertex with the same structure as the ZWW vertex, but with a different coupling constant $g' = g_{Z'WW}$, then the partial decay widths into the $f\bar{f}$ and W^+W^- final states are [7]

$$\Gamma^{f\bar{f}} = \frac{M_{Z'}}{12\pi}((v_f')^2 + (a_f')^2)$$

$$\Gamma^{W^+W^-} = \frac{M_{Z'}}{192\pi}g'^2 x^{-2}(1-4x)^{3/2}(1+20x+12x^2)$$

$$\sim g'^2 M_{Z'}^3, \qquad \text{for } M_{Z'} \gg M_W \qquad (8)$$

where $x = M_W^2/M_{Z'}^2$. Naively, then, the W^+W^- decay mode becomes more and more dominant as the mass of the Z' becomes larger, i.e. for *fixed* coupling g', $BR(Z' \to W^+W^-) \to 1$ as $M_{Z'} \to \infty$. However it is important to note that in most extended gauge models the coupling actually decreases as the mass increases, i.e. $g' \sim 1/M_{Z'}$. In this case the fermionic and W^+W^- decay channels scale asymptotically with $M_{Z'}$ in the same way. To give a specific numerical example, if

$$v_f' = v_f, a_f' = a_f, \qquad \frac{g_{Z'WW}^2}{g_{ZWW}^2} = 1/10, \qquad M_{Z'} = 300\ GeV/c^2, \qquad (9)$$

then

$$\Gamma_{Z'} = 14\ GeV$$

$$BR(Z' \to WW) = 33\%$$

$$BR(Z' \to \sum f\bar{f}) = 67\% \qquad (10)$$

If there *is* a substantial branching ratio for $Z' \to W^+W^-$ then one must address the question as to whether there is any possibility for detection in this channel. At this

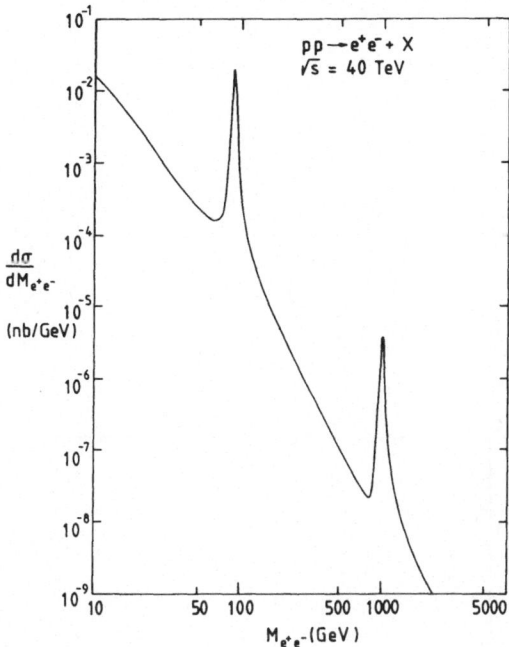

Fig. 7 . Electron pair mass distribution in pp collisions at $\sqrt{s} = 40\ TeV$, in a model with a heavy Z' boson of mass $1\ TeV$ which has the same couplings to light fermions as the standard Z.

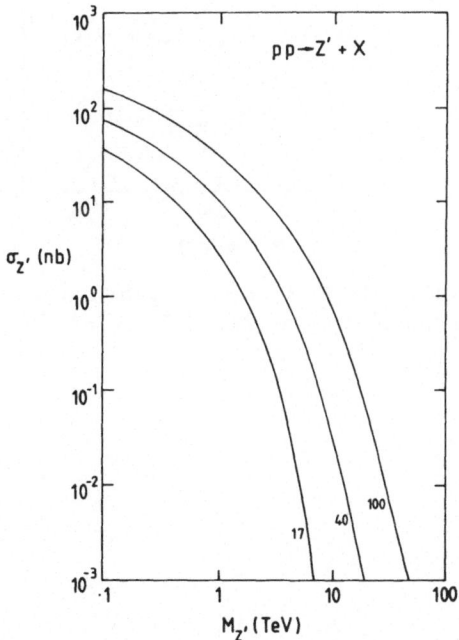

Fig. 8 . Production cross-sections as a function of mass for a heavy Z' gauge boson in pp collisions at three different energies, assuming the same couplings to light fermions as the standard Z.

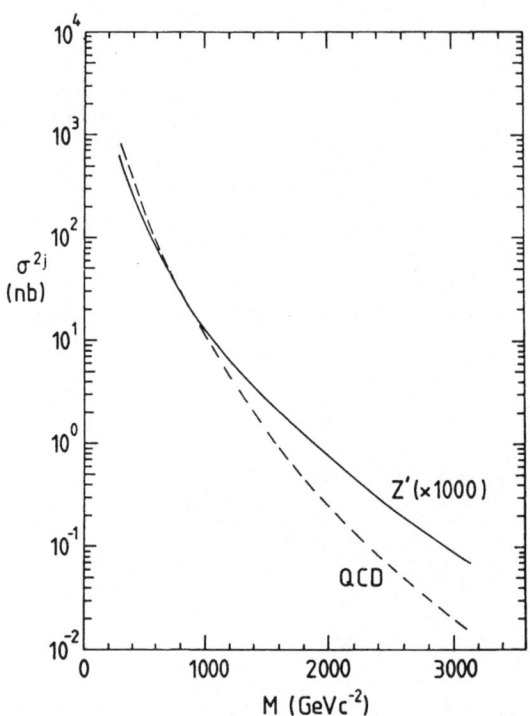

Fig. 9 . The Z' cross-section of Fig. 8 at 17 TeV compared with the QCD two jet background integrated over a mass window $M_{Z'} \pm 5GeV/c^2$.

point the analysis parallels that for a heavy (standard) Higgs boson which decays predominantly into W^+W^- and ZZ in a ratio of 2:1 . Much of the Higgs discussion applies directly to the Z', except of course that there are more undetermined parameters for the latter. In general one can say that for comparable Z' and Z fermionic couplings, $\sigma_{Z'} \gg \sigma_H$. Because of the different spins of the parent particle, there will in addition be different angular correlations for the final state weak bosons. A complete study of the various final states and backgrounds is beyond the scope of the present review. It will be generally true however that the QCD 4-jet background will prevent detection in the $Z' \to WW \to 4\,jet$ channel [7], but that the $Z' \to WW \to 2\,jet + l + \nu$ channel will offer a potential signature. The literature should be consulted for more details in particular models [5].

5. CONCLUSIONS

Drell-Yan physics will continue to be an interesting and important topic for study at future high energy hadron-hadron colliders. The main interest will centre on the production of standard W and Z weak bosons. The enormous production rates at very high energy will allow a detailed search for exotic final states. At large transverse momentum, weak boson production will continue to be an important background for new physics. If there are new heavy gauge bosons like the W and Z then they will again be produced by the Drell-Yan mechanism. An interesting possibility is the availability of the W^+W^- decay channel, which will give signatures similar to those of the standard heavy Higgs boson.

ACKNOWLEDGEMENTS

I am very grateful to the Workshop Director, Professor Ahmed Ali, and to the Project Director and staff at the "Ettore Majorana" Centre for Scientific Culture, Erice, Sicily for organising a most interesting and enjoyable workshop.

REFERENCES

1. E. Eichten, I. Hinchliffe, K. Lane and C. Quigg, Rev. Mod. Phys. 56 (1984) 579.

2. UA1 collaboration: E. Locci, Int. Europhysics Conference on High Energy Physics, Uppsala 1987.

3. UA2 collaboration: R. Ansari et al., Phys. Lett. 194B (1987) 158.

4. Z. Kunszt and W.J. Stirling, Durham University preprint DTP/87/16 (1987), to be published in Phys. Rev. D.

5. See for example: P. Langacker et al., Phys. Rev. D30 (1984) 1470; R. Ruckl, Proc. Int. Conf. on Proton Anti-proton Physics, Aachen 1986; P. Chiapetta and J.-Ph. Guillet, preprint CERN-TH.4628/86 (1986); P. Chiapetta, J.L. Kneur, S. Larbi and S. Narison, preprint CERN-TH.4670/87 (1987); F. del Aguila, M. Quiros and F. Zwirner, preprint CERN-TH-4535/86 (1986); Nucl. Phys. B284 (1987) 530; B287 (1987) 419; LBL preprint LBL-23694 (1987); J.L.

Hewett and T.G. Rizzo, Iowa State preprint IS-J-2702 (1987); T. Rizzo, Phys. Lett. 187B (1987) 169; V. Barger et al., Oregon University preprint OITS-362 (1987) and references therein.

6. S. Geer, Int. Europhysics Conference on High Energy Physics, Uppsala 1987.

7. R. Kleiss and W.J. Stirling, Phys. Lett. 180B (1986) 171.

PROTON-PROTON AND ANTIPROTON-PROTON SCATTERING AT THE HIGHEST ENERGIES

R.H. Dalitz

Department of Theoretical Physics
Oxford University, England

1. INTRODUCTION

Proton-proton and antiproton-proton scattering in Colliders have a special importance in hadronic physics because they bring us so much closer to Asymptotia than do any other hadron-hadron scattering processes we can study at present. The present Collider data at $\sqrt{s}$ = 900 GeV corresponds to $\bar{p}p$ interactions at antiproton momentum p_L = 405 TeV/c in the target proton laboratory frame. It will still be a long time before we can do experiments even at p_L = 1 TeV/c for the πp, Kp, Kp and γp systems.

Since the Collider experiments give us our closest view of Asymptotia, it is important that we have all the kinds of data, which can be obtained, available for both pp and $\bar{p}p$ systems. Their respective amplitudes F(s,t) and $\bar{F}$(s,t) are linked by crossing and by analyticity, and they are jointly constrained by the Dispersion Relations, especially through the forward Dispersion Relations. Their joint availability is so important that this requirement must be taken into account in the design of accelerators in the future, such as the Eloisatron, optimising the convenience with which all the necessary experiments can be carried out. It is not enough to study $\bar{p}p$ interactions alone.

The possibility of spin dependence in these interactions is an unfortunate complication in studies of Asymptotia. We believe that spin effects play little role in pp and $\bar{p}p$ physics at ISR energies - there are quantitative arguments to support this belief[1] - but we do not actually know this for the interactions at Collider energies. At some point, this belief will have to be tested directly, by using polarized p and/or $\bar{p}$ beams in Collider experiments. We shall not refer to this question of spin-dependence again here.

The models we shall discuss here have no clear relationship with QCD at present - although those stemming from the work of Cheng and Wu[2] derive some motivation from it - and certainly none with the New Physics of the Standard Model and beyond (i.e. Higgs particles, very massive quarks, W´ and Z´ mesons, etc.), which has been the main topic discussed at this Workshop. To establish their roots in QCD must certainly be our long-term aim, even

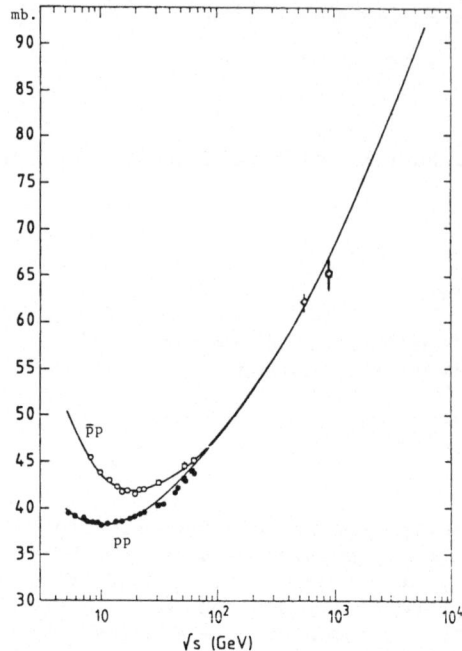

Fig. 1. σ_{tot} and $\bar{\sigma}_{tot}$ vs. $\sqrt{s}$(GeV). Fit is by BSW[8]

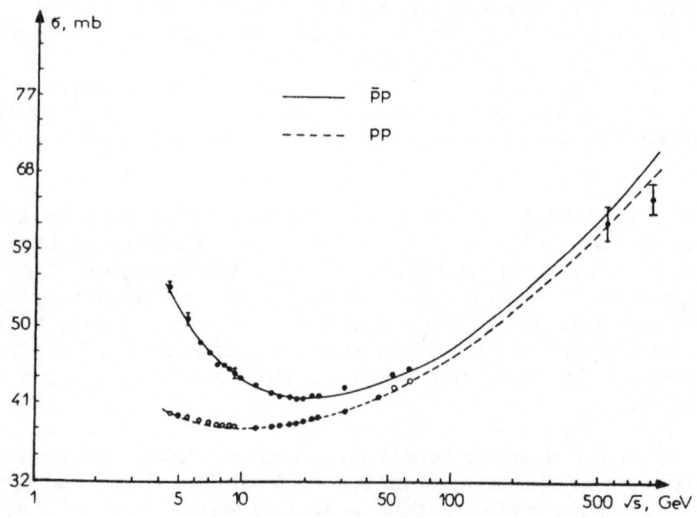

Fig. 2. σ_{tot} and $\bar{\sigma}_{tot}$ vs. $\sqrt{s}$(GeV). Fit is by GNL[18].

though there has been little progress in this direction so far.

We shall first discuss the excellent $\bar{p}p$ data at $\sqrt{s}$ = 546 GeV, and then move on to its most striking aspect in relation to the earlier data available up to $\sqrt{s}$ = 60 GeV, namely the energy-dependence of the pp and $\bar{p}p$ total cross sections, shown in Fig. 1. The earlier indications of a quadratic dependence on (ln s) for both pp and $\bar{p}p$ total cross-sections are strongly re-inforced by the 546 GeV pp data. With geometric models, this requires an explicit energy-dependence for the basic parameters and the Cheng-Wu work provides one plausible prescription for this. This energy dependence also requires the re-consideration of our past theorems and beliefs about the nature of Asymptotia. We shall end with the phenomenological discussion based on the hypothesis of a maximal Odderon contribution to these cross sections, as well as to other hadron-hadron cross-sections, leading to our conclusion that it is essential to have pp and $\bar{p}p$ cross sections in the same energy range if we are to understand clearly what is going on.

2. THE GLAUBER MODEL

For the interaction of a parton cluster (representing a proton) and an antiparton cluster (representing an antiproton), Glauber's multiple scattering theory leads to the following expression for the elastic scattering amplitude $\bar{F}(t)$ for near-forward directions:

$$\bar{F}(t) = i\int J_0(b\sqrt{(-t)}\,(1-e^{-\Omega(b)})b\,db \tag{2.1}$$

where the opacity function $\Omega(b)$ for impact parameter b is related with the parton-antiparton scattering amplitude $\bar{f}(t)$ by the relationship

$$\Omega(b) = \frac{1}{2\pi i} \int e^{-\underline{q}\cdot\underline{b}}\, G(t)^2 \bar{f}(t)d^2q \tag{2.2}$$

where t = (four-momentum transfer)$^2 \approx -q^2$, and G(t) is the form factor for the density distribution of the partons (antipartons) in a proton (antiproton). Denoting the forward parton-antiparton amplitude as

$$\bar{f}(0) = i\bar{K}(1-i\bar{\alpha}) \tag{2.3}$$

where K and α are real, we have

$$\Omega(b) = \bar{K}(1-i\bar{\alpha}) \int J_0(qb)G(t)^2\, \frac{\bar{f}(t)}{\bar{f}(0)}\, qdq \tag{2.4}$$

the amplitude $\bar{f}(t)$ being defined such that for parton-antiparton scattering

$$\frac{d\bar{\sigma}}{dt} = \pi|\bar{f}(t)|^2 \tag{2.4}$$

In the application of these expressions to the $\bar{p}p$ elastic scattering data[3,4] at $\sqrt{s}$ = 546 GeV, Glauber and Velasco[5] took the form factor G(t) from the electric form factor measured by the scattering of electrons by a proton, using for example that reported by Borkowski et al.[6]. The free variables are K, α and the form factor $\bar{f}(t)/\bar{f}(0)$, which they parameterised as $1/\sqrt{(1-at)}$, with a real. In general, this form factor may have a complex t-dependent phase factor but it was taken to be real in this work because the effect of including this factor was found to be significant only in regions where the data was not well determined. The above expressions differ from the geometric model of Chou and Yang[7] only in that the form factor $\bar{f}(t)/\bar{f}(0)$ is taken to be unity in the latter. Glauber and Velasco

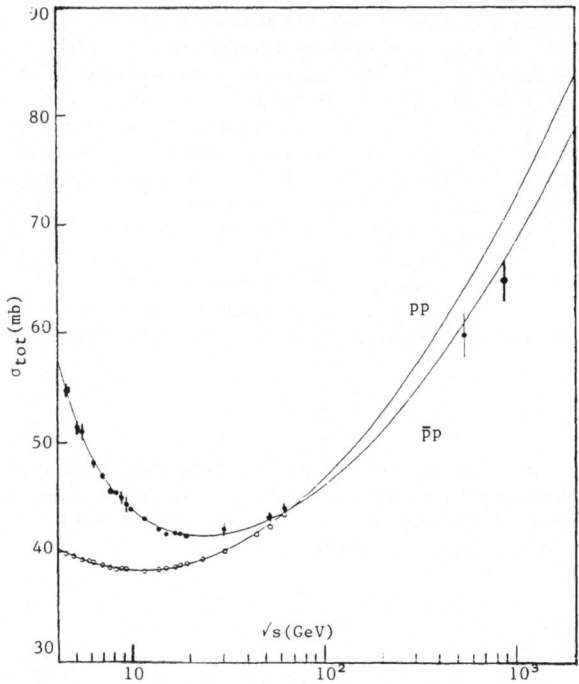

Fig. 3. σ_{tot} and $\bar{\sigma}_{tot}$ vs. $\sqrt{s}$ (GeV). Fit is by BNG[24].

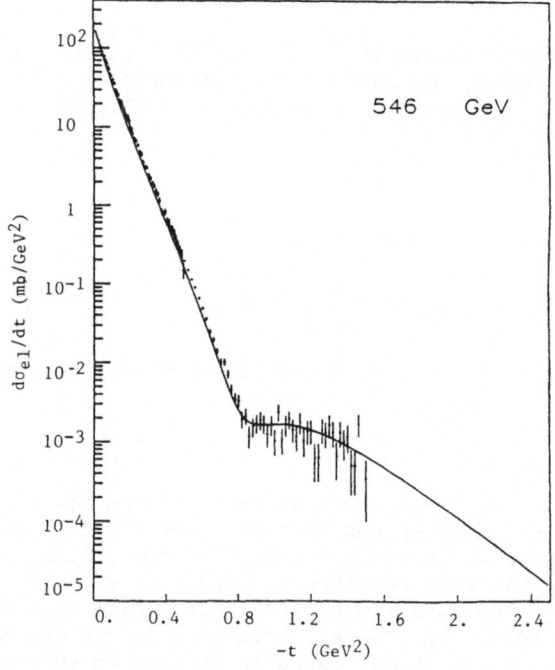

Fig. 4. $d\bar{\sigma}_{el}/dt$ vs. t at $\sqrt{s} = 546$ GeV. Fit is by Glauber & Velasco[5].

found it essential to take a = 0 if a reasonable fit to the data were to be obtained. With a = 0, there is too little freedom available to allow a good fit. As shown in Fig. 1, the Glauber-Velasco fit is quite good, although it is difficult to fit precisely both the fall in $d\sigma(\bar{p}p)/dt$ over the range $t = 0$ to $-t = 0.8$ GeV^2 and the shape of the shoulder in it beyond $-t = 0.8$ GeV^2. Their best-fit parameter values were a = 1.8 GeV^{-2}, α = 0.19, and K = 81.3 mb.

This model has several limitations, of course:

i)   there are no means available for the comparison of $\bar{p}p$ and pp at the same energy, since there is no necessary relation between f(t) for parton-parton scattering and $\bar{f}$(t) for antiparton-parton scattering.

ii) there is no way to discuss the $\sqrt{s}$ dependence of the $\bar{p}p$ fit, since $\bar{K}$, $\bar{\alpha}$ and a are free parameters to be fitted to the data at each energy considered.

At the same time, one must exclaim it as remarkable that it is possible to fit cross-section data ranging over 5 orders-of-magnitude in the forward peak by such a simple model. Further, the rapid t-dependence of $(d\sigma_{el}(\bar{p}p)/dt)/\sigma_{el}(\bar{p}p)$ observed by Bozzo et al.[3] in the range $-t = 0.03$ to 0.50 GeV^2 is well reproduced, without any further assumptions.

3. THE CHENG-WU IMPACT SPACE PARAMETRISATION

The expression for the $\bar{p}p$ amplitude $\bar{F}(s,t)$ is here given by the expression (2.1) with opacity function $\Omega(s,\underline{b})$ explicitly dependent on s. The spin-independent form adopted by Bourrley, Soffer and Wu[7] after some earlier trials was

$$\Omega(s,\underline{b}) = S(s)\Omega(\underline{b}) + R(s,\underline{b}), \qquad (3.1)$$

where Ω(b) is given by (2.2), S(s) is the crossing symmetric term

$$S(s) = \frac{s^c}{(\ln s)^d} + \frac{u^c}{(\ln u)^d} \qquad (3.2)$$

where u is the third Mandelstam variable, here such that

$$s+t+u = 4M^2 \qquad (3.3)$$

and R(s,b) represents the contribution from Reggeon exchanges.

In two-dimensional transverse momentum space, the Fourier transform of (b) is given by

$$\Omega(t) = \bar{f}(0)G(t)^2(\bar{f}(t)/\bar{f}(0)) \qquad (3.4)$$

where $\bar{f}(t)/\bar{f}(0) = (1+t/a^2)/(1-t/a^2)$ is the form adopted by Bourrely et al.[8], quite similar to that used later by Glauber and Velasco (see above). The new inputs are

i)   the term (3.2) which stems from the analysis of high-order graphs generated by QCD, carried out by Cheng and Wu[2].

ii) the Regge contribution R, which they took to have the form

$$R(s,t) = (C_+ + C_- e^{-i\pi\alpha(t)})s^{\alpha(t)} \tag{3.5}$$

with an exchange degenerate trajectory $\alpha(t) = \alpha_0 + \alpha_1 t$.

In this model, the difference between pp and $\bar{p}p$ is due to two effects:

a) the Regge contributions, which are different for pp and $\bar{p}p$ interactions, but which die out as s increases, relative to the first term of (3.1). Hence this model predicts that $\sigma = \sigma(pp)$ and $\bar{\sigma} = \sigma(\bar{p}p)$ should become equal, at sufficiently large energy $\sqrt{s}$.

b) the Coulomb interaction, which, to a sufficient approximation, adds to F(s,t) the term

$$F_c(t) = \pm 2\alpha \frac{G(t)^2}{|t|} \exp(\mp i\alpha\phi) \tag{3.6}$$

where this α is the fine structure constant and ϕ is a known phase angle dependent on t. In (3.6) the upper sign refers to the $\bar{p}p$ system, the lower sign to the pp system. This Coulomb amplitude is, of course, negligible except at very small momentum transfers. Nevertheless, if measurements of $d\sigma_{el}/dt$ can be made at sufficiently small angles, the presence of the Coulomb amplitude (3.6) leads to two observable effects important for our knowledge of the pp and $\bar{p}p$ interactions, as we shall now indicate.

In a phenomenological spirit, the $\bar{p}p$ nuclear amplitude may be conveniently written, for small t, in the form

$$\bar{F}_N = \frac{\bar{\sigma}_{tot}}{4\pi} \cdot (\bar{\rho}+i) e^{\frac{1}{2}\bar{b}t}, \tag{3.7}$$

where the parameters $\bar{\rho}$ and b are real. The first factor stems from the unitarity condition at $0°$, and thus for $t = 0$,

$$\text{Im } \bar{F}_N = \frac{1}{4\pi} \bar{\sigma}_{tot} \tag{3.8}$$

while $\bar{\rho}$ denotes the ratio

$$\bar{\rho} = \text{Re}\bar{F}_N(0)/\text{Im}\bar{F}_N(0)). \tag{3.9}$$

The exponential factor is added to correspond to the physical data, since the differential cross section $d\bar{\sigma}/dt$ does fall at a rate comparable with exponential, with increasing t. The net elastic scattering amplitude is therefore, to sufficient accuracy

$$\bar{F}(t \approx 0) = \frac{\bar{\sigma}_{tot}}{4\pi} \cdot (\bar{\rho}+i) \frac{2\alpha}{|t|} \tag{3.10}$$

These nuclear and Coulomb amplitudes lead separately to cross sections of equal magnitude for

$$|t|_c = \frac{8\pi\alpha}{\bar{\sigma}_{tot}} \cdot \frac{1}{\sqrt{(1+\bar{\rho}^2)}} \tag{3.11}$$

With $\bar{\rho}^2 \ll 1$, and $\bar{\sigma}_{tot} \approx 62$ mb at $\sqrt{s} = 546$ MeV, this expression leads to $|t|_c = 1.1 \times 10^{-3}$ GeV^2, a value such that the neglect of the exponential factor in (3.7) is reasonable, since values $b \approx 15$ GeV^{-2} are typical. With $|t| = \bar{\theta}\sqrt{s}/2$ in this small angle regime, expression (3.11) corresponds to scattering angle

$$\bar{\theta}_c = \frac{16\pi\alpha}{\sigma_t\sqrt{s}} \cdot \frac{1}{\sqrt{(1+\rho^2)}} \qquad (3.12)$$

which is about 120 µrad at $\sqrt{s}$ = 546 GeV.

In the angular region around (3.12), the interference between $\bar{F}_N$ and $\bar{F}_C$ in (3.10) is significant and is proportional to $\bar{\rho}$, thus determining the phase of $\bar{F}_N(0)$, a quantity important for the detailed testing of any model and vital for the use of forward dispersion relations. All of the remarks made above, from (3.7) on, apply equally to pp elastic scattering, when the bar is removed from all the symbols. The values which have been determined in this way for ρ and $\bar{\rho}$ are given in Fig. 2.

Further, if it is possible to make measurements at still smaller angles than (3.12), say at angles as small as $\bar{\theta}_c/n$, the cross section becomes dominated by the Coulomb scattering, then being larger than the nuclear scattering by a factor n^2. Since the Coulomb scattering has a known cross section, its observation provides an independent normalization of all the measured cross sections, in other words it constitutes an absolute measurement of the luminosity of the two interacting beams.

In 1979, Bourrely, Soffer and Wu (referred to as BSW below)[8] made a detailed analysis of all the pp and $\bar{p}$p elastic scattering and total cross section data available above $\sqrt{s}$ = 5.5 GeV at that time. In particular, we note here their determination of the parameters associated with the Regge contribution in (3.1), namely

$$C_+ = 39, \quad C_- = 1.8, \quad \alpha(t) = 0.352 + 0.694t. \qquad (3.13)$$

Their later analyses include the Collider data on $\bar{p}$p interactions up to $\sqrt{s}$ = 546 GeV. Since the parameters (3.13) stem from the differences between pp and $\bar{p}$p interactions, inclusion of the more recent data has little effect on these parameters since there is no pp data above the ISR energies. Further, the values of these parameters imply that the Regge contributions to the pp and $\bar{p}$p elastic amplitudes become negligibly small (relative to the experimental errors) well before $\sqrt{s}$ = 546 Gev.

The later analyses of BSW[9-11] have included all of the data on $\bar{p}$p and pp total cross sections, $\bar{\rho}$ and ρ-values, and differential elastic cross sections. The proton form factor they have adopted is

$$G(t) = \{(1-t/m_1^2)(1-t/m_2^2)\}^{-1} , \qquad (3.14)$$

which they have fitted to the proton magnetic moment form factor $G_M(t)/\mu_p$, obtaining m_1 = 0.586 GeV and m_2 = 1.704 GeV. The remaining parameters of their 1984 fit are as follows:

$$c = 0.167, \quad d = 0.748, \text{ with } a = 1.953 \text{ GeV and } f(0) = 7.115 \text{ GeV}^{-2} \qquad (3.15)$$

It is worth noting that these parameter values are quite comparable with their 1979 values, which were (0.151, 0.756, 2.257, 8.125) in the same sequence, and that the cross section BSW predicted at that time for $\sqrt{s}$ = 546 GeV was 62 mb, in excellent agreement with the value 61.9 ± 1.5 mb measured by Bozzo et al.[12] in 1984. The difference $\Delta\sigma = (\bar{\sigma}_{tot} - \sigma_{tot})$ calculated for $\sqrt{s}$ = 546 GeV with these earlier parameters is difficult to determine precisely from their published curves[8] but appears to be no more than 0.2 mb. For parameters (3.15), both total cross sections are plotted on Fig.1.

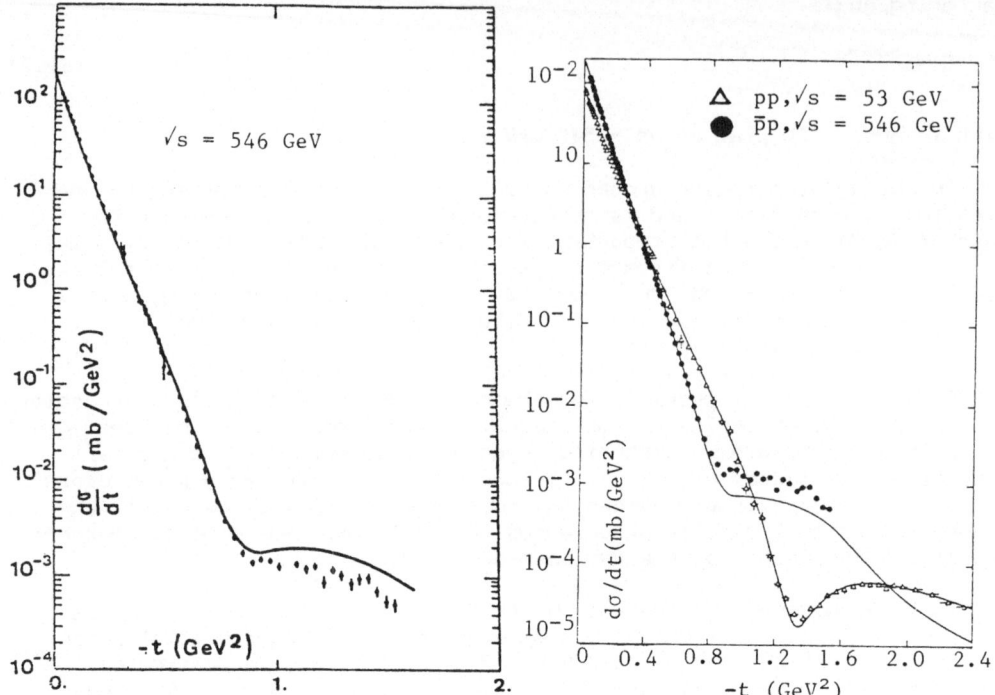

Fig. 5. $d\bar{\sigma}/dt$ at $\sqrt{s}$ = 546 GeV is
compared with the fit of BSW[8]. The
fit is excellent out to $|t|$ = 0.8
GeV[2], but is far above the data beyond.

Fig. 6. $d\sigma/dt$ at 53 GeV and
$d\bar{\sigma}/dt$ at 546 GeV are compared
with calculations of DL[37].

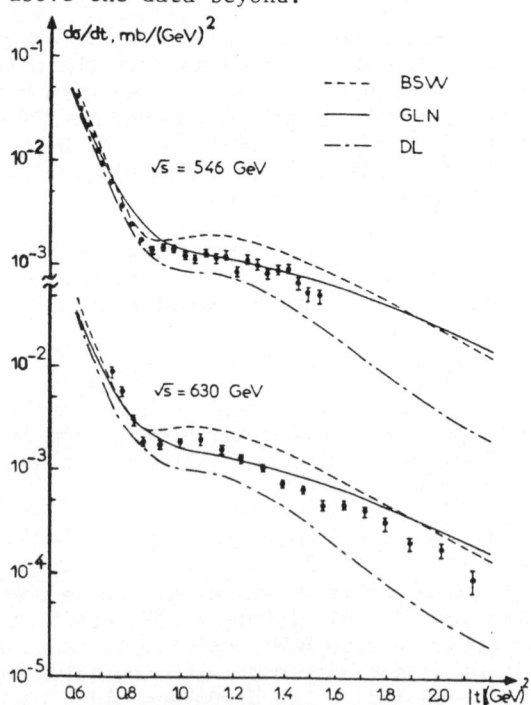

Fig. 7. $d\bar{\sigma}/dt$ for large $|t|$ is plotted for
$\sqrt{s}$ = 546 and 630 GeV and compared with the fits
of BSW[8], GNL[18] and DL[37].

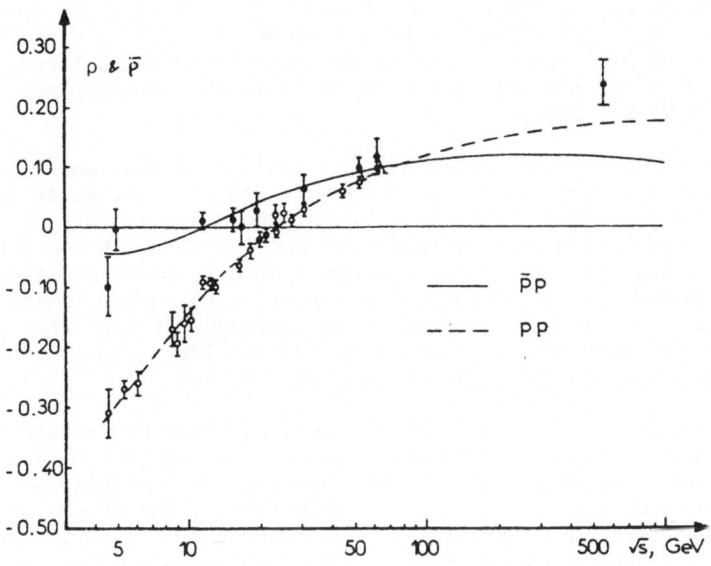

Fig. 8. The data on ρ and ρ̄ are compared with the fit of GNL[18]

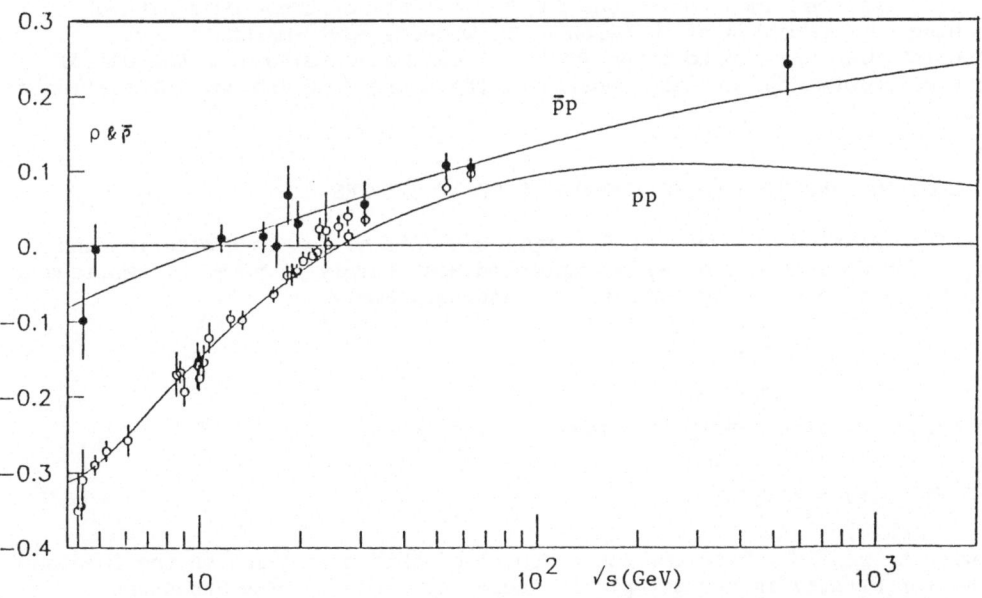

Fig. 9. The ρ and ρ̄ values available are compared with the fit of BGN[24]

The differential cross sections calculated with the parameters (3.15) at √s = 546 and 630 GeV are shown on Figs. 5 and 7; the former may be compared with that of Glauber and Velasco given in Fig. 4. Fig. 5 shows that the BSW fit[11] is much better over the range -t = 0 to 0.7 GeV^2. The location of the break and the general shape of the shoulder are correct but the calculated cross section values beyond about -t = 0.9 GeV^2 are systematically too large by a factor of about 2. Fig. 7 plots the large $|t|$ data of Bernard et al.[13] at 630 GeV, where we see that this discrepancy persists out to -t = 2 GeV^2 and beyond.

The total cross sections calculated by BSW[11] at higher energies are 74.8 mb for √s = 1.8 TeV, the Fermilab energy, and 121.2 mb at 40 TeV, the proposed SSC energy. Over this energy range, the shape of $d\bar{\sigma}_{el}/dt$ changes gradually with increasing √s, the small dip seen at about -t = 0.9 GeV^2 for √s = 546 GeV in Fig. 11 before the shoulder gradually moves to smaller $|t|$ values, and deepens; at √s = 40 TeV, the dip is quite marked, being at -t = 0.3 GeV^2, the ratio (height of secondary peak)/(value at dip) then being about 2. Over the same energy range, a second shoulder gradually develops, its break being at about -t = 1.7 GeV^2 for √s = 40 TeV.

A determination of $\bar{\rho}$ for 546 GeV has been reported by Bernard et al.[14] since these calculations were completed, with the result $\bar{\rho}$ = 0.24±0.04. Its determination was from observations over the range $2\times10^{-3} < |t| < 35\times10^{-3}$ GeV^2 on the upper edge of the Coulomb interference region, and it was considered essential in the fitting to include the exponential factor in the amplitude $\bar{F}_N$ given by (3.7), the value required for b by these data being 15.5±0.8 GeV^{-2}, in good agreement with the value b = 15.3±0.3 GeV^{-2} found earlier from cross section measurements[2] over a wider momentum range, out to 0.1 GeV^{-2}. This value for $\bar{\rho}$ at 546 GeV is much larger than the value 0.13 predicted by the BSW model (cf. Fig. 10). The reason for this discrepancy is not clear. It may be that a more elaborate analysis of the data is required, involving for example a t-dependence for $\bar{\rho}$ or a t-dependent factor in (3.7) more complicated than the exponential factor used there. Bernard et al.[14] explored such variations in the analysis of their data without reaching any reduction of $\bar{\rho}$ below 0.20, whereas BSW[15] maintain that agreement with their prediction $\bar{\rho}$ = 0.13 can be achieved with the use of their calculated F_N, but the details of their argument are not yet available to us.

4. CROSSING AND HIGH-ENERGY BEHAVIOUR OF AMPLITUDES

The relationship between the amplitudes F(s,t) for pp scattering and $\bar{F}$(s,t) for $\bar{p}$p scattering may be expressed most simply in terms of amplitudes $F_\pm$(s,t) which are even or odd under crossing, namely

$$\text{(a) } F = F_+ + F_-, \qquad \text{(b) } \bar{F} = F_+ - F_- \qquad\qquad (4.1)$$

The crossing relationship is expressed as follows:

$$F_\pm(-s,t) = \pm(F_\pm(s,t))^* \qquad\qquad (4.2)$$

We wish to consider here the consequences of this constraint on the leading terms for $F_\pm$(s,t) in the asymptotic region of large s. The Froissart theorem tells us that their leading terms cannot increase faster than s(ln s)^2. The data at present suggest that the asymptotic amplitudes F^{as} and F^{as} may indeed increase as fast as this limit allows.

If we consider the asymptotic form $As(\ln s)^2$ for $F_+(s,t)$ and insert it in the crossing relation (4.2), remembering that $s \to -s = se^{i\pi}$ implies that $\ln s \to \ln(-s) = (\ln(s)+i\pi)$, we conclude from the leading terms on either side that $A^* = -A$, and we write $A = iF_1$, where F_1 is real, so that

$$F_+^{as} \to iF_1 s(\ln s)^2. \qquad (4.3a)$$

With the form $Bs(\ln s)^2$ for $F_-(s,t)$, we conclude $B^* = B$, and we write $B = 0_1$, where 0_1 is real, so that

$$F_-^{as} \to 0_1 s(\ln s)^2. \qquad (4.3b)$$

The even asymptotic term (4.3a) is referred to as the Froissaron contribution; the odd asymptotic term (4.3b) is referred to as the Odderon contribution.

We note that

i)   the elastic scattering cross sections for pp and $\bar{p}$p are

(a) $d\sigma/dt = |F|^2/16\pi s^2$,   (b) $d\bar{\sigma}/dt = |\bar{F}|^2/16\pi s^2$ $\qquad (4.4)$

ii)   The total cross sections are given by the unitarity relation for t=0:

(a) $\mathrm{Im}F(s,t=0) = \frac{s}{8\pi}\sigma_{tot}$ (b) $\mathrm{Im}\bar{F}(s,t=0) = \frac{s}{8\pi}\bar{\sigma}_{tot}$ $\qquad (4.5)$

Using the relationship (4.1) and the asymptotic forms (4.3), we conclude that the leading term of the asymptotic cross section is the same for σ_{tot} and $\bar{\sigma}_{tot}$, both being given by

$$\sigma_{tot} \approx \bar{\sigma}_{tot} \to 8\pi F_1(\ln s)^2 \qquad (4.6)$$

Hence, the Odderon does not contribute to the leading $(\ln s)^2$ term in these total cross sections. However, if we extend (4.3b) to the form $(0_1 s(\ln s)^2 + B_1 s(\ln s))$, the crossing relation leads us to conclude $\mathrm{Im}B_1 = \pi 0_1$, so that the Odderon necessarily contributes to the $(\ln s)$ term in these cross sections. We should add that an arbitrary term $iG_1 s(\ln s)$ with G_1 real can be added to (4.3a) without constraint by the crossing relation, so that the Froissaron also contributes a $(\ln s)$ term equally to the pp and $\bar{p}$p cross sections.

(iii) The cross section difference is given by

$$\Delta\sigma = \bar{\sigma}_{tot} - \sigma_{tot}$$
$$= \frac{8\pi}{s} \mathrm{Im}\{\bar{F}(s,\ t=0) - F(s,t)\} = -\frac{16\pi}{s} \mathrm{Im}(F_-(s,t=0) \qquad (4.7)$$

It follows from (4.3b) that this has no asymptotic $(\ln s)^2$ term, but the discussion of the last paragraph implies that its asymptotic form is dominated by the Odderon, giving the leading term

$$\Delta\sigma \to 16\pi 0_1(\ln s) \qquad (4.8)$$

5.   THEOREMS OF ASYMPTOTIC BEHAVIOUR

The statements we have become accustomed to about the $s \to \infty$ asymptotic forms permissible for amplitudes and cross sections are no longer valid when we have to accept the possibility that total cross sections do not approach constant values in this limit but may increase indefinitely like $(\ln s)^2$. This matter has been discussed by a number of authors recently[16-19].

(i) Pomeranchuk theorem

The general statement of this theorem is that, as $s \to \infty$,

$$\bar{\sigma}_{tot}(s)/\sigma_{tot}(s) \to 1. \tag{5.1}$$

As we have just remarked above, it is now possible that

$$\Delta\sigma = \bar{\sigma}_{tot}(s) - \sigma_{tot}(s) \to 16\pi\, 0_1(\ln s). \tag{5.2}$$

This is not inconsistent with the theorem (5.1), the ratio $\bar{\sigma}/\sigma$ differing from unity only by a term of order $1/(\ln s)$ with coefficient $-20_1/F_1$. In this situation, integrals like

$$\int^\Lambda \frac{\bar{\sigma}(s) - \sigma(s)}{s} \cdot ds \tag{5.3}$$

are not convergent as $\Lambda \to \infty$. This may be the case even if the maximal odderon 0_1 is absent, since the crossing relation for $F_-(s,t)$ still allows a term $-(\pi i/2)\mathrm{Re}B_1/s$ and this leads to an asymptotically constant term in $\Delta\sigma$, which still leaves the integral (5.3) logarithmically divergent.

The reason why (5.1) remains valid even when there is an Odderon contribution 0_1 is that the F_+ contributes equally to σ and $\bar{\sigma}$ in the leading order, while the Odderon contributes to neither σ or $\bar{\sigma}$ in leading order because 0_1 is required to be real by crossing.

(ii) The Froissart-Martin bound

This states that

$$\sigma_{tot}(s) \leq \frac{\pi}{m_\pi^2} \, (\ln(s/s_0))^2 \tag{5.4}$$

for some s_0. The value of the coefficient (π/m_π^2) is two orders of magnitude greater than the value obtained from the $(\ln s)^2$ terms detected to date in total cross sections.

(iii) The Cornille-Martin theorem

This states that, for a fixed value of t within the diffraction peak, we have

$$\left(\frac{d\bar{\sigma}(s,t)}{dt}\right) / \left(\frac{d\sigma(s,t)}{dt}\right) \to 1 \qquad \text{as } s \to \infty. \tag{5.5}$$

However the diffraction peak lies within

$$|t| < t_0/(\ln s)^2 \to 0 \qquad \text{as } s \to \infty . \tag{5.6}$$

Hence it can really hold only for $t = 0$ and so we must conclude that

$$\frac{d\bar{\sigma}(s,t)}{dt} - \frac{d\sigma(s,t)}{dt} \nrightarrow 0 \qquad \text{as } s \to \infty. \tag{5.7}$$

In other words, for a fixed t value, the two differential cross sections do not necessarily approach equality as s → ∞ .

(iv) The Fischer-Martin theorem

The conditions under which the limit of $\rho_- = \text{Re}F_-/\text{Im}F_-$ is zero as s → ∞ are no longer necessarily satisfied but Fischer[19] has recently reconsidered this theorem in the light of the possibility of a maximal Odderon, re-formulating it under a set of conditions which include the case of such an Odderon.

(v) The AKM theorem of Auberson, Kinoshita and Martin[20]

This states that

$$F_\pm(s,t) \rightarrow F_\pm(s,0)g_\pm(\tau) \qquad\qquad \text{as } s \rightarrow \infty \qquad\qquad (5.8)$$

where the $g_\pm$ are analytic functions of the variable $\tau = C\sqrt{-t}.(\ln s)$.   From this result, if follows that the diffraction peak should asymptotically shrink in width like $(\ln s)^{-2}$. Although this is the general trend of model calculations and of what data exist, this asymptotic behaviour might not become effective until energies are reached much higher than are available at present. Of course, for dimensional reasons, (ln s) should be understood as $\ln(s/s_0)$ for some constant s_0; although it has been conventional to take the scale energy to be $s_0 = 1$ GeV, it could well be much larger than this value, putting off the regime of Asymptotia until correspondingly larger s-values.

6. THE ODDERON ANSATZ OF GAURON, NICOLESCU AND LEADER (GNL)

Gauron, Nicolescu and Leader[18] have recently made a phenomenological fit to all pp and p̄p data above about √s = 10 GeV using assumed amplitude forms which include Froissaron and Odderon components up to the maximally possible terms, the standard Pomeron and Reggeon exchange terms, as well as those terms which arise from the joint exchange of two Pomerons or of a Pomeron and a Reggeon, and which are constrained to obey all of the above theorems which are relevant. We then have for each amplitude the sum

$$F = F^{as} + F^P + F^{PP} + F^R + F^{RP} \qquad\qquad (6.1)$$

The construction of $F_\pm^{as}$ amplitudes which satisfy the AKM constraint (5.8) is a key point and is quite complicated. It is carried out in the J-plane, where, for t=0, the Froissaron behaviour corresponds to a triple pole, and the Odderon behaviour to a double pole, at J=1 and the Sommerfeld-Watson in J space is used to return to the F(s,t) amplitudes. The simplest non-trivial behaviour is assumed at all points in their construction; where (ln s) occurs below, it means $\ln(s/s_0)$ with the choice $s_0 = 1$ GeV^2. To indicate the nature of these amplitudes, we specify them schematically:

(i)   the Froissaron term,

$$F_+^{as}/(is) = F_1\ln^2(s)\phi_1 e^{\frac{1}{2}b_1^+ t} + F_2\ln(s)\phi_2 e^{\frac{1}{2}b_2^+ t} + F_3\phi_3 e^{\frac{1}{2}b_3^+ t} \quad , \qquad (6.2)$$

where the ϕ_i are functions of $(R_+\tau)$, of known form. The numerical parameters in (6.2) are F_1, F_2, F_3; b_1^+, b_2^+, b_3^+; and R_+.

(ii) the Odderon term,

293

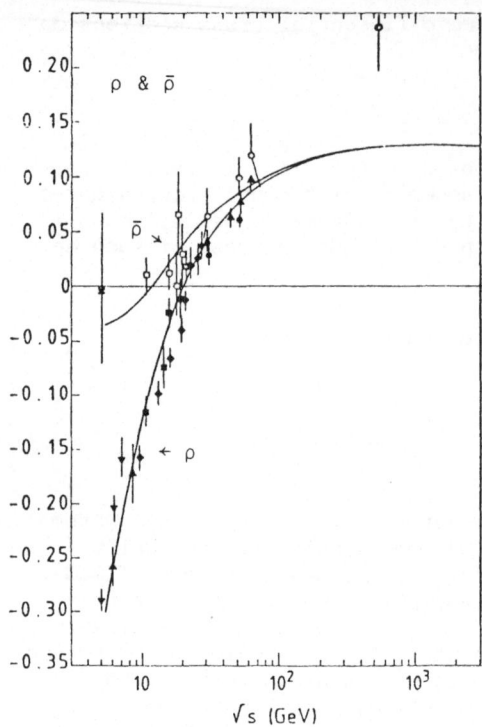

Fig. 10. The data on ρ and ρ̄
compared with the fit of BSW[8].

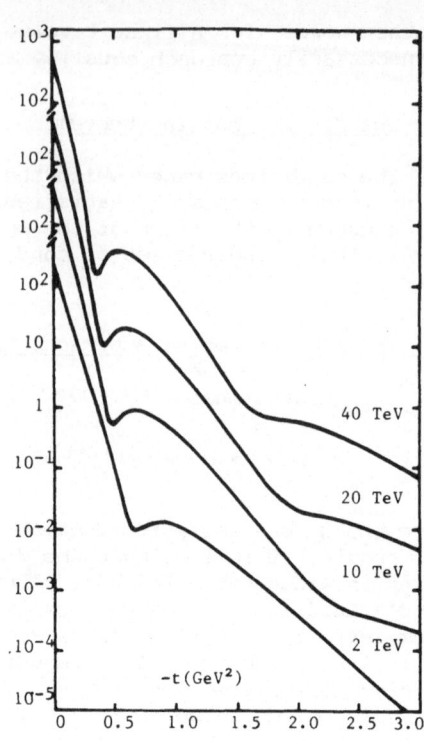

Fig. 11. dσ̄/dt for √s from
2 to 40 TeV, from BSW[8]

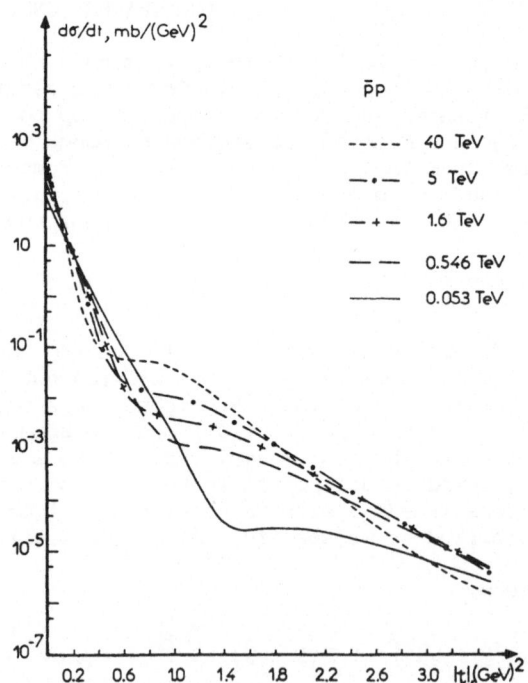

Fig. 12. The angular distributions dσ̄/dt predicted
by the GNL[18] fit are plotted for a range of √s values,
from the ISR energy to the projected SSC energy.

$$F_-^{as}/s = 0_1 \ln^2(s)\psi_1 e^{\frac{1}{2}b_1^- t} + 0_2 \ln(s)\psi_2 e^{\frac{1}{2}b_2^- t} + 0_3 \psi_3 e^{\frac{1}{2}b_3^- t} , \qquad (6.3)$$

where the ψ_i are functions of $(R_-\tau)$, of known form. The numerical parameters in (6.3) are $0_1, 0_2, 0_3$; b_1^-, b_2^-, b_3^-; and R_-.

These forms $F_\pm^{as}$ are introduced in a purely phenomenological spirit, in the same way as the Pomeron has been introduced in the past twenty years or more. No connection with QCD is known, except that three-gluon exchange does lead to a term of the form 0_3 (cf. Sec. 7 below). For fixed t, these functions $F_\pm^{as}$ have the asymptotically leading terms:

$$F_+^{as}(s,t) = is\sqrt{(\ln s)} \cdot h_+(s,t), \qquad (6.4a)$$

$$F_-^{as}(s,t) = s(\ln s) \cdot h_-(s,t), \qquad (6.4b)$$

where the $h_\pm$ are bounded oscillating functions as $s \to \infty$.

The Reggeon and Pomeron terms are introduced as usual:

P and PP: these terms require the introduction of four parameters, (C_P, β_P) and (C_{PP}, β_{PP}), their slope parameters being taken to have the values $\alpha'_P = 2\alpha'_{PP} = 0.25$ GeV^{-2}.

$R_\pm$ and $(RP)_\pm$: each of these require a set of parameters $\alpha(0)$, C and β, a total of 12 parameters. The Regge terms turn out to play little role in the top ISR energy range and essentially no role at all at the Collider energies.

The total number of parameters to be determined from the data is very considerable, and we shall not discuss their best-fit values here. The asymptotic behaviour is determined almost entirely by $F_\pm^{as}(s,t)$ and the leading terms require the values

$$F_1 = 0.29 \text{ mb}, \quad b_1^+ = 8.5 \text{ GeV}^{-2}; \qquad F_2 = -1.70 \text{ mb}, \quad b_2^+ = 10.80 \text{ GeV}^{-2}$$

$$0_1 = 0.031 \text{ mb}, \quad b_1^- = 14.04 \text{ GeV}^{-2}; \qquad 0_2 = -0.30 \text{ mb}, \quad b_2^- = 14.04 \text{ GeV}^{-2}$$

We note the following points only:

(i) the curves for $\bar\sigma_{tot}$ and σ_{tot} predicted by this fit are given as function of (ln s) on Fig. 1, where they are compared with the data available, to indicate the quality of the fit.

(ii) $\Delta\sigma = \bar\sigma_{tot} - \sigma_{tot} \to 2\pi 0_1(\ln s)$, as $s \to \infty$. For the ISR energies, Fig.13 shows that $\Delta\sigma$ is falling as function of s, but this analysis predicts that it will ultimately rise with s, as shown in Figs. 2 and 13.

(iii)The ρ values predicted for pp and $\bar{p}p$ from this analysis are, in the limit $s \to \infty$

$$\rho = \frac{\text{Re}F}{\text{Im}F} \to \frac{\text{Re}F_+^{as}(s,0)}{\text{Im}F_+^{as}(s,0)} \to \frac{0_1}{F_1} , \qquad (6.5a)$$

$$\bar\rho = \frac{\text{Re}\bar{F}}{\text{Im}\bar{F}} \to \frac{-\text{Re}F_-^{as}(s,0)}{\text{Im}F_+^{as}(s,0)} \to \frac{-0_1}{F_1} , \qquad (6.5b)$$

when we retain only the $(\ln s)^2$ terms. The conclusion from (6.5) is, remarkably, that asymptotically

$$\rho(\infty) = -\bar\rho(\infty) \neq 0 \qquad\qquad (6.6)$$

ρ being positive and $\bar\rho$ being negative, as $s \to \infty$. The values predicted for ρ and $\bar\rho$ by GNL from their analysis are plotted on Figs. 8 and 14, for energies up to $\sqrt{s} = 100$ TeV. Their trend is in accord with (6.5) but Asymptotia is clearly very far away from experiment today since $\bar\rho$ is still positive at $\sqrt{s} = 100$ TeV; indeed $\bar\rho$ is predicted to change sign only at $\sqrt{s} = 5 \times 10^4$ TeV, and the asymptotic equality $\rho = -\bar\rho$ is still far from being reached at the Planck mass.

(iv)

$$\left(\frac{d\bar\sigma(s,0)}{dt} - \frac{d\sigma(s,0)}{dt}\right) \to \frac{1}{8}\,(F_2 0_1 - F_1 0_2)(\ln s)^2 + o(\ln s)$$

$$\approx 0.045\,(\ln s)^2 \text{ mb}. \qquad\qquad (6.7)$$

The following general remarks may be made about the GNL model:

(a) for $\sqrt{s} \approx 10$ GeV, the Pomeron and Reggeon terms dominate, with some corrections from the amplitudes $F_\pm{}^{as}$,

(b) for large $\sqrt{s} \gtrsim 500$ GeV, the terms $F_\pm{}^{as}$ dominate strongly over the Pomeron and Reggeon terms.

(c) for $-t < 0.6$ GeV^2, Im$F_+{}^{as}$ dominates and it is the same for pp and $\bar{p}p$. for $-t > 1.2$ GeV^2, Re$F_-{}^{as}$ dominates and this has the opposite sign for pp and $\bar{p}p$. In this fit, this Odderon term is the origin of the rise in magnitude of $d\bar\sigma/dt$ beyond the diffraction dip as s increases from 10 to 546 GeV. The $\sqrt{s}$ dependence of $d\bar\sigma/dt$ predicted by GNL[18] is plotted on Fig. 12.

It is of interest to note here that Gauron, Nicolescu and Szymanowski[21] have outlined a field theoretical realization of the GNL ansatz, starting from a Yang-Mills field theory with a broken symmetry. The asymptotic amplitudes $F_+{}^{as}$ and $F_-{}^{as}$ correspond to the exchange of an axial-vector meson trajectory $\alpha_V(t)$ and a vector meson trajectory $\alpha_A(t)$, respectively. The intercept $\alpha_V(0)$ is close to 1 if the mass M_V is small, $M_V{}^2 \ll 1(\text{GeV})^2$; the mass M_A is not required to be small. The calculations were made only in the leading-logs approximation and the nature of the singularity at $J = 1$ is not settled, except that it is clear that there is no real meson which corresponds to this pole.

This Odderon structure, if it exists in the pp-$\bar{p}p$ system, would be expected to manifest itself in other hadron-hadron systems, such as the $\pi^- p$ system. The amplitude for the charge-exchange process $\pi^- p \to \pi^0 n$ is odd under crossing, so that the Odderon pole can contribute in addition to the exchange of the ρ and ρ' mesons. The Odderon interaction does not involve spin-flip, whereas ρ and ρ' are known to have a strong magnetic coupling with the nucleon. Hence, in principle at least, the interference between these two kinds of amplitude can lead to substantial polarization effects in this charge exchange reaction. In fact, such effects have been reported for this reaction, notably at pion momentum $p_L = 40$ GeV/c by Apokin et al.[22], as shown in Fig. 15, and GNL[23] have found that the introduction of a maximal Odderon term does lead to a striking agreement with these data, the polarization being predicted and observed to change sign twice in the $|t|$ range investigated. Even more important, these calculations predict that the polarization should grow in magnitude with increasing p_L, leading to effects of order 50% in the range $-t = 0.4$ to 0.6 GeV^2 for pion momentum $p_L = 100$ GeV, as shown in Fig. 16.

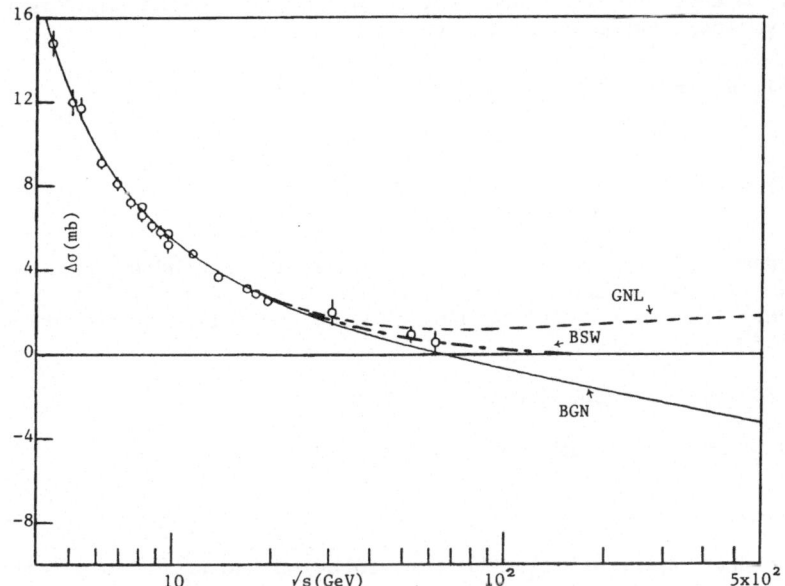

Fig. 13. The difference $\Delta\sigma = (\mathrm{o} - \sigma)$ is plotted
vs. $\sqrt{s}$ for the models of BSW[8], GNL[18] and BGN[23],
and compared with the avaiable data.

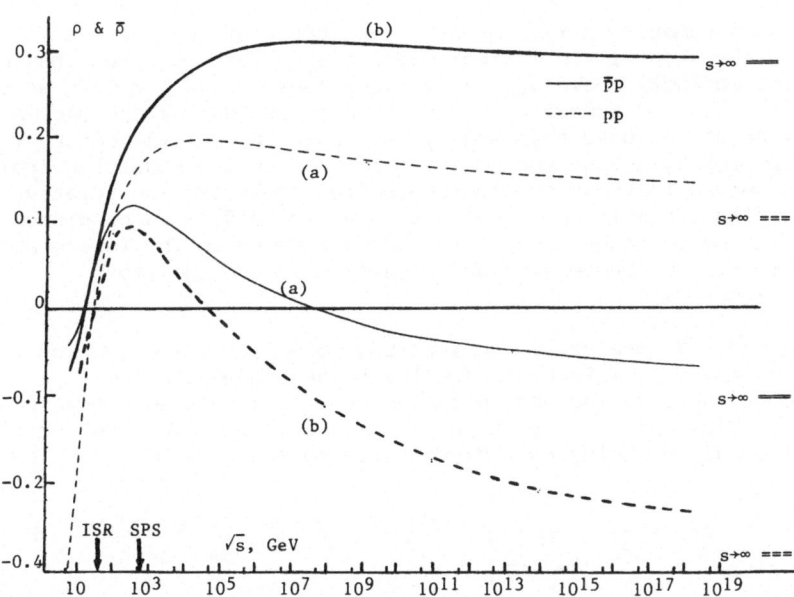

Fig. 14. Calculated ρ and $\bar{\rho}$ values are plotted vs. $\sqrt{s}$ for (a) the GNL[18]
parameters, and (b) the BGN[23] parameters. Note the limiting values for
$s \rightarrow \infty$ plotted on the right. Asymptotia is reached only very slowly - the
$\sqrt{s}$ scale stops at the Planck mass.

On the other hand, the prediction of GNL for $\bar{\rho}$ at $\sqrt{s}$ = 546 GeV is about 0.13, far from the observed value 0.24±0.04. This has recently been considered in terms of the Odderon ansatz by Bernard, Gauron and Nicolescu (BGN)[24]. Since they are concerned only with the t=0 forward amplitude, they write the asymptotic terms (6.2) and (6.3) in the form

$$F_+^{as}(s,0)/is = F_1(\text{in}(s/s_+) - i\pi/2)^2 + f_+ , \tag{6.8}$$

and

$$F_-^{as}(s,0)/s = 0_1(\text{in } s/s_-) - i\pi/s)^2 + f_- , \tag{6.9}$$

where the parameters s_+, f_+ and s_-, f_- are constants. Including Regge terms as in the GNL fit above, BGN fitted all of the t = 0 data (including total cross sections) from $\sqrt{s}$ = 5 GeV to 546 GeV with the following parameter values for $F_\pm^{as}$:

F_1 = 0.38 mb, s_+ = 100 GeV^2, f_+ = 39 mb

0_1 = +0.11 mb, s_- = 2995 GeV^2, f_- = -1.0 mb.

BGN point out that qualitatively the same solution had been found (with an additional constraint $f_- = 0_1\pi^2/4$ and with Regge amplitudes more restricted) by Kang and Nicolescu[25] in 1975 from the limited t = 0 data available then, in the first serious investigation of the possibility of a maximal Odderon amplitude. Indeed, the Kang-Nicolescu fit predicted $\bar{\sigma}_{tot}$ = 63 mb and $\bar{\rho}$ = 0.27 at $\sqrt{s}$ = 546 GeV, in remarkable agreement with the data obtained almost a decade later.

These BGN parameters differ greatly from those of GLN, of course, most strikingly in the sign of the maximal Odderon 0_1. This requires that σ_{tot} should lie increasingly above $\bar{\sigma}_{tot}$ as $\sqrt{s}$ increases beyond 100 GeV, as shown in Fig. 3, i.e. that $\Delta\sigma$ should change sign at about 100 GeV and become increasingly negative above that energy, as shown in Fig. 13, contrary to our natural expectation that the additional final states resulting from pp annihilation should increase the $\bar{p}p$ interaction cross section relative to that for pp. The situation re ρ and $\bar{\rho}$ is now reversed; ρ is predicted to lie below $\bar{\rho}$ in the $\sqrt{s}$ range accessible for experiment in the foreseeable future, ultimately to change sign and finally to reach the limit $\rho(s = \infty) = -\bar{\rho}(s = \infty)$.

Igi and Kroll[26] have raised an important quantitative objection to all these Odderon models. This runs parallel to an argument by Igi[27] a long time ago, about the crossing-odd πN amplitude. This related the πN scattering lengths at threshold to a pion-pole term and an integral over $(\sigma(\pi^+p) - \sigma(\pi^-p))$, following the Goldberger-Miyazawa-Oehme sum rule,

$$\text{Re}(F^-(m)-F^+(m)) = \frac{m}{M^2} \cdot \frac{1}{(1-\frac{m^2}{4M^2})} \cdot \frac{G^2}{4\pi} + \frac{m}{2\pi^2}\int_m^\infty \frac{\sigma^-(E')-\sigma^+(E')}{(E'^2-M^2)} dE' \tag{6.10}$$

where M,m are the masses of the proton and the pion, respectively, and E denotes the pion energy in the laboratory frame. This sum rule is obtained under the assumption that $\int^E(\Delta\sigma(E'))dE'/E'$ is finite as $E \to \infty$. It is well satisfied experimentally, if $\Delta\sigma$ is assumed to fall like $E^{-\gamma}$ for some positive γ for E above the available measurements, and has even been used as one means of determining the πNN coupling constant $G^2/4\pi$. Igi argued that if the integral were not convergent, the sum rule would not hold, whereas it

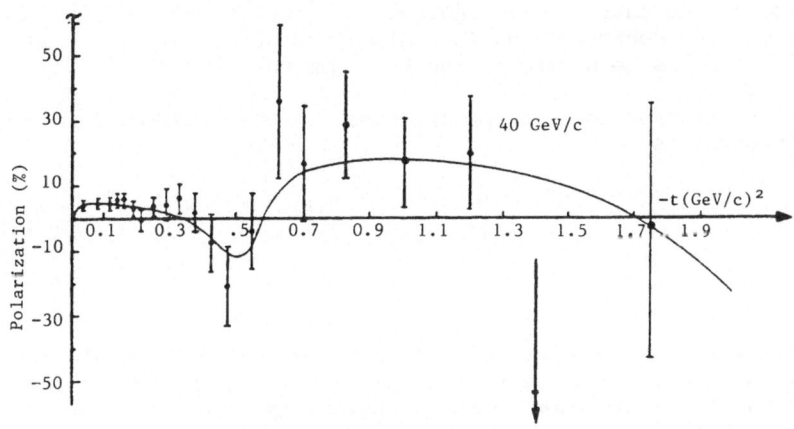

Fig. 15 (above).
Polarization measured
in $\pi^- p \rightarrow \pi^0 n$ at $p_L = 40$
GeV.c[22]. For predicted
curve, see GNL[23].

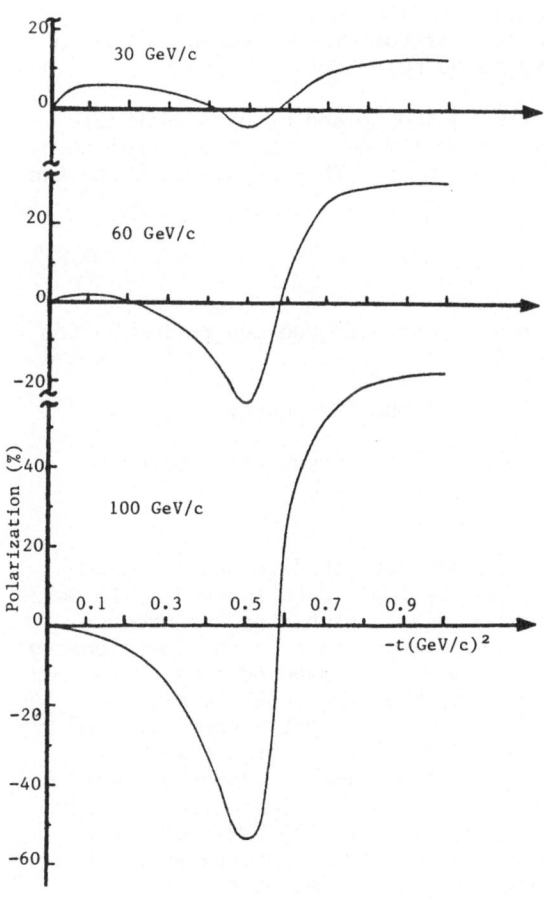

Fig. 16. Predicted
polarizations in
$\pi^- p \rightarrow \pi^0 n$, as function
of $-t$ for $p_L = 30$, 60
and 100 GeV/c, as given
by GNL[23], with the in-
clusion of an Odderon
term. For data, see
Fig. 15 above.

is well satisfied by the data. This argument is circular, but it does raise doubt about any Odderon contribution, for this would break the simple and happy agreement which has been reached for this sum rule (6.10).

The analogous relation for the ($\bar{p}p$-pp) case, in the laboratory frame for the target proton, is

$$\text{Re}F_-(M) = +\frac{1}{4m} \cdot \frac{1}{(1-\frac{m^2}{4M^2})} \cdot \frac{G^2}{4\pi} - \frac{M}{2\pi} \int_{\epsilon(2m)}^{M} \frac{\text{Im}\bar{F}(E')dE'}{(E'^2-M^2)} - \frac{M}{4\pi^2} \int_{M}^{\infty} \frac{\Delta\sigma(E')dE'}{\sqrt{(E'^2-M^2)}}$$

(6.11)

where the integrals are to be understood as principal value integrals. The empirically unknown regions are the unphysical region ranging from $E' = \epsilon(2m) = (-M+2m^2/M)$, the antiproton energy at which $\bar{p}+p \to$ the threshold 2π-state, to $E' = M$, where the incident $\bar{p}$ has zero momentum, and the high energy region above pp c.m. energy 62 GeV (i.e. for lab. energy E' above $\epsilon(62 \text{ GeV}) \approx 1.9$ TeV), where $\Delta\sigma$ is not known. Igi and Kroll evaluate the integral over the unphysical-region according to the best theoretical advice, and determine the other terms from empirical data, concluding that, if this sum rule is valid, the unknown high-energy integral is well given simply by extrapolating $\Delta\sigma(E)$ by the power law $26(E/\text{GeV})^{-0.58}$ mb, a canonical Regge behaviour. In other words, they conclude that the high energy integral $\int \Delta\sigma dE'/E'$ is convergent, which excludes any Odderon term. Again, the argument is circular. An Odderon contribution would nullify the unsubtracted sum rule; it would be necessary to consider a suitably subtracted sum rule instead and the outcome of this is uncertain, although Igi and Kroll consider this possibility to be ruled out.

Block and Cahn[28] have also made fits to the $\bar{p}p$ and pp data available through the ISR energy range and below (to $\sqrt{s} = 5$ GeV) with the inclusion of an Odderon amplitude as well as a Froissaron term. They modulated these two amplitudes by the energy-dependent factor

$$\{1 + a|\ln(s/s_0)-i\pi/2|^2\}^{-1}$$

(6.12)

which limits the total cross section as $s \to \infty$ to a degree controlled by the parameter a. Their conclusions were that:

(i) there was no need for any Odderon term in the ISR energy range,

(ii) there was no firm conclusion that σ_{tot} should necessarily approach

$\approx (\ln s)^2$ as $s \to \infty$.

Their later work[29-32] included the 546 GeV and 900 GeV data and focussed attention on high energy behaviour, using only input data above $\sqrt{s} = 15$ GeV. In this work[29,30] they omitted any Odderon contribution - on the grounds that it had not been found necessary in fitting the data in the lower energy regime, a somewhat dubious argument - and their fit favoured a non-zero value for the parameter a in (6.12), implying that the total sections approach constant values as $s \to \infty$. This work has recently been refined[31,32] and their best fit parameters now give $a = 7.4 \times 10^{-3}$, in which case they predict $\sigma_{tot} = 94$ mb at $\sqrt{s} = 40$ TeV, an estimate almost a factor 2 lower than the value 175 (s.d. +40,-27) mb, and well below the lower limit of 130 mb, which Gaisser, Sukhatme and Yodh[33] obtain for $\sqrt{s} = 30$ TeV from their analysis of Cosmic Ray data. On the other hand, if the constraint $a = 0$ were imposed, Block and Cahn would predict $\sigma_{tot} = 138$ mb at 40 TeV.

7. THREE-GLUON EXCHANGE: THE MODEL OF DONNACHIE AND LANDSHOFF

The analysis of pp and p̄p scattering must be linked with the theory of QCD, and this might be most convincingly possible in the regime of large momentum transfers. It has been known for quite a long time that, over the energy range √s from 30 to 65 GeV in ISR experiments and for large momentum transfer $-t > 3.5$ GeV^2, the pp elastic scattering cross section obeys the relation

$$d\sigma/dt = C(t)^{-8} \qquad\qquad (7.1)$$

where C is a constant, with value 0.09 mb(GeV)^{14}. In 1979, Donnachie and Landshoff (referred to as DL, below) pointed out[34] that this t-dependence is in accord with the process of three-gluon exchange between the two protons, in which each quark in the proton emits or absorbs one gluon, as is illustrated in Fig. 17 below. The three-gluon system exchanged is necessarily a colour singlet and has charge-conjugation parity C = -1. The empirical result (7.1) made it appear natural to DL that this three-gluon exchange may be the dominant C = -1 exchange in the large $|t|$ domain.

This term is not dominant in the small $|t|$ domain. Empirically, the forward pp elastic scattering amplitude is dominantly imaginary, whereas this ggg term is real. Naturally, according to QCD arguments, the small $|t|$ domain is dominated by multi-gluon exchange, and DL represent the amplitude in this domain by means of Pomeron exchange. Apart from an even-signature Regge factor, DL take the Pomeron-proton coupling to have the form[35]

$$3\beta(\gamma_\mu F_1^S(t) + i\sigma_{\mu\nu}q^\nu F_2^S(t)) \quad , \qquad\qquad (7.2)$$

where β denotes the qqP coupling amplitude and the $F_i^S(t)$ are the isoscalar components of the electromagnetic form factor for the nucleon. In other words, DL assume that the Pomeron couples with the nucleon like a C = +1 photon. This possibility is not readily excluded. The nucleonic form factor $F_2^S(t)$ is well known to be small and with (7.2) this implies that there is very little spin-flip in the assumed Pomeron-nucleon coupling; we know well empirically that this is indeed the case. The form factor $F_1^S(t)$ is taken by DL from the form factors known for the proton and neutron, and this turns out to work quite well. The t-dependence of single Pomeron exchange then results both from $F_1^S(t)$ and from the slope α_1 of the Pomeron trajectory, which is assumed to be of the form

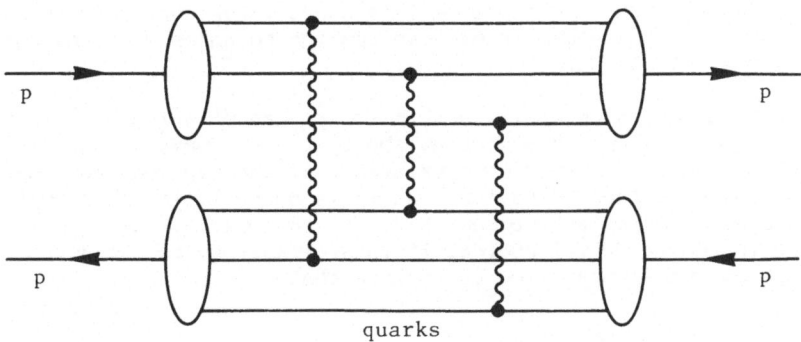

quarks

Fig. 17. Three-gluon-exchange graphs for pp scattering, of the kind considered by DL[34] to be dominant for large $|t|$, which lead to the amplitude F_{ggg} and the cross section (7.1).

$$\alpha(0) = \alpha(0) + \alpha_1 t \quad . \tag{7.3}$$

The increase of total cross section with increasing s is accounted for by assuming that

$$\alpha(0) = 1 + \epsilon , \tag{7.4}$$

where ϵ is positive but small (with value about 0.1). The argument is that this will not necessarily violate the Froissart-Martin bound in Asymptotia since P-exchange will become modified by the branch cuts due to multi-Pomeron exchanges, and that we are concerned here only with a limited energy range, over which the observed rise in $\sigma_{tot}(s)$ with s can be mocked up by this device. For small $|t| \lesssim 0.3$ GeV^2, the P-exchange amplitude leads to the cross section

$$4\pi d\sigma/dt = (3\beta F_1^S(t))^4 (s/M^2)^{2(\epsilon + \alpha_1 t)}, \tag{7.5}$$

which fits the data rather well and determines the value of β, ϵ and α_1, the latter being $\alpha_1 = 0.25$ GeV^{-2}, as we used earlier above.

Single Pomeron-exchange is not sufficient beyond -t = 0.3 GeV^2, since it cannot account for the dip structure observed at about -t = 0.9 GeV^2 in the pp differential cross section, as shown in Fig. 6. DL have suggested that this may be accounted for as the result of a cancellation between the one-Pomeron exchange amplitude and the other multi-gluon exchange terms, a number of which they have calculated[36]. More recently, they have focussed on the two-Pomeron exchange process[37], in order to make their point clear; its imaginary part automatically has the sign needed for cancellation with the P-exchange term. There are then three essential amplitudes to combine, to make the pp elastic scattering amplitude,

$$F = F_P + F_{PP} + F_{ggg} , \tag{7.6}$$

where F_P and F_{PP} contribute to F_+, since they represent C = +1 exchange in the t-channel. F_P is imaginary and dominant for small $|t|$, F_{PP} is always complex, and F_{ggg} gives the only contribution to F_-, being real but having a rather singular dependence on $|t|$, as (7.1) shows, so that DL[36] found it necessary to suppress the $\bar{q}qg$ coupling for momentum transfers -t < (300 MeV)^2. However, the precise details are not important; the essential point is to show how these amplitudes can combine to agree with the outstanding features of the data.

The dip in $d\sigma/dt$ is generally attributed to cancellation between P and PP exchange, giving Im($F_P + F_{PP}$) ≈ 0 at about -t = 0.9 GeV^2 in the ISR-Collider energy range. However, the existence of the dip requires that ReF should also vanish in the dip region, and it is quite implausible that Re($F_P + F_{PP}$) should also vanish at the same t-value. However, since F_{ggg} is real and dominant in the large $|t|$ region, it is available to cancel Re($F_P + F_{PP}$) in this t-region; thus we are led to suppose that

$$\text{Re}(F_P + F_{PP}) + F_{ggg} \approx 0 \tag{7.7}$$

near -t = 0.9 GeV^2, in order to account for the observed dip in the $d\sigma/dt$ cross section for pp.

What check do we have on this mechanism? When we turn to consider $\bar{d}\sigma/dt$ for $\bar{p}p$ scattering, we realise that

$$\bar{F} = F_p + F_{p\bar{p}} - F_{ggg} \ , \tag{7.8}$$

since F_{ggg} is odd under the reversal of C. $\text{Im}\bar{F}$ is $\text{Im}(F_P+F_{PP})$, which must vanish in the dip region, as we have just learned above, but $\text{Re}\bar{F}$ is now non zero in this region, since $\text{Re}(F_P+F_{PP})$ and (F_{ggg}) have the same sign, so that they fill in the dip arising from the vanishing of $\text{Im}F$. This pre- diction concerning the absence of a dip in $d\bar{\sigma}/dt$ was made before both the ISR data and the Collider data were known and was therefore a significant qualitative success. It was then easy for DL to obtain a good quantitative fit to all the $d\sigma/dt$ and $d\bar{\sigma}/dt$ data in the ISR energy range but the para- meters thus obtained still understate the cross sections $d\bar{\sigma}/dt$ in the shoulder region around $-t = 1.2$ GeV^2 by about a factor 2, as shown on Figs. 6 and 7.

This model may appear to owe its success to rather detailed cancell- ations or constructive interferences resulting from the mechanisms invoked. However, we must emphasize that this is the situation for any of the models considered in this review. To illustrate this point, we have plotted on Fig. 18(a) the magnitudes of the real and imaginary parts of the amplitudes $F_+(s,t)$ and $F_-(s,t)$ for the GLN model as function of $|t|$ for $\sqrt{s} = 546$ GeV. This shows that $\text{Re}F_-$ is the dominant contributor to $d\bar{\sigma}/dt$ in the shoulder region from about $-t = 1.2$ GeV^2 onward, while $\text{Im}F_+$ is, as expected, the dominant contributor in the forward peak, out to about $-t = 0.6$ GeV^2. In the dip region, all four amplitudes have comparable magnitudes, and the net result is a matter of some detail. As $\sqrt{s}$ increases, the contributions change in their characteristic ways. The net $d\bar{\sigma}/dt$ is shown as function of $|t|$ for a series of $\sqrt{s}$ values from 2 TeV to 40 TeV on Fig. 11 for the BSW[8] fit and on Fig. 12 for the GNL[18] fit. The magnitudes of $\text{Re}(F_\pm)$ and $\text{Im}(F_\pm)$ at 40 TeV are plotted against $|t|$ on Fig. 18(b). The F_+ amplitude displays the well-known characteristics of a diffraction amplitude, with dips in $|\text{Im}F_+|$ where $|\text{Re}F_+|$ has its maxima, and vice versa. With the GNL model, the F_- terms are dominant for $|t|$ greater than about 0.5 GeV^2. However, the GNL ansatz is rather extreme. It will be of the greatest interest, in due course, to learn how pp and $\bar{p}p$ scattering behave at the LHC, SSC and ELOISATRON energies.

8. CONCLUSION

The models used above for comparison with experiment are rather diverse in their nature and their predictions. The GNL model[18] boldly assumes a maximal Odderon, which implies in turn differences between pp and $\bar{p}p$ cross sections which increase with increasing energy. The BSW model[9] would require the pp and $\bar{p}p$ cross sections to be equal, apart from Coulomb effects, well before the Collider energy of 546 GeV. However, we see from Fig. 1 that, although the $\bar{\sigma}_{tot}$ values lie on a smooth curve, the σ_{tot} values do not appear to join smoothly with the calculated curve which fits the $\bar{\sigma}_{tot}$ values, in the way which BSW require. These data suggest that there might still be quite an appreciable difference $\Delta\sigma$ for energies as high as 200 GeV. There is a strong case, even though it can be done only for $\bar{p}p$ at present, to have total cross section, differential elastic cross section and real forward elastic amplitudes measured for some intermediate energies between 62 and 546 GeV, in order to know the precise $\sqrt{s}$ dependence of these quan- tities in the readily accessible range, since these will give us clues concerning the values these quantities may have beyond this accessible range. Of course, we also look forward now to data on these questions for

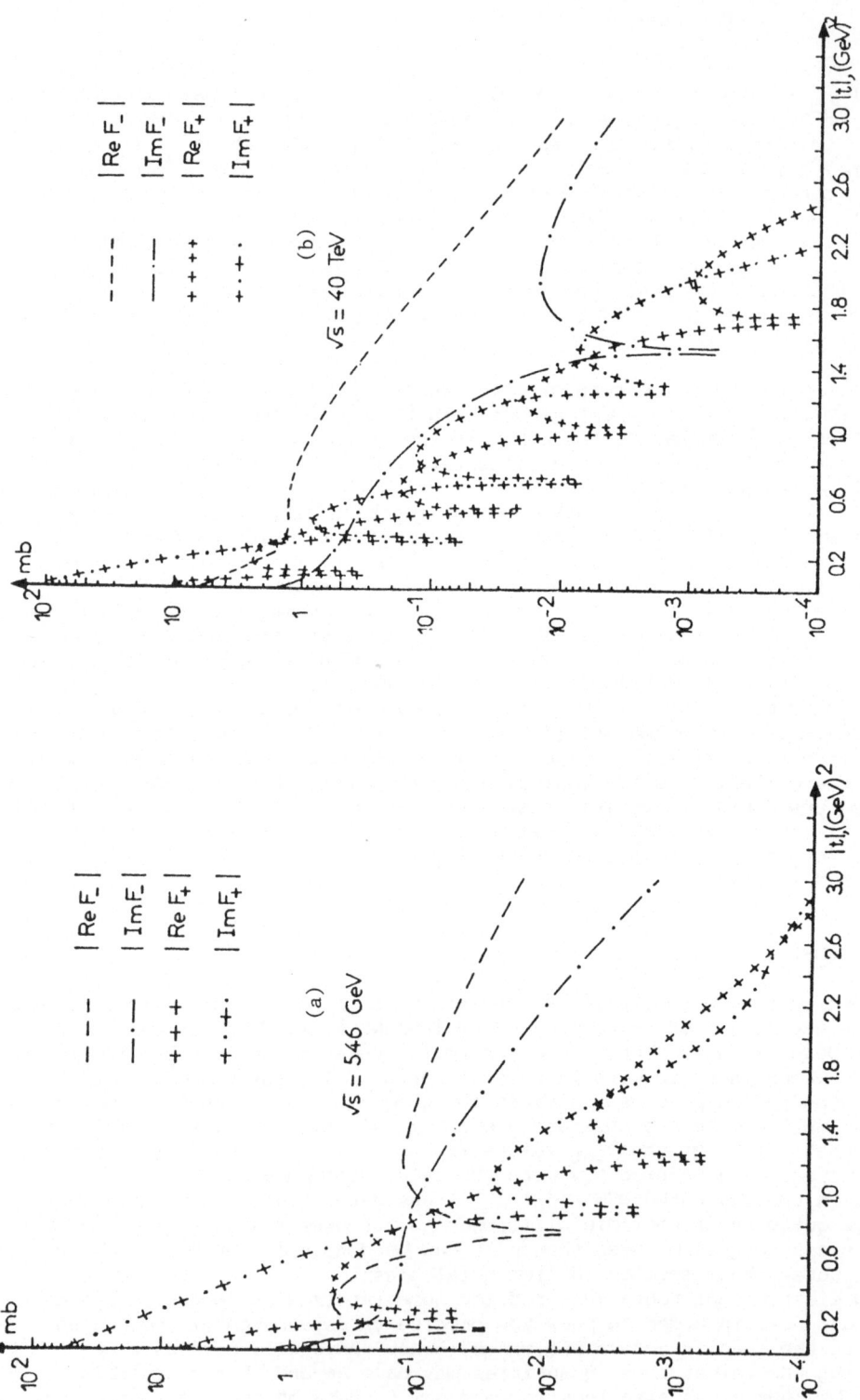

Fig. 18. The magnitudes of the real and imaginary parts of the amplitudes $F_{\pm}(s,t)$ are plotted as function of the momentum transfer $|t|$ for (a) $\sqrt{s} = 546$ GeV, and (b) $\sqrt{s} = 40$ TeV, as calculated with the model of GNL18.

p̄p, both at 900 GeV with the CERN collider and at 1800 GeV with the Tevatron at Fermilab. The BGN model[24] also assumes a maximal Odderon, but the inclusion of the new ρ̄ value[14] at 546 GeV in the fit causes such a large change in its parameters that the difference $\Delta\sigma = (\bar{\sigma}_{tot} - \sigma_{tot})$ may well have opposite sign to that for the GNL parameters. Even if there is no maximal Odderon contributing, there can still be amplitudes generated by the exchange of other C = odd systems. The economical and dynamical DL model[37] is an example, and this predicts quite large differences between dσ/dt and dσ̄/dt in the dip region for the former, even though these cross sections may be equal elsewhere. All of these models also make quite different predictions for these differential cross sections in the regime of large |t| values, beyond -t = 1.5 GeV^2.

These remarks make it clear how vital it is to have both p̄p and pp scattering and reaction data in the same energy range, preferably at the same values of √s. When only one of these two sets of data are available, the analysis can leave quite crucial questions without an answer. A knowledge of the pp cross sections at √s = 546 GeV would allow us to test the theoretical models proposed, and this might exclude some of the models from further consideration. Any further colliders must have two rings to allow measurement of both pp and p̄p cross sections.

Data on elastic scattering for both p̄p and pp in the region of small t is also essential. We have seen above how important the recent measurements of ρ̄ at 546 GeV has been for the assessment of current models; the dust has not yet settled concerning the questions raised and it will not until we have more p̄p data, and preferably more pp data, especially in the √s range up to 546 GeV, although the latter are necessarily far in the future, perhaps not until the mid-1990s, with the SSC and/or the ELOISATRON.

The small t data are needed for:

(i) the determination of ρ = ReF/ImF and similarly ρ̄, for t ≈ 0. This requires measurements of elastic scattering in the Coulomb-nuclear interference region. For a 40 TeV Collider, this means measurements on the outgoing p or p̄ for rather small angles of deflection, namely

$$\theta \approx 8 \times 10^{-7} \text{ radians,} \tag{8.1}$$

to cover fully this interference region. The possibility of such measurements has been considered carefully by Orear[38], with the SSC in mind. If the detector is placed 10 km downstream, the displacement of the proton or antiproton from the path it would have followed if it had not been scattered, is only 0.8 cm. The measurement of this displacement is certainly difficult to achieve and it will need tracking of individual protons or antiprotons before, in, and after the p̄p collision region. Orear argues that this may be considered quite feasible, especially if the Collider is built with a straight section of length of order 10 km provided specifically for this purpose, and DeSalvo et al.[39] have published a design proposal for a detector suitable for these measurements, a multitube drift chamber which can be inserted directly into the beam pipe.

(ii) the Coulomb scattering, observed at still smaller angles of deflection, say

$$\theta \approx 5 \times 10^{-7} \text{ radians ,} \tag{8.2}$$

for √s = 40 TeV, where the Coulomb scattering cross section dominates the nuclear scattering cross section by more than an order of magnitude. Its observation would allow a direct and absolute determination of the luminosity L, which would be of immense value. The calibration of L in this way need not be done for every experimental run. At a definite $\sqrt{s}$, the measurement of the rate for some more readily observed reaction can be made relative to the Coulomb scattering rate at that energy, in order to calibrate absolutely the cross section of the former, in which case the former can then be used in routine operation of the Collider, to serve as a means to measure the luminosity effective during other experiments.

This facility for such small angle measurement is of major importance and must be built into future Colliders, such as the ELOISATRON, right from their first design stage.

The importance of the measurements of ρ and $\bar{\rho}$ has been mentioned many times, most recently by Martin[40] in connection with the measurement of $\bar{\rho}$ at $\sqrt{s}$ = 546 GeV by the UA4 collaboration[14], namely that ρ and $\bar{\rho}$ are connected by forward amplitude dispersion relations with the pp and $\bar{p}p$ total cross sections at energies beyond those physically accessible at the time and can therefore give some indications of the $\sqrt{s}$ dependence of σ_{tot} and $\bar{\sigma}_{tot}$ in that physically inaccessible region. Martin[40] and Leader[41] have both emphasized that a large rise in the $\bar{p}p$ total cross section around 1 TeV could account for the large $\bar{\rho}$ value observed at 546 GeV. Such a rise could occur through the presence of some resonance-like phenomenon, beyond which energies the cross section might settle down again to a slow $(\ln s)^2$ rise with increasing $\sqrt{s}$, or because of some threshold for new reaction processes not possible at present energies. The Collider measurement of $\bar{\sigma}_{tot}$ = 65.3±1.7 mb at 900 GeV by the UA5 collaboration[42] shows no sign of any such increase but lies well on (or even below) the calculated curve for $\bar{\sigma}_{tot}$ extrapolating beyond the earlier measurement at 546 GeV. In this respect, the $\sqrt{s}$ = 1.8 TeV total cross section being measured now at the Tevatron I will be of particular interest, for it could well rule out this hypothesis of a sudden and quite large rise in $\bar{\sigma}_{tot}$ as an explanation for the large value measured at 546 GeV.

It will also be of particular interest to measure $d\bar{\sigma}/dt$ and $d\sigma/dt$ for large $|t|$ values, first of all for the present $\sqrt{s}$ range, since the various models discussed above differ widely in their predictions for $-t > 1.5$ GeV^2, but also for higher $\sqrt{s}$ values. This appears quite feasible at Colliders, and it may even become a little easier as $\sqrt{s}$ goes up, because the large-$|t|$ differential elastic cross sections are predicted by the Odderon models to increase with $\sqrt{s}$. Such an increase with $\sqrt{s}$ is already seen in the comparison between the ISR and the CERN Collider measurements over the $|t|$ region of 1 to 2 (GeV/c)^2.

To conclude, we may sum up by remarking that the measurements made on total, Coulomb interference elastic and high-$|t|$ elastic cross sections to date have been immensely fruitful, but that the absence of pp data above $\sqrt{s}$ = 62 GeV to match the $\bar{p}p$ data up to 900 GeV has limited greatly the conclusions which can be drawn from this data. This difficulty must be rectified in the final ELOISATRON proposal, irrespective of what its energy may be, or of what form it may take. The study of the pp-$\bar{p}p$ system will then take us far into Asymptotia, a region which we physicists have always wished to explore particularly in the hope that some simplicity in the fundamental interactions might emerge there beyond the already known and remarkable degree of unification which has become established as the SU(3)$_C$ x SU(2)$_W$ x U(1) symmetry scheme, a hope whose fulfillment in this way has proved quite elusive to date.

REFERENCES

1. A. Martin, Proc. 6th Int. Conf. High Energy Spin Physics, Marseille 1984 (ed. J. Soffer), in J. Physique $\underline{46}$, Coll. 2 suppl.,p. C2-727 (Fevrier 1985).
2. H. Cheng and T.T. Wu, Phys. Rev. Lett. $\underline{22}$, 666 (1969); Phys. Rev.$\underline{182}$, 1852, 1868 & 1899 (1969).
3. UA4 Collaboration. M. Bozzo et al. Phys. Lett. $\underline{B147}$, 385 (1984).
4. Ibid, Phys. Lett. $\underline{B155}$, 197 (1985).
5. R.J. Glauber and J. Velasco, Phys. Lett. $\underline{B147}$, 380 (1984).
6. F. Borkowski, G. Simon, V. Walther and R. Wendling, Nucl. Phys. $\underline{B93}$, 461 (1975).
7. T.T. Chou and C.N. Yang, Phys. Rev. Lett. $\underline{128B}$, 457 (1983).
8. C. Bourrely, J. Soffer and T.T. Wu, Phys. Rev. $\underline{D19}$, 3249 (1979).
9. Ibid, Nucl. Phys. $\underline{B247}$, 15 (1984)
10. Ibid, Phys. Lett. $\underline{196B}$, 591 (1987).
11. J.Soffer, Impact-picture predictions for pp and $\bar{p}p$ elastic scattering at high energies, Orsay preprint CPT-87/P.2057 (December 1987). {Proc. 2nd Intl. Conf. on Elastic and Diffractive Scattering, held on 15 - 18 October 1987 at Rockefeller Univ. New York (in Press).}
12. UA4 Collaboration. M. Bozzo et al., Phys. Lett. $\underline{B147}$, 392 (1984).
13. UA4 Collaboration. D. Bernard et al., Phys. Lett. $\underline{B171}$, 142 (1986).
14. UA4 Collaboration. D. Bernard et al., Phys. Lett. $\underline{B198}$, 583 (1987).
15. C. Bourrely, J. Soffer and T.T. Wu, preprint CPT-87/P.2032 (August, 1987), Z. Phys.C (in press).
16. A. Martin, Z. Phys, $\underline{C15}$, 185 (1982).
17. J.Fischer, R. Saly and I. Vrkoc, Phys. Rev. $\underline{D18}$, 4271 (1978).
18. P. Gauron, B. Nicolescu & E. Leader, Nucl. Phys. $\underline{B299}$, 640 (1988).
19. J. Fischer, Z. Phys. $\underline{C36}$, 273 (1987).
20. G. Auberson, T. Kinoshita and A. Martin, Phys. Rev. $\underline{D3}$, 3185 (1973).
21. P.Gauron, B. Nicolescu and L. Szymanowski, A possible field theoretical description of the Odderon, Orsay preprint IPNO/TH87-53 (June 1987).
22. V.D. Apokin et al., Z. Phys, $\underline{C15}$, 293 (1982).
23. P.Gauron, B. Nicolescu and E. Leader, Phys. Rev. Lett. $\underline{52}$, 1952 (1984).
24. D. Bernard, P.Gauron and B. Nicolescu, Phys. Lett. $\underline{B199}$, 125 (1987)
25. K. Kang and B. Nicolescu, Phys, Rev, $\underline{D11}$, 2461 (1975).
26. K. Igi and P. Kroll, A sum rule for the (pp-$\bar{p}$p) forward amplitude - evidence against the odderon, preprint CERN-TH.4891/87 (November, 1987).
27. K. Igi, Phys. Rev. $\underline{130}$, 820 (1963).
28. M.M. Block and R.N. Cahn, Rev. Mod, Phys. $\underline{57}$, 563 (1985).
29. Ibid, Phys. Lett. $\underline{B168}$, 151 (1986).
30. Ibid, Proc. XXIst Rencontre de Moriond on Strong Interactions and Gauge Theories (ed. J. Tran Thanh Van, Edition Frontieres, Gif sur Yvette, France, 1987).
31. Ibid, Phys. Lett. $\underline{B188}$, 143 (1987).
32. Ibid, High energy predictions for pp and $\bar{p}p$ elastic scattering and total cross sections, Rept. LBL-23472 (May, 1987).
33. T.K. Gaisser, U.P. Sukhatme and G.B. Yodh, Phys.Rev. $\underline{D36}$, 1350 (1987).
34. A. Donnachie and P.V. Landshoff, Z. Phys. $\underline{C2}$, 55 (1979).
35. Ibid, Phys. Lett. $\underline{123B}$, 345 (1983).
36. Ibid, Nucl. Phys. $\underline{B231}$, 189 (1984).
37. Ibid, in Elastic and Diffractive Scattering at the Collider and beyond (ed. B. Nicolescu and J. Tran Thanh Van, Editions Frontieres, Gif sur Yvette, France, 1985) p.209.

38. J. Orear, Proc. 1984 Summer Study on Design and Utilization of the Superconducting Super Collider (eds. R. Donaldson and J.G. Morfin, Div. Particles & Fields, Amer. Phys. Soc.) p.743.

39. R. DeSalvo, J. Orear, R. Maleyran and S. Zuchelli, A small angle elastic scattering detector for the LHC and/or SSC, Cornell Univ. preprint CLNS 86/724 (April 1986).

40. A. Martin, Are we seeing a new threshold in $\bar{p}p$ scattering?, preprint CERN-TH.4852/87 (September 1987). {Proc. 2nd. Intl. Conf. on Elastic and Diffractive Scattering, held on 15 – 18 October 1987 at Rockefeller Univ. New York (in press)}.

41. E. Leader, Phys.Rev. Lett. 59, 1525 (1987)

42. UA5 Collaboration, G.J. Alner et al., Z. Phys. C32, 153 (1986).

ROLE OF MULTIPLE PARTON PROCESSES AT COLLIDERS

Nello Paver

University of Trieste, Italy
INFN, Sezione di Trieste, Italy

ABSTRACT

Multiple parton interactions are expected to occur in high energy hadron-hadron inelastic collisions in the so-called small-x region of the kinematical variables, and to have interesting physical effects there. We discuss, in particular, the role of such processes in the production of four (mini) jets, and in the unitarization of the total "hard" hadron- hadron cross section.

INTRODUCTION

High p_T hadronic jet studies [1,2,3] at collider energies have lead to spectacular confirmations of the QCD-parton model description of highly inelastic hadron-hadron collisions, based on hadronic structure and elementary "hard" interactions among the quark and gluon hadron constituents [4,5]. The hadronic cross sections are large enough that, besides the dominant two-jet events, considerable statistics is available also for three- and four-jet events, which are expected, with rates suppressed by powers of α_s, from higher order QCD processes, and which accordingly should allow detailed tests of the underlying dynamics.

As well-known, the leading QCD mechanism for high p_T jet production in hadron-hadron collisions results in the following expression for the inclusive jet cross section [6]:

$$d\sigma^{n-jets} \sim \sum_{a,b} F_a(x_1,Q^2) \, F_b(x_2,Q^2) \, d\hat{\sigma}_{ab}(2 \to n) . \tag{1}$$

In eq.(1) $d\hat{\sigma}_{ab}$ (with a,b the partons species) are the elementary, "hard" $2 \to n$ parton-parton cross sections, to be computed using perturbative QCD. Analytic expressions for all subprocesses are available in the cases n=2[7], n=3[8] and n=4[9]. Calculations for more than four jets are only possible, at present, by Monte Carlo simulations [10]. In all cases the gluon-gluon and the gluon-quark subprocesses appear to be the dominant ones.

The F's in eq.(1) are the partons longitudinal fractional momentum distributions in the colliding hadrons, and thus characterize the incoming parton flux. The variable Q^2, on which they depend in addition to x, denotes the "large" mass, or equivalently the typical virtualities involved in the process under consideration, of the order $Q \sim P_T \gg M_{hadron}$. There exist different possible parametrizations of quark and gluon densities [11], consistent with the Q^2-evolution equation [12].They all have the common important feature of sharply increasing as $x \to 0$, and, moreover, gluons by far dominate this large parton flux for sufficiently small x.

The point now is that the values of the incoming partons fractional momenta, relative to jet events with given total transverse jet energy E_T, are fixed by kinematics to be of order $x_{1,2} \sim 2E_T/\sqrt{s}$, with $\sqrt{s}$ the collider total center of mass energy. One then expects from eq. (1) the jet yields, σ^{n-jets} $(E_T, \sqrt{s})$, to sharply increase with $\sqrt{s}$, as the consequence of parton distributions being probed at smaller and smaller x, to the point of becoming an appreciable fraction of the total hadronic cross section $\sigma_{tot}(\sqrt{s})$. Also, as the parton flux composition changes, gluon-gluon physics should dominate for large enough $\sqrt{s}$.

This is represented qualitatively in Fig.1 [13], which shows multijet cross sections, as given by eq.(1),integrated from a lower p_T cutoff.

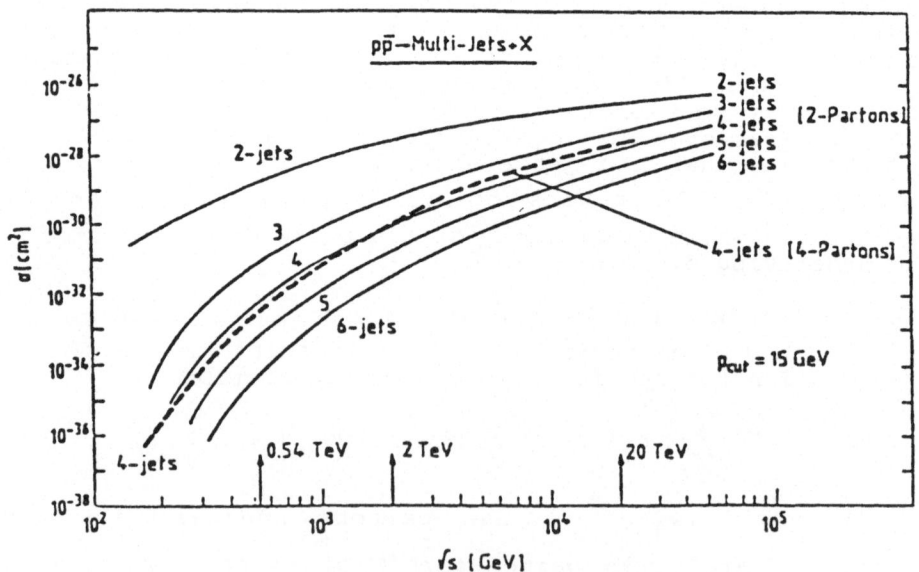

Fig.1. Cross sections for multijet events

This behavior is confirmed experimentally, and actually is the reason why the observable jet rates have become so copious in going from the ISR ($\sqrt{s}$ =63 GeV) to the S$\bar{p}$pS collider ($\sqrt{s}$ = 630 GeV), allowing such spectacular progresses in this field.

Hadronic jet analyses have been extended down to the "moderate" p_T region, i.e. to jets with $p_T \gtrsim p_T^{min}$ ~5 GeV [14]. These "minijets" seem to represent a sizeable fraction of the totality of inelastic events, which steadily increases with the collider energy $\sqrt{s}$. The corresponding values of incoming partons fractional momenta, of order $x \gtrsim 10^{-2}$ at the Collider, would become for such a value of p_T^{min} as small as $x \sim 10^{-4}$ at a supercollider with $\sqrt{s}$ = 100 TeV. In this small-x region the transition from partonic physics to Regge-dominated dynamics should take place.

On the other hand, a p_T^{min} of 5 GeV or so looks still large enough, compared to M_{proton}, to eventually justify the use of perturbative QCD to estimate quark and gluon interactions, and to continue to speak the parton model language in the above mentioned region also. In so doing, however, modifications to the basic, leading order QCD parton model emerge, giving rise to a new class of "semihard" phenomena [15,16].

Firstly, at small x terms of order $\alpha_s(p_T) \ln x$ can become of order unity, in spite of $\alpha_s(p_T)$ still being small, and eventually spoil the use of perturbation theory. Such terms have to be resummed, in order that the perturbative expansion is reset. Recent estimates [17] seem to indicate that, owing to partial cancellations, these ln x corrections should be relatively unimportant in the range of x and Q^2 relevant to minijets.

The other, interesting effect is the "overcrowding" of partons at small x, as the consequence of parton densities being increasingly large as $x \to 0$ [15,16]. The emergence of physical consequences of this parton overcrowding depends on the effective behaviour of parton densities for small x (notice that in practice only gluons matter in this limit). For example, the Q^2-evolution equation for the gluon distribution, with the effect of quarks neglected for simplicity, admits an analytic solution which behaves for $x \approx 0$ as:

$$g(x,Q^2) \propto \frac{1}{x} \exp\left[\sqrt{\frac{144}{33-2N_f} \ln(1/x) \ln\left(\frac{\ln Q^2/\Lambda^2}{\ln Q_o^2/\Lambda^2}\right)} \right] , \qquad (2)$$

and thus is more singular than 1/x. In the general case, where $g(x,Q^2)$ is determined by solving numerically the coupled evolution equations of quarks and gluons, a behavior faster than 1/x is similarly obtained.

One can define a relative gluon density within the proton:

$$W(x,Q^2) = \alpha_s \ (Q^2) \ \frac{xg(x,Q^2)}{Q^2R^2} \quad , \tag{3}$$

where R is a typical hadronic radius, of the order of 1 fm. Eq.(3) simply states that in a proton-proton highly inelastic collision partons have an effective cross section area of order α_s/Q^2, with $Q\sim p_T$, while all longitudinal dimensions are squeezed by Lorentz contraction.

With $g(x,Q^2)$ steeper than $1/x$, for values of x sufficiently small $W(x,Q^2)$ can approach unity and partons spatially overlap. In that region of x the simple QCD parton model description of eq.(1), based on independent hard parton-parton collisions, stops being the dominant one, and there should be the onset of new kinds of phenomena, generated by more complicated, semihard parton processes, favoured by parton overcrowding. The smaller the x, the more likely those effects should be, and as such they should represent an interesting issue in hadronic physics at the multiTev colliders.

For example, gluon coherence and recombination effects should "shadow" the $x\to0$ increasing behaviour of $g(x,Q^2)$ [18].

Similarly, multiple parton scattering should show up with considerable rates in multijet production; in addition, it should prevent the indefinite rise of the "hard" hadronic cross section with $\sqrt{s}$, which would be implied by the leading QCD alone, in apparent contrast with unitarity. These last two points will be the subject of the following Sections.

MULTIPARTON SCATTERING

The simplest example is represented by the double-disconnected parton scattering (DDS) depicted in Fig.2. In this process two partons of the projectile have independent, hard collisions with two different partons of the target, giving origin to four high p_T final jets (actually four minijets as we are at low x).

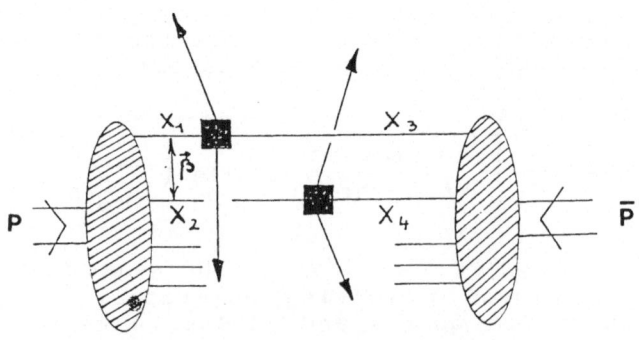

Fig.2. Double parton scattering

The $4 \to 4$ partonic amplitude is disconnected, namely it is factorized into the product of two $2 \to 2$ hard parton amplitudes, separated by a distance, within the hadron, much larger than $1/E_T$.

The DDS mechanism has been extensively studied in the literature [19]. The corresponding four-jet cross section can be written as

$$d\sigma_{DDS}^{4jet} \sim \sum_{\substack{a,b \\ c,d}} \int d^2\beta \; F_{ab}(x_1,x_2;\beta) \, F_{cd}(x_3,x_4;\beta) \, d\hat\sigma_1 d\hat\sigma_2 \,. \qquad (4)$$

In eq. (4) $d\hat\sigma_1$ and $d\hat\sigma_2$ are the cross sections of the two independent hard parton scatterings, to be computed by perturbative QCD. $F_{ab}(x_1,x_2;\beta)$ are the two-parton distributions with fractional momenta x_1 and x_2 and with transverse distance within the hadron β. As such, they govern the flux of ingoing parton pairs.

We have thus identified a new fundamental quantity in QCD, namely the double- (in general the multi-) constituent distribution within the hadron [20]. Clearly these must contain more detailed information on the hadronic structure, with respect to the familiar, single parton distributions which appear in eq. (1). This simply reflects the fact that, in going down with $x \sim 2E_T/\sqrt{s}$, one starts to feel the hadronic wavefunction. In fact, the dependence on the variable β, in addition to x_1 and x_2 (*), which can be ascertained by careful inspection of the kinematics of Fig. 2, is directly related to the dimensionality $1/R^2$ carried by the double distributions, with R the hadronic size. Single parton distributions are instead dimensionless.

On account of this new scale R the DDS jet cross section of eq. (4) will be of order $1/R^2 E_T^2$ with respect to the leading QCD (LQCD) four-jet cross section from eq. (1) (notice that the two cross sections are of the same order in α_s). The DDS jet events should accordingly have a steeper dependence on E_T (in particular on the cutoff E_T^{min} used to define the jets) than the LQCD ones. This is one distinctive character of DDS.

The second, important qualitative aspect concerns the behaviour with the collider energy $\sqrt{s}$. As remarked previously, the LQCD cross section of eq. (1) increase with $\sqrt{s}$ at fixed E_T, driven by the increasing parton flux. Intuitively, the flux of parton pairs should grow even faster, for decreasing

---

(*) Of course, implicit in eq. (4) there also is the dependence of double parton distributions on the short distance scale $Q^2 \sim p_T^2$.

$x_{1,2} \sim 2E_T/\sqrt{s}$, and correspondingly the rise of the DDS cross section of eq.(4) with $\sqrt{s}$ is expected to be steeper.

Changing these qualitative observations into a more quantitative statement requires the knowledge of the double distributions $F(x_1,x_2;\beta;Q^2)$. In principle, these should be measured, as it is for the single parton densities $F(x,Q^2)$. Lacking this information, it should be reasonable to assume, as a first approximation, the simple, factorized $x \ll 1$ ansatz:

$$F_{a,b}(x_1,x_2;\beta;Q^2) = F_a(x_1,Q^2)\ F_b(x_2,Q^2)\ \Phi(\beta), \qquad (5)$$

where $\Phi(\beta)$ is a profile in the relative partons transverse distance. For the latter one may choose, for example

$$\Phi(\beta) = \frac{1}{\pi R^2}\ \exp(-\beta^2/R^2)\ , \qquad (6)$$

where R has been introduced before. With the parametrization of eqs.(5) and (6) R becomes the key parameter, which controls the normalization of the DDS relative to the LQCD.

Using eqs.(5) and (6) in eq.(4) one easily derives the following expression:

$$\frac{d\sigma_{DDS}^{4jet}}{dE_T} \sim \int_{p_T^{min}}^{E_T-p_T^{min}} dE_1\ dE_2\ \frac{d\sigma^{2jet}}{dE_1}\ \frac{d\sigma^{2jet}}{dE_2}\ \frac{\delta(E_1+E_2-E_T)}{\sigma_{eff}}\ , \qquad (7)$$

where $\sigma_{eff} = \pi R^2$. From eq.(7) one can finally obtain the integrated cross section $\sigma_{DDS}^{4jet}(E_T^{min}, \sqrt{s})$. The latter is represented, for the choice $R \sim 1$ fm, by the dashed line in Fig.1, which shows the expected rapid rise with $\sqrt{s}$, essentially like the square of the two-jet cross section. Although qualitative, Fig.1 indicates that multiparton processes should become quite common at the multiTeV colliders. Of course, evidencing them in practice would depend very sensitively on the experimental cuts.

SIGNATURES OF DDS

According to the previous discussion, signatures of multiparton processes should appear in multijet events with $E_T \ll \sqrt{s}$ (i.e. in minijets). The problem is to distinguish in practice the DDS jets from the LQCD ones, as the observed jet sample will contain jets from both sources:

$$d\sigma^{4jet} = d\sigma_{LQCD}^{4jet} + d\sigma_{DDS}^{4jet} \ . \tag{8}$$

The qualitative features pointed out above, namely the different behaviour with E_T and with $\sqrt{s}$, do not by themselves represent a sufficient criterion, because the relative rates of the DDS and the LQCD cannot be assessed theoretically with enough accuracy. In fact, as well-known, the normalization of eq.(1) is greatly affected by the different possible choices of $F(x,Q^2)$ and by the definition of the variable Q^2 itself. On the other hand the DDS rate of eq.(4), even in the simplest model for double distributions of eq.(5), has a large uncertainty associated to the range of values a priori possible for R.

The very distinctive feature is that the LQCD and the DDS give rise to quite different topologies of the four-jet events. The former mechanism, dominated by double gluon bremsstrahlung, leads to configurations where one leading-p_T jet is balanced by three recoiling jets in the transverse plane, and the jets are correlated in a typical way. The DDS instead prefers configurations where the jets balance their transverse momenta in pairs (as evident from the two separate transverse momentum conservations in Fig.2), and those jet pairs should be almost uncorrelated.

One should thus try to define some simple variables, sensitive to correlations among the jets, therefore to the different mechanisms, and to study the jets differential distributions with respect to these variables. As the two mechanisms will populate the phase space differently, one can attempt to evidence the DDS by looking for excess of events in those ranges of the above mentioned variables where the LQCD is unlikely. Once such an excess is eventually found, important checks can be provided by the E_T and by the $\sqrt{s}$ behaviours mentioned above.

For a quantitative analysis it is absolutely necessary to take into account the cuts which are applied to the final hadronic state to experimentally resolve the jets. In the four-jet analyses performed so far they have been formulated, at the partonic level, as follows:

$$\sum_{i}^{4} | p_{T_i} | > E_T^{min} \quad ; \quad p_{T_i} > p_T^{min} \tag{9}$$

$$\theta^{ij} > \theta^{min} \quad\quad\quad ; \quad |\eta_i| < \eta^{max} \quad , \tag{10}$$

where θ^{ij} are the opening angles among the jets and η_i are the jets pseudorapidities.

Among various possibilities [21], the correlation variable which looks particularly convenient, in order to distinguish the DDS from the LQCD four-jet events, is the transverse momentum umbalance:

$$P_{umb} = \{2 \min_{i \neq max} \{(\mathbf{p}_T^{max} + \mathbf{p}_{T_i})^2\}\}^{1/2} \, , \qquad (11)$$

where transverse momenta are taken in the four-jet center of mass system. This variable essentially indicates how "back-to-back" are the leading-p_T jet and its recoil jet.

An example, for given cuts, is represented in Fig.3. As one can see,

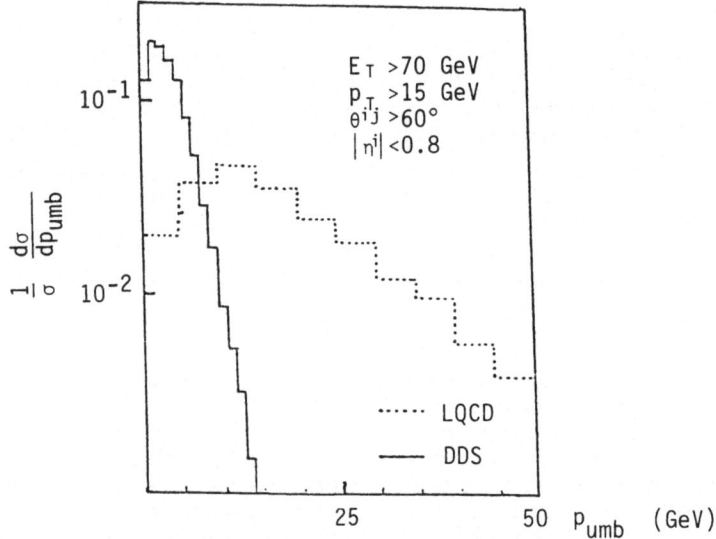

Fig.3. P_{umb} distribution

the DDS and the LQCD markedly differ: the former priviledges small values of P_{umb} (of the order of those observed in two-jet events), while much larger P_{umb} is accessible to the latter.

Turning now to experimental searches of multiparton processes, the present situation is somewhat controversial. Evidence of double parton scattering has been claimed recently as the result of the analysis, following the lines above, of four-jet events at the ISR ($\sqrt{s}$ =63 GeV)[22].The distribution in the umbalance variable[*] of those data is shown in Fig.4, which displays the same kind of situation as prospected in Fig.3, namely an excess of events at smaller umbalance, with respect to the LQCD mechanism represented by the solid line [23]. Agreement with the data is obtained by adding to the LQCD a contribution from DDS (represented by the dashed line), estimated via eqs.(5) to (7) with σ_{eff}=11 mb or equivalently R=0.6 fm.

---

[*] Notice that this is defined in Ref.[22] slightly differently from eq.(11).

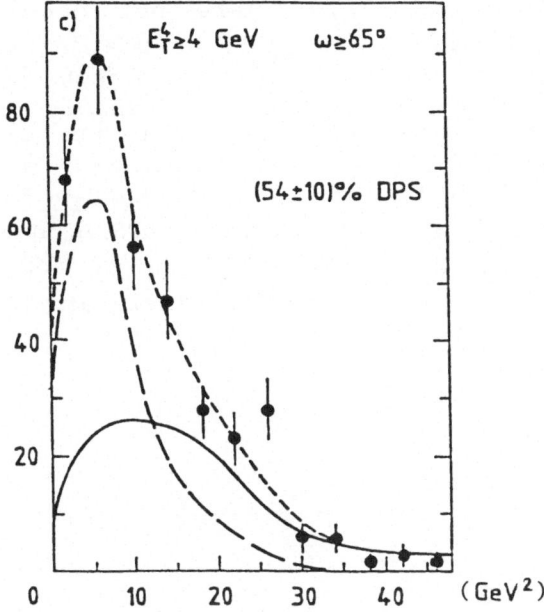

Fig.4. Umbalance distribution of Ref.[22]

Moreover, the DDS candidate "pairwise balanced" events, which populate such an excess, can be selected from the total sample and studied separately. They show the characteristic dependences, e.g. on the transverse energy E_T and on the interjet angles, qualitatively expected for the DDS jets.

Support to the above results comes from double lepton pair Drell-Yan production, as it has been discussed in Ref.[24]. This process should be sensitive to the double disconnected quark-antiquark annihilation into e^+e^- or $\mu^+\mu^-$ pairs. Although suppressed by α_{QED}, it has the advantage of being unaffected by fragmentation effects, and of a somewhat better theoretical control over double quark distributions, hence on the rates. The interesting result is that the existing multilepton production data can be explained by double quark annihilation, if the value of R, characterizing double distributions, is taken equal to 0.7 fm. This consistency with the value of R implied by the four-jet analysis of Ref.[22] looks really encouraging.

On the other hand multiple parton processes have not been observed by the UA2 collaboration in their sample of four-jet events at the S$\bar{p}$pS collider ($\sqrt{s}$ =630 GeV) [3$_b$]. Those jets are found quite consistent with the LQCD mechanism alone. As an example the corresponding umbalance distribution is shown in Fig.5. In fact,by comparing the data with eq.(7), the lower limit on the characteristic cross section area σ_{eff}>10 mb (or equivalently R >0.5 fm) has been obtained [25].

317

The results of Ref.[3_b,25] are not necessarily in contradiction with those of Ref.[22], as the jet samples are in different regions of Q^2 and $\sqrt{s}$, and a direct comparison is not so obvious. Indeed the simplest explanation why multi-parton processes have not been observed at the Collider, in spite of the larger $\sqrt{s}$, could be, as already suggested in Ref.[22], that the SppS jets have larger E_T ($E_T > 70$ GeV) than the ISR ones ($E_T \sim 30$ GeV). With the ratio DDS/LQCD expected to rapidly fade away as $1/R^2 E_T^2$, the E_T^{min} cut of Ref.[3_b] could simply be too high to evidence the DDS. There are however more

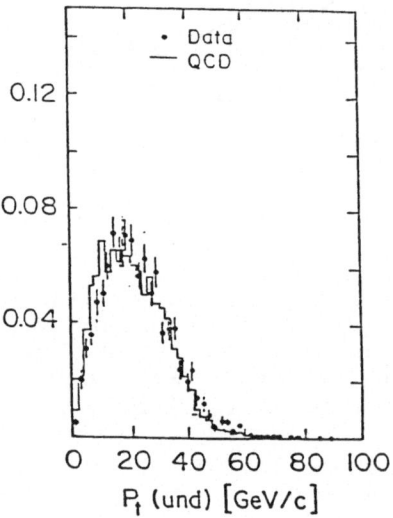

Fig.5. Umbalance distribution of Ref.[3_b]

elements which can potentially complicate such a comparison. For example, as they are in different ranges of x, ISR jets should be mostly induced by quark-gluon and quark-quark processes, while the $S\bar{p}pS$ ones should dominated instead by gluon-gluon, and the scale R in the double distributions might well depend on the parton species (as well as on $\sqrt{s}$).

The search of DDS processes should thus be pursued further, both at the present and at future colliders, to hopefully bring the present situation to a full clarification. Clearly, in addition to establishing multiparton processes, this would allow considerable progress in the understanding of the partonic structure of the proton.

The observation that the jet integrated cross section:

$$\sigma_{jet}(p_T^{min}, \sqrt{s}) = \int_{p_T^{min}}^{\sqrt{s}/2} dp_T \frac{d\sigma_{jet}}{dp_T} \tag{12}$$

rapidly rises with $\sqrt{s}$, to become an appreciable fraction of the $p\bar{p}$ total inelastic cross section [14], has aroused great theoretical interest [26].

Indeed, the onset of such an increased jet activity has been advocated as a possible explanation of a variety of non-scaling (with $\sqrt{s}$) effects observed in high energy pp and $p\bar{p}$ collisions, such as the KNO scaling violation and the increase of the $<p_T>$ with the particle multiplicity [27].

As repeatedly stressed, the leading order QCD integrated cross section:

$$\hat{\sigma} \equiv \sigma_{jet}^{QCD}(p_T^{min}, \sqrt{s}) \tag{13}$$

keeps on growing, essentially like a power of $\sqrt{s}$, due to the increasing parton densities as $x \to 0$. With p_T^{min} of the order of few GeV, as appropriate to minijets, $\hat{\sigma}$ can equal the total $p\bar{p}$ cross section at the energies of the planned colliders [28].

One should also notice that $\hat{\sigma}$, by itself defining the total cross section for partons with $p_T > p_T^{min}$, has the disappointing feature of wildly depending on the value chosen for the parton p_T^{min}, not necessarily identical to the jet p_T^{min} and thus somewhat arbitrary. This reflects the divergence of $\hat{\sigma}$ for $p_T^{min} \to 0$.

It has been recently pointed out that multiparton processes could play a significant role as a mechanism to regularize the behaviour of $\hat{\sigma}$ mentioned above. In fact they should "shadow" the indefinite rise of $\hat{\sigma}$ with $\sqrt{s}$, thus reconciling the "hard" total inelastic hadron-hadron cross section with unitarity, and in so doing offer at the same time the possibility to compensate the unphysical dependence on the p_T^{min} [29].

The starting point is that the leading order QCD parton model cross section is an inclusive cross section, so that $\hat{\sigma}$ should be interpreted in fact as the jets multiplicity, or equivalently as the average number of parton-parton collisions with $p_T > p_T^{min}$ per proton-proton collision:

$$\hat{\sigma} \rightarrow <n> \sigma_{tot}^{inel} \ . \tag{14}$$

From this point of view, $\hat{\sigma}$ becoming comparable with σ_{tot} simply means the possibility of several hard parton-parton scatterings in a hadron-hadron interaction.

It is useful to characterize each proton-proton collision with (in addition to $\sqrt{s}$ and p_T) an impact parameter **b** in the transverse plane, as depicted in Fig.6. For small b the

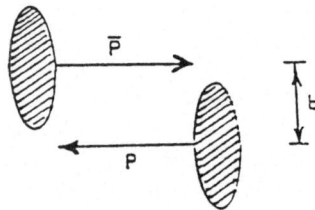

Fig.6. Impact parameter

colliding hadrons almost overlap, and the corresponding probability for parton scattering will be large. For b larger than the typical hadronic radius R such a probability would be vanishingly small, as partons have no way to make a hard interaction. To take this into account one can define an overlap function $\Phi(b)$, similar to eq.(6), such that $\Phi(b) \sim 0$ for b>R. The integrated QCD cross section will then involve the integration over all values of b, weighted by $\Phi(b)$. Thus the average number of parton interactions with $p_T > p_T^{min}$, per hadron-hadron collision at center of mass energy $\sqrt{s}$ and impact parameter b, can be expressed as:

$$\chi \equiv <n(b,\sqrt{s})>\Big|_{p_T>p_T^{min}} = \Phi(b)\ \frac{\sigma(p_T^{min},\sqrt{s})}{\sigma_{tot}^{inel}} \tag{15}$$

Assuming now the number of parton-parton collisions to follow a Poisson probability distribution, $P_n(b)=\chi^n e^{-\chi}/n!$, as somehow implicit in the parametrization of eq.(5), with average χ given by eq.(15), the series of multiple disconnected parton scatterings can be explicitly summed.

The "hard" cross section, i.e. the contribution of all hard parton processes to the total inelastic hadronic cross section, is simply given by the probability of having at least one hard parton interaction. The summation of multiple parton scatterings gives:

$$\sigma_H(p_T^{min},\sqrt{s}) = \int_0^R d^2\mathbf{b}\,(1-e^{-\chi})\ , \tag{16}$$

where χ depends on b, p_T^{min} and $\sqrt{s}$ according to eq.(15). Notice that the **b** integration, in principle from zero to infinity, in practice has R as upper endpoint due to Φ(b)~0 for b>R. Also, the perturbative QCD cross section $\hat{\sigma}$, which appears (via eq.(15)) in the exponential of eq.(15) representing the effect of multiparton scattering, does not necessarily need to be the cross section integrated over the whole rapidity range. It can be as well the QCD cross section integrated within the cuts which define the criterion to select the hard events.

In eq.(16) one recognizes two distinct regimes. The first one is the regime where $\hat{\sigma}$ is small compared to σ_{tot}. By expanding the exponential one verifies that the "hard" hadron-hadron cross section σ_H essentially coincides there with the lowest order perturbative QCD cross section $\hat{\sigma}$. In the other regime the p_T^{min} and/or the $\sqrt{s}$ are such that $\hat{\sigma}$ is comparable or even larger than σ_{tot}. In that case the exponential of eq.(16) is vanishingly small, and $\sigma_H \sim \pi R^2$, i.e. the hadron would behave like a "black-disc" with radius R. One should also notice that the p_T^{min}-divergence affecting $\hat{\sigma}$ is regularized by the exponential. The definition of σ_H in eq.(16) should thus be stable with respect to that parameter.

The shadowing corrections such as in eq.(16), modifying the leading QCD description, are expected to have considerable phenomenological implications at the energies of the planned supercolliders [30].

CONCLUDING REMARKS

Briefly summarizing the preceding discussion, multiparton processes have many points of interest for strong interaction phenomenology at the collider and at future supercolliders.

As we have seen, such processes should manifest themselves with characteristic signatures, related to minijet production, and with direct bearing on the multiconstituent structure of hadrons. These signatures should be either peculiar jet event topologies, or shadowing corrections to the hadronic hard cross section, or finally screening effects on small-x parton densities from parton recombination . They all would signal the physical p_T cutoff where, as the consequence of parton overcrowding, the leading QCD parton model description of deep inelastic hadron collisions becomes insufficient, and hard, perturbative dynamics starts merging into Regge physics.

Thus clearly future experimental searches for multiparton processes, improving the present situation, will allow significant progress of our understanding of the properties of QCD.

ACKNOWLEDGEMENTS

We wish to thank Prof. A.Zichichi and Prof. A.Ali for a very interesting and enjoyable meeting.

Thanks are also due to Mrs. E.Talocchi for the careful preparation of the manuscript.

REFERENCES

[1] T.Akesson et al.:Phys.Lett 118B(1982)185,193; 123B(1983) 133.

[2a] M.Banner et al. (UA2 Collaboration): Phys.Lett. 118B (1982) 203;

[2b] G.Arnison et al.(UA1 Collaboration): Phys.Lett. 123B (1983) 115.

[3] For most recent reviews see:

[3a] F.Ceradini(UA1 Collaboration): in Proceedings of the 23rd International Conference on High Energy Physics, Berkeley 1986;

[3b] P.Bagnaia (UA2 Collaboration): in Proceedings of the 16th International Conference on pp Physics, Aachen 1986; and references there.

[4] R.P.Feynman: Photon Hadron Interactions, Benjamin (1972).

[5] For reviews see e.g.:
G.Altarelli: Phys.Rep. 81 (1982)1;
A.J. Buras: Rev.Mod.Phys. 52 (1980)189.

[6] For reviews see e.g.:
M.Jacob and P.V.Landshoff: Phys.Rep. 48(1978) 285;
R.Horgan and M.Jacob: Physics at Collider Energy,
CERN-DESY School of Physics (1980), CERN 81-04.

[7] B.Combridge, J.Kripfganz and J.Ranft: Phys.Lett.70B (1977) 234;
R.Cutler and D.Sivers: Phys.Rev. D16 (1977) 679.

[8] T.Gottshalk and D.Sivers: Phys.Rev. D21 (1980) 102;
Z.Kunszt and E.Pietarinen: Nucl.Phys. B164 (1980)45;
F.A.Berends, P. De Causmaecker, R.Gastmans, R.Kleiss and
T.T.Wu: Phys.Lett. 103B(1981)124.

[9] J.F.Gunion and Z.Kunszt: Phys.Lett 159B(1985)167;176B (1986) 163,477;
S.Parke and T.Taylor: Fermilab-Pub-85/118-5; Fermilab-Pub-85/162-T;
Z.Kunszt: Nucl.Phys. B271(1986)333.

[10] R.Odorico: Nucl.Phys. B228(1983)381; Comp.Phys.Comm. 32 (1984)139;
For a review see for instance:
T.D.Gottshalk: in Proceedings of the 1986 Madison Workshop on Physics Simulations at High Energy.

[11] D.W.Duke and J.F.Owens: Phys.Rev. D30 (1984) 461;
E.Eichten, I.Hinchliffe, K.Lane and C.Quigg: Rev.Mod.Phys. 56(1984)579;
M.Diemoz et al.: Phys.Rep. 130(1986)293;
and references there.

[12] G.Altarelli and G.Parisi: Nucl.Phys. B126(1977)248.

[13] B.Humpert and R.Odorico: in Proceedings of the 4th Topical Workshop on Proton-Antiproton Collider Physics, Bern 1984, CERN 84-09.

[14] G.Arnison et al.: Phys.Lett. 172B(1986)461;
F.Ceradini: in Proceedings of the International Europhysics Conference on High Energy Physics, Bari 1985.

[15] For an extensive review of low-x physics see:
     L.V.Gribov, E.M.Levin and M.G.Ryskin: Phys.Rep.
     100(1983)1.
[16] J.C.Collins: in Proceedings of the Oregon Workshop on
     Super High Energy Physics, Eugene 1985.
[17] J.Kwiecinski: Krakow report 1328/PH(1986); Zeit.Phys. C29
     (1985)561.
[18] A.H.Mueller and J.Qiu: Nucl.Phys. B268(1986)427;
     A.H.Mueller: in Proceedings of the 1985 Annual DPF
     Meeting, Eugene, Oregon.
[19] N.Paver and D.Treleani: Il Nuovo Cimento 70A(1982) 215;
     73A(1983); Phys.Lett. 146B(1984)252; Zeit.Phys. C28(1985)
     187;
     B.Humpert and R.Odorico: Phys.Lett. 154B(1985)211;
     M.Jacob: CERN TH-3639(1983);
     P.V.Landshoff and J.C.Polkinghorne: Phys.Rev. D18(1978)
     3344.
[20] H.D.Politzer: Nucl.Phys. B172(1980)349;
     H.R.Gerhold: Nuovo Cimento 59A(1980)373;
     R.K.Ellis, R.Petronzio and W.Furmanski: Nucl.Phys.
     B207(1982)1;
     M.Mekhfi: Phys.Rev.D32(1985)2371,2380.
[21] Ll.Ametller: in Proceedings of the 7th International
     Symposium on Multiparticle Dynamics, Seewinkel 1986.
[22] T.Akesson et al.: Zeit.Phys. C34(1987)163.
[23] Z.Kunszt and J.W.Stirling: Phys.Lett. 171B(1986)307.
[24] F.Halzen, P.Hoyer and J.W.Stirling: Phys.Lett. 188B
     (1987)375.
[25] F.Pastore: in Proceedings of the International
     Europhysics Conference on High Energy Physics, Uppsala
     1987.
[26] For a review see:
     M.Jacob and P.V.Landshoff: Mod.Phys.Lett. A1(1986)657.
[27] G.Pancheri, Y.N. Srivastava and M.Pallotta: Phys.Lett.
     151B(1985)657;
     T.K.Gaisser, F.Halzen, A.D.Martin and C.J.Maxwell:
     Phys.Lett. 166B (1986)219.
[28] G.Pancheri and Y.N.Srivastava: Phys.Lett: 182B(1986)199;
     T.K.Gaisser and F.Halzen: Phys.Rev.Lett. 54(1985)1754.
[29] L.Durand and P.Hong: Phys.Rev.Lett. 58(1987) 303;
     T.Sjöstrand and M.van Zijl: Phys.Lett. 188B(1987)149;
     Ll.Ametller and D.Treleani: SISSA preprint 13/87/EP
     (1987);
     J.Kwiecinski: Phys.Lett. 184B(1987)386;
     A.Capella, J.Kwiecinski and J.Tran Than Van:preprint LPTHE
     86/51 (1986).
[30] D.Treleani: SISSA preprint 19/87/EP (1987).

JET PHYSICS: THEORETICAL ISSUES IN SIMULATING VERY HIGH ENERGY COLLISIONS[1]

B. R. Webber

Cavendish Laboratory
University of Cambridge
Cambridge CB3 0HE, U.K.

INTRODUCTION

Understanding jet physics – that is, the production and development of jets of hadrons at large transverse momenta in hard collisions – is of paramount importance for the design of future particle accelerators and detectors. Most of the new objects we hope to find at higher energies – the Higgs boson, new quarks and leptons, supersymmetric partners of known particles, and so on – will produce jets. At the same time, there will be a large multijet background from familiar processes like gluon-gluon scattering. The more we understand about jets, the more chance we have of distinguishing a new signal from the background. In addition, there is a lot of interesting physics in the jets themselves.

The generally accepted picture of jet production in a hard hadron-hadron collision is illustrated in Fig. 1[2]. In the lowest order of perturbative QCD, two constituent partons of the incoming hadrons interact to produce some outgoing partons. In higher orders, additional partons are radiated from the incoming and outgoing lines of the hard subprocess (initial- and final-state QCD bremsstrahlung, respectively). There are also loop corrections involving additional virtual partons. These higher-order contributions can get very large, but we shall never have the computing power to evaluate them exactly beyond the first few orders. The best we can do is to identify the largest terms (the leading singularities or leading logarithms) in each order and sum them up to all orders.

There are two important classes of leading singularities: collinear and soft. Collinear singularities occur when parton momenta are parallel, forcing internal lines onto mass shell. Since they enhance diagrams in which cascades of partons are emitted close to the directions of the hard-scattered partons, they are responsible for the very existence of jets. In a suitable gauge, the enhanced diagrams do not interfere significantly and may be summed by means of the 'jet calculus' [4], leading to the familiar Altarelli-Parisi [5] evolution equations.

[1] Research supported in part by the U.K. Science and Engineering Research Council.
[2] For reviews of the QCD basis for this picture, see for example Refs. [1 – 3].

Because interference can be neglected in the summation of leading collinear logarithms, it proves simple and convenient to carry it out by means of the Monte Carlo technique, generating the parton cascades by a classical Markov branching process. This is the basis of most of the available simulation programs for jet physics [6].

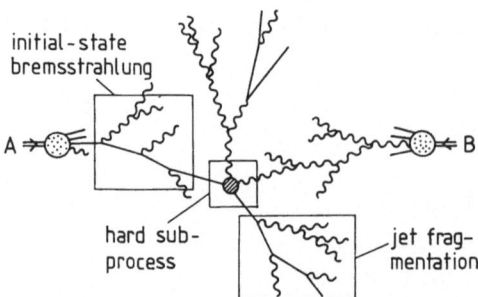

Figure 1. Parton-level structure of a hard hadron-hadron collision

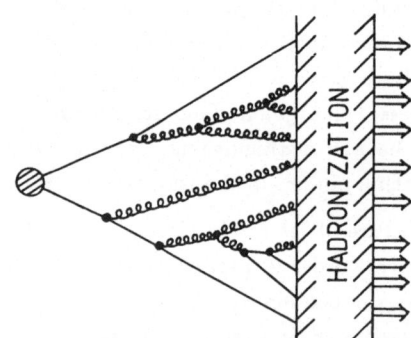

Figure 2. QCD cascade picture of jet development

The leading soft singularities are more difficult to deal with. They come from the emission of gluons with energies that are small compared with those of the hard-scattered partons. As the energy increases, more and more of the emitted partons fall into this category. Furthermore, there is no choice of gauge in which interference between the enhanced diagrams becomes negligible. This leads to important soft-gluon coherence effects within and between jets that have to be taken into account. It turns out that many of these effects can be incorporated in the Monte Carlo programs by imposing angular ordering constraints on the parton branching process [7].

The resulting picture of jet development is shown in Fig. 2, for the case of jets coming from outgoing partons (final-state bremsstrahlung). The momentum transfer scale Q in the hard subprocess sets an upper limit on virtual masses in the parton cascade. Those virtual partons nearest to the hard subprocess tend to be farthest from mass shell. As the cascade proceeds, the typical virtual mass q falls as the hard parton momenta are shared between more and more lines. As long as $q \gg \Lambda_{QCD}$, it is possible to describe the cascade using perturbation theory in $\alpha_S(q^2)$. At some scale $Q_0 \gtrsim \Lambda_{QCD}$, however, nonperturbative effects become important and the process of 'hadronization' sets in, converting the parton jets into hadron jets. Our theoretical understanding of this process is as yet very limited, although some quite successful models have been developed[3]. We shall assume that hadronization involves only momentum transfers of the order of Q_0 or smaller, for some fixed Q_0. Then for sufficiently large Q the momentum and quantum number structure of jets will be determined mainly in the parton branching regime $Q \gg q \gg Q_0$, where perturbative methods are applicable.

In the case of jets from incoming partons (initial-state bremsstrahlung), the picture is broadly similar, except that the hard virtual partons have spacelike momenta and the simulation is simplest if the cascade proceeds backwards from the hard subprocess to the incident hadrons [11].

In this review, I shall discuss some recent work on refining the way the above QCD-based picture is incorporated in jet simulation programs. First, I shall outline some new results on quantum angular correlations in jets [3, 12 − 15]. These correlations, which come from the coherent sum over gluon helicities in the parton cascade, are formally of leading collinear logarithmic order but have usually been neglected by treating the cascade as an essentially classical process. The problem of going beyond the classical approximation raises interesting fundamental questions in the theory of simulation and computability. Also, of course, it is of practical importance for the correct simulation of angular distributions in jets. Iterative algorithms are now under development [15, 16] and they can be compared with new explicit results [14] up to order α_S^5.

The other main topic I shall cover is recent work on the simulation of soft gluon coherence. Coherence has been found to be important for the understanding of several aspects of e^+e^- annihilation, such as the average multiplicity [17] and the 'string effect' [18]. In hard hadronic collisions, it should lead to a rich diversity of analogous phenomena involving coherent emission from initial and final partons [19, 20]. As mentioned above, many features of coherence can be simulated by imposing appropriate angular constraints on the parton cascades. This is the basis of a new hadron-hadron simulation program [21], from which I shall present some preliminary results.

SPIN CORRELATIONS IN JETS

The type of correlation I shall be discussing is illustrated in Fig. 3. Here ϕ_i and ϕ_j represent the *azimuthal* angles of two arbitrary branchings i and j in some parton cascade. In the leading collinear approximation, the azimuths may be defined with respect to some convenient overall plane such as that of the hard scattering subprocess. After integrating over

---

[3]See, for example, Refs. [8, 9] and the recent review by Gustafson [10].

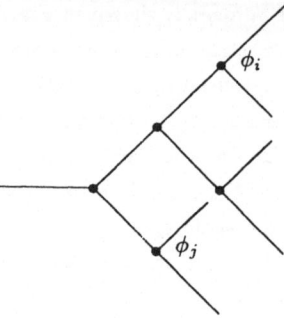

Figure 3. Azimuthal angles in a parton cascade

all other azimuths[4], the joint probability distribution of ϕ_i and ϕ_j will be (up to an overall normalisation)

$$P(\phi_i, \phi_j) = 1 + C_{ij} \cos 2(\phi_i - \phi_j) \tag{1}$$

where C_{ij} is some function of the momenta of all the lines in the cascade that connect vertices i and j. The correlation coefficients C_{ij} are of interest because:

- They are a manifestation of the Einstein-Podolsky-Rosen 'paradox' [22], *i.e.*, the existence of acausal correlations in quantum mechanics. In this case, the 'paradox' is that the measured value of ϕ_i influences the probability distribution of ϕ_j, however far apart the vertices may be and independent of their temporal sequence. Furthermore this correlation cannot be correctly computed by ascribing it to a common cause or a hidden variable in any classical formulation of the branching process [5].

- They are a leading collinear effect that should be included in jet simulation programs for consistency and could substantially affect angular distributions in jets.

- Their calculation is a computational problem that appears initially to be insoluble.

The last point requires some amplification. Suppose $M_{\lambda \to \lambda' \lambda''}$ is the amplitude for the branching of a parton with helicity λ into those with λ' and λ''. For the branching $g \to gg$ with all helicities positive, for example, we have (omitting factors of $g_S/2\pi$, *etc.*)

$$M_{+ \to ++} = [\sqrt{z/(1-z)} + \sqrt{(1-z)/z}]e^{-i\phi} \tag{2}$$

where z is the momentum fraction in the branching. Now the probability distribution for the cascade $\lambda \to \lambda' \lambda'', \lambda' \to \lambda''' \lambda'''', \cdots$, summed over final helicities $\{\lambda_f\}$, will be of the form

$$P = \sum_{\{\lambda_f\}} \left| \sum_{\{\lambda_i\}} M_{\lambda \to \lambda' \lambda''} M_{\lambda' \to \lambda''' \lambda''''} \cdots \right|^2 \tag{3}$$

---

[4]If other azimuths are not integrated out, there will be correlations involving more than two vertices.

[5]This is of course a peculiarity rather than a genuine paradox: no information can be transmitted between observers by means of the correlation.

Table I. Computation of correlation coefficients using Eq. 3.

Vertices	Terms	Time	Ref.
2	512	1 sec	[3, 12]
3	16k	30 sec	[13]
4	500k	15 mins	[14]
5	17M	8 hours	[14]
6	500M	10 days	—
7	17G	10 months	—

where $\{\lambda_i\}$ represents the helicities of internal lines. If each factor M consists of two terms, then P for a tree diagram with n_v vertices will be a sum of $32^{n_v}/2$ terms. As shown in Table I, if we allow a notional time of one second for an algebraic computer to calculate the $n_v = 2$ case, the time and memory requirements[6] become prohibitive by about $n_v = 7$.

What we have here is an *exponential algorithm* for computing the correlations: the number of operations increases exponentially with the number of steps. Of course, many terms vanish and once they have all been computed there are many cancellations, so the final expressions are not that huge. For example, the result obtained by Knowles [14] for the 5-vertex gluon cascade in Fig. 4 is

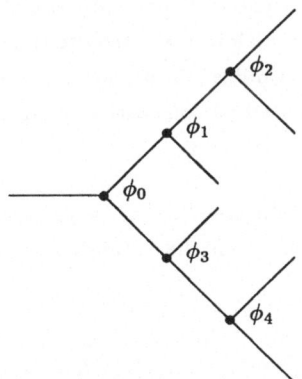

Figure 4. Five-vertex gluon cascade

[6]These correspond roughly with experience using Veltman's M68000 SCHOONSCHIP algebra package.

$$P^{(Fig.4)}(z_0, z_1, z_2, z_3, z_4; \phi_0, \phi_1, \phi_2, \phi_3, \phi_4) = P(z_0)\, P(z_1)\, P(z_2)\, P(z_3)\, P(z_4)$$

$$+\, u(z_0)\, w(z_1)\, P(z_2)\, P(z_3)\, P(z_4)\, \cos 2(\phi_1 - \phi_0)$$

$$+\, u(z_0)\, v(z_1)\, w(z_2)\, P(z_3)\, P(z_4)\, \cos 2(\phi_2 - \phi_0)$$

$$+\, v(z_0)\, P(z_1)\, P(z_2)\, w(z_3)\, P(z_4)\, \cos 2(\phi_3 - \phi_0)$$

$$+\, v(z_0)\, P(z_1)\, P(z_2)\, v(z_3)\, w(z_4)\, \cos 2(\phi_4 - \phi_0)$$

$$+\, P(z_0)\, u(z_1)\, w(z_2)\, P(z_3)\, P(z_4)\, \cos 2(\phi_2 - \phi_1)$$

$$+\, w(z_0)\, w(z_1)\, P(z_2)\, w(z_3)\, P(z_4)\, \cos 2(\phi_3 - \phi_1)$$

$$+\, w(z_0)\, v(z_1)\, w(z_2)\, w(z_3)\, P(z_4)\, \cos 2(\phi_3 - \phi_2)$$

$$+\, w(z_0)\, w(z_1)\, P(z_2)\, v(z_3)\, w(z_4)\, \cos 2(\phi_4 - \phi_1) \qquad (4)$$

$$+\, w(z_0)\, v(z_1)\, w(z_2)\, v(z_3)\, w(z_4)\, \cos 2(\phi_4 - \phi_2)$$

$$+\, P(z_0)\, P(z_1)\, P(z_2)\, u(z_3)\, w(z_4)\, \cos 2(\phi_4 - \phi_3)$$

$$+\, u(z_0)\, w(z_1)\, P(z_2)\, u(z_3)\, w(z_4)\, \cos 2(\phi_1 - \phi_0) \cos 2(\phi_4 - \phi_3)$$

$$+\, u(z_0)\, v(z_1)\, w(z_2)\, u(z_3)\, w(z_4)\, \cos 2(\phi_2 - \phi_0) \cos 2(\phi_4 - \phi_3)$$

$$+\, v(z_0)\, u(z_1)\, w(z_2)\, w(z_3)\, P(z_4)\, \cos 2(\phi_3 - \phi_0) \cos 2(\phi_2 - \phi_1)$$

$$+\, v(z_0)\, u(z_1)\, w(z_2)\, v(z_3)\, w(z_4)\, \cos 2(\phi_4 - \phi_0) \cos 2(\phi_2 - \phi_1)$$

$$+\, P(z_0)\, u(z_1)\, w(z_2)\, u(z_3)\, w(z_4)\, \cos 2(\phi_2 - \phi_1) \cos 2(\phi_4 - \phi_3)$$

where

$$u(z) = (1 - z)/z, \quad v(z) = z/(1 - z), \quad w(z) = z(1 - z),$$

$$P(z) = u(z) + v(z) + w(z). \qquad (5)$$

Since the parton cascades in a multi-TeV hard process may contain a hundred or more vertices, we clearly need a *linear* algorithm if we want to include azimuthal correlations in the general case. A recent paper by Collins [15] proposes a novel approach which offers this possibility. In this approach, a simulated parton cascade with all the correct correlations can be generated as follows:

i) An arbitrary branch of the cascade (the 'leader') is generated first, with azimuthal angles correlated with earlier ones on that branch but independent of those on side-branches;

ii) Working backwards from the end of the leader, side-branches are generated with distributions that depend on a 'decay matrix' that feeds back information about later branchings on the leader. This introduces the acausal element needed to reproduce the EPR phenomenon.

Fig. 5a shows the sequence of generation of azimuths in the cascade of Fig. 3 when the uppermost branch is taken to be the leader.

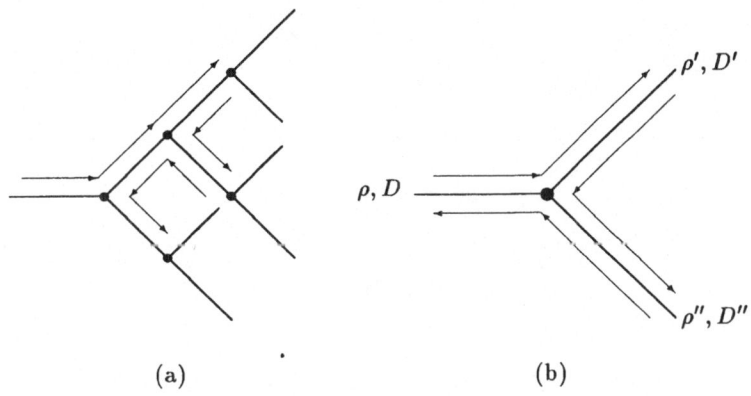

(a)   (b)

Figure 5. Generation of azimuths via Collins algorithm

The flow of information at a general vertex is shown in Fig. 5b. Here ρ represents a spin density matrix and D a decay matrix. The algorithm is defined by the density matrix of the parton initiating the cascade, the decay matrices of the final partons, and the following equations applied recursively in the above sequence:

$$\rho'_{\mu\mu'} \sim \rho_{\lambda\lambda'} M_{\lambda\to\mu\nu} M^*_{\lambda'\to\mu'\nu'}$$

$$\rho''_{\nu\nu'} \sim \rho_{\lambda\lambda'} M_{\lambda\to\mu\nu} M^*_{\lambda'\to\mu'\nu'} D'_{\mu\mu'} \tag{6}$$

$$D_{\lambda\lambda'} \sim M_{\lambda\to\mu\nu} M^*_{\lambda'\to\mu'\nu'} D'_{\mu\mu'} D''_{\nu\nu'}$$

where '$\sim$' here denotes equality after dividing out the trace.

It proves convenient to express Eqs. 6 in terms of polarisation and decay vectors ρ and D, and a transverse vector $\mathbf{n}$, where

$$\rho \sim 1 + \boldsymbol{\sigma} \cdot \boldsymbol{\rho}$$

$$D \sim 1 + \boldsymbol{\sigma} \cdot \mathbf{D} \tag{7}$$

$$\mathbf{n} = (\cos 2\phi, \sin 2\phi, 0)$$

For a purely gluonic cascade averaged over initial and summed over final polarisations, for example, we can write Eqs. 6 simply as

$$\rho' = \mathbf{M}(z; 0, \rho, \mathbf{n})$$

$$\rho'' = \mathbf{M}(z; \mathbf{D}', \mathbf{n}, \rho) \tag{8}$$

$$\mathbf{D} = \mathbf{M}(z; \mathbf{n}, \mathbf{D}', \mathbf{D}'')$$

where

$$\mathbf{M}(z; \mathbf{a}, \mathbf{b}, \mathbf{c}) = [u(1 + \mathbf{a} \cdot \mathbf{b})\mathbf{c} + v(1 + \mathbf{c} \cdot \mathbf{a})\mathbf{b} + w(1 + \mathbf{b} \cdot \mathbf{c})\mathbf{a}$$
$$+ 2(\mathbf{a} \wedge \mathbf{b}) \wedge \mathbf{c} + 2z(\mathbf{b} \wedge \mathbf{c}) \wedge \mathbf{a}]/(P + u\mathbf{a} \cdot \mathbf{b} + v\mathbf{c} \cdot \mathbf{a} + w\mathbf{b} \cdot \mathbf{c}) \tag{9}$$

u, v, w and P being as defined in Eq. 5.

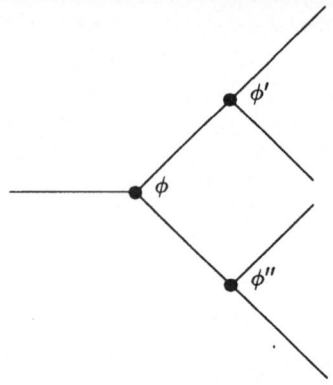

Figure 6. Three-vertex gluon cascade

Let's see how this method works by computing the azimuthal correlations for the three-vertex gluon cascade in Fig. 6. The average over initial and sum over final spins means that ρ for the incoming line and $\mathbf{D}$ for all outgoing lines are zero. Therefore, in an obvious notation,

$$\boldsymbol{\rho}' = \mathbf{M}(z; \mathbf{0}, \mathbf{0}, \mathbf{n}) = u\mathbf{n}/P$$

$$\mathbf{D}' = \mathbf{M}(z'; \mathbf{n}', \mathbf{0}, \mathbf{0}) = w'\mathbf{n}'/P'$$

$$\mathbf{D}'' = \mathbf{M}(z''; \mathbf{n}'', \mathbf{0}, \mathbf{0}) = w''\mathbf{n}''/P''$$

$$\boldsymbol{\rho}'' = \mathbf{M}(z; \mathbf{D}', \mathbf{n}, \mathbf{0}) = (v\mathbf{n} + ww'\mathbf{n}'/P')/(\mathcal{J} + uw'\mathbf{n} \cdot \mathbf{n}'/P')$$

(10)

and so the overall probability distribution is

$$P(z, z', z''; \mathbf{n}, \mathbf{n}', \mathbf{n}'') \sim 1 + \boldsymbol{\rho}'' \cdot \mathbf{D}''$$

$$\sim PP'P'' + uw'P''\mathbf{n} \cdot \mathbf{n}' + vw''P'\mathbf{n} \cdot \mathbf{n}'' + ww'w''\mathbf{n}' \cdot \mathbf{n}''$$

(11)

as found in [13].

To test the validity of the algorithm in a more complicated case, 10^7 Monte Carlo gluon cascades like Fig. 4 were generated with momentum fractions z_i set to one-half at each vertex. As shown in Table II, the azimuthal correlations in the Monte Carlo 'data' agree with the algebraic results of Eq. 4, within the expected statistical errors (shown in parentheses for the last digit).

We see from Table II that for purely gluonic cascades the correlations are numerically small — a few percent at most — when the momentum fractions at all vertices are comparable. For momentum fractions $z_i \neq \frac{1}{2}$ they are typically of the same order or smaller. For certain momentum configurations, there are correlations of a few percent between arbitrarily widely separated vertices — see Ref. [14].

When quarks are included, as illustrated in Fig. 7, there are large correlations associated with the branching $g \to q\bar{q}$. The values of C_{ij} given are for $z_i = z_j = \frac{1}{2}$. They rise to -1 when

Table II. Correlation coefficients for the branching process shown in Fig. 4

i	j	C_{ij}	Value	Monte Carlo
0	1	$\frac{4}{81}$	0.0494	0.0492(2)
0	2	$\frac{16}{729}$	0.0219	0.0219(3)
1	3	$\frac{1}{729}$	0.0014	0.0016(3)
2	3	$\frac{4}{6561}$	0.0006	0.0004(3)
2	4	$\frac{16}{59049}$	0.0003	0.0001(3)

the connecting gluon line becomes soft. For vertices connected by a quark line, on the other hand, helicity conservation in the branching $q \to qg$ implies that there are no correlations at all. Thus we expect the inclusion of azimuthal correlations to be most important in processes that depend on the branching $g \to q\bar{q}$, such as charmed particle production in jets at hadron colliders.

In the Monte Carlo programs of Refs. [9, 21], azimuthal correlations are partially taken into account: the 'nearest-neighbour' correlations (those shown in Fig. 7 and C_{01} in Table II) are included but not the more 'long-range' ones like C_{02}. In this approximation, there is no evidence yet for any dramatic correlation effects that survive hadronization, at least not in global event structures for e^+e^- annihilation at present energies [13]. Work is now under way to incorporate spin effects in full, including the parton polarisations induced in hard scattering, which also give rise to interesting correlations [16].

SOFT GLUON COHERENCE

As mentioned earlier, many aspects of soft gluon interference can be included in a Monte Carlo simulation by ensuring that the opening angles of successive branchings in which gluons are emitted are ordered. Suppose for example we want to compute the coherent soft gluon emission from two partons i and j (either incoming or outgoing) participating in some hard subprocess (Fig. 8).

The soft gluon distribution will be of the form

$$
\begin{aligned}
dP &= \frac{d^3k}{k} \frac{(p_i \cdot p_j)}{(p_i \cdot k)(p_j \cdot k)} \\
&= \frac{d^3k}{k^3} \frac{(1 - \cos\theta_{ij})}{(1 - \cos\theta_i)(1 - \cos\theta_j)} \\
&= \frac{dk}{k} d\cos\theta_i \, d\phi_i \, W_{ij} + (i \leftrightarrow j)
\end{aligned}
\tag{12}
$$

where

$$
W_{ij} \equiv \frac{1}{2} \left[\frac{(1 - \cos\theta_{ij})}{(1 - \cos\theta_i)(1 - \cos\theta_j)} + \frac{1}{(1 - \cos\theta_i)} - \frac{1}{(1 - \cos\theta_j)} \right]
\tag{13}
$$

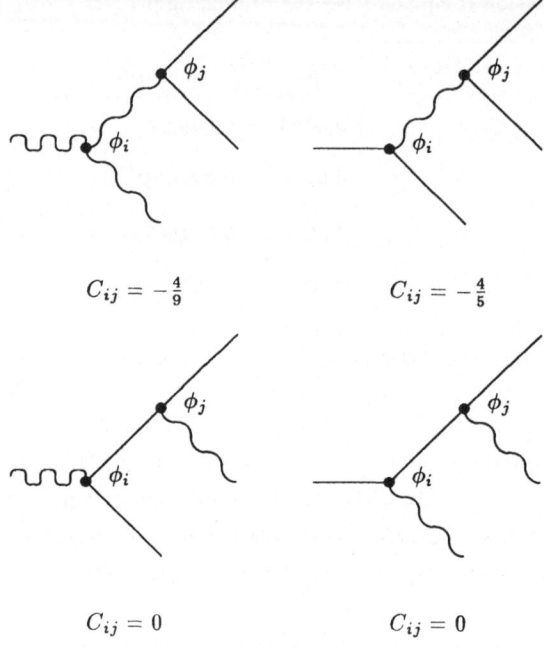

Figure 7. Quark-gluon correlations

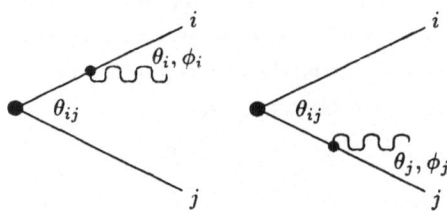

Figure 8. Coherent soft gluon emission

The point about splitting the radiation pattern into W_{ij} and W_{ji} is that only the former has the pole at $\theta_i = 0$, so this piece can be thought of as 'belonging to' parton i. Furthermore, if W_{ij} is averaged with respect to the azimuthal angle ϕ_i around the direction of p_i we get

$$\langle W_{ij} \rangle \equiv \int \frac{d\phi_i}{2\pi} W_{ij} = \Theta(\theta_{ij} - \theta_i)/(1 - \cos\theta_i) \tag{14}$$

In other words, the azimuthal average of W_{ij} is just the *incoherent* radiation from parton i, confined to the cone $\theta_i < \theta_{ij}$. Similarly W_{ji} azimuthally averaged about p_j gives radiation from j into the cone $\theta_j < \theta_{ij}$. This is the origin of angular ordering.

Generalizing to multigluon emission, we may consider each gluon as emitted coherently from a subgraph consisting of those lines that are nearer to the hard subprocess. In Fig. 9, for example, we obtain the angular ordering $\theta_0 > \theta_1 > \theta_2 > \theta_3 \ldots$ for radiation from the incoming quark and $\theta_0 > \theta_1' > \theta_2' \ldots$ for that from the outgoing one.

For quantities that are integrated over all azimuths, such as parton energy spectra and

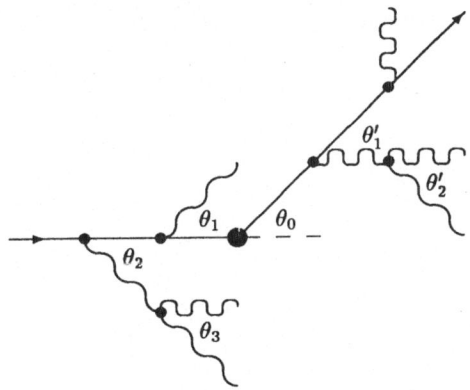

Figure 9. Angular ordering

multiplicity distributions, the substitution

$$W_{ij} \rightarrow \langle W_{ij} \rangle = \Theta(\theta_{ij} - \theta_i)/(1 - \cos\theta_i), \tag{15}$$

which we shall call the *angular-ordering approximation*, gives the correct leading and next-to-leading logarithmic behaviour in the soft region, i.e. for gluon energies k such that $Q \gg k \gg Q_0 \gg \Lambda$, where Q is the hard process scale and Q_0 is the infrared cutoff, as discussed in Sect. 1. For other quantities it provides an approximation scheme that seems to be rather accurate in many cases but needs to be investigated more systematically.

The angular-ordering prescription is very well suited to Monte Carlo simulation because all the terms involved are non-negative. Parton distributions can be generated by a Markov process that looks like a classical branching process, while implicitly including the destructive interference effects that, on the average, cancel the probabilities of disordered configurations.

As a refinement that still keeps everything non-negative, instead of simply replacing W_{ij} by $\langle W_{ij} \rangle$ as in Eq. 15 we may use the *improved angular-ordering approximation*

$$W_{ij} \rightarrow W_{ij} \, \Theta(\theta_{ij} - \theta_i), \tag{16}$$

thus retaining the full coherence effect inside the cone $\theta_i < \theta_{ij}$ while still treating it only on the average outside the cone. As we shall see, this prescription seems to reproduce quite accurately many of the phenomena due to soft gluon interference.

Angular ordering occurs for any colour gauge group and indeed is well known in QED as the Chudakov effect [23]. Although the colour algebra for SU(N) is not very complicated, a particularly simple picture emerges if we work to leading order in $1/N$, which turns out to be a surprisingly good approximation. Then gluons may be represented by a colour and an anticolour line, with half the radiation coming from each, and only colour-connected partons interfere. For example, in $e^+e^- \rightarrow q\bar{q}g$ the full soft gluon radiation pattern is of the form

$$W_{q\bar{q}g} = C_F(W_{qg} + W_{\bar{q}g}) + \tfrac{1}{2}C_A(W_{gq} + W_{g\bar{q}}) + \tfrac{1}{2}N^{-1}(W_{qg} - W_{q\bar{q}} + W_{\bar{q}g} - W_{\bar{q}q}) \tag{17}$$

where $C_F = (N^2-1)/(2N) = 4/9$ and $C_A = N = 3$ are respectively the quark and gluon colour-charges squared.

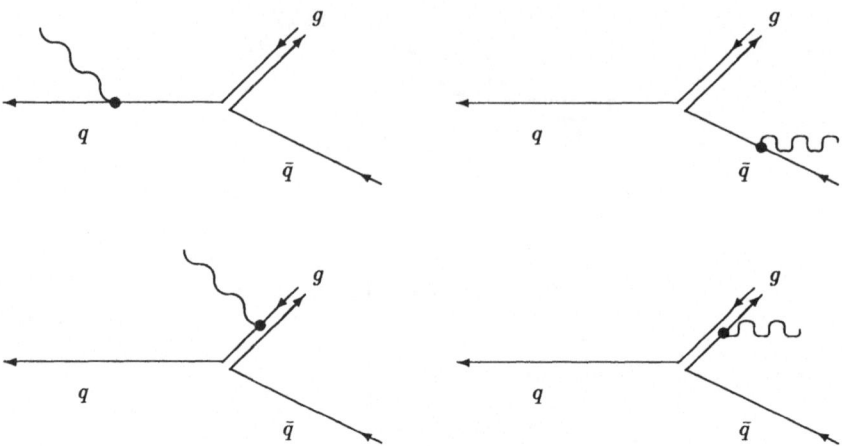

Figure 10. Soft gluon radiation in $e^+e^- \to q\bar{q}g$

The first two terms in Eq. 17 correspond to the contributions in Fig. 10, where only the qg and $\bar{q}g$ lines interfere. The final term is not only of relative order $1/N^2$ but also dynamically suppressed: it has no collinear poles and averages to zero throughout much of phase space (c.f. Eq. 14). Thus the error involved in neglecting it turns out to be only a few percent. Note that this is only true if the different colour factors C_F and $\frac{1}{2}C_A$ are used for radiation from a quark and from a colour line in a gluon, respectively.

Fig. 11 shows the projection onto the hard parton plane of the full soft gluon radiation pattern of Eq. 17 (full line), and the angular-ordered (dashed) and improved (dot-dashed) approximations of Eqs. 15 and 16. The hard quark, antiquark and gluon directions were fixed at $\theta = 0°$, $155°$ and $230°$ respectively. The histograms were generated from 50,000 random points in phase space. For the two approximate curves the last term in Eq. 17 was dropped. We see that both approximation schemes reproduce the 'string effect', i.e. the large dip around $\theta = 80°$, between the quark and antiquark jets. This is due to destructive interference between the gluon jet and the emission from the quark and antiquark. Not surprisingly, the improved approximation is better than simple angular ordering overall, particularly in the other two dip regions, where the interference is constructive.

Fig. 12 shows that the string effect is also seen in the hadron distribution of real three-jet events: the data are from the JADE Collaboration [18] and the solid and dashed curves show the predictions of QCD Monte Carlo simulations with [9] and without [24] angular ordering, respectively.

A similar analysis of soft gluon radiation in $2 \to 2$ hard parton scattering can be performed, comparing the two approximation schemes with the soft limit of the corresponding $2 \to 3$ cross section [19]. A complication here is that there may be more than one possible colour configuration of the hard partons. In quark-gluon scattering, $qg \to q'g'$, for example, the colour of the incoming quark may be connected to either the initial or the final state gluon

(Fig. 13). Correspondingly, to leading order in $1/N$ we expect the radiation pattern to be of the form

$$W_{qg \to q'g'} \simeq (H_a W_a + H_b W_b)/(H_a + H_b) \qquad (18)$$

where H_a and H_b are the contributions of the two colour configurations and

$$W_a = C_F(W_{qg} + W_{q'g'}) + \tfrac{1}{2}C_A(W_{gq} + W_{g'q'} + W_{gg'} + W_{g'g}),$$
$$W_b = C_F(W_{qg'} + W_{q'g}) + \tfrac{1}{2}C_A(W_{gq'} + W_{g'q} + W_{gg'} + W_{g'g}). \qquad (19)$$

There will be a quark-exchange pole in the s-channel for configuration a and in the u-channel for b, and in both cases a t-channel gluon pole. Up to an overall normalization, we find

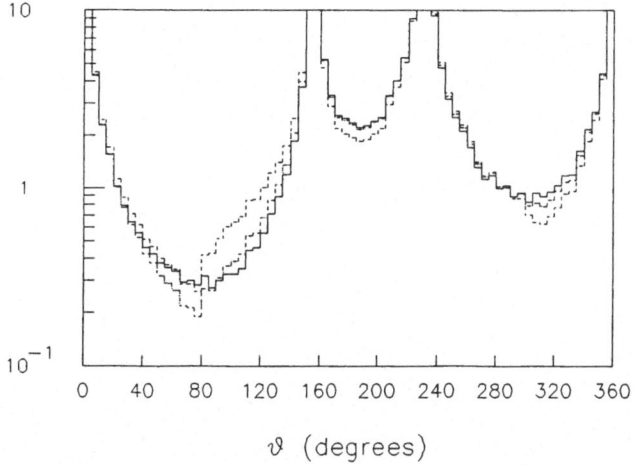

Figure 11. Soft gluon radiation pattern from $q\bar{q}g$ system, projected onto $q\bar{q}g$ plane, as a function of angle θ relative to quark jet.

$$H_a \simeq \left(\frac{s^2 + u^2}{t^2} - \frac{1}{N^2}\right)\frac{u}{s},$$
$$H_b = \left(\frac{s^2 + u^2}{t^2} - \frac{1}{N^2}\right)\frac{s}{u} \qquad (20)$$

where s, t and u are the Mandelstam variables for the hard subprocess.

The inclusion of the $1/N^2$ terms in Eqs. 20 is somewhat arbitrary since we are working to leading order in $1/N$, but it ensures that $(H_a + H_b)$ reproduces the exact lowest-order $qg \to qg$ differential cross section. Furthermore, as in the case of $e^+e^- \to q\bar{q}g$, one then finds that the approximate expression in Eq. 18 is generally accurate at the few percent level. This is illustrated in Fig. 14, which compares the exact soft gluon radiation pattern from 90° hard quark-gluon scattering (solid line, formula given in Ref. [19]) with the angular-ordered (dashed) and improved (dot-dashed) approximations based on Eq. 19. The incoming and outgoing quark jets are at 0° and 90°, the incoming and outgoing gluon jets at 180° and 270°. Again we see that the interference effects, destructive in the angular region between the quark jets (0°–90°)and constructive in the other dip regions, are well reproduced by the improved approximation scheme of Eq. 16.

Preliminary results from full Monte Carlo simulations of hadron-hadron hard scattering, using the program of Ref. [21], indicate that interference between jets should be a signifi-

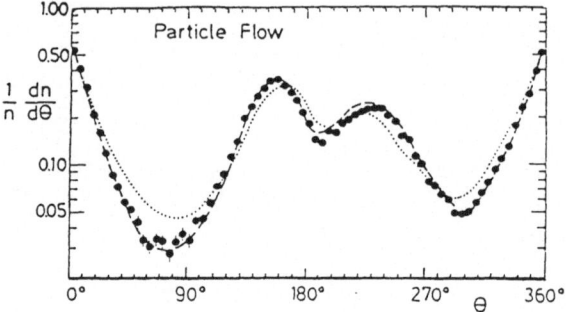

Figure 12. Hadron flow in $e^+e^- \to 3$ jets

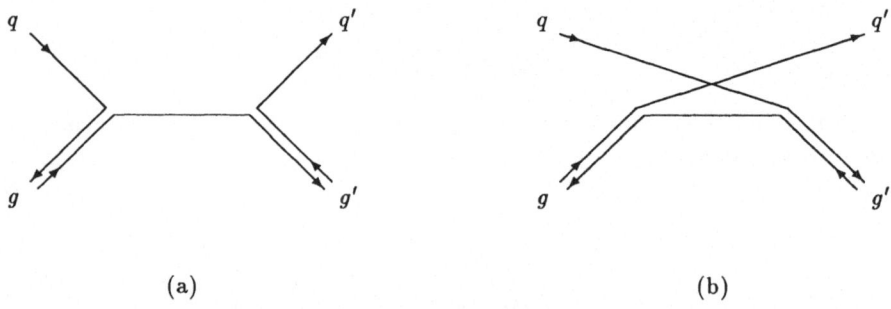

(a)   (b)

Figure 13. Colour configurations in $qg \to q'g'$

cant, and potentially useful, phenomenon. As an illustration, the solid line in Fig. 15 shows the predicted particle distribution in the hard scattering c.m. frame for $W^\pm$ production and hadronic decay at 90°. The dashed line shows that for the process considered above ($qg \to qg$), again at 90°, with a lower limit of 80 GeV on the sum of the jet transverse energies. In both cases parton cascades from incoming and outgoing hard partons are treated in the improved angular-ordering approximation, with nearest-neighbour spin correlations. For hadronization, a cluster model similar to that of Ref. [9] is used. The soft underlying event is included, using a multi-cluster model based on the UA5 Monte Carlo [25].

We see that, even after hadronization and inclusion of the underlying event, the strong destructive interference in the region between the quark jets remains clear in quark-gluon scattering. In going from the process $q\bar{q} \to W^{\pm} \to q\bar{q}$ to $qg \to qg$, we replace two quark jets by gluon jets, but contrary to intuition this *decreases* the hadron flow in the region opposite the gluon jets. This is a kind of 'string effect' like that in $e^{+}e^{-}$ annihilation, where an analogous decrease in hadron production opposite the gauge boson is seen [26] when one goes from $e^{+}e^{-} \to q\bar{q}\gamma$ to $e^{+}e^{-} \to q\bar{q}g$.

It would be exciting if the differences between the two radiation patterns displayed in Fig. 15 could be used to enhance the signal to background ratio for hadronic $W^{\pm}$ decays [27],

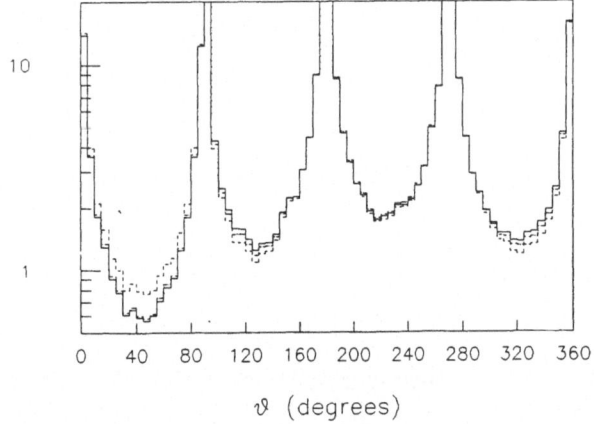

Figure 14. Soft gluon radiation pattern from $qg \to qg$, projected onto the scattering plane, as a function of angle θ relative to incoming quark jet

for which the principal background at the SPS Collider is indeed quark-gluon hard scattering. For example, one could reject events with a large asymmetry in the populations of the dip regions. Unfortunately, the Monte Carlo studies suggest that event-to-event fluctuations are too great for this type of criterion alone to be very useful. However, it may prove valuable if used in conjunction with other selection criteria.

CONCLUSIONS

In this review I have outlined some recent work in two main areas that should help to improve the simulation of jet physics at high energies:

- A full treatment of azimuthal correlations due to gluon spin now seems feasible. This is surely an important goal even though many of the phenomenological consequences may turn out to be small. After all, the vector nature of the gluon is central to the concept of a gauge theory of strong interactions. In addition, there may be large effects in some important rare processes, such as heavy quark production in gluon jets.

- The treatment of interference within and between jets has been improved, by extending the angular-ordering technique to initial-state jets and comparing the results with the soft limits of $2 \rightarrow 3$ hard scattering cross sections. In addition, an improved angular-ordering

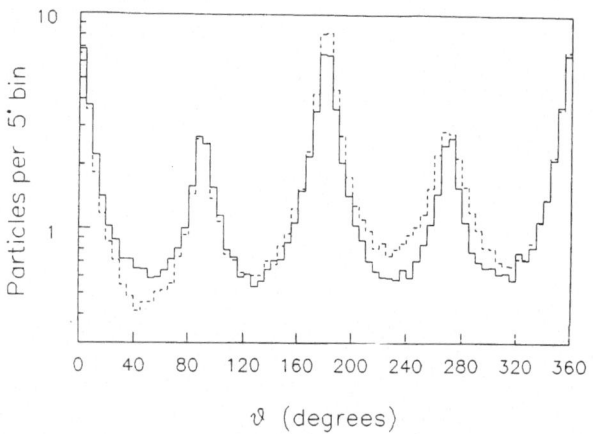

Figure 15. Simulated particle distributions in $p\bar{p}$ collisions at 630 GeV. Solid: $q\bar{q} \rightarrow W^{\pm} \rightarrow q\bar{q}$. Dashed: $qg \rightarrow qg$.

prescription that approximates the full coherent radiation pattern more closely has been introduced. Taking into account the effects of hadronization and the underlying soft event, significant interference between jets is expected to be apparent in hard hadron-hadron collisions. Interference effects could prove a valuable additional tool in the task of extracting important signals from the QCD background.

In summary, our understanding of jet physics is steadily increasing. This understanding is slowly finding its way into the Monte Carlo simulation programs that will play a crucial role in the design of future experiments and the analysis of their data.

It is a pleasure to acknowledge many helpful conversations with J.C. Collins, R.K. Ellis, I.G. Knowles and G. Marchesini.

REFERENCES

1. A. Bassetto, M. Ciafaloni and G. Marchesini, Phys. Reports **100** (1983) 201.

2. A.H. Mueller, in *Proc. Int. Symp. on Lepton and Photon Interactions at High Energies, Kyoto, 1985*, p.162.

3. B.R. Webber, Ann. Rev. Nucl. Part. Sci. **36** (1986) 253.

4. K. Konishi, A. Ukawa and G. Veneziano, Nucl. Phys. **B157** (1979) 45.

5. G. Altarelli and G. Parisi, Nucl. Phys. **B126** (1977) 298.

6. See, for example, *Proc. UCLA Workshop on Observable Standard Model Physics at the SSC: Monte Carlo Simulation and Detector Capabilities*, ed. H.-U. Bengtsson, C. Buchanan, T. Gottschalk and A. Soni (World Scientific, 1986).

7. G. Marchesini and B.R. Webber, Nucl. Phys. **B238** (1984) 1.

8. B. Andersson, G. Gustafson, G. Ingelman and T. Sjöstrand, Phys. Reports **97** (1983) 31.

9. B.R. Webber, Nucl. Phys. **B238** (1984) 492.

10. G. Gustafson, Lund preprint LU TP 87–15, to appear in *Proc. Int. Europhysics Conf. on High Energy Physics, Uppsala, Sweden, June 1987*.

11. T. Sjöstrand, Phys. Lett. **157B** (1985) 321.

12. M.P. Shatz, Nucl. Phys. **B224** (1983) 218.

13. B.R. Webber, Phys. Lett. **193B** (1987) 91.

14. I.G. Knowles, Cambridge preprint Cavendish-HEP-87/5, June, 1987.

15. J.C. Collins, Illinois Institute of Technology preprint, June, 1987.

16. J.C. Collins, I.G. Knowles and B.R. Webber, work in progress.

17. B.R. Webber, Phys. Lett. **143B** (1984) 501.

18. JADE Collaboration: W. Bartel *et al.*, Phys. Lett. **134B** (1984) 275; ibid. **157B** (1985) 340.
    TPC Collaboration: A. Aihara *et al.*, Z. Phys. **C28** (1985) 31.

19. R.K. Ellis, G. Marchesini and B.R. Webber, Nucl. Phys. **B286** (1987) 643.

20. Yu.L. Dokshitzer, V.A. Khoze, S.I. Troyan and A.H. Mueller, Columbia preprint CU–TP–374 (1987).

21. *HERWIG, a New Monte Carlo Event Generator for Simulating Hadron Emission Reactions With Interfering Gluons*, G. Marchesini and B.R. Webber, in preparation.

22. A. Einstein, B. Podolsky and N. Rosen, Phys. Rev. **47** (1935) 777.

23. A.E. Chudakov, Izv. Akad. Nauk. SSSR, ser. fiz. **19** (1955) 650.

24. T.D. Gottschalk, Nucl. Phys. **B214** (1983) 201.

25. UA5 Collaboration: G.J. Alner *et al.*, Nucl. Phys. **B291** (1987) 445.

26. TPC Collaboration: H. Aihara *et al.*, Phys. Rev Lett. **57** (1986) 945.
    Mark II Collaboration: P.D. Sheldon *et al.*, Phys. Rev Lett. **57** (1986) 1398.

27. UA2 Collaboration: R. Ansari *et al.*, Phys. Lett. **186B** (1987) 452.

SOFT PROCESS IN VERY HIGH ENERGY PROTON-PROTON COLLISIONS

G. Ingelman
Deutsches Elektronen-Synchrotron DESY
Notkestrasse 85, D-2000 Hamburg 52, FRG

Abstract

Soft processes in terms of fragmentation of high-$p_\perp$ and low-$p_\perp$ jets are considered in relation to the high and low momentum transfer mechanisms that produce them. The evolution of high-$p_\perp$ jets can be reliably described based on perturbative QCD radiation and hadronization models tuned at present energies and we give realistic predictions for the properties of such jets at very high energies. The modelling of low-$p_\perp$ spectator jets and minimum bias events, on the other hand, involve important uncertain elements regarding the interaction mechanism and the resulting colour field structure which lead to large uncertainties in the resulting predictions. Diffractive scattering is discussed in terms of the nature of the pomeron and its interactions, in particular the possibility of parton constituents in the pomeron and 'hard diffractive scattering' is considered.

1 Introduction

The notion of 'soft' process is not quite well defined and when extrapolating to the enormous energies that the ELOISATRON would provide one has to first discuss its meaning, e.g. whether 'soft' is to be understood on an absolute or relative scale. Usually a limited absolute scale in transverse momentum exchange of the order 1 GeV is thought of, and this will also be our attitude. A scale increasing with total energy, $\sqrt{s}$, would mean that many processes which are presently successfully described as hard would be classified as soft at future high energy machines. One should note, however, that the requirement of a low *transverse* momentum exchange does not exclude the occurence of substantial *longitudinal* momentum transfers. There are in fact models for minimum bias interactions that involve such longitudinal momentum exchanges, either explicitly by construction or implicitly. Another aspect is that processes or objects that occur as hard processes at lower energies may also arise through soft processes at larger energies. An example is provided by charm production which seems to be explained by hard QCD fusion processes at fixed target energies, but may have an important diffractive component at ISR energies. Another case is W and Z bosons which appear in hard processes at present collider energies, but in some circumstances can be treated by a Weizäcker-Williams 'equivalent boson' approximation at supercollider energies. The recent discussion on 'minijets' and their possible relation to the total cross-section rise, also illustrate the vague borderline between 'hard' and 'soft' physics with increasing energy.

There is thus an important interplay between the 'hard' and 'soft' physics that we might associate with the parton and hadron 'worlds', respectively. In the ultimate theory these aspects will be related, presumably by colour confinement which today is a major unsolved problem in high energy hadron physics. It is only due to our present ignorance and limited calculational techniques, e.g. perturbation theory, that 'hard' and 'soft' physics are treated separately. To improve our understanding it is, therefore, interesting to investigate the relations between our present theories and models for hard and soft interactions. This is of importance also for practical reasons of planning future experiments, namely in order to make realistic predictions of what events will look like at supercollider energies.

Although soft processes also include important issues like total cross-sections, elastic and diffractive scattering, this will not be discussed here. The only exception is the possibility of a parton structure in the pomeron and its consequences in terms of so-called 'hard diffractive scattering'. The organization of the paper is as follows. The properties of high-$p_\perp$ jets resulting from perturbative jet evolution and non-perturbative hadronisation is discussed in Section 2, whereas Section 3 considers low-$p_\perp$ jets in the sense of beam jets and minimum bias interactions. In Section 4 some ideas on pomeron structure and interactions are discussed and, finally, we end with some concluding remarks in Section 5.

2   Hard and soft components in high-$p_\perp$ jets

The properties of high-$p_\perp$ jets are influenced both by hard, or semihard, parton processes that can be calculated in perturbative QCD and by soft hadronization processes of a non-perturbative QCD origin which cannot be calculated from fundamental principles at present. Phenomenologically successful models for the latter exists and to the extent that these are not just parametrizations of data, but rather based on more or less elaborate physics ideas, they can have a large predictive power leading to useful tests of the assumptions involved in the models. The use of a separate treatment of the pertubative and non-perturbative phases is based on the assumption that they occur sequentially with two sufficiently well separated space-time scales. Nevertheless, the two parts must be properly connected and our attempts to understand high-$p_\perp$ jets in detail thus involve an interplay of 'hard' QCD theory and 'soft' models and can thereby serve as a suitable test bench for our purposes.

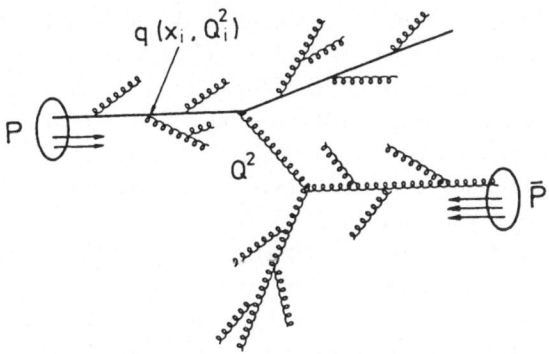

Figure 1. *Initial and final state parton radiation in high-$p_\perp$ hadron-hadron scattering*

2.1 Perturbative jet evolution

The calculation of matrix elements in QCD perturbation theory is well defined, but quickly becomes very complicated for higher order processes. In high-$p_\perp$ hadron-hadron scattering exact matrix elements have been calculated up to order α_s^4 [1], corresponding to at most four partons in the final state (spectators not counted). Still higher order diagrams, giving rise to multiparton final states, are important not only at higher energies, but also at present energies in order to understand detailed jet evolution properties. This requires the use of approximation schemes and the parton cascade approach (for a review see [2]) is particularly useful since the complete final parton state can be dynamically simulated on a computer giving access to all possible variables. The basic idea is that partons emitted in a large transverse momentum process can be off mass shell and emit bremsstrahlung gluons, which may in turn split into gluon or $q\bar{q}$ pairs, leading to a shower of partons as illustrated in Fig. 1.

Parton cascade models have been developed to a high degree over the last few years [3,4,5]. The common feature of such models is that first order QCD matrix elements in the leading logarithm approximation is used for each separate branching, i.e. the Altarelli-Parisi equations [6] for the basic perturbative splitting processes $q \rightarrow qg$, $g \rightarrow gg$ and $g \rightarrow q\bar{q}$ are used in an iterative procedure which is stopped when all parton virtualities, m_i^2, are below a cutoff, t_{cut}, usually chosen of order 1 GeV^2. Together with Λ_{QCD} this cutoff regulates the amount of partons radiated. One should bear in mind that this approximation is not expected to work properly for hard gluon emission at large angles where interferences between diagrams with the gluon emitted from different parton lines are important. These models are rather intended for studies of e.g. jet broadening due to the emission of several but not very hard gluons at large angles. Nevertheless, these models may be used also for multiple jet phenomena simply because better higher order calculations are lacking at present; the general features will certainly be adequately described although the rates and some distributions will not be exactly the correct ones.

In a high-$p_\perp$ proton-proton collision, Fig. 1, the scattered partons may have a virtual mass squared up to order Q^2 and produce final state parton emission of a timelike character, i.e. all partons have $m^2 \geq 0$, which can be described by models developed for e^+e^- annihilation. This usually involves an angular ordering to take interferences between soft gluons into account [4] and very good agreement with experimental data can be obtained [7]. In hadron collisions, however, not only the scattered parton, but also the partons entering the hard scattering process can emit radiation, Fig. 1. In this case, the radiating parton develops a negative virtuality, i.e. has a space-like 4-vector giving $m^2 < 0$. This initial state cascade evolution may be viewed as a quantum fluctuation which can only be realized in a large momentum transfer process which put the parton back on shell or to a positive virtuality, i.e. $m_i^2 < 0 \rightsquigarrow m_f^2 \geq 0$. For practical reasons it is better to start with the hard scattering, given by the $2 \rightarrow 2$ exact matrix elements, and perform the initial cascade evolution backwards in time. Such a scheme, developed in [8], must also take into account constraints from the structure functions, since at each intermediate step one should have the correct probability of finding a parton with momentum fraction x_i at the proper momentum transfer scale Q_i^2 as indicated in Fig. 1. This leads to a modified splitting probability

$$dP_b = |dt| \, \frac{\alpha_s(Q^2)}{2\pi} \sum_a \int \frac{dx'}{x'} \frac{f_a(x',t)}{f_b(x,t)} \, P_{a \to bc}\left(\frac{x}{x'}\right) \qquad (1)$$

with the ratio of the structure functions, f, before and after splitting as a weight. Otherwise the procedure is similar to that of final state radiation. The structure functions, however, has the effect of reducing the amount of radiation as compared to the case for final state radiation. For the properties of high-$p_\perp$ jets the initial state radiation is of less importance, but it does influence the underlying event and also generate a transverse momentum of the hard scattering system.

Although the parton shower approach is phenomenologically very useful, one should realize its limitations. Not only does it involve the QCD leading log approximation, but also some details which are not theoretically well-defined. It cannot, therefore, replace exact matrix element calculations for fundamental tests of QCD and the determination of Λ_{QCD} in a well-defined renormalization scheme. Ideally one would use exact matrix elements up to the order available and then add the higher order effects by a cascade model. This give rise, however, to problems of a proper joining of the two methods such that double counting is avoided; so far this is only solved for order α_s matrix elements in e^+e^- annihilation [5]. Furthermore, α_s used in each separate branching depends on the momentum transfer in that vertex, and therefore becomes larger the further the cascade is evolved and the perturbative approximation will break down at some point when the parton virtualities become small. The value for the parameter t_{cut}, which determines the border line between the region where perturbative QCD can be considered trustworthy and the following non-perturbative region, is not given by theory but is basically a free parameter which is obtained from comparison with data after a method for the final hadronization step has been included.

2.2 Non-perturbative hadronization

The simplest fragmentation model where all partons from the shower hadronize independently of each other using, e.g., the Field-Feynman parametrization [9] is not stable with respect to changes of this t_{cut} value [10] and is therefore not suitable. Being essentially a parametrizatiuon it is furthermore not so interesting since it does not

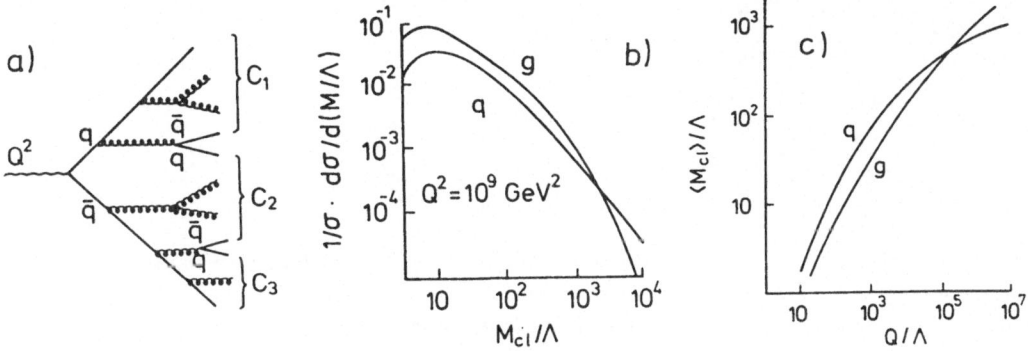

Figure 2. *(a) Preconfinement of perturbatively produced quanta into colour singlet clusters. (b) Cluster mass distribution at asymptotic energy. (c) Average mass versus energy scale. Energies and masses are scaled with Λ_{QCD}; q and g refer to quark and gluon jet, respectively. From [12].*

provide much understanding. Physically more interesting is the possibility of cluster formation from the partons and, in particular, the idea of preconfinement [11]. Given the well-defined colour ordering of the planar graph the partons can be associated with colour singlet clusters (Fig. 2a) which could form a link to the hadronic final state. Considering e^+e^- annihilation for simplicity, the original $q\bar{q}$ colour singlet system can give rise to more than one cluster only if additional $q\bar{q}$ pairs are formed in the perturbative shower evolution. Consequently, the cluster multiplicity and mass spectrum depends on the frequency of the $g \to q\bar{q}$ branching. Analytical calculations in an asymptotic limit indicates that the typical cluster mass is close to the shower cutoff, i.e. close to the hadronic mass scale, and, moreover, essentially independent of Q^2 [11]. More detailed investigations based on Monte Carlo simulations show that this is not quite correct [12]. The mass spectrum of such clusters, Fig. 2b, has indeed a peak at small masses, but also a long tail to large masses which makes the average mass quite appreciable and also increasing with increasing Q^2, Fig. 2c. The cluster masses are therefore in reality significantly above the hadron mass scale and become even more so at higher energy scales. Unfortunately, this prevents an easy connection between perturbatively produced clusters and the final state hadrons.

Nevertheless, it is possible to construct phenomenological cluster models [4] by first splitting the gluons into $q\bar{q}$ pairs and then let colour-connected $q\bar{q}$ pairs form lower mass clusters, which are finally decayed into ordinary hadrons (including resonances) using pure phase space. The continous cluster mass spectrum obtained will obviously depend on the parton shower cutoff value and hence a low cutoff is preferable in order to get cluster masses not too much above the hadron mass scale. Even with a small cutoff, however, large mass clusters will occur and their isotropic phase space decay will produce too spherical events compared to e^+e^- data. This is usually solved by splitting heavy clusters, with a mass larger than 3–4 GeV, into lower mass clusters using a longitudinal (string-like) decay.

Another approach is to connect the perturbatively produced partons, whose colour ordering is given by the shower evolution, with a colour string force field and apply the Lund fragmentation model [13] for the final hadronization step. A colour triplet and antitriplet charge, e.g. q and $\bar{q}$, are here represented by the endpoints and a

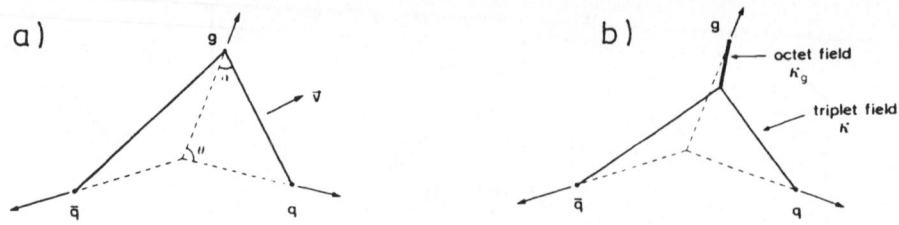

Figure 3. *(a) Representation of a $q\bar{q}g$ system using a triplet string stretched via the colour octet gluon. The velocity of a string piece is $v = \cos\theta/2$. (b) Alternative gluon model with a colour octet string joining the two triplet strings at a junction.*

gluon colour octet charge by an energy-momentum carrying kink on the string, Fig. 3a. Thus, a rather complicated string topology arise when many gluons have been emitted. The string model provides a desired stability [10,14], of the final hadron state properties with respect to variations of the arbitrary t_{cut} parameter since the extra gluons emitted with a lower cutoff will only produce small disturbances on the string configuration obtained without them. In e^+e^-, e.g., more than one colour string system will only occur if $q\bar{q}$ pairs have been produced in the perturbative evolution; these strings thus correspond to the 'preconfinement clusters' in Fig. 2a. The string decays into pieces which are usually forced to correspond directly to final hadrons (including resonances), but in the symmetric Lund model [15] one could equally well have had an intermediate step with 'larger' pieces that would resemble the clusters in the previous model. Thus, the two models are not completely ortogonal, although they are constructed rather differently.

Representing a gluon with a kink, Fig. 3a, means that no additional assumptions and parameters are needed for the gluon fragmentation model, since it is determined by the basic break-up of the colour triplet field as given by quark jets and constrained by, e.g., e^+e^- data. Gluon jet fragmentation is, however, experimentally not well measured and other models can certainly be conceived of. One possibility [16], which can be included in the string framework, is that the gluon stretches a colour octet field which is split into two triplet fields at a junction, Fig. 3b. The position of the junction is determined by the ratio κ_g/κ_q of the string tensions in the octet and triplet fields and if it is larger than two, as suggested by the ratio 9/4 obtained from the Casimir operator magnitudes for the octet and triplet representations of QCD, it is energetically favourable for the octet field to collapse to zero length; thus giving the gluon as a kink on the string. The hadronization of an octet field is, furthermore, unknown and additional assumptions would have to be made to construct a model. On general grounds one would expect an octet field to break by gluon pair creation resulting in the production of glueballs (if existing) and isoscalar particles, but such a model [17] has been found inconsistent with results on η and ϕ production in Υ decays, whereas the Lund gluon-kink model does provide a good description [18].

2.3 Jets at present energies

The importance of the higher order QCD effects included in the parton shower approach has been demontrated by detailed studies of the rate of multijet events, event

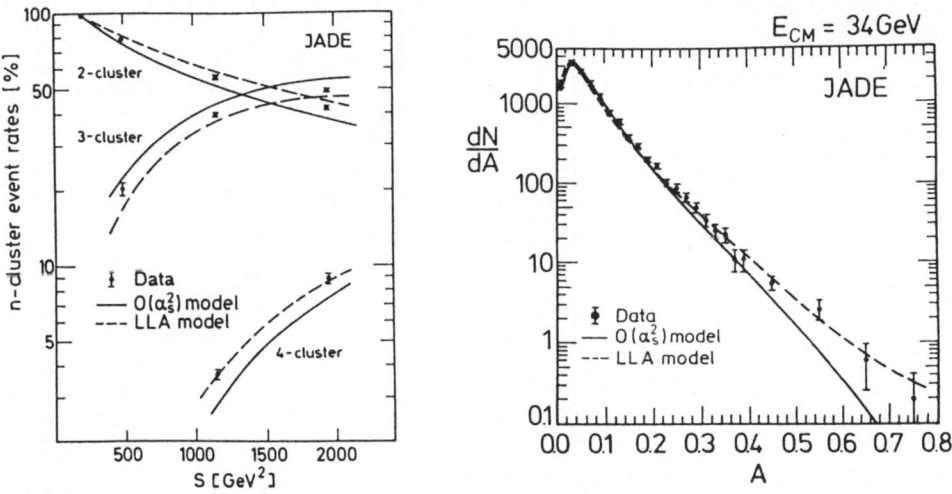

Figure 4. *n-jet event rates versus cms energy and acoplanarity distribution in e^+e^- data compared to model calculations [19].*

shape measures and internal jet properties. Using a cluster algorithm to find jets in e^+e^- annihilation, the JADE collaboration [19] at PETRA observes an excess of 4-jet events as compared to the expectations from order α_s^2 matrix elements. The parton shower model can, however, reproduce the jet rates observed in the data, Fig. 4a, as well as the acoplanarity distribution, Fig. 4b, where the tail of more spherical events is properly generated by the additional gluon radiation compared to the $\mathcal{O}(\alpha_s^2)$ model, which fails in this region. These deficiencies of the $\mathcal{O}(\alpha_s^2)$ model cannot trivialy be cured by an increased α_s since the 3-jet rate will then be overestimated, but the use of an optimized scale choice [20] may improve the agreement [21]. Nevertheless, the higher order effects are still needed for other observables and at higher energies.

The properties of high-$p_\perp$ jets in pp and $p\bar{p}$ collisions are also influenced by parton radiation processes. Thus, the jets measured by the UA1 collaboration [22] show clear evidence for these effects [23] in, e.g., the inclusive fragmentation function

$$D(z) = \frac{1}{N_{jet}} \cdot \frac{dN_{ch}}{dz} \quad ; \quad z = \frac{\bar{p}_{track} \cdot \bar{p}_{jet}}{|\bar{p}_{jet}|^2} \tag{2}$$

of charged particles. This is shown in Fig. 5 together with ISR data from the AFS collaboration [24] for comparison. The reason for the collider jets to be considerably softer is two-fold. Firstly, they are dominated by the intrinsically softer gluon jets, $\approx 60\%$ according to the model, whereas the ISR jet sample contains $\approx 70\%$ quark jets. Secondly, the harder interaction at the collider, resulting in $< p_{\perp jet} > \approx 39$ GeV compared to 13 GeV at the ISR, leads to more parton radiation; an effect which is also more accentuated by the colour octet charge of a gluon jet. The parton radiation also generate significantly enhanced transverse momenta with respect to the jet axis as seen in Fig. 6a. The low-$p_\perp$ part of the distribution depends rather sensitively on the cut applied to remove soft particles from the underlying event as illustrated with the two z-cuts shown for the data. Therefore, a mismatch between data and model calculation for the effective z-cut used may cause the observed difference at low $p_\perp$.

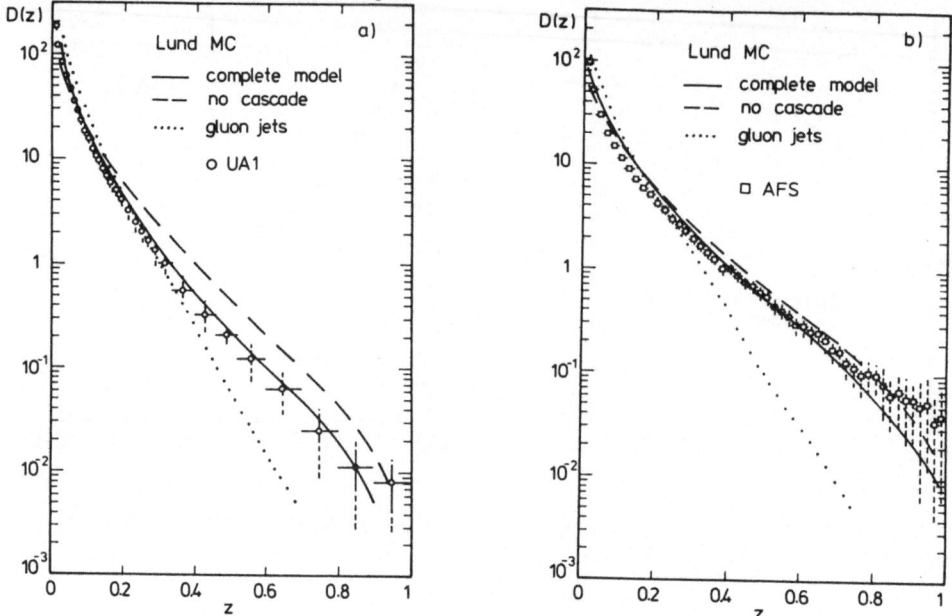

Figure 5. *Fragmentation functions, eq. (2), of high-$p_\perp$ jets at SPS collider (a) and ISR (b) energies. Data from UA1 [22] and AFS [24] collaborations with statistical (full) and systematic (dashed) error bars. The curves represent the model with the parton cascade included (full) and excluded (dashed), with quark and gluon jets mixed according to their relative cross-sections. Pure gluon jets including cascade are also shown (dotted curve) for comparison.*

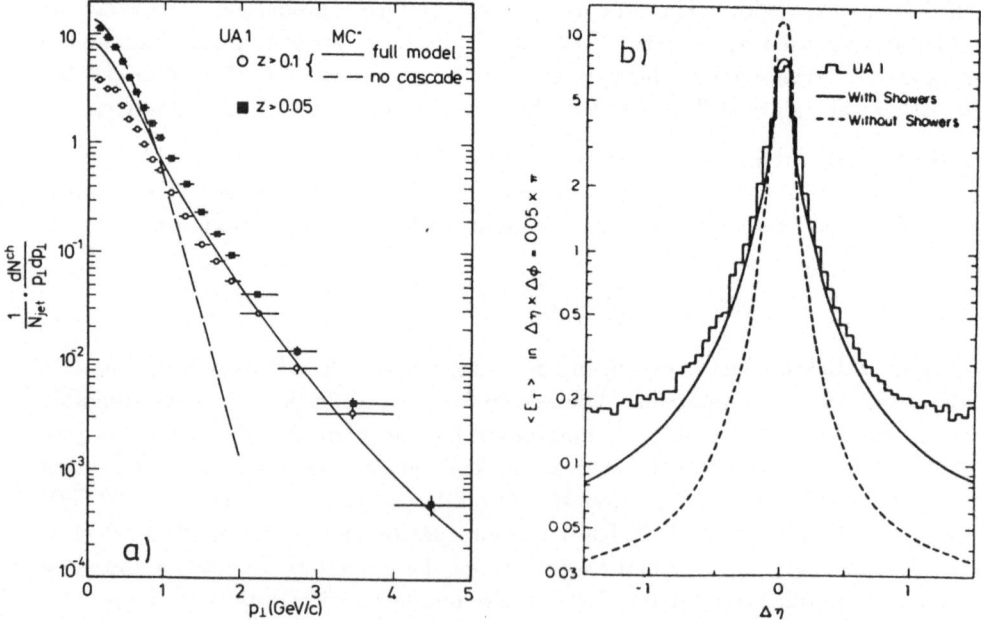

Figure 6. *(a) Transverse momentum distribution of charged particles with respect to the jet axis [23]. (b) Transverse energy flow of high-$p_\perp$ jets versus the rapidity distance to the jet axis [8]. Model curves including (full) and excluding (dashed) parton radiation (in (a) only final state emission).*

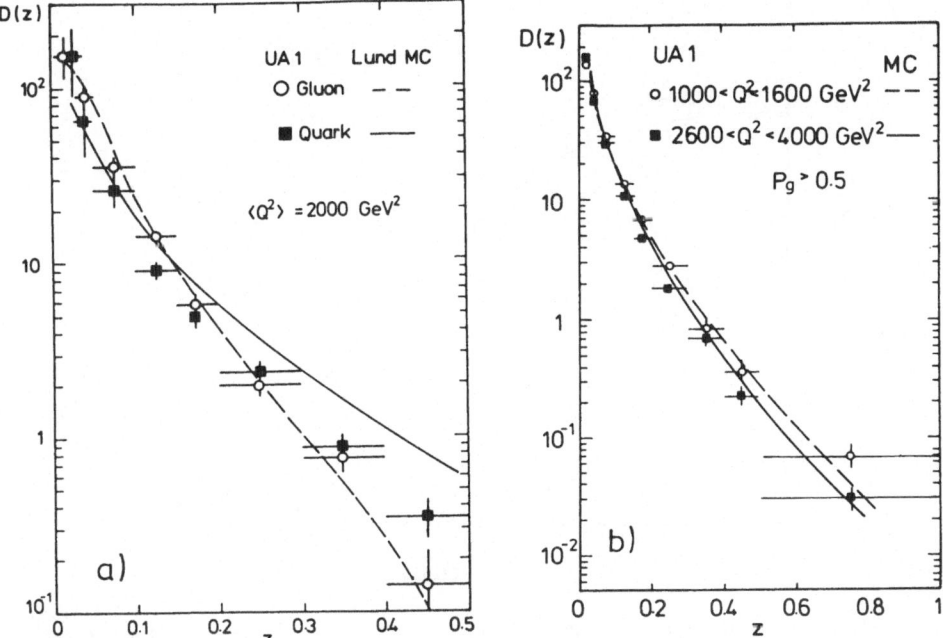

Figure 7. *(a) Fragmentation function for separated quark and gluon jet samples in data compared to pure quark and gluon model predictions. (b) Fragmentation function for gluon enriched jet samples at different Q^2 scales showing scaling violations in data and model.*

The width of a jet, e.g. defined in terms of the energy flow versus rapidity around the jet axis as shown in Fig. 6b, can be rather well described with the inclusion of the final state parton radiation model, whereas without it a much too narrow jet is obtained with non-perturbative fragmentation alone. Although the rapidity distribution away from the jet, $|\eta| \stackrel{>}{\sim} 1$, is considerably raised by the initial state parton radiation this is not sufficient to describe the observed energy flow plateau in Fig. 6b. This shows that the underlying event contains more physics than parton shower evolution and simple fragmentation.

In [23] the properties of quark and gluon jets are investigated and a fair agreement between data and model is found. As expected, gluon jets are softer and wider than quark jets, Fig. 7a. There are, however, some tendency of a smaller quark-gluon difference in the data as compared to the model. This could indicate an inadequacy of the model, but since they could also follow from a non-complete separation of quark and gluon jets or other systematic uncertainties in the data the model can be considered satisfactory. The variation with the momentum transfer Q^2 of the longitudinal and transverse jet properties are also found to be essentially the same in data and model, Fig. 7b. In the limited range covered by the UA1 data, the the Q^2 variation is rather small as expected from the dominant leading log Q^2 dependence in QCD.

Current state-of-the-art models for the perturbative jet evolution and the non-perturbative fragmentation, either in terms of strings or clusters, are thus able to reproduce present-day data on jets quite well.

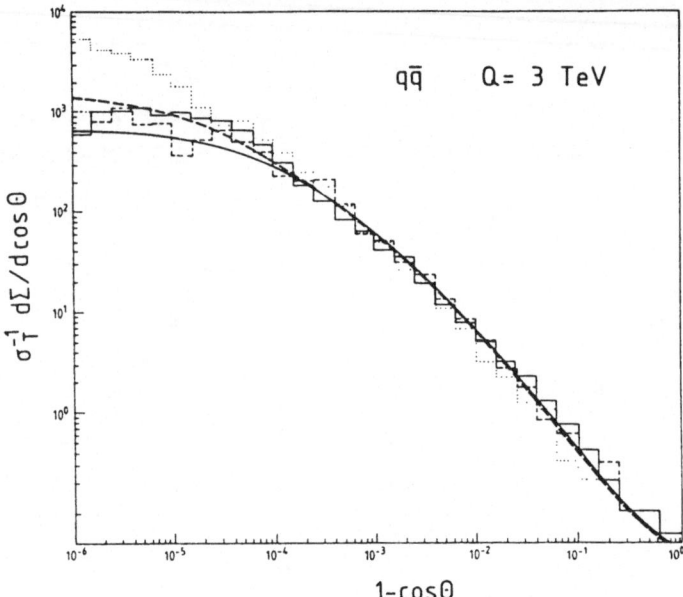

Figure 8. *The energy-energy correlation function, eq. (3), for a quark-antiquark system at $Q = 3$ TeV obtained from Monte Carlo simulation (histograms) of coherent parton shower evolution [4] with $t_{cut} = 0.5$ (full) and 5 GeV^2 (dashed), and without soft gluon interference ($t_{cut} = 5$ GeV^2, dotted) compared to analytical calculations (full and dashed curves) using two separate fits to data of non-perturbative effects [25].*

2.4 Jets at supercolliders

Given that the jet models are based on sound physics input, and not just parametrizations of data, together with their ability to describe present-day jet data it becomes meaningful to make extrapolations to the TeV energy scale that may become available at future colliders; such as LHC, SSC and the ELOISATRON as the most extreme case. There are also theoretical cross-checks that can be made to further increase the confidence in such extrapolations, e.g. by comparing Monte Carlo and analytical calculations in QCD [25]. Such a case is provided by the angular energy-energy correlation function defined by

$$\frac{1}{\sigma}\frac{d\Sigma}{d\cos\theta} = \frac{1}{\sigma}\frac{1}{4}\sum_{A,B}\int_0^1 dx_A dx_B \frac{d\sigma(e^+e^- \to A+B+X)}{dx_A dx_B d\cos\theta} \cdot x_A x_B \qquad (3)$$

where the sum is over all particle pairs A, B with angle θ between $\vec{p}_A$ and the negative of $\vec{p}_B$. This function is shown in Fig. 8 for $e^+e^- \to q\bar{q}$ at an invariant mass of 3 TeV. The peak centered at the back-to-back direction $\theta = 0$ arise from two-jet final states in which one hadron is detected from each jet and the width of the peak is thus a measure of the angular width of a jet and is strongly affected by soft and collinear gluon emission. The two curves correspond to the analytical QCD calculation using two different approximations for the non-perturbative fragmentation based on fits to present data. Since this calculation is based on quite different assumptions and approximations compared to the Monte Carlo, histograms in Fig. 8, it is reassuring

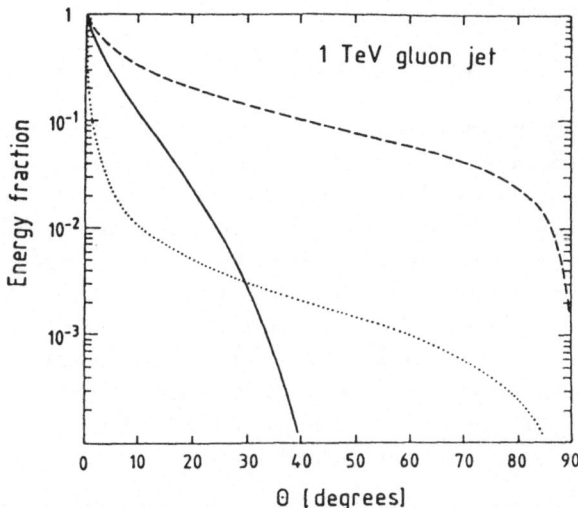

Figure 9. *Energy fraction outside a cone with half-opening angle* θ, *around the overall jet axis, i.e. parton direction, (dashed curve) and around the reconstructed (sub)jet axes (full curve) obtained from a gluon-gluon system with parton shower at* $Q = 2$ *TeV and, for comparison, without shower with* θ *relative to overall jet axis (dotted curve).*

to see the good agreement between the two approaches even when extrapolating to energies much higher than presently available, where they are both tuned to the data. It is also interesting to note that a parton shower algorithm without the coherence among soft gluons taken into account gives a different result compared to the coherent shower and the analytic calculation. This, and other considerations, make the incoherent cascade theoretically disfavoured, although such models can also be tuned to fit most aspects of current data which are, however, in a rather restricted energy region.

Although the bremsstrahlung nature of the QCD radiation predominantly results in soft and collinear parton emission, there will also be occasional hard emission at large angles resulting in the splitting of a jet into a sub-jet structure. Between these two extremes there is, of course, a continous distribution which makes the concept of a jet rather arbitrary from the theoretical point of view. The angular energy flow arising from a 1 TeV gluon jet is illustrated in Fig. 9. A very narrow jet is obtained with pure fragmentation, whereas a significant energy flow also at very large angles (with respect to the gluon direction) arise with the inclusion of parton radiation. The measured jet width will, however, be much smaller and depend on the resolution of sub-jet structures. This illustrates the importance of applying experimentally realistic jet definition criteria to the Monte Carlo generated events when predicting jet properties at TeV colliders. This can be done from the energy flow pattern, e.g., as follows. An idealized 'calorimeter' covers the full azimuthal angle around the beam axis and the pseudo-rapidity region $|\eta| \leq 3$ and is divided into cells of size $\Delta\eta \times \Delta\phi = 0.1 \times 5^0$ in each of which the particle energies of a Monte Carlo generated event are summed. Starting from the cell with largest transverse energy, $E_\perp$, the transverse energy of nearby cells within

$$\Delta R = \sqrt{\Delta\eta^2 + \Delta\phi^2} \le 0.7 \qquad (4)$$

are summed. If $\sum E_\perp$ exceeds a certain cut-off value, E_c, then all particles/cells within the cone are said to form a jet with axis given by the $E_\perp$-weighted center of the cells. This procedure is iterated until all jets with $E_\perp$ larger than E_c, typically 10–20 GeV, are found. At TeV energies the details of this procedure make no difference. Thus, alternatives like using total energy rather than transverse energy or summing cell 4-vectors obtained from the energy deposited in and the location of the calorimeter cells, give essentially the same results. A coarser grained calorimeter, with cell size $\Delta\eta \times \Delta\phi = 0.2 \times 10^0$, also give similar results. The size of the cone used for the jet definition is, however, important for the jet properties since it regulates not only how many soft, wide angle particles that are included in the jet, but also the experimental resolution to separate nearby jets as indicated above.

In order to give definite predictions of high-$p_\perp$ jet properties at the TeV energy scale [26], we generate events using the Lund PYTHIA program [27] based on the $2 \to 2$ QCD matrix elements combined with initial and final state parton cascade evolution and followed by hadronization using the Lund string model [13,28]. A cutoff to assure hard scattering is applied by requiring the partons emerging from the simple $2 \to 2$ process to have

$$x_\perp \equiv \frac{2p_\perp}{\sqrt{s}} \ge 0.1 \qquad (5)$$

(Due to the subsequent gluon radiation a final jet with this minimum $p_\perp$ need not emerge.) To illustrate the dependence on centre-of-mass energy results are shown in the following for $p\bar{p}$ collisions at $\sqrt{s}=$ 0.63, 2, 18 and 100 TeV, with the $x_\perp$ cutoff corresponding to $p_\perp = 30$, 100 GeV and 1, 5 TeV, respectively. The events are analysed with the above jet algorithm keeping the parameters fixed with increasing energy, since they correspond to fixed detector properties. The jet multiplicity distributions are given in Fig. 10a and show the increasing number of jets with increasing cms energy. Naturally, most of the increase at TeV energies are 'lower' energy jets, since the cutoff is kept fixed at $E_{\perp jet} > 20$ GeV which can be expected to be enough for the jets to stand out clearly above the underlying event. Seen on a TeV energy scale, however, the events are often quite clean with only a few, narrow energy clusters surrounded by additional activity at a reduced energy, Fig. 10b. Of course, the number of jets depend rather strongly on the basic jet resolution as illustrated in Fig. 10a by the curve for $\Delta R = 0.2$, which is about the smallest value possible for the given calorimeter granularity before fluctuations become important when only a few cells are used to define a jet.

The characteristics of 'typical high energy' jets at each centre-of-mass energy is extracted by only taking jets in a 'window' around $x_\perp \simeq 0.1$ and separating quark and gluon jets [26]. The longitudinal distribution within the jets is illustrated in Fig. 11a by the fraction of jet energy carried by particles with a minimum fractional energy $z = E_{particle}/E_{jet}$, i.e. the integral of the energy-weighted fragmentation function

$$I(z) = \int_z^1 dz' \, z' \cdot D^{all}(z') \, , \quad D^{all}(z) = \frac{1}{N_{jets}} \cdot \frac{dN^{all}}{dz} \qquad (6)$$

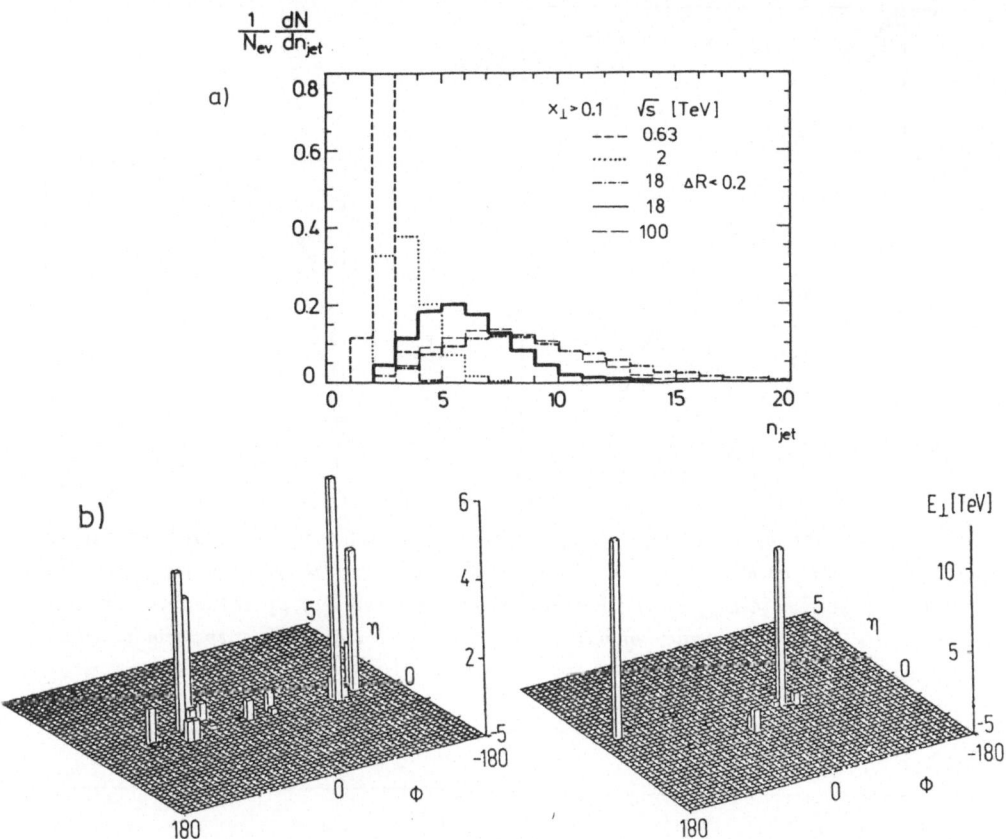

Figure 10. *(a) Jet multiplicity in hadron collider high-$p_\perp$ events. Standard jet cone size $\Delta R = 0.7$ in all cases except the one marked.*
(b) Transverse energy distribution in the pseudorapidity-azimuthal angle space for two Monte Carlo events at $\sqrt{s} = 100$ TeV with a hard $2 \to 2$ QCD scattering with $p_\perp > 10$ TeV.

The much softer nature of the gluon jets compared to quark jets is evident as is the increased softness (in the scaling variable) for higher energy jets; e.g. 10% of the quark jet energy is carried by particles with $z > 0.45$, 0.3 at $\sqrt{s} = 0.63$, 100 TeV corresponding to particle energies of $\sim$ 15, 1500 GeV. The decreasing width of a jet with increasing energy is shown in Fig. 11b in terms of the integrated energy flow inside a cone with half-angle θ around the jet axis; e.g. 90% of the quark jet energy is inside a cone of 18^0, 9^0 for a quark jet of 30 GeV and 5 TeV, respectively. The particle flow is less collimated compared to the energy flow due the occurence of lower energy particles at wider angles [26].

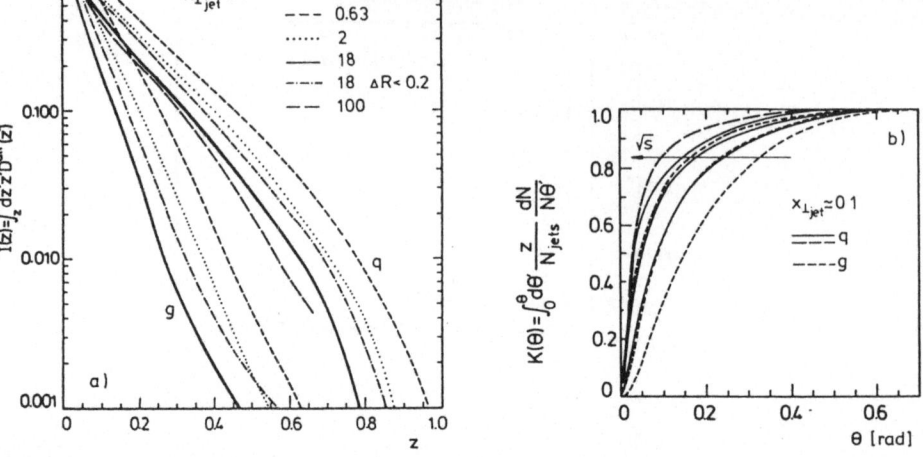

Figure 11. *Fraction of jet energy (a) carried by particles with fractional jet energy larger than z, and (b) within a cone of half-angle θ around the jet axis; for quark (q) and gluon (g) jets with $x_\perp \simeq 0.1$ at different cms energies of $p\bar{p}$ collisions. In (b), the curves are for cms energies of 630 GeV, 2, 18 and 100 TeV, respectively, when going towards θ = 0.*

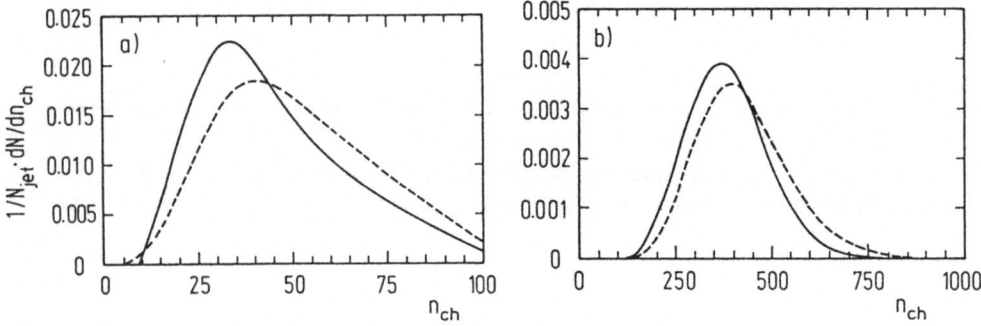

Figure 12. *(a) Charged particle multiplicity for jets with $E_\perp \approx 5$ TeV (full) and 15 TeV (dashed). (b) Total charged particle multiplicity in events with a QCD 2 → 2 hard scattering with $p_\perp > 10$ TeV, for two somewhat different models of the underlying event.*

The expected charged particle multiplicity in high-$p_\perp$ jets and in total for high-$p_\perp$ events at the ELOISATRON is shown in Fig. 12. The total multiplicity in particular, but also the jet multiplicity, should only be taken as an indication since they depend on the model used for the underlying event. As mentioned above, see Fig. 6b, the underlying event in high-$p_\perp$ events is not well understood making extrapolations rather uncertain. This is related to the absence of a proper understanding of the spectator partons and minimum bias event properties in general.

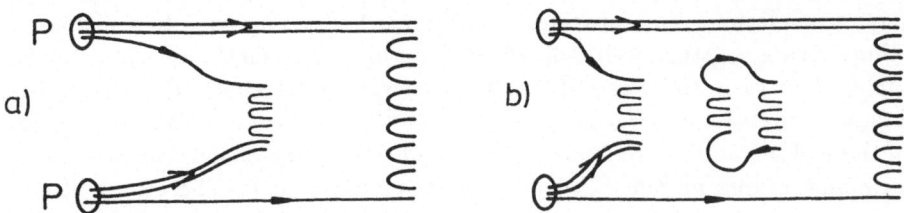

Figure 13. *Low-p$_\perp$ proton-proton interaction in the DTU model. (a) The simplest case giving two quark-diquark strings, hadronizing independently of each other. (b) Next order correction with sea quark interactions giving two additional quark-antiquark strings.*

3   Minimum bias physics and beam jets

The low-p$_\perp$ interactions of minimum bias events provide most of the inelastic hadron scattering cross-section and in that sense constitute the dominant soft interaction. Although high statistics data on very detailed properties of the final multiparticle state are available our understanding of the interaction mechanism is far from being understood. The hadronization models discussed, which successfully describe e^+e^- annihilation, deep inelastic scattering and the high-p$_\perp$ jets in hadron scattering, could in principle work also for low-p$_\perp$ interactions. The problem is that we have no calculable underlying parton level process which can define the structure of the colour fields to which the fragmentation models can be applied. Nevertheless, models to describe the production of the final hadron state in minimum bias events have been developed and some of them work rather well in describing the data.

3.1   Minimum bias models

The simplest model is provided by the 'old' Lund low-p$_\perp$ model [29] where it is assumed that the main features can be understood by the fragmentation of a single colour triplet field, which hadronizes similar to the one in e^+e^-. Thus, a string is stretched between the interacting hadrons with the valence quarks distributed along it according to a phenomenological recipe. It is remarkable how well this simple model can be tuned to fit data concerning the beam and target fragmentation regions at fixed target energies [30], but with increasing energy the model fails badly with too little activity in the central region and is therefore not useful at collider energies.

Another approach is given by the dual topological unitarization (DTU) scheme [31], which is based on Regge theory, with 'cut pomerons', to give the probablity for a certain number of strings to be formed as illustrated in Fig. 13. In the lowest order process, a quark from one beam proton is connected with a diquark from the other and the remaining diquark with the corresponding quark, thus forming two colour triplet fields. Higher order effects give additional strings between sea quarks and antiquarks from the interacting protons. Quark structure functions are used to assign momenta to the quarks, antiquarks and diquarks, thereby specifying the invariant masses of the strings, which can then be hadronized using normal models from e^+e^- annihilation. This provides a non-perturbative approach to multiparton interactions giving a rather well defined model which is in fact very successfull in reproducing data even up to the SPS collider energy with respect to multiplicities, central rapidity plateau, long range correlations and $p_\perp$ increase with multiplicity [31]. Some of these observables were, however, not predicted by the model, but reproduced by tuning and adding new detailed ingredients to the model. There may also be problems of a more fundamental kind. Firstly, the model is founded on Regge theory and one may question whether this is the proper soft limit of QCD. Secondly, there are no dynamical gluons in the model. This is in my view disturbing since we know that in perturbative QCD, gluon scattering become dominating as the momentum transfer is decreased. Based on the assumption of a smooth transition to the soft limit one might therefore expect the gluons to play an important dynamical role in minimum bias interactions. Also the connection to diffractive scattering, although covered by Regge theory, may be problematic if the exchanged pomeron is a gluonic object as discussed below. Although this model works effectively in describing many aspects of the data, it need not reveal the true interaction mechanism. In any case, alternative models are important tools to investigate other ideas of the interaction mechanism.

A new approach based on multiparton scattering in perturbative QCD has recently been developed [32,33], which leads to a natural connection to high-$p_\perp$ scattering. The starting point is the cross-section in leading order QCD

$$\sigma(p_\perp) = \int_{p_{\perp min}} dx_1 dx_2 d\hat{t} \sum_{i,j} f_i(x_1, Q^2) f_j(x_2, Q^2) \frac{d\sigma}{d\hat{t}} \qquad (7)$$

which increases strongly for decreasing $p_{\perp min}$ and in fact becomes equal to the total cross-section at the SPS collider for $p_{\perp min} \simeq 1.5$ GeV. This is not necessarily unphysical since this cross-section is not the proton-proton cross-section, but a parton-parton one. Viewing the proton as a beam of partons one can interpret the ratio $\sigma(p_{\perp min})/\sigma_{tot}$ as the number of parton-parton scatterings per proton-proton collision. Thus, a multiparton scattering situation naturally arise. One may still worry whether a perturbative calculation is really applicable at such small momentum transfers, but this is not a conceptual problem but rather a technical one. In particular, higher order QCD corrections may be large making the numerical results unreliable.

In a first attempt this probability for multiple parton scatterings, using a fixed value of $p_{\perp min}$, was introduced in the Lund hadron scattering Monte Carlo [32]. The additional scatterings have the effect of producing a more complicated string topology with either extra strings or more kinks on the existing ones, which in turn lead to an increased hadron multiplicity. In comparison to observed minimum bias event multiplicities, Fig. 14a, the result is not satisfactory, but encouraging when compared to the result of the previous case, including single $2 \to 2$ hard QCD processes and initial plus final state parton cascade evolution as discussed above. The result is, however, very sensitive to the exact value of the new basic parameter $p_{\perp min}$, as seen in Fig. 14a. In an improved version of the model an impact parameter picture is introduced [33]. For a large overlap between the colliding hadrons, associated with a small impact parameter, an enhanced probability for multiple interactions occur, resulting in 'more strings' and consequently increased multiplicity. An event with a high-$p_\perp$ process, i.e. a manifest parton-parton scattering, should then involve a

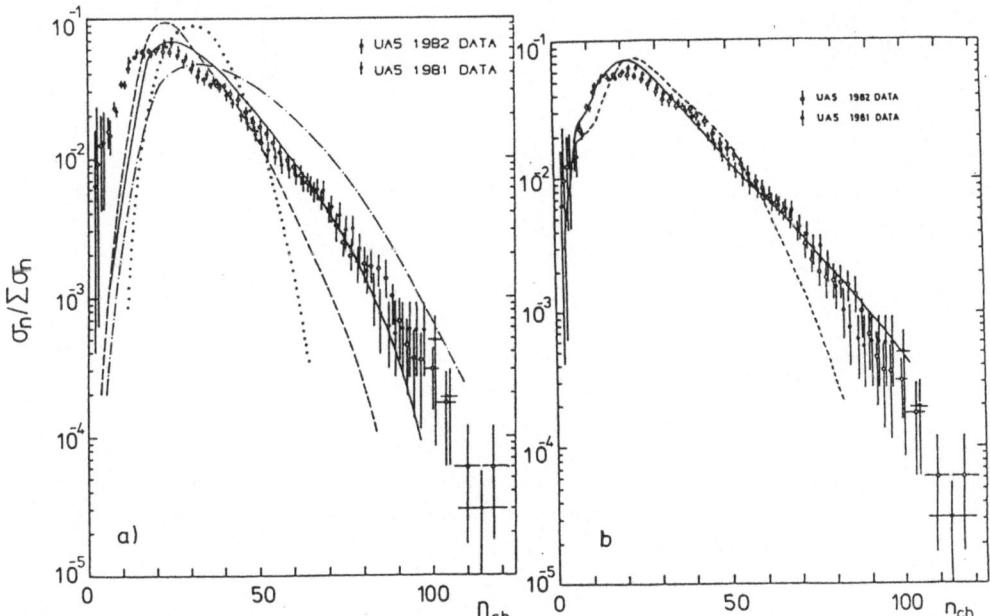

Figure 14. *Charged particle multiplicity distribution in 540 GeV pp̄ collisions [34] in comparison to models. (a) Simple multiparton scattering model with cutoff $p_{\perp min} = 2$ GeV (dashed), 1.6 GeV (full) and 1.2 GeV (dash-dotted); excluding multiple scattering (dotted). (b) Multiparton scattering in an impact parameter model with varying (full) and fixed (dashed) impact parameter. In all cases hard scattering processes with initial and final state parton radiation is included. From [33].*

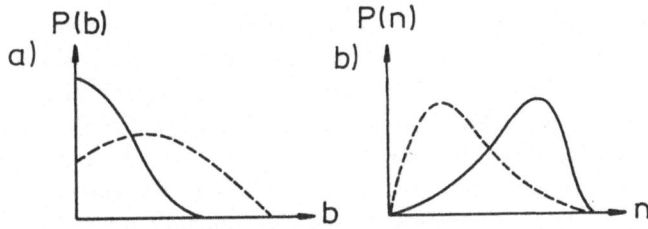

Figure 15. *Probability distributions for (a) impact parameter and (b) multiplicity in minimum bias events (dashed) and high-$p_\perp$ events (full).*

larger overlap, compared to the average minimum bias event, and therefore give rise to a larger multiplicity, Fig. 15.

To construct this model in detail the distribution of matter in the colliding hadrons has to be specified and a few simple functions was considered to describe the effective parton distributions. A solid sphere, a Gaussian or an exponential distribution give similar results that are not able to properly describe the pedestal effect, i.e. the increased rapidity plateau in high-$p_\perp$ jet events. A better result is obtained with a

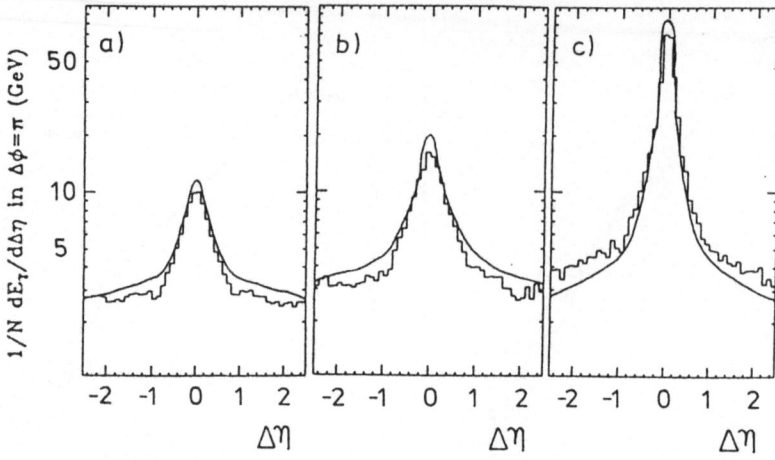

Figure 16. *Jet energy profiles in rapidity* ($\Delta\eta = \eta - \eta_{jet}$) *in the 'same side' region,* $|\phi - \phi_{jet}| < 90^0$. *UA1 data [35] in comparison to multiparton scattering model [33] for (a)* $E_{\perp jet} > 5$ *GeV, (b)* $E_{\perp jet} > 10$ *GeV and (c)* $E_{\perp jet} > 30$ *GeV.*

double Gaussian of the form

$$\rho(r) = \frac{1-\beta}{a_1^3} e^{-r^2/a_1^2} + \frac{\beta}{a_2^3} e^{-r^2/a_2^2} \tag{8}$$

corresponding to a small core of radius a_2 with a fraction β of the matter in the hadron, which has radius a_1. One may speculate whether this two-component structure is related to the pion cloud or a simple representation of a more complicated structure, e.g. with three cores corresponding to the valence quarks. Numerically, $\beta = 0.5$ and $a_1/a_2 = 5$ was found appropriate. The problem remains with the divergence for $p_\perp \to 0$ of the QCD $2 \to 2$ matrix elements, which are used for calculating the multiparton interaction probability. A regularization was enforced by replacing the $1/p_\perp^4$ dependence with $1/(p_\perp^2 + p_{\perp 0}^2)^2$ in the matrix element and using $\alpha_s(p_\perp^2 + p_{\perp 0}^2)$, such that the basic cut-off parameter is $p_{\perp 0}$ which is assigned the value 2 GeV in order that the model can reproduce the collider multiplicity distribution, Fig. 14b, rather well. The increasing rapidity plateau with energy can be reproduced and its behaviour in high-$p_\perp$ events is better described, although the agreement is not perfect, see Fig. 16. Correlations in forward-backward multiplicity and $p_\perp$-multiplicity gets a natural explanation in terms of increased 'activity' from additional parton scatterings [33].

This model is thus able to give a rather satisfactory description of basic observations of soft interactions. The smooth connection to hard scattering processes throught its modern parton language basis is theoretically appealing. This may, however, also be a weakness in the sense that such a description may be inadequate, or even invalid, at small momentum transfers. In any case, calculational problems are certainly present in terms of possibly large higher order corrections in the perturbative expansion and a rather strong dependence on the basic $p_\perp$ cutoff. Although this cutoff is introduced in a technically different way by the parameters $p_{\perp min}$ and $p_{\perp 0}$ in the simple model and the impact parameter model, respectively, they represent the same problem giving rise to a similar kind of uncertainty. Additional uncertainties

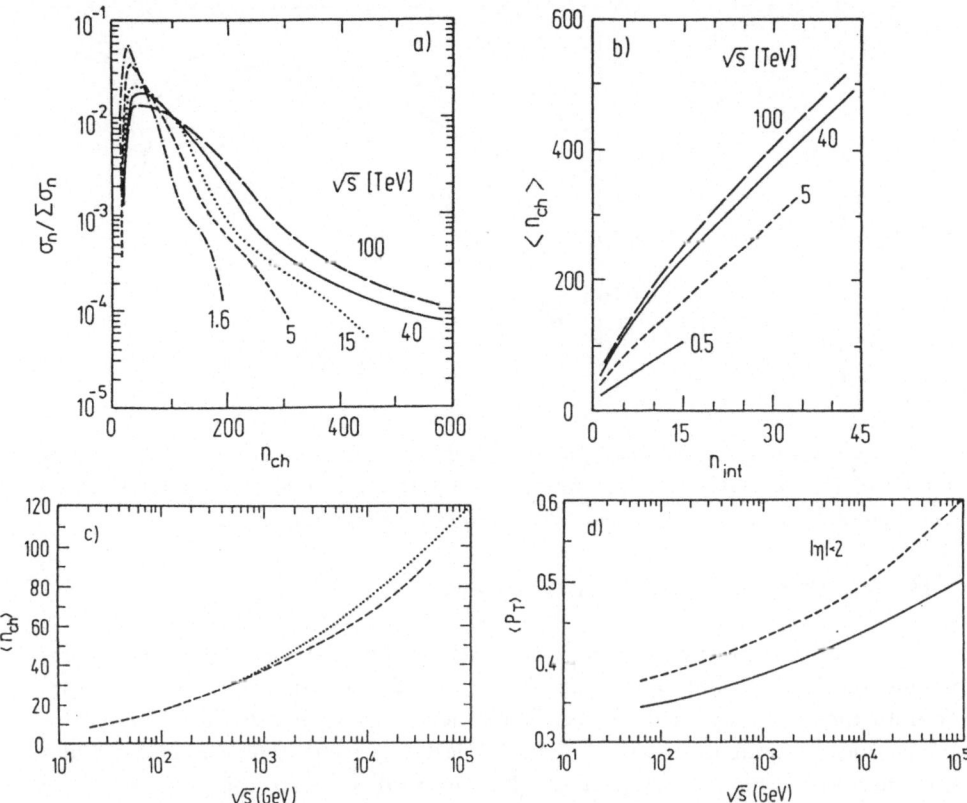

Figure 17. *Expectations for p$\bar{p}$ collisions at TeV energies based on the model including impact parameter dependent multiparton scatterings and high-$p_\perp$ processes with parton cascade evolution. (a) Charged particle multiplicity distribution. (b) Dependence of mean charged multiplicity on the number of parton-parton interactions. (c) Energy dependence of mean charged multiplicity for a fixed $p_{\perp 0} = 2$ GeV (dotted) and $p_{\perp 0} = 2 + 0.08\,ln(\sqrt{s}/540)$ (dashed). (d) Energy dependence of mean transverse momentum of central (dashed) and all particles (full).*

arise through the choice of string configurations which is not quite clear in case of multiple interactions. Nevertheless, we use this new interesting model for extrapolations to higher energies, but stress that the results should not be taken too literally but rather as an educated guess of what can be expected at TeV colliders. In particular, an unknown energy dependence of the $p_{\perp 0}$ parameter cannot be excluded, although it can be argued to be essentially a constant or only varying logarithmically with energy.

The multiplicity distribution, Fig. 17a, develops a very long tail with increasing energy. This arises through events having many parton-parton collisions, Fig. 17b, producing additional strings and 'longer' strings with gluon kinks. The dependence of the mean multiplicity with energy is given in Fig. 17c, for a fixed regularization $p_{\perp 0} = 2$ GeV and for a mildly energy dependent one, but a stronger variation is not excluded. The increasing number of parton collisions with increasing energy also give rise to an increased transverse activity as illustrated by the mean $p_\perp$ in Fig. 17d; in particular the central rapidity region is affected.

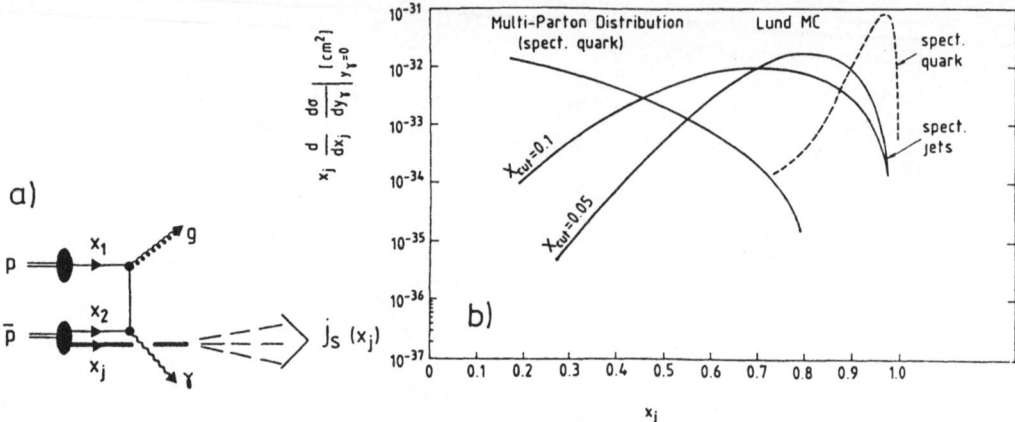

Figure 18. *(a) Example of process for the study of correlations between the hard scattering and the spectator jet. (b) Momentum distribution of the spectator jet in the Lund approach and using multiparton momentum distributions, see text.*

3.2 Beam jets in high-$p_\perp$ processes

The properties of the underlying event in hard scattering processes is given by the remaining partons in the colliding hadrons. These are often considered to be spectators that do not experience any hard scatter. Nevertheless, they provide the dynamics for generating the colour field structure of the underlying event, which is in the Lund model connected to the high-$p_\perp$ partons. It is natural to relate the problem of the underlying event to that of minimum bias event structure and thus try to solve both with the same model.

The properties of beam jets in high-$p_\perp$ events has hardly been investigated experimentally; in particular the correlations with the high-$p_\perp$ process would be interesting. In a first study [36] the consequences of two simple and orthogonal models were considered for the process $p\bar{p} \rightarrow \gamma g + j_s + X$ in Fig. 18a. The distribution of scaled momentum, x_j, of the spectator jet is shown in Fig. 18b, for a prompt photon with $p_\perp > 10$ GeV at zero rapidity. In one model, used in the previous Lund Monte Carlo [37], all spectator partons are assumed to act as a single colour charge with momentum fraction $x_s = 1 - x_2$, reflecting the hard scattering directly, and stretching a single string which hadronizes in the usual way. A hard spectator jet will thus arise, the exact distribution depending on the detailed definition of the jet concept. A completely different, much softer, spectrum is obtained in a model using 2-parton momentum distributions $V(x_2, x_j)$ to admit that the remaining energy is distributed on different spectator partons that hadronize independently. The division into several spectators in the latter model may seem unphysical but illustrates the main point, i.e. whether the spectator can be considered as a whole or not which is not obvious as will become clear in the following.

To illustrate the large uncertainties concerning the beam jets we use the various options available in the Lund hadron scattering Monte Carlo [27]. Fig. 19 show the angular energy flow, $x_E/\sigma \cdot d\sigma/d\log\theta$, of the final hadrons, with $x_E = 2E_h/\sqrt{s}$, in $p\bar{p}$ collisions at $\sqrt{s} = 100$ TeV having a hard $2 \rightarrow 2$ QCD scattering with $p_\perp > 10$

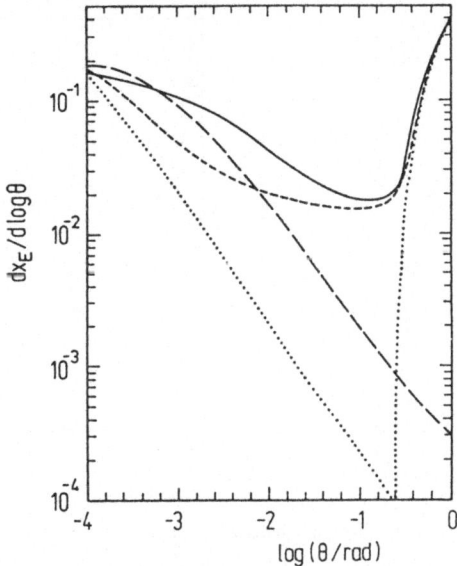

Figure 19. *Scaled angular energy flow with respect to the beam axis in $p\bar{p}$ collisions at $\sqrt{s} = 100$ TeV having a hard $2 \to 2$ QCD scattering with $p_\perp > 10$ TeV. The curves are for simple process (dotted), including parton radiation in initial (and final) state (dashed) and, in addition, multiparton scatterings (full). Result for minimum bias model, based on multiple scatterings, is included for comparison (long-dashed).*

TeV. Although the transverse partons and the forward spectator partons are usually connected by a string, the final jets are very clearly separated in the simplest model, but less so when the initial state parton radiation is taken into account. This produces a wider energy flow in the beam jet, which is also strongly increased by the multiparton scattering mechanism. Even if the parton radiation process can be considered as rather well controlled, the multiple scattering model introduces significant uncertainties of the beam jet properties. For comparison, the result of the minimum bias model is also given. Concentrating on the beam jet by selecting a forward cone of 100 mrad half-angle, the longitudinal and transverse particle spectra are shown in Fig. 20. For high-$p_\perp$ events the rapidity distribution is strongly influenced by the gluon radiation and also to some extent by multiparton scattering. The transverse momentum distribution is similarly influenced, although the large-$p_\perp$ tail is totally dominated by the parton cascade effects. The multiplicity distribution is particularly sensitive to the inclusion of these effects, Fig. 21, and the predictions correspondingly uncertain.

In summary it is evident that the beam jet properties are strongly dependent on variations within the model. This is not particular for this model, but other models would tend to add to the 'theoretical' uncertainty and the predictions for TeV energies will therefore span a rather wide region. It would be very valuable to have data from the present colliders to constrain the models, or even rule some out, in order to obtain a better understanding of the interaction mechanisms and be able to make more solid predictions to be tested at higher energies.

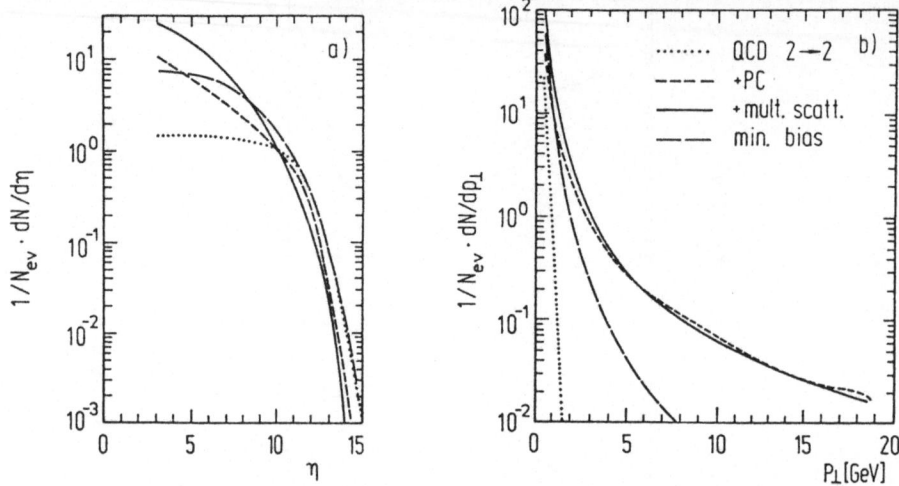

Figure 20. *Distribution in (a) pseudorapidity and (b) transverse momentum of charged particles in a forward cone of 100 mrad half-angle to the beam axis of $p\bar{p}$ collisions at $\sqrt{s} = 100$ TeV having a hard $2 \to 2$ QCD scattering with $p_\perp > 10$ TeV. Model curves as in Fig. 19.*

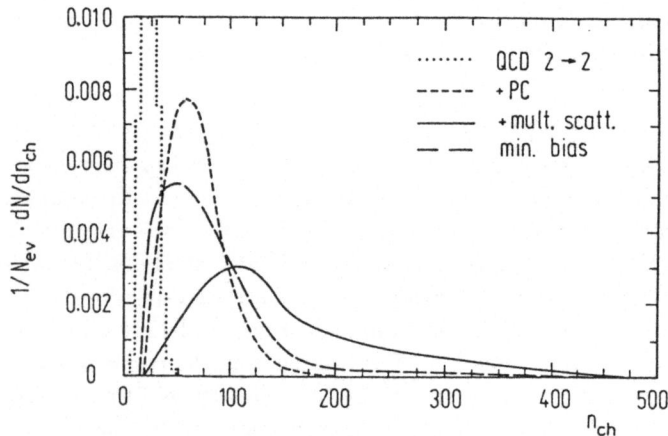

Figure 21. *Beam jet charged particle multiplicity distribution under the same conditions as in Fig. 20.*

3.3 Speculative discussion

Given the lack of a fundamental understanding of soft interaction mechanisms one has to consider modifications of the discussed models and, of course, be open to completely new ideas. The hadron remnant in a high-$p_\perp$ interaction has conventionally been treated as a genuine spectator, i.e. non-interacting and more or less as a single system, Fig. 22a. Some of the remnant partons may, however, interact via additional essentially independent scatterings as in the multiparton scattering model and other interaction mechanisms are also possible. Given the additional gluon radiation in

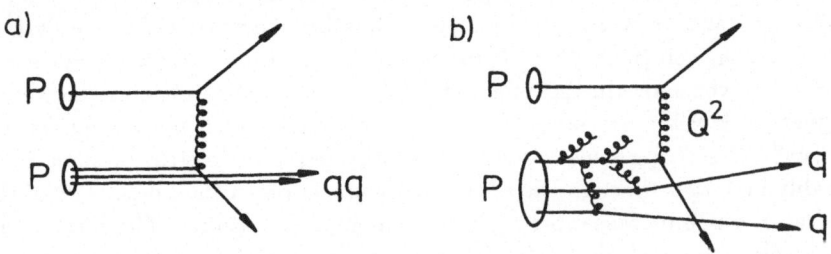

Figure 22. *Illustration of (a) 'true' spectator and (b) 'interacting spectator'.*

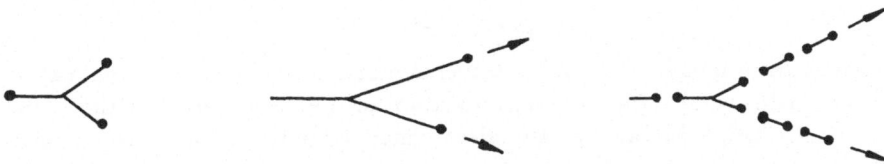

Figure 23. *Stretching of a Y-shaped baryon string due to a hard scattering and additional spectator interactions resulting in a non-leading baryon.*

the process, gluon exchanges like in Fig. 22b may occur giving rise to 'interacting spectators' which are connected to the large Q^2 process. This would be a kind of higher twist effect, but there are certainly severe calculational problems, such as what structure functions and q^2 scale to use and how to obtain the colour field structure in order to apply a hadronization model, but the qualitative effects are clear. With increasing Q^2 one would expect the forward jet to become softer and wider. The leading baryon may furthermore disappear, namely as illustrated in Fig. 23 where the baryon is considered as a Y-shaped string and the remaining valence quarks are separated through this additional gluon exchange giving rise to leading mesons before the baryon is formed. In $p\bar{p}$ collisions the baryons may totally disappear, but only if separate strings are formed between the valence quarks and antiquarks which is presumably very unlikely.

More generally one may consider whether separate strings formed in a collision can interact with each other if they are 'overlapping' in space-time. In particular with multiple interactions, in the DTU model or the pertubative QCD approach, such cases arise. The degree of overlap depends on the transverse size of such colour flux tubes, which is usually taken as a typical hadron diameter of ~ 1 fm. Also very important is, however, the internal field configuration. One possibility [38] is a thin core surrounded by a more extended field, similar to a vortex line in a type II superconductor. In this case the most essential properties are given by the thin core, e.g. the dynamics via the massless relativistic string, and the effective overlap for interactions between strings would be reduced and the assumption on non-interacting strings be better founded.

The possibility remain, however, that overlapping strings interact and could coalesce into a field of higher colour representation. Alternatively, the original interaction, e.g. via multiple gluon exchange, can generate higher colour charges on the separating hadron remnants leading to a colour field of a higher rank in between them. The probability for the generation of such different fields is unclear, but a model based on a random walk in colour space has been proposed for the case of relativistic heavy ion collisions [39]. A model based on gluon exchange dynamics in QCD would be even more interesting. The most important question is, however, whether higher rank fields are more likely to be produced with increasing centre-of-mass energy? The string constant κ, i.e. energy per unit length, of such fields can presumably be larger through the advocated connection to the magnitude of the Casimir operator which increases for higher colour representations. The increase is, however, not very strong as shown in Table 1 for the next few higher representation of the SU(3) colour group in comparison to the basic triplet field. In the Lund model the string breaking is treated as a tunneling process giving a quark production probability

$$P \sim e^{-\frac{\pi}{\kappa}m_q^2} e^{-\frac{\pi}{\kappa}p_\perp^2} \tag{9}$$

which is Gaussian in quark mass and transverse momentum. A simple application to higher rank fields would thus give an indication of the dependence on the string constant, κ, as illustrated in Table 1. A moderate increase in mean transverse momentum, $<p_\perp> \sim \sqrt{\kappa}$, can thus be be expected and a significant increase of strangeness production. Charm production, which is normally absent, may be dramatically increased to an observable level.

Table 1

rank of field	3	6,$\bar{6}$	8	10,$\overline{10}$	15	15'
κ/κ_3	1	10/4	9/4	9/2	4	7
$<p_\perp>$ (GeV)	0.4	0.64	0.6	0.84	0.8	1.0
$s\bar{s}/u\bar{u}$	0.3	0.6	0.6	0.77	0.74	0.84
$c\bar{c}/u\bar{u}$	10^{-11}	$4 \cdot 10^{-5}$	10^{-5}	$4 \cdot 10^{-3}$	$2 \cdot 10^{-3}$	$3 \cdot 10^{-2}$

A major problem is, however, that the decay or hadronization of such a field is not known. Given the fundamental colour charges 3, $\bar{3}$ and 8 of quarks, antiquarks and gluons the screening of the field in the breaking process would have to proceed by pair production of composite parton states, e.g. a 15-plet field could be broken by $qg - \bar{q}g$ pair production. Alternatively, $q\bar{q}$ and/or gluon pair production can be made in a stepwise manner that successively reduce the rank of the intermediate field. Thus, the fragmentation function of such fields is basically unknown, but the larger energy per unit length implies a concentration of the energy and presumably a correspondingly higher central rapidity plateau. Correlations of forward-backward multiplicities and $p_\perp$-multiplicity would arise naturally through the occurence of different event classes in terms of different fields.

Although these speculations are not very quantitative they illustrate the fact that confinement is an unsolved problem and completely different approaches are certainly possible. One such is based on hydrodynamical concepts applied to a quark-gluon plasma [40], but is not yet developed to the extent that definite predictions for TeV colliders can be made.

4   Pomeron structure and diffractive scattering

The differential cross-section for elastic and diffractive scattering of hadrons has been well measured by the detection of the (quasi-)elastically scattered particle and can

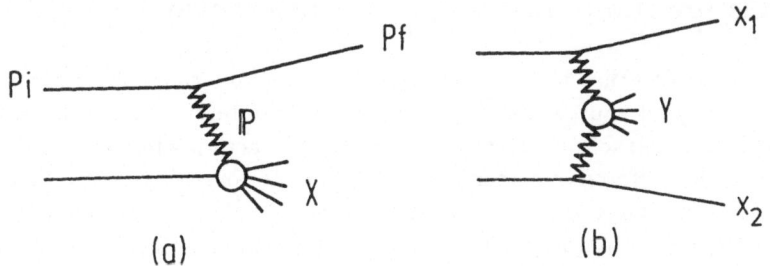

Figure 24. *(a) Pomeron exchange in single diffractive scattering. (b) Double pomeron exchange process.*

be 'understood' in Regge theory in terms of pomeron exchange [41]. Unfortunately, this has not lead to any real understanding of the true nature of the pomeron and its interactions. Since they involve hadrons and have cross-sections of order millibarn, they are clearly strong interaction processes. It is therefore natural to ask whether they can be understood in a 'modern language' based on an underlying parton level process. In particular, the pomeron itself may have a parton substructure, perhaps mainly of a gluonic nature [42].

In the following we will concentrate on the single diffraction process

$$h_1 + h_2 \rightarrow h_1 + X, \tag{10}$$

with one beam hadron quasi-elastically scattered with a remaining large momentum, $x = p_f/p_i > 0.9$, and the other excited to a system X (Fig. 24a). This process has been studied with various combinations of incident particles and at different energies. Even up to the highest available SPS collider energies [43], it shows a consistent behaviour of the cross-section in terms of the excitation mass, M_X, and momentum transfer, t, to the system X. However, the X system itself has not been much studied and its properties are largely unknown. The mass and rapidity of the excited system is given by

$$M_X^2 = (1-x)s \ , \quad y = \ln \frac{\sqrt{s}}{M_X} = \ln \frac{1}{\sqrt{1-x}} \ (= 1.5 \ for \ x = 0.95) \tag{11}$$

such that very large masses can be excited in a rather central region at TeV colliders, Table 2. Double pomeron exchange, Fig. 24b, provide pomeron-pomeron collisions giving excited states of mass

$$M_Y^2 = (1-x_1)(1-x_2)s \tag{12}$$

that can also be quite substantial at future colliders. The study of these systems should provide important information about the pomeron, its possible structure and interaction mechanisms which are essentially unknown at present.

Table 2

	M_X $(x = 0.95)$		M_Y $(x_1 = x_2 = 0.95)$	
ISR	13	GeV	3	GeV
S$p\bar{p}$S	140	GeV	30	GeV
LHC	4	TeV	0.9	TeV
ELOISATRON $\sqrt{s} = 100$ TeV	22	TeV	5	TeV

4.1 Models for the pomeron-proton interaction

Based on the observation that factorisation (between the upper and lower vertices in Fig. 24a) holds to a good approximation, the system X can be said to be the outcome of a pomeron-proton interaction. The nature of this interaction will clearly influence the properties of the diffractive system, e.g. whether its overall shape is isotropic in its centre-of-mass or longitudinal along some prefered axis. At high enough mass M_X, high-$p_\perp$ jets may even emerge [44] to signal the occurence of a hard scattering at the parton level!

A coherent pomeron-proton interaction could conceivably excite the proton wave function as a whole without any prefered direction. This could lead to a decay into a spherically symmetric state, if angular momentum effects are not important. Even for a system which is overall isotropic there can be internal asymmetries. The momentum vector of the final baryon may, for example, be aligned with the direction of the incoming proton. A mechanism for this is not easily constructed within a fireball type model, but one could perhaps think of 'hard' valence quarks which go forward with less disturbance, whereas the softer sea and glue components 'thermalize' with the many soft gluons that could make up the pomeron.

A very interesting model [45] for the pomeron interaction is based on the apparent similarity with photon interactions, which is deeper than the obvious diagrammatic one (in Fig. 24a replace the upper proton line with a lepton and the exchanged pomeron with a photon). The pomeron is thus assumed to have a pointlike coupling to quarks with a strength, β, that can be determined from elastic scattering. Within this framework the differential cross-section for single diffraction can be written as (when the approximation $\alpha_P(t)=1$ is made for simplicity) [45]

$$\frac{d^2\sigma}{dt dM_X^2} = \frac{1}{4\pi}[3\beta F_1(t)]^2 \frac{1}{M_X^2}(1 - \frac{M_X^2}{s})\beta^2 \tilde{F}_2(x, Q^2) \tag{13}$$

where F_1 is the proton formfactor (upper vertex in Fig. 24a) and $\tilde{F}_2$ the (lower vertex) proton structure function (normal F_2 but without quark charges) evaluated at $x = \frac{Q^2}{M_X^2 + Q^2} \approx \frac{Q^2}{M_X^2}$ and $Q^2 = -t$. The resemblence with deep inelastic scattering appears clearly in the cross-section formula, which for deep inelastic electron or muon scattering is

$$\frac{d^2\sigma}{dQ^2 dM_X^2} \approx \frac{1}{4\pi}[\frac{e^2}{Q^2}]^2 \frac{1}{M_X^2}(1 - \frac{M_X^2}{s})F_2(x, Q^2) \tag{14}$$

in the kinematic region ($y \ll 1$, $|Q^2 - m_p^2| \ll M_X^2$) of interest for the comparison with single diffractive scattering.

The amazing thing with eq. (13) is that it has no free parameters and yet it does reproduce data quite nicely [45]. Furthermore, this model has very clear consequences for the properties of the system X. Viewed in the pomeron-proton cms, a proton quark (or antiquark) would be back-scattered by the pomeron, leaving forward-going spectator quarks (in the simplest case a diquark). The separated colour charges are then expected to stretch a colour field which hadronize in the same way as is obseved in deep inelastic scattering. Obviously, this leads to a longitudinal event structure with a leading baryon in the direction of the initial proton. Based on the pomeron-photon analogy one can thus construct a detailed model for the production of the diffractive system [46]. It is still unclear, however, why the pomeron should couple in a pointlike way, or perhaps only effectively appears to. Furthermore, why should it couple to quarks only and not to the gluon component of the proton?

In another approach the pomeron is considered to have hadronic properties and gluonic interactions could then be of great importance. Being a strongly interacting object, one could expect the pomeron to contain coloured parton constituents. In such a pomeron-hadron analogy the pomeron-proton interaction should resemble a normal hadron-hadron interaction and typically give rise to a minimum bias topology, i.e. longitudinal event shape with a leading baryon effect. It was suggested more than ten years ago that the pomeron could be a gluonic system [42]. However, the simplest case of a two-gluon system, where the two gluons could couple to the same or different quarks, has theoretical problems with gauge invariance and factorization breaking [47]. A more complicated many-gluon system may work, but is difficult to calculate and is still an open question. Some non-perturbative developments have, however, been made recently [48] in this context.

Experimentally there has been progress by the clear observation of a longitudinal structure of the X system (in its cms) both at ISR [46] and the SPS collider [49]. Although the diffractive mass is only a few GeV at the ISR, a detailed comparison with model simulations show a clear longitudinal structure along the pomeron-proton collision axis. The final state proton momentum is, furthermore, found to be aligned along the incoming proton, an effect which becomes more pronounced with increasing diffractive mass [46]. At collider energies, the longitudinal structure is directly revealed in the rapidity distribution of the hadrons in the diffractive system [49]. Spherical symmetry is thus clearly ruled out and thereby all kinds of 'fireball' models. A further interesting result is the evidence for pomeron single-quark interactions obtained by a study of the fully reconstructed exclusive diffractive reactions $pp \rightarrow (\Lambda^0 \phi^0 K^+)p$ and $pp \rightarrow (\Lambda^0 \bar{\Lambda}^0 p)p$ by the R608 Collaboration at the ISR [50]. As seen in the proton-pomeron cms, there is in each case a forward Λ^0 in correlation with a backward K^+ or p, respectively, whereas the ϕ^0 and $\bar{\Lambda}^0$, which do not contain any proton valence quarks, show no such behaviour but are centrally produced. In particular the first reaction is a very suitable analyzer of the interaction mechanism since the fate of the proton valence quarks are well defined; a valence ud diquark goes forward and the remaining u valence quark is apparently back-scattered leading to a quark-diquark separation which hadronizes into the forward Λ^0, backward K^+ and central ϕ^0 by the creation of two $s\bar{s}$ pairs. This result is certainly in qualitative agreement with point-like pomeron-quark coupling used in the pomeron-photon analogy, but it may also be consistent with a pomeron-hadron picture having a $q\bar{q}$ component such that the $\bar{q}$ can annihilate a proton valence quark and the observed backward quark be viewed as a spectator from the pomeron. Independently of the details of these interpretations, however, the conclusion of an underlying partonic process in the pomeron-proton interaction seems unavoidable.

4.2 Hard diffractive scattering

If the pomeron has a parton substructure it may be possible to probe it in a hard scattering process [44] as, for example, between a gluon in the pomeron and a parton

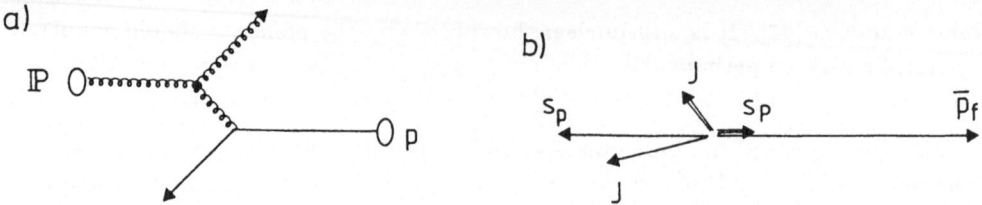

Figure 25. *(a) The hard parton-parton scattering subprocess in the pomeron-proton system, i.e. the lower vertex (or "blob") in Fig. 24a. (b) The topology of the scattering process as seen in the overall $p\bar{p}$ cms; $\bar{p}_f$ is the quasi-elastically scattered antiproton, J denotes the jets from the hard scattering and S denotes spectator jets. Double arrows are jets originating from the pomeron, while simple arrows are those from the proton.*

in the proton (Fig. 25a). This would give rise to a very characteristic event topology, as illustrated in Fig. 25b, with a quasi-elastically scattered particle and an opposite (in rapidity) system having two high-$p_\perp$ jets and two low-$p_\perp$ spectator jets from the non-interacting partons in the pomeron and proton respectively. Of course, for this jet-structure to be observable, the diffractive mass has to be large enough, i.e. colliders are needed (Table 2).

The process proceeds via the emission of a pomeron with a small momentum transfer (upper vertex in Fig. 24a), and one of its constituent partons then experience a hard scatter against a proton quark or gluon. Although the pomeron is often thought to interact coherently in the dominant low momentum transfer processes, one may assume an individual coupling of its constituents at the lower vertex because of the large momentum transfer at that vertex. Factorization allows the diffractive two-jet cross-section to be written [44]

$$\frac{d^2\sigma_{jj}}{dtdM_X^2} = \frac{d^2\sigma_{sd}}{dtdM_X^2} \cdot \frac{\sigma_{Pp\to jj}}{\sigma_{Pp\to X}} \tag{15}$$

The single diffractive cross-section, σ_{sd}, has been measured at the collider by UA4 [43] and can be parametrized as $\frac{d^2\sigma_{sd}}{dtdM_X^2} = \frac{6.8}{M_X^2}[e^{5.6t} + 0.04 \cdot e^{2t}]$. A 'pomeron-proton total cross-section' of $\sigma_{Pp\to X} \approx 1$ mb can be extracted from data using Regge analysis [51]. The pomeron-proton hard scattering into jets can be calculated in QCD

$$\sigma_{Pp\to jj}(M_X) = \int dx_1 dx_2 d\hat{t} \sum_{i,k} f_i(x_1, Q^2)G(x_2)\frac{d\hat{\sigma}_i^k}{d\hat{t}} \tag{16}$$

if a pomeron structure function, $G(x)$, is provided. This function is clearly unknown, but with the assumption of only gluons in the pomeron one may try different possibilities to investigate the sensitivity. A function like $xG(x) = 6x(1-x)$ would be expected if only two gluons share the pomeron momentum, or $xG(x) = 6(1-x)^5$ if the pomeron were a many-gluon system. The gluon distribution is in principle experimentally measurable because a change in its shape shifts the parton-parton cms with respect to the X cms, which in turn changes the distribution of the high-$p_\perp$ jets.

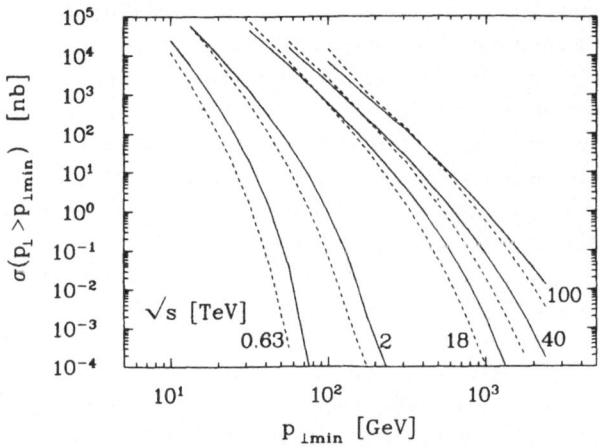

Figure 26. *Cross-section for diffractive events with jets having $p_\perp > p_{\perp min}$ in the region $|\eta| < 3$. The curves correspond to different $p\bar{p}$ collider energies and gluon structure functions of the pomeron given by $xG(x) = 6x(1 - x)$ (full) and $6(1 - x)^5$ (dashed).*

The resulting cross-sections for diffractive jet production are found to be quite large [44]. At the SPS collider the total cross-section for jets with $p_\perp \geq 8$ GeV is between 4 and 6×10^{-29} cm^2 depending on the pomeron structure function. Extrapolating to TeV energy colliders one obtains cross-sections as shown in Fig. 26. Since there are several uncertainties in this approach, see below, this naive extrapolation should not be taken too seriously, but rather as a first order of magnitude estimate. The expected major background process is given by normal QCD high-$p_\perp$ scattering where the spectator partons hadronize into a leading, diffractive-like proton with $x_F \geq 0.9$. At SPS energies, this process was estimated to have a cross-section two orders of magnitude below the diffractive jet signal [52]. This is, however, rather an upper limit since the uncertain elements of the beam jet fragmentation, as discussed above, tend to soften the beam jet.

These ideas are the starting point for 'hard diffraction' which has seen interesting developments since. First, heavy flavour production was considered [53] through gluon-gluon fusion into $Q\bar{Q}$ as the hard subprocess in the pomeron-proton interaction. The charm cross-section is, however, found to be far below the experimentally observed ones at the ISR. Due to the softness of the exchanged pomeron this is natural, since the available gluon-gluon cms energy is much lower than that for the same subprocess in non-diffractive proton-proton interactions, which is already below the data. The explanation for the large charm cross-sections at the ISR has to be sought elsewhere!

The theoretical basis for parton scatterings in the pomeron-proton interaction has furthermore been strengthened using Regge theory [54]. The important factorisation property is valid to the extent that Regge factorisation is valid in normal diffractive scattering. Triple Regge theory and counting rules suggest a gluon structure function of the pomeron of the form $xg(x) = (0.18 + 5.46x)(1 - x)$ which fulfills the momentum sum rule $\int_0^1 xg(x)\,dx = 1$. One should note that this is very close to the 'hard' structure function used as Ansatz and that the resulting phenomenology therefore will be the same. Based on the pomeron-photon analogy a different scheme has been proposed [55], in which the pomeron structure function is more similar to the photon structure function. Thus, a dominant $q\bar{q}$ structure is advocated with a structure function $xq(x) = 0.2x(1-x)$, which is normalised to $\int_0^1 xg(x)dx = 1/30$. Although the same shape as the simple 'hard' Ansatz is used, the hard diffractive cross-sections will be correspondingly reduced by this normalisation. The dominance of quarks rather than gluons would reduce the jet cross-sections further, due to the smaller colour charge, and exclude heavy flavour production in leading order, but would on the other hand give rise to other interesting processes like, e.g., Drell-Yan production of muon pairs and weak bosons at high enough diffractive masses. A quark substructure in the pomeron could also be directly measured in ep collisions at HERA, whereas a gluon structure would involve a more indirect, higher order process like photon-gluon fusion [44,56].

The idea of probing a possible pomeron parton structure by hard scattering processes has thus been well established. From a theoretical point of view it is hard to judge which of these scenarios is preferable and hence experimental information is needed to make progress. The UA8 collaboration has given preliminary results [57] which clearly hint at the existence of high-$p_\perp$ jets emerging from diffractively excited systems. The first indication of a smaller rate than expected should not be taken too literally since experimental jet finding problems were not included in the comparison with theoretical parton level calculations. A proper study is, however, in progress and the results will be most interesting, whichever way they will point!

5   Conclusions

Some important soft processes have been discussed in connection with the hard processes aiming at an understanding of the soft processes also in terms of a modern parton framework. This interplay between hard perturbative processes and soft hadronization is well developed for high-$p_\perp$ jets, which are well under control based on their production through leading order QCD Feyman diagram calculations, their perturbative evolution described by the parton cascade approach and finally their hadronization in terms of elaborate phenomenological models. The extrapolations to TeV colliders presented should therefore be taken seriously. Smaller discrepancies in comparison with future data may be cured by improving the models, but large ones would presumably indicate the occurrence of new physics phenomena!

Beam 'spectator' jets and minimum bias event structure cannot be considered understood, although models exist which are able to describe present energy data rather well. The increased activity, in terms of multiplicity and transverse momenta, with increasing energy seem to require some kind of multiple interactions, such as in the DTU approach or the new model based on perturbative QCD which can both be tuned to obtain good agreement with present data. The latter, although having some problems related to the cut-off against low-$p_\perp$ divergences, seems promising since it provides a smooth transition to high-$p_\perp$ scattering. When extrapolating to future TeV collider energies a rather wide range of results are obtained due to the unknown energy dependence of some parameters in the models. Multiplicities are notoriously difficult to predict, whereas energy flows are more stable because of their smaller sensitivity to very soft particles.

Although the presently used fragmentation models work quite well in describing the experimental observations, the confinement problem is unsolved and therefore unconventional models are not excluded and some possibilities were mentioned. Therefore, this topic certainly leaves room for surprises!

In the field of diffractive scattering the discussion was concentrated on the new topic of 'hard diffraction' which was invented recently and had interesting developments. The notion of partons in the exchanged pomeron which can participate in hard scattering processes has become a theoretically realistic possibility and the first experimental indication for its occurence has been seen in terms of transverse energy clusters. This may lead to an exploration of the pomeron structure and its interactions in a parton level framework which provides an alternative, or perhaps complementary, treatment compared to the previous Regge analysis.

In conclusion, soft processes have for a long time been considered 'dirty' in the sense that a fundamental theory for proper calculations has been lacking. Nevertheless, they have many interesting aspects and are certainly very important for a general understanding of high energy interactions. New ideas and models are an important tool for improving our knowledge and crucial tests of their predictions will certainly involve, or even require, the higher energy interactions at future TeV colliders.

Acknowledgements
I am grateful to Dr. A. Ali for organizing an interesting workshop and a pleasant week in Erice.

References

[1] Z. Kunszt, E. Pietarinnen, Nucl. Phys. B164 (1980) 45, Phys. Lett. 132B (1983) 453
Z. Kunszt, W.J. Stirling, Phys. Lett. 171B (1986) 307 (and references therein).

[2] G. Ingelman, DESY 87–145, and in proc. XVth International Winter Meeting on Fundamental Physics, Sevilla, Spain, 1987.

[3] G.C. Fox, S. Wolfram, Nucl. Phys. B168 (1980) 285
R.D. Field, S. Wolfram, Nucl. Phys. B213 (1983) 65
T.D. Gottschalk, Nucl. Phys. B214 (1983) and CALT-68-1083
R. Odorico, Nucl. Phys. B228 (1983) 381

[4] G. Marchesini, B.R. Webber, Nucl. Phys. B238 (1984) 1
B.R. Webber, Nucl. Phys. B238 (1984) 492

[5] M. Bengtsson, T. Sjöstrand, Nucl. Phys. B289 (1987) 810

[6] G. Altarelli, G. Parisi, Nucl. Phys. B126 (1977)298
G. Altarelli, Phys. Rep. 81 (1982) 1

[7] A. Petersen et al., SLAC-PUB-4290
W. Hoffman, LBL 23922 and proc. Int. Symposium on Lepton and Photon Interactions at High Energies, Hamburg 1987.

[8] T. Sjöstrand, Phys. Lett. 157B (1985) 321

[9] R.D. Field, R.P. Feynman, Nucl. Phys. B136 (1978) 1

[10] G. Ingelman, Physica Scripta 33 (1986) 39

[11] D. Amati, G. Veneziano, Phys. Lett. 83B (1979) 87
A. Bassetto, M. Chiafoloni, G. Marchesini, Phys. Lett. 83B (1979) 207; Nucl. Phys. B163 (1980) 477

[12] C.-K. Ng, Phys. Rev. D31 (1985) 469

[13] B. Andersson, G. Gustafson, G. Ingelman, T. Sjöstrand, Phys. Rep. 97 (1983) 31

[14] T. Sjöstrand, Phys. Lett. 142B (1984) 420

[15] B. Andersson, G. Gustafson, B. Söderberg, Z. Phys. C20 (1983) 317

[16] I. Montvay, Phys. Lett. 84B (1979) 331

[17] C. Peterson, T.F. Walsh, Phys. Lett. 91B (1980) 455

[18] A. Drescher, Thesis, Inst. f. Physik der Universität Dortmund, January 1987
U. Matthiesen, Thesis, Inst. f. Physik der Universität Dortmund, January 1987

[19] W. Bartel et al., JADE collaboration, Z. Phys. C33 (1986) 23

[20] G. Kramer, B. Lampe, DESY 87-106

[21] N. Magnussen, private communication.

[22] G. Arnison et al., UA1 collaboration, Nucl. Phys. B276 (1986) 253

[23] P. Ghez, G. Ingelman, Z. Phys. C33 (1987) 465

[24] T. Åkesson et al., AFS collaboration, Z. Phys. C30 (1986) 27

[25] G. Ingelman, D.E. Soper, Phys. Lett. 148B (1984) 171

[26] P.N. Burrows, G. Ingelman, Z. Phys. C34 (1987) 91

[27] H.-U. Bengtsson, T. Sjöstrand, Comput. Phys. Commun. 46 (1987) 43

[28] T. Sjöstrand, Comput. Phys. Commun. 39 (1986) 347
T. Sjöstrand, M. Bengtsson, Comput. Phys. Commun. 43 (1987) 367

[29] B. Andersson, G. Gustafson, I. Holgersson, O. Månsson, Nucl. Phys. B178 (1981) 242

[30] E.A. De Wolf, in proc. XV International Symposium on Multiparticle Dynamics, Lund, June 1984. Eds. G. Gustafson, C. Peterson, World Scientific, p. 2

[31] P. Aurenche, F.W. Bopp, Phys. Lett. 114B (1982) 363
    A. Capella, J. Tran Than Van, Phys. Lett. 114B (1982) 450; Z. Phys. C18 (1983)
    85; ibid. C23 (1984) 165

[32] T. Sjöstrand, FNAL Pub-85/119-T

[33] T. Sjöstrand, M. van Zijl, Phys. Rev. D36 (1987) 2019.

[34] G.J. Alner et al., UA5 Collaboration, Phys. Lett. 138B (1984) 304

[35] C. Albajar et al., UA1 Collaboration, in proc. workshop on Physics Simulations
    at High Energies, Madison, Wisconsin 1986.

[36] B. Humpert, G. Ingelman, Int. conference 'Hadron Structure 83', Smolenice,
    Czechoslovakia 1983.

[37] H.U. Bengtsson, G. Ingelman, Comput. Phys. Commun. 34 (1985) 251

[38] G. Gustafson, Phys. Lett. 175B (1986) 453

[39] T.S. Biro, H.B. Nielsen, J. Knoll, Nucl. Phys. B245 (1984) 449

[40] L. van Hove, Z. Phys. C27 (1985) 135

[41] U. Amaldi, M. Jacob, G. Matthiae, Ann. Rev. Nucl. Sci. 26 (1976) 385
    G. Alberi, G. Goggi, Phys. Rep. 74 (1981) 1
    K. Goulianos, Phys. Rep. 101 (1983) 169

[42] F.E. Low, Phys. Rev. D12 (1975) 163
    S. Nussinov, Phys. Rev. Lett. 34 (1975) 1286; Phys. Rev. D14 (1976) 246

[43] M. Bozzo et al., UA4 collaboration, Phys. Lett. 136B (1984) 217

[44] G. Ingelman, P.E. Schlein, Phys. Lett. 152B (1985) 256

[45] A. Donnachie, P. V. Landshoff, Nucl. Phys. B244 (1984) 322

[46] A.M. Smith et al., R608 Collaboration, Phys. Lett. 167B (1986) 248

[47] D.G. Richards, Nucl. Phys. B258 (1985) 267

[48] P.V. Landshoff, O. Nachtman, Z. Phys. C35 (1987) 405

[49] D. Bernard et al., Phys. Lett. 166B (1986) 459

[50] A.M. Smith et al., R608 Collaboration, Phys. Lett. 163B (1985) 267

[51] K.A. Ter-Martirosyan, Phys. Lett. 44B (1973) 179
    A.B. Kaidalov, K.A. Ter-Martirosyan, Nucl. Phys. B75 (1974) 471
    D.P. Roy, R.G. Roberts, Nucl. Phys. B77 (1974) 240

[52] G. Ingelman, proc. first Int. workshop on Elastic and Diffractive Scattering at
    the Collider and Beyond, Chateau Blois, France 1985, Eds. B. Nicolescu, J. Tran
    Than Van, Editions Frontières, p. 135

[53] H. Fritzsch, K.H. Streng, Phys. Lett. 164B (1985) 391

[54] E.L. Berger, J.C. Collins, D.E. Soper, G. Sterman, Nucl. Phys. B286 (1987) 704

[55] A. Donnachi, P.V. Landshoff, DAMTP 87/2, DAMTP 87/16

[56] N. Arteaga-Romero, P. Kessler, J. Silva, Mod. Phys. Lett. vol 1, No. 3 (1986) 211

[57] P.E. Schlein, in proc. XXIII International Conference on High Energy Physics, Berkeley 1986, Ed. S.C. Loken, World Scientific, vol II, p. 1331

THE PATTERN OF LEPTON AND QUARK MASSES

Harald Fritzsch

Sektion Physik der Universität München and
Max-Planck-Institut für Physik und Astrophysik
- Werner Heisenberg Institut - München, Germany

ABSTRACT

The observed hierarchy of the lepton-quark masses and the emerging hierachical pattern of the weak interaction mixing matrix is discussed, as well as an underlying theoretical framework based on a step-wise breaking of chiral symmetries. The flavor mixing angles are functions of quark mass ratios and must vanish if the associated generation is massless. A specific pattern of quark mass matrices emerges, which can be used to extrapolate to higher energies and to deduce the properties of ultra-heavy quarks.

Everyone attending this meeting is interested in the possibility that the spectrum of leptons and quarks may extend beyond the heavy b and t quarks and the τ-lepton discovered (or in the case of t anticipated firmly) thus far. And everyone will agree that such an insight into the lepton-quark spectrum would have immediate consequences with respect to our theoretical understanding of the particle physics universe.

Thus far all particle physics phenomena have been found to be in accordance with the standard model - the well-known superposition of QCD and the electroweak theory. No doubt, the standard model exhibits signs of eternal truth. Our understanding of, in particular, the basic interaction of photons with elektrons and quarks (QED) and of the gluons with themselves or with quarks (QCD), and to a somewhat lesser extent the interaction of the weak bosons with the leptons and quarks show the simplicity and beauty which one expects from a basic and correct theory.

It is surprising that on the one hand a dramatic breakthrough has been made in understanding the structure of the basic interactions, but on the other hand very little progress has been made in order to understand the rather bizarre looking tower of mass scales in the standard model, especially those of the mass spectrum of the leptons and quarks. The "explanation" of these mass scales given in the standard model by relating them to a set of Yukawa coupling constants describing the strengths of the couplings of the leptons and quarks to the scalar Higgs field is nothing but a convenient parametrization. It offers no insight into the dynamical details of the mass generation and is probably wrong. A deeper understanding of the problem can only be reached by going beyond the standard model, which probably means also going beyond some of the apparent beauty and truth of the standard model and replacing some of its features by something else.

Sometimes during this talk I shall consider a world in which more than three lepton-quark families exist, specifically four families. The fourth family is denoted as follows:

$$\begin{pmatrix} \nu_\sigma & \vdots & h \\ \sigma^- & \vdots & l \end{pmatrix}.$$

The observed mass spectrum shows an interesting feature, which is not understood: the mass ratios m_i/m_{i+1} (m_i: mass of a quark or a charged lepton belonging to family i; m_{i+1}: mass of the corresponding quark or lepton of family $i+1$) are all small and, very roughly speaking, of similar order of magnitude such that one is tempted to connect the mass values by straight lines on a logarithmic scale. This works well for the quarks of charge $-1/3$, since $m_d/m_s \approx m_s/m_b \approx 1/20$. If this trend continues, the fourth quark of charge $-1/3$, the l-quark, should have a mass of about $20 \cdot m_b \approx 105$ GeV. But this is the mass value, obtained at an energy scale of 1 GeV, where the other quark mass values are defined. The physical l-mass, relevant for the spectroscopy of $\bar{l}l$-onia or l-flavored mesons, is reduced by factor of about 0.6 such that $m_l \approx 60 \ldots 65$ GeV.

An important parameter to extrapolate to the fourth family is the t-quark mass. But one might expect that the ratios m_h/m_l, m_t/m_b and m_c/m_s are of similar order of magnitude, in which case we would have $m_h \approx 5 \ldots 10 \cdot m_l \approx 250 \ldots 600$ GeV.

The ratio m_b (1 GeV) / m_τ is about three, in accordance with the expectation within the minimal SU(5) theory, and a reasonable guess would be to take a similar value for the ratio m_l (1 GeV) / m_σ, in which case we have $m_\sigma \approx m_b(1 \text{ GeV})/3 \approx 35$ GeV.

These values are, of course, nothing but rough guesses, with little theoretical foundations, but they might provide some preliminary guidance for the experimentalists searching for phenomena due to the presence of a fourth family.

The observed lepton-quark spectrum is of an exponential type, i.e., the mass values increase roughly like exp (const. $\cdot$ i) [i: family quantum number]. Thus this spectrum is unlike any other spectrum of states observed in physics before, e.g. the spectrum of atoms, nuclei or hadrons. Atomic levels are most suitably displayed on a linear energy scale, likewise nuclear levels. The spectrum of the light hadrons, on the other hand, should be displayed on a scale quadratic in the energy; this reflects the relativistic dynamics of the light quarks.

It looks like as if the masses of the lighter families are generated by a perturbative mechanism involving the masses of the members of the heavier families, a mechanism, which at the same time could lead to the phenomenon of weak interaction mixing.

Suppose there exists a fourth generation, and suppose that the masses of the h and l quarks serve as driving terms for the generation of the lighter quark masses by a mixing phenomenon which at the same time leads to the observed flavor mixing. In this case the mass matrix for the up and down type quarks could be[2]:

$$M = m_{h,l} \cdot \begin{pmatrix} 0 & a & 0 & 0 \\ a & 0 & b & 0 \\ 0 & b & 0 & c \\ 0 & 0 & c & 1 \end{pmatrix} \tag{1}$$

It is well-known that such an ansatz, in the case of three generations, describes well the observed flavor mixing angles in terms of quark mass ratios. In the case of four generations the masses m_h and m_l would be the dominating and most important mass terms. The ratio m_h/m_l might be of special relevance, since this ratio would presumably not be influenced by the flavor mixing phenomenon, due to the large mass values. It could be close to four[1], i.e. given by the ratio of the squares of the quark charges, as suggested by simple composite models in which the lepton and quark masses are of electromagnetic origin. (If the ratio were four, the h-mass would be $m_h = 4\ m_l \approx 250$ GeV, using $m_l \approx 60 \ldots 65$ GeV, as mentioned above).

Independent of any particular model for the mass matrices, the idea of connecting the mass eigenvalues and the mixing angles leads one to the possibility of estimating the order of magnitude of the t-quark mass. Since the u and d masses are very small compared to m_t or m_b, we can neglect the first family and look solely at the interplay between the second and the third one. The flavor mixing between these two families is observed to be rather small, the relevant mixing angle being of the order of 0.04. Thus it would be a good approximation to start from the limit in which there is no flavor mixing. In this case one might expect that the various masses are proportional: $m_t/m_b = m_c/m_s$, i.e. m_t(1 GeV) $\approx m_b \cdot$ (1 GeV) $\cdot (m_c/m_s)$. The ratio m_c/m_s is not well determined, due to the rather uncertain value of m_s, which may vary between about 120 MeV and 220 MeV. The ratio m_c/m_s correspondingly varies between 5 and 10, thus m_t (1 GeV) $\approx$ 26 GeV ... 55 GeV. Taking into account the QCD rescaling factor (about 0.6) to obtain the physical t-mass, one obtains $m_t \approx$ 15 GeV ... 34 GeV. The lower part of this spectrum is already ruled out by the e^+e^--annihilation experiments, while the upper part may still be consistent with the present experimental constraints.

Taking into account the flavor mixing between the second and third family, a larger range of t-quark values is obtained.

The exact range depends on the specific ansatz, which determines the interplay between the masses and the mixing angles. In general we expect:

$$\frac{m_c}{m_t} \cong \frac{m_s}{m_b} - c_1\theta - c_2\theta^2 \ldots \qquad (2)$$

Such an expansion makes sense $\sin\theta = \theta_{cb}$ is observed to be rather small: $\theta \approx 0.045$, and the second order term $c_2\theta^2$ can already be neglected. If we start from a mass matrix of the type (1) (with a specific choice of the phase parameters), one finds, for example,

$$\frac{m_c}{m_t} \cong \frac{m_s}{m_b} - 2\,\theta\,\sqrt{\frac{m_s}{m_b}} \quad . \qquad (3)$$

Given the uncertainty of the various quark mass values (especially the uncertainty of m_s), one obtains in this case a rather broad range for the physical t-quark mass: 30 GeV $< m_t <$ 90 GeV.

It is a well-known, however yet unexplained fact that the quark states for which the charged weak currents are diagonal are not eigenstates of the quark mass matrix. If all quark masses were zero or had a common nonzero value, the phenomenon of weak interaction mixing would not exist since in this case the eigenstates, defined by the weak interactions, and the ones given by the strong interactions, would coincide. Thus the phenomenon of weak interaction mixing and the breaking of the flavor symmetry in the internal space spanned by the various quark families are expected to be related to each other.

We shall adopt the usual convention and suppose that the charge 2/3 quarks are unmixed, and the weak mixing is expressed in terms of a unitary n x n matrix V (n: number of families) relating the weak eigenstates q' to the mass eigenstates q[3)]

$$\begin{pmatrix} d' \\ s' \\ b' \end{pmatrix} = \begin{pmatrix} V_{ud} & V_{us} & V_{ub} \\ V_{cd} & V_{cs} & V_{cb} \\ V_{tb} & V_{ts} & V_{tb} \end{pmatrix} \begin{pmatrix} d \\ s \\ b \end{pmatrix} \qquad (4)$$

Taking into account the strengths of observed weak transitions as well as the unitarity constraints, the 3 x 3 unitary flavor mixing matrix has to obey, one finds[8)] for the matrix (4).

$$V = \begin{pmatrix} 0.9743 - 0.9757 & 0.219 \quad 0.225 & 0.000 - 0.008 \\ 0.219 \quad - 0.225 & 0.9733 - 0.9748 & 0.039 - 0.050 \\ 0.002 \quad - 0.017 & 0.037 \quad - 0.048 & 0.9987 - 0.9993 \end{pmatrix} \qquad (5)$$

(only the ranges of the absolute values of the matrix elements are indicated).

Thus far it is unknown whether the elements of the mixing matrix V depend on the quark masses or might even be specific functions of quark mass ratios (for various attempts in this direction see e.g., refs. (2,4-7).

I shall discuss a general approach to the problem using a set of chiral symmetries which provides strong arguments for such a dependence. Under rather general circumstances we derive stringent constraints for the weak mixing angles.

We shall suppose that there are three doublets of quarks:

$$
\left\{
\begin{array}{ccc}
\text{I} & \text{II} & \text{III} \\
u \ [5.1] & c \ [1\ 350] & [\ ?\ ] \\
d \ [8.9] & s \ [175] & [5300]
\end{array}
\right\}
\tag{6}
$$

(the numbers in the brackets denote the quark mass values, normalized at an energy of 1 GeV; for a review see ref. (9)).

With one possible exception, the t-quark, whose mass is not yet known, all quark masses are relatively small compared to the typical weak interaction energy scale of the order of 0.3 TeV.

Furthermore a clear hierarchical pattern has emerged: the masses of the first generation (u, d) are small compared to the corresponding masses of the second generation (c, s), and an analogous hierarchy exists if we compare the masses of the third generation (t, b) to the masses of the second generation. Thus it seems useful to consider the various stages of mass generation given in the subsequent table, following the arguments given in ref. (10).

Stage	Masses different form zero
I	none.
↓(t,b massive)	
II	t,b.
↓(c,s massive)	
III	t,b; c,s.
↓(u,d massive)	
IV	t,b; c,s; u,d.

If all six quark masses vanish, the Lagrangian of the standard model is invariant under the chiral symmetry $U(3)_L \times U(3)_R$, where U(3), where U(3) is the symmetry group connecting the three generations U(3) stands as usual for the direct product SU(3) x U(1), and the index L(R) refers to the left-handed or right-handed quark fields q_L (q_R)).

At the present time it is completely unknown what dynamical mechanism is responsible for generating the masses of the quarks (and leptons). Nevertheless it is a fact that the chiral symmetry U(3)$_L$ x U(3)$_R$ in the space of the generations is exact only if all quark masses vanish, and vice versa. Furthermore in this limit all weak mixing angles vanish. Of course, in the real world this symmetry is broken. Nevertheless the chiral generation symmetry does play a relevant rôle. It guarantees that the symmetry limit m_q = 0 (q = u,d,..) is a natural limit, i.e. it can be realized without posing artifical constraints on the underlying yet unknown dynamical framework (whatever this may be).

Stage II: $m_u = m_d = 0$, $m_c = m_s = 0$, m_t, $m_b \neq 0$.

The (u,d)-quarks as well as the (c,s)-quarks are massless. The (t,b)-quarks are massive (their masses assume the values realized in nature).

Obviously this stage is not very far from the case given in nature, since the masses of the (u,d; c,s)-quarks are indeed very small compared to the t and b masses and might be regarded as perturbations. In terms of mass eigenstates, the mass matrices for the (2/3) and (-1/3) charged quarks are proportional to the matrix

$$\begin{pmatrix} 0 & 0 & 0 \\ 0 & 0 & 0 \\ 0 & 0 & 1 \end{pmatrix} . \qquad (7)$$

This situation can be obtained in a natural way only if it follows from a specific breaking of the original chiral generation symmetry. Since only the t and b quarks are massive, the original chiral symmetry U(3)$_L$ x U(3)$_R$ is not completely violated, but it is reduced to the subsymmetry U(2)$_L$ x U(2)$_R$ acting on the (u,d; c,s)-system.

What is the weak interaction mixing in the stage II ? In principle the stage II mass pattern allows a non-trivial weak mixing described by just one angle, as described by the following weak doublets:

$$\begin{pmatrix} u & \vdots & c & \vdots & t \\ d & \vdots & s\,\cos\alpha + b\,\sin\alpha & \vdots & -s\,\sin\alpha + b\,\cos\alpha \end{pmatrix} \qquad (8)$$

It can easily be seen that any parametrization of the weak interaction mixing at the stage II can be reduced to the one given in eq. (8). Note that here the (u,d)-pair remains unmixed; this can always be arranged by a suitable unitary transformation among the two massless doublets.

The weak interaction mixing given in eq. (8) must be the consequence of a particular structure of the quark mass matrix.

The quark mass term describing such a mixing is as follows:

$$\overline{(d_0, \; s_0, \; b_0)}_R \begin{pmatrix} 0 & 0 & 0 \\ 0 & 0 & 0 \\ 0 & B & A \end{pmatrix} \begin{pmatrix} d_0 \\ s_0 \\ b_0 \end{pmatrix}_L \qquad + \text{h.c.} \qquad (9)$$

(d_0, s_0, b_0: quark states in which the weak currents are diagonal). The parameters A and B can be arranged to be real and positive.

It is obvious from the matrix eq. (9) that the mixing parameter does not influence the mass of the s-quark; the determinant of the mass matrix in eq. (9) is zero, independent of the magnitude of the mixing. The mass term can be written explicitly as

$$B \cdot \overline{b}_{0_R} s_{0_L} + A \cdot \overline{b}_{0_R} b_{0_L} + \text{h.c.} \qquad (10)$$

We should like to interpret the mass pattern at stage II as a result of a specific breaking of the chiral $U(3)_L \times U(3)_R$ symmetry which is broken down to $U(2)_L \times U(2)_R$. This reduction of the symmetry establishes a hierarchical pattern of the fermion masses (hierarchical symmetry breaking).

The masslessness of the quarks of the first two generations in the stage II can only be guaranteed if the subgroup $U(2)_L \times U(2)_R$ of the original group $U(3)_L \times U(3)_R$ is preserved. However the mass term given in eq. (10) does not respect this symmetry if the mixing term $A \cdot \overline{b}_{0_R} s_{0_L}$ is present; this term transforms under $U(2)_L \times U(2)_R$ as $(1/2, 0)$. For example, as a result of the chiral transformation $s_{0_L} \rightarrow e^{i\alpha} s_{0_L}$, $s_{0_R} \rightarrow e^{-i\alpha} s_{0_R}$ the first term in eq. (10) is multiplied by $e^{i\alpha}$, violating the chiral $U(2)_L \times U(2)_R$ symmetry. The weak interaction mixing and the chiral symmetry cannot both be implemented. Thus there is no weak interaction mixing in the limit of stage II. The weak interaction doublets are simply given by the mass eigenstates:

$$\begin{pmatrix} u & \vdots & c & \vdots & t \\ d & \vdots & s & \vdots & b \end{pmatrix} \qquad (11)$$

In reality it is observed that the (t,b) system mixes only very slightly with the other quarks, which in our view is related to the fact that the (c,s)-masses are small compared to the (t,b)-masses, and the hierarchical symmetry limit of stage II is not far away.

Stage III: $m_u = m_d = 0$, m_c, m_s; m_t, $m_b \neq 0$.

In order to arrive at stage III, the chiral symmetry is broken down to $U(1)_L \times U(1)_R$. In terms of weak eigenstates the general mass term for the charge $(-1/3)$-quarks respecting this symmetry has a structure such that only the submatrix connecting the second and third family has entries different from zero.

The mass term can be diagonalized by a rotation in the (s_o, b_o)-plane. Only one mixing angle α enters, and we arrive at the same weak doublets as given in eq. (4). The weak mixing matrix at stage III is orthogonal - no complex phases are present. We conclude that in the limit $m_u = m_d = 0$ there is no CP violation. All CP violating effects must be proportional to positive powers of the u or d quark masses. Furthermore the u and d quarks remain unmixed in the limit $m_u = m_d = 0$. The Cabibbo angle vanishes in this limit as well as the transitions $u \leftrightarrow b$ or $t \leftrightarrow d$.

This result is, of course, not valid in general, but follows in our special approach from the underlying hierarchical structure.

Stage IV: All quark masses different from zero.

At the last stage IV the (small) masses of the u and d quarks are generated. The mixing matrix V takes its final form. Since the u and d masses are small compared to the other masses, the weak mixing matrix V can be regarded as a perturbation of the mixing matrix obtained at stage III. This implies that the elements V_{us}, V_{ub}, V_{cd} and V_{td}, which vanish in the limit $m_u = m_d = 0$, can be expanded in positive powers of the ratios m_d/m_s, m_d/m_b, m_u/m_c and m_u/m_t. Of course, the corresponding expansion coefficients cannot be predicted in the general approach discussed here. This can only be done if one has a specific algebraic structure for the mass matrix. For example, in the ansatz of ref. (2) the chiral symmetry constraints discussed here are fulfilled. One finds:

$$V_{us} \approx -\sqrt{m_d/m_s} + e^{i\alpha}\sqrt{m_u/m_c}$$

$$V_{ub} \approx \sqrt{m_d/m_b}\ m_s/m_b + \sqrt{m_u/m_c}\ (e^{i\alpha}\sqrt{m_s/m_b} - e^{i\beta}\sqrt{m_c/m_t})$$

$$V_{cb} \approx \sqrt{m_u/m_c} + e^{i\alpha}\sqrt{m_d/m_s} \qquad\qquad (12)$$

$$V_{td} \approx \sqrt{m_u/m_t}\ m_c/m_t + \sqrt{m_d/m_s}\ (e^{i\alpha}\sqrt{m_c/m_t} - e^{i\beta}\sqrt{m_s/m_b})$$

(α, β phase parameters).

All these matrix elements vanish in the limit m_u, $m_d \to 0$, in accordance with the general conclusions drawn above.

As the u and d quarks acquire their masses, the b-s mixing coefficients (V_{cb}, V_{ts}) are essentially not changed (apart from very small corrections due to unitarity).

The considerations made avove can easily be extended to include further quark doublets, provided the masses of the new quarks fit into the hierachical mass pattern as observed for the first three doublets. Suppose there exists a fourth doublet denoted by (h,l) h: "high", l: "low"). If the masses of these quarks are large enough, e.g. $m_l > 80$ GeV, $m_h > 200$ GeV, the set of chiral symmetries discussed above can be extended by another stage at which all masses except the h and l masses are zero. At this stage the (h,l)-doublet remains unmixed. The mixing of this doublet with the (t,b)-doublet would arise once the t and b masses are introduced.

A UNIQUE PARAMETRIZATION OF THE WEAK MIXING

In general the flavor mixing of the quarks is described by a unitary n x n matrix V. In the case of the observed three generations the matrix elements of V are parametrized by three rotation angles and one phase parameter[11...15]. For an arbitrary number n of generations one has n(n-1)/2 angles and (n-1)·(n-2)/2 phases, i.e. altogether (n-1)² parameters.

It is well known that there exist many equivalent ways to choose these parameters[16-29]. Here we should like to point out that a prescription for constructing the general mixing matrix given by Harari and Leurer and by Plankl and the author follows the chiral expansion of the mixing matrix mentioned above and considered in more detail subsequently. Instead of deriving our result directly for an arbitrary number of generations, we prefer to study first the case of three generations explicitly.

THREE GENERATIONS

The three weak doublets are denoted by

$$\begin{pmatrix} u \\ d \end{pmatrix}_L \begin{pmatrix} c \\ s \end{pmatrix}_L \begin{pmatrix} t \\ b \end{pmatrix}_L.$$

We consider first the limit where m_t, $m_b \neq 0$ and all other masses are zero as a result of an underlying $SU(2)_L \times SU(2)_R$-symmetry acting on the first two generations. In this limit the mixing matrix is the unit matrix. Subsequently the masses of the quarks of the second generation are introduced. The mixing matrix is given by:

$$V = \begin{pmatrix} 1 & 0 & 0 \\ 0 & c_{23} & s_{23} \\ 0 & -s_{23} & c_{23} \end{pmatrix} \qquad (13)$$

($s_{23} = \sin\theta_{23}$, $c_{23} = \cos\theta_{23}$, θ_{23}: angle referring to the mixing of the second and third generation). As discussed above the angle θ_{23} must be a function of the ratios m_c/m_t and m_s/m_b. The parametrization is uniquely fixed.

Finally the masses of the u and d quarks are introduced, and the mixing matrix takes its final form:

$$V = \begin{pmatrix} V_{ud} & V_{us} & V_{ub} \\ V_{cd} & V_{cs} & V_{cb} \\ V_{td} & V_{ts} & V_{tb} \end{pmatrix}. \qquad (14)$$

The elements V_{us}, V_{ub}, V_{cb}, and V_{td} are functions of the ratios m_d/m_s, m_u/m_c, m_c/m_b and m_u/m_t. One description to parametrize the matrix is singled out by the following argument. The transition elements V_{cb} and V_{ts} depend in the limit $m_u = m_d = 0$ on mass ratios, involving the masses of the quarks of the second and third generation. Once the u and d masses are introduced, these matrix elements which do not involve the u and d quarks will remain essentially unchanged, apart from very small corrections due to unitarity. If we request that the parametrization of the matrix reflects this inertia of the V_{cb} and V_{ts} matrix elements, we must describe them in terms of the same angle θ_{23}, and not in terms of a combination of angles. One way to achieve this is to take

$$V = \begin{pmatrix} 1 & 0 & 0 \\ 0 & c_{23} & s_{23} \\ 0 & -s_{23} & c_{23} \end{pmatrix} \begin{pmatrix} c_{13} & 0 & s_{13}e^{-i\delta_{13}} \\ 0 & 1 & 0 \\ -s_{13}e^{i\delta_{13}} & 0 & c_{13} \end{pmatrix} \begin{pmatrix} c_{12} & s_{12} & 0 \\ -s_{12} & -c_{12} & 0 \\ 0 & 0 & 0 \end{pmatrix}$$

$$= \begin{pmatrix} c_{12}c_{13} & s_{12}c_{13} & s_{13}e^{-i\delta_{13}} \\ -c_{23}s_{12} - c_{12}s_{23}s_{13}e^{i\delta_{13}} & c_{12}c_{23} - s_{12}s_{23}s_{13}e^{i\delta_{13}} & c_{13}s_{23} \\ s_{12}s_{23} - c_{12}c_{23}s_{13}e^{i\delta_{13}} & -c_{12}s_{23} - c_{23}s_{12}s_{13}e^{i\delta_{13}} & c_{13}s_{23} \end{pmatrix}$$

$$(15)$$

386

where s_{ij}, c_{ij} $(i,j = 1,2,3)$ stand for $\sin\theta_{ij}$, $\cos\theta_{ij}$ respectively, and θ_{ij} can be viewed as the mixing angles describing the mixing between generation i and j. The index (13) of the phase δ_{13} indicates that this phase appears always in association with the angle θ_{13}. The parametrization eq. (15) has been discussed by Chau and Keung[14] and the author[15], following earlier work by Maiani[12]:

Introducing the symbols $(\tilde{R})_{ij}$ for the 3 x 3 (complex) rotation matrices in eq. (15), we can write:

$$V = \tilde{R}_{23} \cdot R_{13} \cdot R_{12}. \qquad (16)$$

Besides this order of multiplication one can consider five other permutations e.g. $R_{23} R_{12} R_{13}$. Furthermore the phase parameter δ could appear in any of the three rotation matrices.

All these representations have the property that the chiral limit $m_u = m_d = 0$ can be realized simply by turning off the rotations R_{12} and R_{13}, and leaving the matrix R_{23} unchanged.

The parametrization given by Kobayashi and Maskawa[11] may serve as an example of a description which shows a "bad" behaviour in the chiral limit $m_u = m_d = 0$. Therefore the specific KM representation should not be used to describe the weak mixing.

FLAVOR MIXING FOR N GENERATIONS

The (lefthanded) weak doublets are denoted by

$$\begin{pmatrix} U_1 \\ D_1 \end{pmatrix}_L \qquad \begin{pmatrix} U_2 \\ D_2 \end{pmatrix}_L \qquad \cdots \qquad \begin{pmatrix} U_n \\ D_n \end{pmatrix}_L \qquad (17)$$

the mass eigenvalues by m_{U_i}, m_{D_i} $(i = 1,2, \ldots, n)$. We suppose that the physical quark masses obey a hierarchical order:

$$\frac{m_{U_{i-1}}}{m_{U_i}} \ll 1 \qquad \frac{m_{D_{i-1}}}{m_{D_i}} \ll 1, \qquad (18)$$

in accordance with observation, as far as the first three generations are concerned.

We consider various hypothetical limits in which the masses of the first k generations $(k = 1,2\ldots n)$ are set to zero. In this case the mass term is required to obey a chiral $U(k) \times U(k)$ symmetry. As a result the flavor mixing matrix, denoted by $V^{(k)}$, is diagonal for the first k generations:

$$V^{(k)} = \begin{pmatrix} & 1 & 2 & 3 & . & . & k & k+1 & . & . & . & . & n & \\ 1 & 1 & 0 & 0 & . & . & 0 & 0 & & . & . & . & . & 0 & \\ 2 & 0 & 1 & 0 & . & . & 0 & 0 & & . & . & . & . & 0 & \\ 3 & 0 & 0 & 1 & . & . & 0 & 0 & & . & . & . & . & 0 & \\ & . & . & . & . & . & . & . & & . & . & . & . & . & \\ & . & . & . & . & . & . & . & & . & . & . & . & . & \\ k & 0 & 0 & 0 & . & . & 1 & 0 & & & . & . & 0 & \\ k+1 & 0 & 0 & 0 & . & . & 0 & V_{k+1,k+1} & . & . & V_{k+1,n} & \\ & . & . & . & . & . & . & . & & . & . & . & \\ & . & . & . & . & . & . & . & & . & . & . & \\ n & 0 & 0 & 0 & . & . & 0 & V_{n,k+1} & & . & . & V_{n,n} & \end{pmatrix} \qquad (19)$$

Thus a nontrivial mixing takes place only among the massive generations $k+1$, ..., n. We arrive at the physical mixing matrix V by a succession of n steps:

STEP 1: The masses of the nth generation are introduced. The chiral symmetry reduces from $U(n)_L \times U(n)_R$ to $U(n-1)_L \times U(n-1)_R$. The flavor mixing matrix $V^{(n-1)}$ is diagonal.

STEP 2: The masses of the $(n-1)^{th}$ generation are introduced. The chiral symmetry reduces to $U(n-2)_L \times U(n-2)_R$. The flavor mixing $V^{(n-2)}$ is simply a rotation among the generations $(n-1)$ and (n), described by one angle $\theta_{n-1,n}$, which must be a function of the ratios $m_{U_{n-1}}/m_{U_n}$ and $m_{D_{n-1}}/m_{D_n}$: $V^{(n-2)} = R_{n-1,n}$.

STEP 3: The $(n-2)^{th}$ generation becomes massive. The chiral symmetry reduces to $U(n-3)_L \times U(n-3)_R$. The flavor mixing is de-described by the 3 x 3 submatrix:

$$V^{(n-3)} = R_{n-1,n}(\tilde{R}_{n-2,n} \, R_{n-2,n-1}). \qquad (20)$$

The mixing angles $\theta_{n-2,n}$ and $\theta_{n-2,n-1}$ are functions involving the masses of quarks of the $(n-2)^{th}$ generation.

The procedure can be repeated a number of steps until we arrive at step k, where the masses of the generation $n-k+1$ are introduced. The mixing matrix is given by

$$V^{(k)} = (R_{n-1,n}) \, (\tilde{R}_{n-2,n} \, R_{n-2,n-1}) \cdots$$

$$\cdots (\tilde{R}_{n-k+1,n} \, \tilde{R}_{n-k+1,n-1} \cdots R_{n-k+1,n-k}). \qquad (21)$$

The new mixing angles $\theta_{n-k+1,n-1} \cdots \theta_{n-k+1,n-k}$ appearing at this are functions involving the masses of the k generation. Finally we arrive at the last step n, where the masses of the first generation are introduced. The full mixing matrix is given by

$$V = V^{(n)} = (R_{n-1,n}) \; (\tilde{R}_{n-2,n} \; R_{n-2,n-1}) \; (\tilde{R}_{n-3,n} \; \tilde{R}_{n-3,n-1} \; R_{n-3,n-2}) \cdots$$

$$\cdots ((\tilde{R}_{1n} \; \tilde{R}_{1\,n-1} \; \cdots \; R_{12}). \tag{22}$$

In eq. (22) we have denoted by the parentheses and the numbers below them those blocks of matrices appearing at each step of the mass generation.

The order of the rotation matrices given in eq. (22) is different from the one proposed in ref. (28,29). However a closer inspection shows that due to the vanishing of the commutators $[R_{ij} \; R_{kl}] (i \neq k, \; j \neq l)$ the two parametrizations are, in fact, identical[30].

CONCLUSIONS

It seems today that the spectrum of the leptons and quarks shows a number of regularities which are worth to be studied in more detail. Especially the observed hierarchy of the masses calls for an approach which resembles the chiral perturbation theory used in the past in order to extract certain properties of the dynamics of hadrons, e.g. of the pions. A cascade-like breaking of chiral symmetries has been used to connect the weak interaction mixing angles and the mass eigenvalues. Step by step the masses of the various generations are introduced. Along with the mass generation goes the evolution of the flavor mixing matrix which starts out as the unit matrix and develops more and more elements as more generations become massive.

The actual parametrization of the mixing matrix is, of course, not of direct relevance of the underlying physics. However if we require that the expansion of the mixing matrix is manifestly described by introducing more mixing parameters and not changing the ones introduced before, unique parametrizations are singled out.

It remains to be seen whether the approaches discussed here may lead to a deeper understanding of the flavor problem. In the case of hadronic dynamics it worked in the past: the chiral symmetry approach combined with the current algebra led eventually to the foundations of QCD.

ACKNOWLEDGEMENT

I would like to thank Dr. Ali and the staff of the Centro Majorana for the excellent organization of the Eloisatron workshop and for the successful attempt to improve the local weather in Erice just at the beginning of the meeting.

REFERENCES

1. U. Baur and H. Fritzsch, Phys. Lett. 134 B (1984) 105.
2. H. Fritzsch, Nucl. Phys. B155 (1979) 189;
   See also: H. Fritzsch, Phys. Lett. 166 B (1986) 423.
3. Review of Particle Properties, Phys. Lett. 170B (1986).
4. B. Stech, Phys. Lett. 130B (1983) 189.
5. G. Ecker, Z.Phys. C24 (1984) 353.
6. H. Georgi, A. Nelson and M. Shin, Phys. Lett 150B (1985) 306.
7. M. Shin, Phys. Lett. 152 B (1985) 83.
8. K. Kleinknecht and B. Renk, Z. Phys. 34 (1987) 209.
9. J. Gasser and H. Leutwyler, Physics Reports 87 (1982) 77.
10. H. Fritzsch, Phys. Lett. 184 B (1987) 3091.
11. M. Kobayashi and K. Maskawa, Progr. Theor. Phys. 49 (1973) 652
12. L. Maiani, Phys. Lett. B 62 (1976) 183, and Proc. Intern.
    Symp. on Lepton and Photon Interactions (Hamburg, 1977)
    p. 867–
13. L. Wolfenstein, Phys. Rev. Lett, 51 (1984) 1945.
14. L.L. Chau and W.Y. Keung, Phys. Rev. Lett. 53 (1984) 1802.
15. H. Fritzsch, Phys. Rev. D 32 (1985) 3058.
16. J. Schechter and J.W.F. Valle, Phys. Rev. D 21 (1980) 309;
    D 22 (1980) 2227.
17. M. Gronau, R. Johnson and J. Schechter, Phys. Rev. D 32
    (1985) 3062.
18. M. Gronau and J. Schechter, Phys. Rev. D 31 (1985) 1668.
19. A.A. Anselm, J.L. Chkareuli, N.G. Uraltsev and
    T.A. Zhukovskaya, Phys. Lett. B 156 (1985) 102.
20. V. Barger, K. Whisnant and R.J.N. Phillips.
    Phys. Rev. D 23 (1981) 2773.
21. R.J. Oakes, Phys. Rev. D 26 (1982) 1128.
22. S. Pakvasa, New particles 85, Conf. at the University of
    Wisconsin (Madison, May 1985), KEK preprint 85-34.
23. X.G. He and S. Pakvasa, Phys. Lett. B 156 (1985) 236;
    University of Hawaii preprint, UH-511-572-85.
24. I.I. Bigi, Z. Phys. C 27 (1985) 303.
25. G. Kramer and I. Montvay, Z. Phys. C 11 (1981) 1128.
26. U. Türke, E.A. Paschos, H. Usler and R. Decker,
    Nucl. Phys. B 258, 313.
27. R. Mignani, Lett. al Nuovo Cim. 28, 529 (1980).
28. H. Harari and M. Leurer, Phys. Lett. 181 B (1986) 123.
29. H. Fritzsch and J. Plankl, Phys. Rev. D 35 (1987) 1732.
30. H. Fritzsch, Phys. Lett. B 189, 191-196 (1987).

NON-ACCELERATOR PHYSICS AT $\sqrt{S} \sim 40$ TEV*

T. K. Gaisser

Bartol Research Institute
University of Delaware
Newark, DE 19716

ABSTRACT

   In this talk I focus on the relation between particle physics and astrophysics in the interpretation of data from high energy air showers, with particular emphasis on the Fly's Eye experiment. This experiment has a dual purpose in this energy range -- to determine the proton cross section and to determine the spectrum and composition of the primary cosmic rays. As with large collider experiments extensive Monte Carlo simulation is needed to interpret the measurements.

I. INTRODUCTION

   I was asked to talk on "Particle Multiplicity Profiles of High Energy Showers" in the context of this meeting on particle physics at future colliders. For such a talk I prefer the title I have used for this written version. My theme is the relation of non-accelerator experiments to high energy physics at accelerators.

   The primary emphasis of high energy non-accelerator experiments is on astrophysical questions such as the origin of cosmic rays and their acceleration to ultra-high energies. An understanding of particle interactions is necessary to interpret the results of cosmic ray measurements, which generally involve cascades of interactions at high energy. This effort can be considered as a branch of applied particle physics. In addition some aspects of the highest energy cosmic ray experiments are still of interest for high energy physics. One of these is the determinations of the proton cross section around $\sqrt{s} \sim 40$ TeV which is the main subject of this talk. I will not discuss here the major non-accelerator efforts to search for nucleon decay, monopoles and extra-terrestrial neutrinos, nor the efforts to observe point astrophysical sources with small air shower experiments.

Contemporary and proposed accelerator experiments focus on new physics, for which the mass scale is high and the corresponding cross sections very low. In contrast, in high energy cosmic ray experiments, the flux is low and therefore only processes with large cross sections play a role. These are the "minimum bias" events that constitute the background for new physics at colliders. I illustrate this point in Table 1 by working out the equivalent luminosity of typical cosmic ray experiments. Integrated luminosities of order inverse microbarns are to be contrasted with the accelerator goals of inverse picobarns and better.

An interesting by-product of the low rates in high energy cosmic ray experiments is that any mistake or fluctuation will be assigned a large cross section. Thus if the probability of mistake per experiment is constant, cosmic ray mistakes will have much larger cross sections than mistakes in accelerator experiments. In addition, their resolution will take longer because of the lower rates. I believe it is this difference that has sometimes led to the mistaken impression that cosmic ray experiments are more error prone than accelerator experiments.

Table 1. Cosmic Ray Rates and Effective "Luminosities"

	~ 2 TeV	~ 40 TeV
$\sqrt{s}$	~ 2 TeV	~ 40 TeV
E_{Lab}	~ 10^{15} eV	~ 10^{18} eV
Flux	3×10^{-6} m^{-2}sec^{-1}	3×10^{-12} m^{-2}sec^{-1}
Typical Area of Detector	~ 10^4 m^2	~ 10^8 m^2
Rate	~ 1/minute	~ 1/hour
$L = \dfrac{\text{Rate}}{\sigma}$	~ 10^{24} cm^{-2}sec^{-1}	~ 3×10^{21} cm^{-2}sec^{-1}
$\int_{\text{year}} L dt$	~ 30 μb^{-1}	~ 100 mb^{-1}

Many of the talks at this workshop have dealt with the use of Monte Carlo simulations for studies of accelerator experiments. Simulations are also essential for design and interpretation of cosmic ray experiments. Cosmic ray experiments have reached the stage where it is desirable to incorporate some of the advanced features of the sophisticated Monte Carlo programs such as PYTHIA and ISAJET, in particular, jet production. At the same time, the cosmic ray cascade calculations emphasize features of hadronic interactions that are relatively unimportant for accelerator simulations, such as the energy distribution of the leading nucleon, which is crucial for determining how a cascade develops. An important technical difference is that the simulation of a hadronic interaction takes place in the context of a cascade, so that the incident particle type and energy cannot be predetermined for a large number of interactions as would be the case for an accelerator simulation. Thus it is not practical simply to take a complex code designed for accelerator analysis and plug it into a cosmic ray calculation where the beam particle and its energy vary from one interaction to another within a cascade. I will return to the question of models and simulations later in the talk when I discuss the model-dependence of the cosmic ray result for the proton cross section.

The cosmic ray flux decreases rapidly with increasing energy. It is so low that direct experiments with detectors flown at the top of the atmosphere become impractical for energies above 10^{14} eV ($\sqrt{s} \sim 400$ GeV). (Above this energy the total flux is less than one particle per square meter per hour.) To compensate for the low flux it is necessary to expose arrays of large effective area for long time periods. These air shower experiments can glimpse physics and astrophysics at very high energies -- but at a very high price: the primary particles are not detected directly. A conventional air shower detector only samples the cascade at one depth. Some arrays are equipped to distinguish and measure separately the hadronic and muonic component of the cascade as well as the dominant electromagnetic component. The most rapidly growing application of conventional air shower experiments at present is the search for and study of astrophysical point sources of PeV and TeV signals. Professor Halzen will refer to these in some detail in his talk. Here I will briefly describe how the Fly's Eye detector has been used to measure the cross section and to study the composition of the primary cosmic ray nuclei around 10^{18} eV ($\sqrt{s} \sim 40$ TeV).

II. PROTON CROSS SECTION IN AIR

The Fly's Eye detector[1] measures atmospheric scientillation light excited by large showers as they traverse the sky. From the timing of the

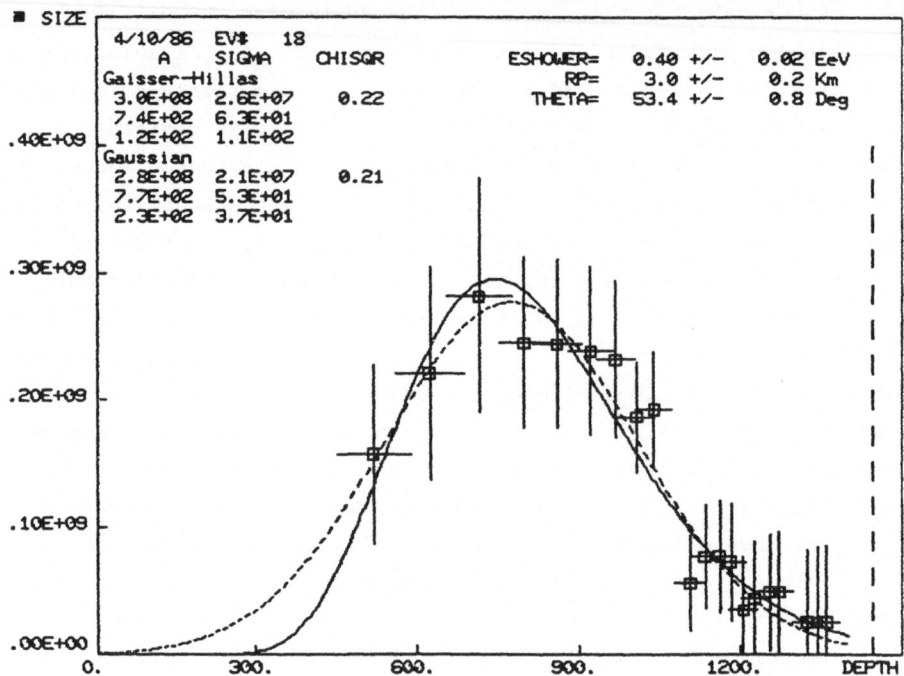

Fig. 1. Example of a measured shower profile fitted with two shapes to obtain depth of maximum for the shower.

signals across the segmented eye the trajectory is reconstructed and from the pulse heights the shower size. Thus the atmosphere acts as a calorimeter in which shower energy and penetration depth can be measured. Fig. 1 shows[2] a typical longitutinal shower profile with two fitted profiles used to determine energy and depth of shower maximum. (Two fits are compared to help insure that the result is independent of the parametrization used to define depth of maximum.)

If it were possible to measure shower starting points and to identify the beam particle as a proton, one would have a direct measure of σ_{p-air}^{inel}. Showers are only visible after they are well developed, however. In addition to the cross section, shower development is affected by the nature of the primary nucleus and by fluctuations in individual hadronic interactions. (For the same total energy, showers initiated by heavy nuclei develop faster than proton showers.) There is not enough information to be able to determine all three aspects uniquely from the

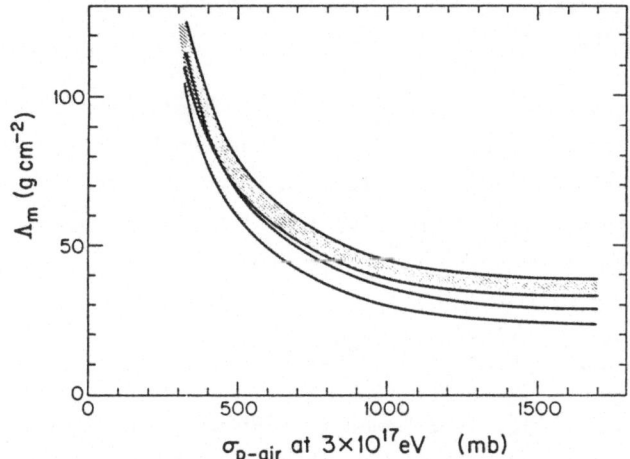

Fig. 2 Λ_m versus $\sigma_{p\text{-air}}$ at $\sqrt{s} \sim 40$ TeV for proton showers chosen from a power law energy spectrum (integral index = 1) with $E_o > 3 \times 10^7$ eV. The upper band represents the scaling model and the lower one represents the scaling violation with $\alpha = 0.09$. Band width indicates statistical uncertainty of the calculation. (From reference 4.)

data. In practice, one assumes a particular form for extrapolation of hadronic interactions beyond machine energies then finds a best fit for cross section and composition. A consistency check is that the fitted composition contains enough protons ($\geq 20\%$) so that the tail of the depth of maximum distribution reflects the proton cross section in air. (In addition one needs to assume that the α/p ratio is not anomalously large, i.e., much larger than the cosmic abundance ratio of 0.1, which translates to 0.4 when p and α are compared at the same energy per nucleus. This is because α and proton showers look quite similar.)

A plot[2] of the experimental depth of maximum distribution shows an exponential tail characterized by a length $\Lambda_m = 70 \pm 6$ gm/cm^2. Simulations are needed to interpret this result. These were first described in ref. 3 and later extended[4] to study the dependence of the result on the model of hadronic interactions. Fig. 2 shows the relation between Λ_m and $\sigma_{p\text{-air}}^{inel}$ for two different extrapolations of hadronic interactions beyond present accelerator energies. From the upper curve one finds $\sigma_{p\text{-air}}^{inel} = 540 \pm 50$ mb.

III. RELATION TO σ_{pp}

In a cosmic ray cascade elastic and quasi-elastic interactions with nuclei have no practical effect. (This includes both quasi-elastic interactions (QE) that leave the nucleus in an excited or continuum state and diffraction dissociation events (D) in which only a target nucleon dissociates.) What is measured is[5]

$$\sigma_{p\text{-}air}^{inel} = \sigma_{p\text{-}air}^{tot} - \sigma_{p\text{-}air}^{elastic} - \sigma_{p\text{-}air}^{QE} - \sigma_D.$$

The p-nucleus cross section is related to σ_{pp}^{tot} with the help of Glauber multiple scattering theory,[6] including the correction[7] for inelastic

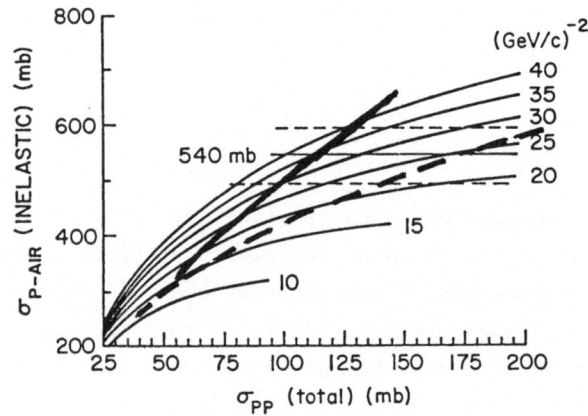

Fig. 3 Relation between proton-air and proton-proton cross sections for various values of slope parameter. The broad solid line is the locus of B for geometrical scaling and the broad dashed curve for the Chou-Yang model. The Fly's Eye result for the proton-air inelastic cross section is also shown.

screening (see ref. 5 for details). In the multiple scattering theory the calculated value of a hadron nucleus cross section depends both on the hadron-nucleon cross section and on the slope parameter $B \equiv d\ln(d\sigma/dt)/dt$. Thus the relation between σ p-air and σ_{pp}^{tot} at $\sqrt{s} \sim 40$ TeV depends on how B_{pp} is extrapolated beyond the energy region where it has been measured. This has important practical consequences for the conclusion to be drawn from the Fly's Eye measurement of $\sigma_{p\text{-air}}$, as shown in Fig. 3. When the Fly's Eye result for the cross section was first reported[2] the geometrical scaling model[8] was used, in which $B_{pp} \propto \sigma_{pp}^{tot}$. This led to an inferred value of $\sigma_{pp}^{tot} = 122 \pm 11$ mb. Results from $\bar{p}p$ collider make it clear that geometrical scaling is not valid, at least up to $\bar{p}p$ collider energies. Block and Cahn[9] suggest that an extrapolation due to Chou and Yang[10] gives a better and somewhat model-independent result for B. Using this extrapolation we[5] found

$$\sigma_{pp}^{tot} = 175^{+40}_{-27} \text{ mb at } \sqrt{s} \simeq 40 \text{ TeV}.$$ I emphasize that this large

difference is for the same value of $\sigma_{p\text{-air}}^{inel}$ and reflects only the dependence of the multiple scattering theory on B.

IV. EXTRAPOLATION OF HADRONIC INTERACTIONS

The model dependence just discussed has no consequences for the cascade development itself. On the other hand, the extrapolation of hadronic interactions does directly effect cascade development. This is illustrated by the two models shown in Fig. 2. Because most ionization in a shower is due to electrons and positrons, cascade development depends on the rate at which energy is transferred from the hadronic to the electromagnetic component of the shower, and also on the extent to which energy is subdivided into small pieces before it enters the electromagnetic component. A large number of low energy neutral pions makes a more rapidly developing shower than a few high energy neutral pions.

In my opinion the conservative way to extrapolate hadronic interactions to high energy is to assume that the inclusive cross sections approximately scale in the fragmentation region, with just enough violation to accommodate a logarithmic increase of the inclusive cross section in the central region of rapidity. In this scheme interactions tend to have a few high energy secondaries, leading to deeply penetrating showers with large fluctuations. Data from the $\bar{p}p$ collider are consistent with this kind of extrapolation of hadronic interactions,[11] but a definitive test is not possible with detectors designed to study the central region around 90° in the CM. In addition we require an extrapolation fully two orders of magnitude in CM energy beyond S$\bar{p}p$S and Tevatron energies.

The second model shown in Fig. 2 is due to Wdowczyk and Wolfendale[12] in which scaling violation is parametrized by

$$\frac{E}{\sigma} \frac{d\sigma}{d^3 p} \longrightarrow \left(\frac{s}{s_0}\right)^{\alpha} f\left(x\left(\frac{s}{s_0}\right)^{\alpha}, P_T\right).$$

$\alpha = 0$ is the scaling limit. The lower band in Fig. 2 corresponds to $\alpha = 0.09$ with $s_0/M_p = 2$ TeV. With this parametrization, the fragmentation region is suppressed in favor of creating an increasingly large number of low energy particles. This leads to more rapid shower development and fewer fluctuations. It therefore leads to a smaller value of $\sigma_{p\text{-air}}$ than the scaling model, 450 ± 50 mb rather than 540 mb. Ref. 11-a shows some scaling violation at $P_T > <P_T>$ but would be inconsistent with the kind of strong violation of Ref. 12.

As indicated in the preceding discussion, the gross energetics of hadronic interactions is of paramount importance for cascade development. Correspondingly, the most important aspect of shower simulations is the treatment of the leading particles and their fragments, i.e., beam jets. In contrast the focus of accelerator simulation is on hard scattered gluon and quark jets. The subject of mini-jets is common to both problems. They comprise the minimum bias background for the central region and they must also be related -- at least by energy momentum conservation -- to the fragmentation region. At $\sqrt{s} \sim 40$ TeV multiple jet production will become a dominant feature of hadronic interactions[13] and should be incorporated in shower simulations.

Jet production is also important for lateral structure of showers. Though the Fly's Eye experiment itself only sees longitudinal structure, most air shower experiments are also sensitive to lateral structure. High energy transverse structure is particularly important for measurements of coincident multiple muons in deep underground detectors of large area. These measurements are capable of giving information about composition of the primary cosmic rays around 10^{15} eV.[14]

V. ASTROPHYSICAL APPLICATIONS

The main astrophysical goal of air shower experiments is to find the origin of the high energy ($> 10^{15}$ eV) cosmic rays. We want to know, for example, whether these high energy particles are primordial (protons and helium nuclei) or the end products of stellar nucleosynthesis, including heavy nuclei such as iron. Showers initiated by heavy nuclei tend to develop high in the atmosphere and to have a rather narrow depth of maximum distribution, as reflected in Fig. 4.[3] The figure also illustrates how the deep tail of the depth of maximum distribution reflects protons and their cross section. To learn about composition, one needs to be able to study the shallow part of the depth of maximum

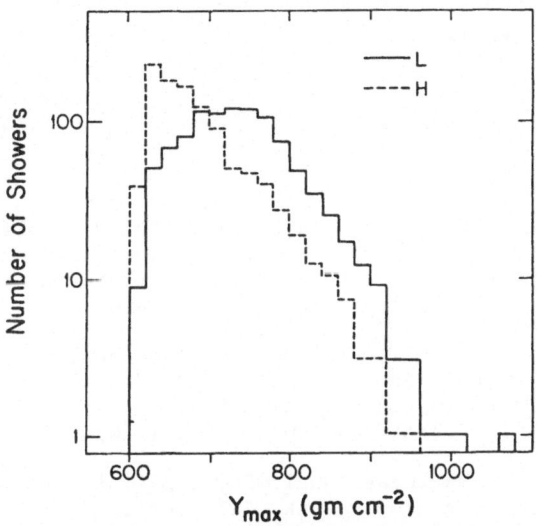

Fig. 4   Distribution of depth of maximum for showers of energy $\geq 3 \times 10^{17}$ eV per nucleus for two compositions: p:α:CNO + Mg:Fe = 0.55:0.21:0.16:0.08 as at low energy (denoted L); and p:α:CNO + Mg:Fe = 0.2:0.08:0.07:0.65, denoted H. (From reference 3.)

distribution. This has only recently become possible because showers that reach maximum high up are somewhat more difficult to reconstruct and interpret. A preliminary analysis, presented at the Moscow Conference[15] shows that a substantial deep tail on the observed X_{max} distribution requires 45 ± 12% protons. The rapid rise of the distribution around 600 g/cm^2 can only be accounted for by the presence of heavy nuclei like iron at the level of 30 ± 10% and the observed distribution is too wide to be accounted for by any single component - i.e., a significantly mixed composition is required.

These conclusions about composition are probably rather insensitive to the hadronic interaction model. Indeed, there is probably some further information about the interaction model to be obtained from a fully self-consistent analysis of the data in which composition, cross section and an inelasticity parameter such as α are all allowed to vary to obtain a best fit to the data. Such a study is in progress.

ACKNOWLEDGMENTS

This work is supported in part by the U.S. Department of Energy. I am grateful to Todor Stanev, Gaurang Yodh and Pierre Sokolsky for helpful discussions and collaboration.

REFERENCES

1. G. L. Cassiday, Ann. Revs. Nucl. & Particle Sci. <u>35</u>, 321 (1985).
2. R. M. Baltrusaitas et al., Proc. XIX Int. Cosmic Ray Conf. (La Jolla) <u>6</u>, p. 5 (1985) and Phys. Rev. Lett. <u>52</u>, 1380 (1984).
3. R. W. Ellsworth, T. K. Gaisser, Todor Stanev and G. B. Yodh, Phys. Rev. <u>D26</u>, 336 (1982).
4. Todor Stanev and T. K. Gaisser, Proc. Workshop on Very High Energy Cosmic Ray Interactions, ed. M. L. Cherry, K. Lande & R. I. Steinberg, U of Pennsylvania p. 125 (1982).
5. T. K. Gaisser, U. P. Sukhatme and G. B. Yodh, Phys. Rev. D (to be published).
6. R. J. Glauber and G. Matthiae, Nucl. Phys. <u>B21</u>, 135 (1970).
7. V. A. Karmanov and I. A. Kondratyuk, JETP Letters <u>18</u>, 266 (1973).
8. J. Dias de Deus and P. Kroll, Acta Phys. Pol. <u>B9</u>, 159 (1978).
9. M. M. Block and R. Cahn, Revs. Mod. Phys. <u>57</u>, 563 (1985).
10. T. Chou and C. N. Yang, Proc. of Second Int. Conf. on High Energy Physics, Rehovoth, Israel, (1967), ed. G. Alexander and Phys. Rev. <u>170</u>, 1591 (1968) and Phys. Letters <u>128B</u>, 457 (1983); F. Hayot and U. Sukhatme, Phys. Rev., <u>D10</u>, 2183 (1974); C. Bourrely, J. Soffer and T. T. Wu, Phys. Rev. Lett., <u>54</u>, 757 (1985); H. Cheng, J. Walker and T. T. Wu, Phys. Lett., <u>44B</u>, 97 (1973); L. Durand and R. Lipes, Phys. Rev. Letters <u>20</u>, 637 (1968).
11. M. Hagenauer et al., Proc. 20th Int. Cosmic Ray Conf. (Moscow) <u>5</u> p. 23 (1987). See also UA5 collaboration (G. J. Alner et al.) CERN-EP/86-126.
12. J. Wdowczyk and A. W. Wolfendale, J. Phys. <u>G10</u>, 257 (1984).
13. See e.g., L. Durand and H. Pi, Phys. Rev. Lett. <u>58</u>, 303 (1987).
14. T. K. Gaisser and Todor Stanev, N.I.M. <u>A235</u>, 183 (1985).
15. R. M. Baltrusaitas et al., Proc. 20th Int. Cosmic Ray Conf. (Moscow) <u>1</u>, 394 (1987).

COSMIC ACCELERATORS: A NEW ERA OF

COSMIC RAY ASTROPHYSICS AND PARTICLE PHYSICS[*]

Francis Halzen

Physics Department
University of Wisconsin
Madison, Wisconsin 53706, USA

and

Department of Physics
University of Tokyo, Tokyo, Japan

ABSTRACT

A new generation of cosmic ray experiments will be able to measure $\gamma N \rightarrow \pi X$ photoproduction for γ-energies exceeding 100 TeV using tagged photon beams emitted by cosmic accelerators. A source such as Cygnus X-3 could emit as many as 10^5 photons/km^2/year with energies exceeding 100 TeV. These experiments, although motivated by astronomy, should be of interest to particle physics as they are unlikely to be ever performed with accelerators. They also avoid the classical pitfalls of present cosmic ray experiments in this energy range as (i) they can achieve reasonable statistics with good signal/noise, (ii) they use a beam of known composition $(i.e.$ photons) and (iii) they observe showers whose development in the air is dictated by QED and therefore calculable so that *unusual phenomena* can be unambiguously interpreted as new physics. They can provide us with a first look at the energy regime probed by future supercolliders. We also briefly touch upon the value of these experiments as telescopes for the study of the origin of cosmic rays and the detection of supernova.

I. COSMIC PARTICLES AND THEIR DETECTION

A crash course in the detection of cosmic particles is given in Fig. 1. Particles in the cosmic ray beam with large interaction cross-section with matter (hadrons, photons, ...) interact near the top of the atmosphere. They initiate an electromagnetic shower (either directly or via $\pi^0(\rightarrow \gamma\gamma)$ production). If the primary energy exceeds 10 GeV this electromagnetic cascade reaches Earth and can be observed

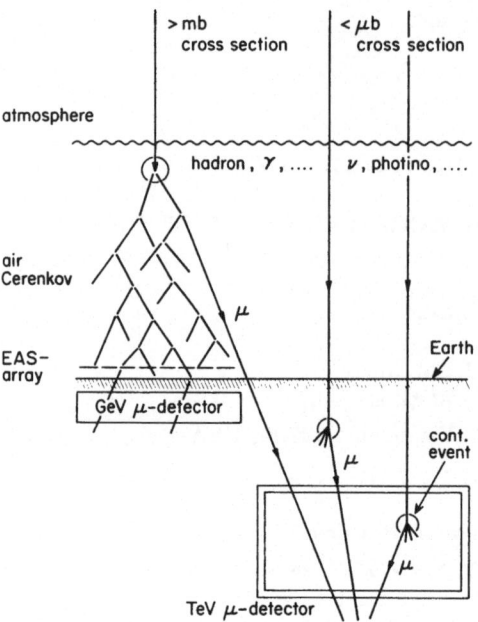

Fig. 1. Summary of observational techniques for cosmic particles with large and small interaction cross sections with matter.

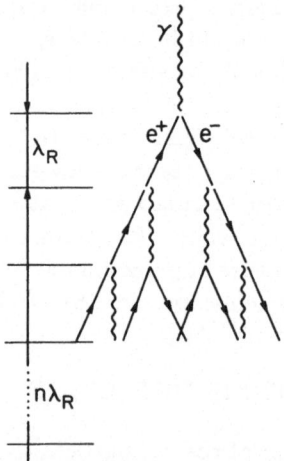

Fig. 2. The γ-ray initiated cascade in the atmosphere. Each layer represents a radiation length λ_R. The number of layers is limited by the maximum energy of the source, *e.g.* 10^5 TeV for Cygnus X-3.

with an array of particle detectors (EAS-array). The shower can also have a muon component from charged π production followed by the decay $\pi \to \mu\nu$. Muons of GeV energy can be detected at the Earth's surface by detectors shielded by e.g., Fe or a few metres of earth or concrete. The TeV muons can reach deep underground *proton-decay* detectors. These detectors can also reveal the presence in the cosmic ray flux of weakly interacting particles such as the neutrino. Neutrinos do not interact with the atmosphere but can produce muons inside or in the rock above or below deep underground detectors via the usual ν + nucleon $\to \mu$ weak interaction; see Fig. 1.

As suggested in Fig. 1 the electromagnetic showers initiated by cosmic ray hadrons are not spectacularly different from those initiated by very high energy γ-rays emitted by an astronomical source, making astronomy in the TeV-band of the electromagnetic spectrum a challenge. A primary photon converts to an e^+e^- pair after 1 radiation length λ_R in the atmosphere which is about 25 radiation lengths thick, with $\lambda_R \simeq 37$ g/cm^2. In subsequent radiation lengths the electromagnetic particles further lose energy by bremsstrahlung $e^\pm \to \gamma e^\pm$ and pair production $\gamma \to e^+e^-$; see Fig. 2. The prominent feature that can set γ-rays apart from background cosmic ray hadrons is their low muon content. The number of muons in a γ-shower is typically a few per cent of that in a hadron shower in which muons are abundantly generated by the decay of the produced $\pi^\pm$. In γ-initiated showers processes resulting in muons are characterized by small cross-sections: γ + nucleon $\to \pi$ photoproduction followed by the decay $\pi \to \mu\nu$, production and subsequent semi-leptonic decay of charm quarks and $\gamma \to \mu^+\mu^-$ pair production which is suppressed by a large factor $(m_e/m_\mu)^2$ relative to $\gamma \to e^+e^-$. The muon production through these channels has been calculated.[1] An illustrative result is shown in Fig. 3 where the number of muons in excess of 1 GeV is plotted as a function of shower size (number of electrons N_e) for γ- and nucleon-initiated showers. The result is qualitatively easy to understand: muons are the progeny of hadrons and the photon is hadronic at $O(\alpha)$, i.e., order 10^{-2}.

To the experimentalist the shower of Fig. 2 looks like a pancake of electromagnetic energy about $10^2 \sim 10^3$ m^2 in size and a few nanoseconds thick, moving down the atmosphere at the speed of light. Showers in excess of 10 TeV reach the ground, as shown in Fig. 1. By recording the arrival time of the pancake in several detectors of an extensive particle array, the arrival direction of the shower can be reconstructed from the timing sequence. Lower energy showers do not reach the ground. A 100 GeV photon will typically produce about 100 electrons at 10 km altitude. Their Čerenkov light does, however, reach earth and can be collected by mirrors viewed by phototubes. The angular spread of the Čerenkov cone is only about 1.5° around the parent photon direction.

Experiments recording the arrival direction of cosmic particles are, of course, as old as cosmic ray physics. The subject emerged on the forefront when in 1983 the Kiel group, after mapping cosmic rays in the sky for five years, reported[2] a 4σ excess coming from the direction of Cygnus X-3. High energy cosmic rays are, after all, not totally isotropic! No other source was found at a significant statistical level. Roughly 20 experiments[3] have by now found evidence for the emission of very high energy γ-rays from the direction and with the characteristic time structure (more about that later) of the binary star Cygnus X-3. Some made repeated observations. An incomplete compilation of results is shown in Fig. 4. The source seen in radio, infra-red, MeV-γ and X-ray experiments is observed to emit TeV photons by Čerenkov telescopes. Ground-based detectors suggest emission all the way up to 10^5 TeV, while the Haverah park array suggest a cut-off at that energy (open circles in Fig. 4). The flux can be approximated by[4]

$$F(> E) = \frac{4 \times 10^{-11}}{E(\text{TeV})} \text{ particles cm}^{-2}\text{sec}^{-1} \qquad (1.1)$$

for $E \gtrsim 0.1$ TeV. We have determined (1.1) from time-averaged data. Ground arrays, such as Kiel, yield for certain epochs fluxes higher by more than one order of magnitude. Some TeV experiments have identified periods of increased activity of a few minutes of duration.[5] The Cygnus beam has a very rich time structure which has not been deciphered.[3] Periodicities of 4.8 hours, 19 days and one year have been suggested. Only the 4.8-hour bunching of the beam is established; high energy emission occurs very sporadically in these 4.8-hour periods. Other very high energy emitters have been reported: Hercules X-1, Crab pulsar, 4U0115 + 63, PSR 1953, LMC X-4, Centaurus A, Vela pulsar, M31, PSR 1802, and the galactic plane. Some of these certainly require confirmation.

It must be emphasized that the surface experiments do not detect the primary particle, only its atmospheric cascade. The primaries are assumed to be photons because they are the only known particle that is neutral, stable (and hence capable of travelling in straight lines in the galactic magnetic field over very long distances) and initiates air-showers. At this point the only cloud in this firmament of exciting results is the observation by the Kiel group[2] that the on-source showers from Cygnus show some deficit in the number of muons with respect to hadron showers; the muon content exceeds however, the expected value by over one order of magnitude. This result has been debated,[6] e.g., on the basis of punch through in the 2 m of concrete used to identify the muons in a shallow detector as in Fig. 1, but never successfully challenged. This started speculations that the particles carrying the radiation from Cygnus X-3 are not photons.[7] We will return to this later.

II. COSMIC ACCELERATORS

As a particle physicist one wonders how an X-ray binary living at the far edge of the galactic plane (Cygnus is at least 12 kpc away) is beaming 10^5 TeV particles at us with a luminosity generating an excess of particles over the isotropic background cosmic ray flux. In Fig. 5 we sketch a possible model[8,9] of a binary star emitting very high energy photons. The system consists of a compact star in a 4.8 hour orbit with a star that has not yet collapsed. The compact partner somehow accelerates protons, perhaps by a pulsar mechanism[10] or through conversion of energy from accretion of matter from the companion star.[11,12] The accelerated particles then interact with the companion or the surrounding gas (see Fig. 5) to produce a cascade of secondaries, the stable end products of which are as in any beam dump experiment photons, neutrinos, protons, antiprotons, electrons and positrons. The charged particles are injected into the galaxy as cosmic rays, though the electrons and positrons especially will be much degraded in the source. Some fraction of photons and neutrinos from

$$\begin{aligned} p &\to \pi^0 \to \gamma \\ p &\to \pi^\pm \to \nu \end{aligned} \qquad (2.1)$$

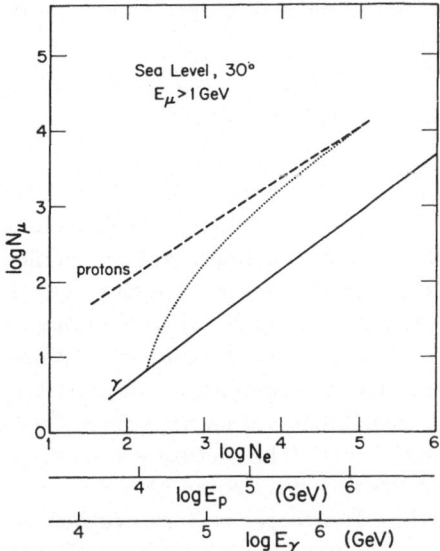

Fig. 3. Number of muons in excess of 1 GeV (N_μ) versus shower size (*i.e.* number of electrons N_e) for γ- (solid line) and proton- (dashed line) initiated showers. The dotted line shows the muon signature in a model where the photon becomes strongly interacting at very high energies. The horizontal energy scales show the conversion of shower size to initial proton or γ-ray energy.

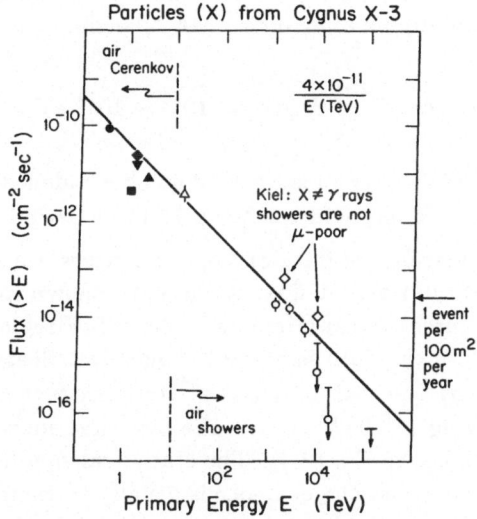

Fig. 4. Integral flux of very high energy particles from the X-ray binary Cygnus X-3. We will call these particles γ-rays although the Kiel experiment challenges this identification. Note the flatter E^{-1} energy dependence compared to the $E^{-1.7}$ fall-off of the cosmic ray flux.

405

will escape the source and travel in straight paths. They may be detected on earth if their production is sufficiently prolific. As in any beam dump any new particle for which the beam is above threshold will be produced and beamed to earth if it is sufficiently stable and neutral, *e.g.*

$$p \to \tilde{g} \to \tilde{\gamma} \,. \tag{2.2}$$

In all but a few experiments (*e.g.*, Kiel[2] and Haleakala[5]) a signal cannot be established without exploiting the fact that on-source showers are a 4.8 hour pulsed signal on a continuous (and typically 100 times larger) cosmic ray background. It is customary to define a phase angle from 0 to 1 which traces the binary revolution as shown in Fig. 5. The 0.0 corresponds to X-ray minimum when the compact star is eclipsed by its companion. In the model shown the beam-on positions are in the vicinity of phases 0.2 and 0.8. A compilation[1] of emission phases is shown in Fig. 6 along with the phase distribution of showers from the Cygnus direction as measured by the Haverah park array. The picture is a warning that the structure of the accelerator is complex. Its structure could be different for different energies, if the accreting matter provides the target material rather than the companion's atmosphere, emission at the other phases is possible; see Fig. 7.

The primary proton beam in Fig. 5 must of course reach energies up to 10^5 TeV to account for the secondary spectrum shown in Fig. 4. Beams of 10^5 TeV sound out of the ordinary. The basic reason why such energies are possibly achieved can, however, be understood on the basis of a dimensional argument. The typical pulsar is shown in Fig. 8. It is 10 km in diameter and accelerates particle beams along the dipole axis of its magnetic field which can reach values of 10^{12} Gauss. Like a lighthouse the beams spin with pulsar periods of $10^{-3} \sim 1$ sec around a rotation axis not coincident with the dipole axis. The EMF of such a system is

$$\mathcal{E} = B\ell v \quad \text{or} \quad B\frac{\partial \phi}{\partial t} = B\frac{\ell^2}{t} = 10^{16} \sim 10^{19} \text{ eV} \,. \tag{2.3}$$

It should be pointed out here that no pulsar period had been identified for Cygnus X-3 until the Durham group[13] found evidence for a 12 msec period.

This and many other features of data on cosmic sources are statistically weak and certainly require confirmation. A flux at the level shown in Fig. 4 has been fairly well-established by Cerenkov experiments in the 1 TeV region. In the energy-band covered by EAS-arrays the data is at best suggestive. Scepticism about this data has been reinforced by the lack of success[14] to construct specific models of cosmic accelerators achieving 10^5 TeV despite the fact that such energies are possible on dimensional grounds; see Eq. (2.3). The statistical significance of the EAS observations is difficult to assess as the evidence is usually extracted from computer searches for time periodicity in the data. In that respect experiments observing a genuine excess of on-source showers irrespective of the time structure of the showers are the most suggestive. Except for the original Kiel experiment, this has only been achieved during short time periods when the source showed increased activity: by the Baksan group[15] during two days in October 1985, the Fly's Eye group[16] during one night in July 1985 and the Los Alamos group[17] during two months in 1986. The Baksan air shower array detected an excess of 100 TeV showers, shown in Fig. 9,

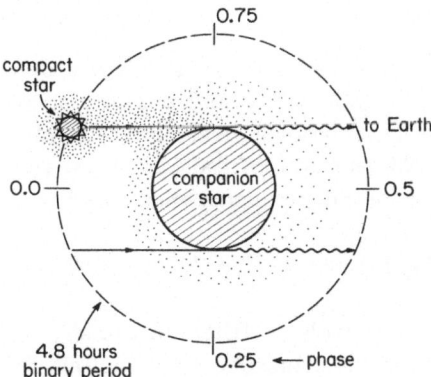

Fig. 5. The Cygnus accelerator beams particles to earth produced in the beam dump at phases 0.2 and 0.8 of the pulsar which accelerates the primary beam. The dots represent matter accreting on the pulsar and material blown off the companion.

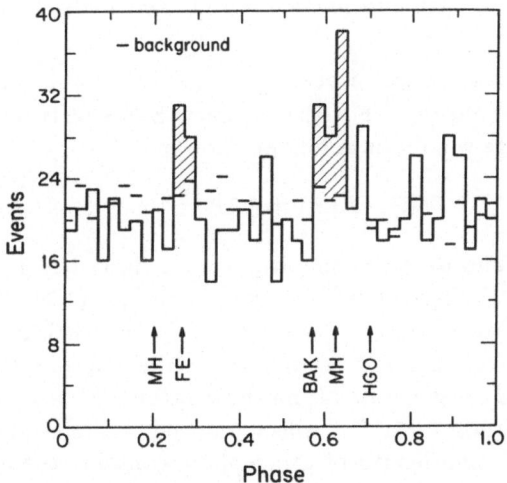

Fig. 6. Phase distribution of the radiation in the direction of Cygnus X-3 as observed by the Haverah park extensive air shower array. Emission is also observed in the general vicinity of $\phi \simeq 0.2$ and $0.6 \sim 0.7$ by Baksan, Fly's Eye, Mount Hopkins and Haleakela (HGO).

in the direction of Cygnus X-3 during a period that coincided with the largest radioburst of this source ever, *i.e.* the largest since monitoring started in 1972. The flux in the 11.1 cm radio band increased by a factor 300 during October of 1985. The Haleakala experiment[5] has evidence for burst of order 1 minute of TeV γ-rays during the same period.

This discussion should remind us that the duty cycle of cosmic accelerators could be small (1 $\sim$ 10% for Cygnus X-3?) and possibly erratic. In that respect other sources might prove to be superior accelerators, *e.g.* Hercules X-1, a source for which both the pulsar (1.24 sec) and orbital (1.7 days) time variability of the flux are well-known. Both can be independently monitored.

III. PARTICLE PHYSICS WITH COSMIC ACCELERATORS[18]

As Fig. 2 suggests, Cygnus is truly a HERA in the sky; primary photons interacting with hadrons populate the atmosphere with an electromagnetic cascade. The interactions of these particles can be studied by ground-based detectors. The potential of such experiments for particle physics should not be judged against conventional cosmic ray experiments which have a distinguished but somewhat checkered record. Here the direction and characteristic time structure of the radiation from a point source can be used to tag a beam of known composition (γ-rays) with well-understood interactions (described by QED) with the target atmosphere. An experiment using a tagged cosmic photon beam, therefore, overcomes the classic hurdles in interpreting cosmic ray experiments. New physics cannot be confused with some *mundane* (at least from the particle physics point of view) change in chemical composition of the primary particles or of their hadronic interaction with atmospheric nuclei.

The structure of the atmospheric beam is extremely simple and can be calculated using linear shower theory.[19] For the primary E^{-1} spectrum of Eq. (1.1) at the top of the atmosphere we obtain a flux of photons

$$N_\gamma(> E, z) = \tfrac{1}{2}AE^{-1} \tag{3.1}$$

as a function of energy and depth in the atmosphere described in terms of linear column density $z(\text{g/cm}^2)$. Here A is the constant in Eq. (1.1). Notice that the number of photons of a given energy E is independent of depth z. This somewhat surprising result can be understood as follows: while photons lose energy with depth, others of energy E are generated from showers with $E' > E$. Energy loss and feed-down are in equilibrium for a E^{-1} spectrum, hence the z-independent result of Eq. (3.1). Every radiation length of atmosphere contains the same number of photons of energy E. The number of radiation lengths is of course limited to some number n_{max} which is determined by the maximum energy of the source (10^5 TeV for Cygnus, see Fig. 4) and by the fact that a photon will lose half of its energy per radiation length. Therefore,

$$n_{\text{max}}(\text{radiation lengths}) \simeq \frac{\ln(E_{\text{max}}/E)}{\ln(2)}, \tag{3.2}$$

where $n_{\text{max}} \simeq 15$ for 1 TeV γ-rays. It is also important to remember that we are observing roughly equal numbers of γ's and electrons interacting above a detector; see Fig. 2.

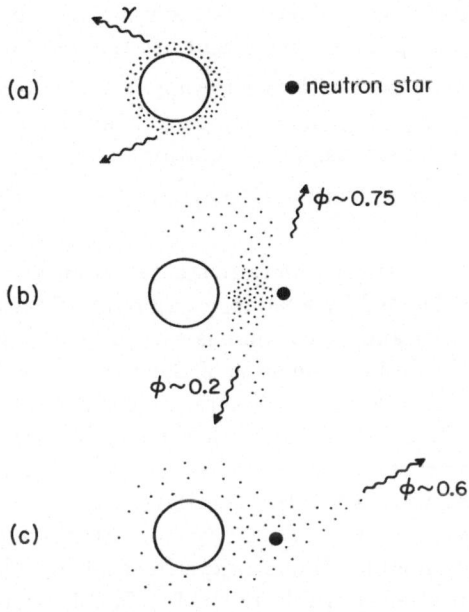

Fig. 7. Models where the accreting matter itself ((b) and (c)) rather than the companion's atmosphere ((a) and Fig. 5) provide the target material for the beam dump. Emission at other phases is possible.

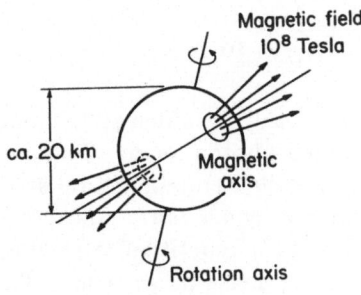

Figure 8. Pulsar

The beam is free. The question is what the detector requirements are to achieve statistics and signal/noise to do a quality photoproduction experiment at energies inaccessible to artificial accelerators and to possibly probe the much-anticipated new structure in particle physics at a scale of $(\sqrt{2}\, G_F)^{-1/2} \simeq 0.25$ TeV. As we have seen the most discriminating signature of the photon-induced shower against the hadron-induced cosmic ray background showers is the scarcity of muons. The two essential components of a tagged photon experiment are therefore obvious:

(i) an array of scintillation counters with fast timing or a lattice of simple single mirror Čerenkov telescopes giving an observation of the electromagnetic component of the shower with good angular resolution on the arrival direction,

(ii) an array of shielded counters identifying the content of (GeV) muons in the shower.

Both goals could be achieved by having vertically stacked counters in coincidence. The lower counters would be shielded by a sufficient amount of material to absorb the e, γ component of the shower and detect muons with $\mathcal{O}$ (GeV) energy. The array can be triggered on specific timing sequences of the counters which correspond to showers from one or even multiple source directions. From Eq. (1.1) we can see that at 1 TeV energy even a "modest" $(100 \text{ m})^2$ array can accumulate 10^5 γ-rays per year. The number of electrons/muons at sea level is, however, insufficient to perform a realistic experiment. With a 100 TeV threshold this problem is solved but it now takes a $(1 \text{ km})^2$ array to accumulate the 10^5/year statistics required to take a detailed look at photoproduction under difficult experimental circumstances. Some relevant information is summarized in Table 1. Such a facility would still constitute a not-too-far extension of present facilities such as the Akeno array[20] in Japan. One should, however, be aware of the possibly severe punch-through problems when identifying GeV muons by their penetration through matter. Not just hard electromagnetic particles could contribute; softer ones have a low probability to punch through but they have very large multiplicity and can also fake muons. For a detailed discussion, see Gaisser et al., in Ref. 6. What about signal-to-background?

Present detectors (e.g. Čerenkov) telescopes achieve a nominal

$$\frac{S}{N} = 10^{-2} \, , \qquad (3.3)$$

as previously mentioned. This is clearly insufficient to hope to detect traces of new particle thresholds in the tagged photon beam. This number can be greatly improved in the future when the detailed emission time structure of sources like Cygnus X-3 are better known. The binary 4.8 hour and candidate 12 msec (pulsar), 19 day, 12 month (?) and 5 year burst (?) repetition rates of the source can be used to enhance S/N possibly by 10^2 using phase information. The μ-poor property of γ showers further enhances the signal by another factor 10^2 but not more as at that level γ showers photoproducing a π in the primary interaction are virtually indistinguishable from a hadron-induced background cosmic ray cascade. With improvements in angular resolution $S/N \simeq 10^2 \sim 10^3$ can possibly be achieved. As discussed in the previous section several experiments suggest increased emission during bursts with a duration of minutes, days and possibly months. During these bursts $S/N \simeq 1$ rather than 10^{-2}. Tagging such bursts could therefore yield data at the $10^4 \sim 10^5$ signal to noise level. It is also very easy to convince oneself that deep underground experiments observing TeV–rather than GeV–muons would not

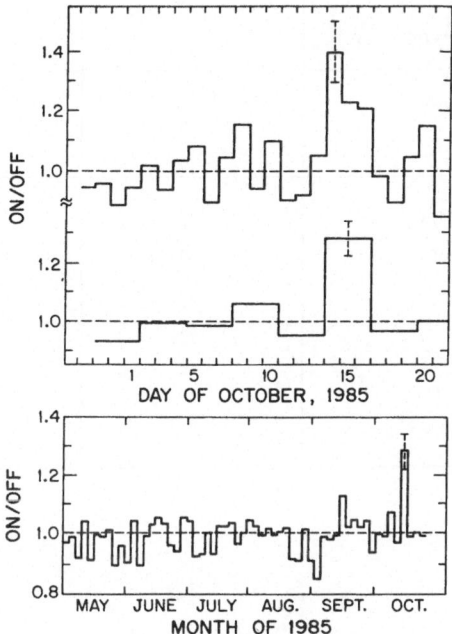

Fig. 9. Baksan observation of a DC signal of 100 TeV showers from Cygnus X-3 during the October '85 radio-flare.

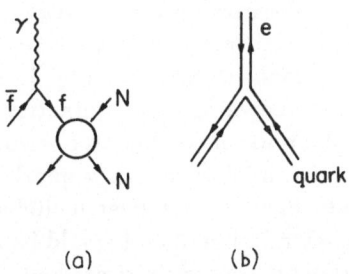

(a)   (b)

Fig. 10. Examples of new physics modifying electromagnetic cascades: (a) photons interacting with matter via a new fermion f, (b) electrons converting into quarks (or muons) via shared constituents.

411

Table 1. Properties of hadron- and photon-induced air showers at Akeno (depth $920\,\mathrm{g/cm^2}$).

Primary energy (TeV)	1	10^2
γ-flux from Cyg X-3 using (1) $(\mathrm{km^{-2}year^{-1}})$	10^7	10^5
Number of electrons in hadron shower	–	3800
Number of muons in hadron shower $(E > 1\,\mathrm{GeV})$	30	1525
Number of muons in γ-shower, (see also Fig. 3)	–	40

be competitive.[21] We will review some specific proposal for next-generation cosmic ray telescopes further on. We next illustrate some physics issues relevant to these experiments. As with all explorations of a new energy regime the "unexpected" discovery will be the most exciting one.

New physics can reveal itself in these experiments in two distinct ways: anomalous interactions of the photons or electrons in the atmospheric cascade (see Fig. 2) or admixture in the γ-beam of new neutral particles produced in the cosmic beam dump where the primary energies reach 10^5 TeV; see Fig. 5. We illustrate both possibilities. In Fig. 10, examples are shown of new interactions of photons and electrons with matter. In Fig. 10a, the photon couples to nucleons via new fermions f. Such interactions are certainly associated with heavy quark photoproduction with $f = c,\, b,\, t,\, \ldots$ and $\alpha_f = \alpha_s$. They have already been included in the calculations of the muon content of γ-showers shown in Fig. 3 and would be observable in the more ambitious versions of the experiment previously described. Speculation that photon interactions become strong at high energies,[22] though unpopular in the context of gauge theories, could easily be investigated. If at some energy $\alpha_f \simeq 1$, photon showers would eventually resemble hadron showers. The dotted line in Fig. 3 shows the result of a Monte Carlo study of the muon number dependence on energy for a new strong coupling process with 0.25 TeV threshold.

It is important to realize that electromagnetic cascades contain roughly equal numbers of electrons and photons. Composite models of quarks and leptons provide us with an example where new physics in the electron-nucleon interaction will also affect the development of the γ-initiated cascade. At very high energy, electrons have a finite size in which pointlike preons move around. They can interact with quarks (hadrons) via the exchange of preons. If these are colored, electrons can turn into quarks and begin to interact like hadrons; see the preon diagram of Fig. 10b. A drastic increase of the muon yield will result.

In both examples shown in Fig. 10, photon showers will turn mixed electro-magnetic-hadronic at high energy, with enhanced muon production signaling the onset of the hadronic component. The Kiel experiment,[2] showing an increase of the muons in the Cygnus showers by at least a factor of 10 over the QED calculation, could be interpreted along these lines.

The photino ($\tilde{\gamma}$) or gluino ($\tilde{g}$) supersymmetric partners of the photon and gluon provide us with postulated particles which definitely have to be produced by cosmic accelerators and could possibly be detected as admixtures in the γ-beam. The generation of particles in the Cygnus beam dump of Fig. 5 can be calculated[18,23,24] with the same techniques[19] used to calculate particle production by the cosmic ray beam in the Earth's (rather than the main sequence star's) atmosphere. For stable $\tilde{\gamma}$'s produced via production and decay of $\tilde{g}$'s (*i.e.*, $\tilde{g} \to \tilde{\gamma} q \bar{q}$) we can obtain the approximate estimate

$$\frac{dN_{\tilde{\gamma}}/dE}{dN_{\gamma}/dE} \sim \frac{1}{3} \frac{\sigma(p \to \tilde{g})}{\sigma_{p,\text{tot}}} . \tag{3.4}$$

For gluinos with $\mathcal{O}$ (GeV) masses, a possibility arguably not ruled out by present accelerators, $\sigma(p \to \tilde{g})$ is $\mathcal{O}$(mb). This results via (3.4) into a 10^{-3}-10^{-2} admixture of stable supersymmetric particles in the photon beams emitted by cosmic accelerators. If the $\tilde{g}$ itself is stable, the $\tilde{g}/\gamma$ ratio would be even larger. They would be bound in gluino-hadrons interacting with the Earth's atmosphere like hadrons rather than photons resulting again into a muon excess in the muon-poor photon beam.

IV. NEXT-GENERATION COSMIC RAY TELESCOPES

It is exciting to contemplate the feasibility of tagging 100 TeV photons in detectors which represent a not too far-fetched extrapolation of present facilities. In the U.S. three efforts are under way to design and construct such experiments. Their main characteristics are summarized in Table 2. Needless to say that many more groups in the U.S. and abroad are considering similar experiments. The experiments all approach the characteristics of the "theoretical" detector described in the previous section.

The Chicago air shower array (CASA)[25] will be operated in Utah at the site of Fly's Eye II. It is made up of 1064 stations separated by 10 m and covers therefore roughly $(300 \text{ m})^2$ area. Each station consists of 4 scintillator counters of 0.37 m^2 size. With fast timing the array hopes to measure shower direction to a precision of 0.01 rad. CASA is operated in conjunction with the Michigan muon array presently under construction.[26] It detects muons of GeV energies in scintillator buried under 10 feet of earth. The scintillators are distributed over 8 sites of roughly 150 m^2 each. A site consists of 64 closely packed counters. As already mentioned one might worry about punchthrough of high energy electromagnetic particles through 10 ft. of earth thus faking muons, a problem brought to the forefront by the Kiel

Table 2. Summary of Some Experiments

Institution	Area m^2	Number of Stations	Electrons	Muons	θ resolution
Chicago*	10^5	1064	* 4 × 0.37 m² scintillator		0.011 rad
Michigan*	1.2×10^3	512		* scintillator under 10 ft. of earth	
Notre Dame	10^4	64	PWC 8 planes 1.2 m²	1.5 inch *steel* + 2 planes	0.25°
Irvine/ Hawaii	6.25×10^4	5×10^3 PMT	PMT plane under 10~20 m of water	PMT plane under additional 30-60 m of *water*	< 1°

* operated in coincidence

experiment. The solution of this group is to ignore underground signals close to the shower core whose position is located by CASA. The Akeno group indeed claims to have demonstrated that punchthrough is under control provided one ignores "muons" within 30 m of the core.[27] Whereas this is a satisfactory solution from the point of view of astronomy (*i.e.*, for rejection of muon-rich background hadron showers), it might make an actual measurement of the muon abundance difficult.

This problem is solved in the ambitious Irvine/Hawaii proposal[28] where two vertically stacked planes of 2500 phototubes each, are spread over $(250 \text{ m})^2$ and immersed in 10 m and 50 m of water, respectively. The actual depths have not been fixed but the idea is that the top and bottom layer identify the electron and muon components, respectively.

The Notre Dame group on the other hand tries a totally novel approach.[29] Their array consists of PWC's of 1.2 m^2 size. The direction of each shower particle is measured individually by 8 planes of wires in each PWC, rather than by timing between stations. The last two planes are shielded by a 1.5 inch steel plate and separates electrons (showering or scattering in the steel) from muons. By high energy physics standards these units are simple and cheap and could be constructed with time in an ever-expanding array at relatively little cost.

V. DISCUSSION

There is very little doubt that these experiments will measure the photoproduction of pions in the 100 TeV energy range. As the π's are identified by μ-decay any other copious source of muons or hadrons in the otherwise purely electromagnetic shower will be detected. Semileptonic decay of heavy quarks is probably within reach of some experiments. As previously discussed, new physics has already been contemplated that could give detectable signals. It should here be pointed out that some experiments have already claimed results on anomalous muons[30] reaching far beyond the imagination of any theorist. Also the muon abundance measured in the on-source showers by the Kiel group not only challenges standard particle physics but is unlikely to be accommodated by any of the new physics contemplated in this talk. It should be mentioned however that their observations come dangerously close to being reproduced by "punchthrough" calculations.[6] These calculations certainly underestimate "punchthrough" as they cannot evaluate the contribution of electromagnetic particles below a few MeV. The claim of observation[30] of TeV muons associated with point sources forces us to essentially abandon the idea that photons carry the radiation; see however Ref. 31. These results defy explanation, they are however controversial and the situation underscores the necessity for a new generation of experiments. The detectors, previously discussed, can confirm or rule out these observations to better than 5σ within days or at best months of operation.

Even if nothing but routine particle physics is done by these experiments, imagine their power as telescopes! The most spectacular results implied by the measurements in Fig. 4 is not the energy but rather the implied flux of the source which is estimated[9,32] to be in excess of 10^{39}ergs sec^{-1}. This luminosity is more than one million times the total energy output of the sun and exceeds the Eddington limit for spherically symmetric accretion onto a one solar mass compact object. This discovery might have solved the old problem of the origin of very high energy cosmic rays. It has been known for some time that the cosmic ray spectrum can be understood[33] up to perhaps 100 TeV in terms of shock wave acceleration in supernova remnants. Although the spectrum shows a kink just above this energy, it steepens but doesn't stop. A straightforward estimate, see *e.g.* Ref. 34, reveals that the Cygnus accelerator's power is more than adequate by itself to supply all the cosmic rays in the galaxy in the interval 10^3-10^5 TeV. Beyond that energy, cosmic rays can be extra-galactic. At present this is a speculation, the new generation of cosmic ray experiments are in a position to quantitatively test the idea that cosmic accelerators are the source of high energy cosmic rays. They will also undoubtedly improve on our knowledge of the chemical composition of cosmic rays at these energies, a measurement of astrophysical importance for understanding the origin of cosmic rays and the structure of our galaxy.

Finally in 1987 it is impossible to conclude without addressing the relevance of these experiments to the observation of supernova. If a star collapses to a young pulsar the supernova becomes a cosmic accelerator.[35] As in the binary shown in Fig. 5 the pulsar is the site of acceleration. The target matter for producing γ-rays from π^0-decay is now provided by the receding shell rather than a companion star. The detectable γ-ray flux will depend on luminosity L of the pulsar and the total column density z_{tot} of the shell along our line of sight to the pulsar. Both can be estimated. The luminosity L from liberation of rotational energy due to the pulsar's dipole magnetic field is[36]

$$\left[\frac{L}{4 \times 10^{43}\text{ergs/s}}\right] \simeq \left[\frac{B}{10^8 T}\right]^2 \left[\frac{R}{10\,\text{km}}\right]^6 \left[\frac{1\,\text{ms}}{P}\right]^4 \left[1 + \frac{t}{4.4\,\text{yr}\left(\frac{P}{1\,\text{ms}}\right)^2 \left(\frac{B}{10^8 T}\right)}\right]^{-2}$$

(5.1)

The maximum energy is

$$E_{\max} \simeq 10^5\ TeV \left[\frac{B}{10^8 T}\right]^{2/3} \left[\frac{1\,\text{ms}}{P}\right]^{4/3}.$$

(5.2)

For a young pulsar $P \simeq 1$ ms the luminosity could exceed that of Cygnus by a factor $[12.59\,\text{msec}/1\,\text{msec}]^4$. The target column density has been calculated by Nakamura *et al.*,[37] it is of course time dependent.

$$\left[\frac{z_{tot}}{5 \times 10^5 \text{g/cm}^2}\right] \simeq \left[\frac{M}{7 M_\odot}\right]^2 \left[\frac{2 \times 10^{51}\text{ergs}}{E}\right] \left[\frac{1\,\text{day}}{t}\right]^2.$$

(5.3)

Here M is the mass and E the explosion energy. For SN1987 the first two factors in (5.3) are believed to be close to unity. The maximum of π^0 production and therefore of γ-emission is expected when the shell's density is comparable to a proton interaction length ($\sim 30\,\text{g/cm}^2$). This occurs after months as can be seen from (5.3). Whereas it is an open question whether SN1987 will be a TeV γ-ray emitter, these estimates are encouraging for the possibility of observing γ-rays from an eventual galactic supernova. Even for SN1987 the handicap that it is 5 times the distance of Cygnus is overcome by its 100% duty cycle compared to a few % for Cygnus X-3. In the absence of observation, limits can be useful. The absence of γ-rays could support evidence for the formation of a black hole. Here $L < 10^{38}$ergs/s, as the Eddington limit applies and detection of a signal is unlikely.

ACKNOWLEDGEMENTS

This research was supported in part by the University of Wisconsin Research Committee with funds granted by the Wisconsin Alumni Research Foundation, and in part by the U. S. Department of Energy under contract DE-AC02-76ER00881.

REFERENCES

1. For a detailed discussion see F. Halzen, K. Hikasa and T. Stanev, Phys. Rev. D34:2061 (1986); T. Stanev, T. K. Gaisser and F. Halzen, Phys. Rev. D32:1244 (1985); T. Stanev and Ch. P. Vankov, Phys. Rev. Lett. 158B:75 (1985); T. Stanev, Ch. P. Vankov and F. Halzen, *19th International Cosmic Ray Conference Papers*, La Jolla, 1985 (NASA, Washington, D.C., (1985)), Vol. 7, p. 219; T. Stanev and Ch. P. Vankov, Com. Phys. Comm. 16:363 (1979).

2. M. Samorski and W. Stamm, Ap. J. 268:L17 (1983).

3. For a complete review, see A. A. Watson, rapporteur paper presented at the *International Cosmic Ray Conference*, La Jolla, 1985 (to be published). See also, F. Halzen, *SLAC 1986 Summer Institute on Particle Physics*, E. Brennan, ed.; T. Weeks, Physics Reports (in print), J. Learned, University of Hawaii Internal Report (unpublished), 1985; J. W. Elbert, R. C. Lamb and T. C. Weeks in *Proceedings of New Particles '85*, Madison, Wisconsin, V. Barger, D. Cline and F. Halzen, eds., World Scientific, Singapore, (1986); B. M. Vladimirskii, A. M. Gal'per, B. I. Luchkov and A. A. Stepanyan, Sup. Riz. Nauk. 145:255 [Sov. Phys. Usp. 28:153 (1985)].

4. M. V. Barnhill III, T. K. Gaisser, T. Stanev and F. Halzen, Nature 317:409 (1985); *19th International Cosmic Ray Conference Papers*, La Jolla, 1985, (NASA, Washington, D.C., 1985), Vol. 1, p. 99.

5. See *e.g.* J. Fry, *Proceedings of the 1986 Aspen Winter Conference on Particle Physics*, M. Block, ed.

6. T. Stanev, T. K. Gaisser and F. Halzen, Phys. Rev. D32:1244 (1985).

7. M. V. Barnhill III, T. K. Gaisser, T. Stanev and F. Halzen, Ref. 4; A. Dar, J. J. Lord and R. J. Wilkes, Phys. Rev. D33:303 (1986); V. S. Berezinsky, J. Ellis and B. L. Ioffe, Moscow preprint ITEP-21 (86). For reviews see *e.g.* A. De Rújula (CERN-TH.4267/85), F. Halzen, L. Maiani (CERN-TH.4326/85), in *Proceedings of the International Europhysics Conference on High Energy Physics*, Bari, 1985, L. Nitti and G. Preparata, eds.; T. Stanev, *Proceedings of New Particles '85*, Madison, V. Barger, D. Cline and F. Halzen, eds., World Scientific, Singapore, (1986).

8. W. T. Verstrand and D. Eichler, Ap. J. 261:251 (1982); V. S. Berenzinsky, *Proceedings of 1979 DUMAND Summer Workshop*, J. G. Learned, ed., p. 245; D. Eichler, Nature 275:725 (1978).

9. A. M. Hillas, Nature 312:50 (1984).

10. D. Eichler and W. T. Vestrand, Nature 307:613 (1984).

11. G. Chanmugam and K. Brecher *et al.*, Nature 313:767 (1985).

12. D. Eichler and W. T. Verstrand, *19th International Cosmic Ray Conference*, La Jolla, 1985; D. Kazanas and D. C. Ellison, Nature, to be published.

13. T. Turver, *19th International Cosmic Ray Conference*, La Jolla, 1985, NASA, Washington, D.C., (1985).

14. D. Eichler, in Johns Hopkins Workshop on Heavenly Accelerators, Baltimore (1987).

15. C. Chudakov, private communication and *European Cosmic Ray Physics Conference*, Bordeaux, France.

16. R. M. Baltrusaitis *et al.*, Utah preprint (1987).

17. G. Yodh, in Johns Hopkins Workshop on Heavenly Accelerators, Baltimore (1987).

18. This is an abbreviated discussion of the results in F. Halzen, K. Hikasa, and T. Stanev, Ref. 1.

19. See *e.g.*, B. Rossi, *High Energy Particles*, Prentice Hall, N.Y., (1952).

20. For a description of the Akeno array see *e.g.*, T. Hara *et al.*, in *Proceedings of the International Symposium on Cosmic Rays and Particle Physics*, Tokyo, 1984, A. Ohsawa and T. Yuda, eds.

21. T. Stanev, Bartol preprint BA-86-7 (1986).

22. See *e.g.*, W. Ochs and L. Stodolsky, Max-Planck Institute (München) preprint MPI PAE/Pth 54/85 (1985).

23. F. Halzen and J. R. Cudell, Phys. Rev. D. 36:346 (1987).

24. V. J. Stenger, Nature 317:411 (1985) and Ap. J. 284:810 (1984); R. W. Robinett, Phys. Rev. Lett. 55:469 (1985); G. Auriemma, L. Maiani and S. Petrarca, Phys. Lett.164B:179 (1985); V. S. Berezinsky and B. L. Ioffe, Moscow preprint ITEP-127 (1985); S. Midorikawa and S. Yoshimoto, University of Tokyo preprint INS-Rep.-560 (1985); K. Hagiwara, private communication. For related suggestions see A. Zee, Phys. Lett. 161B:141 (1985) and R. N. Mohapatra, S. Nussinov and J. W. F. Valle, Phys. Lett. 165B:417 (1985).

25. M. K. Cambell *et al.*, CASA proposal.

26. D. Sinclair *et al.*, private communication.

27. M. Nangano, private communication.

28. M. Svoboda in *Proceedings of the IV Telemark Conference*, World Scientific, Singapore (1987), V. Barger and F. Halzen, eds.; J. G. Learned, private communication.

29. J. Poirier *et al.*, private communication.

30. J. Bartelt *et al.*, (Soudan I Collaboration), Phys. Rev. D32:1630 (1985); M. L. Marshak *et al.*, Phys. Rev. Lett. 54:2079 (1985); *ibid* 55:1965 (1985); G. Battistoni *et al.*, (NUSEX Collaboration), Phys. Lett. 155B:465 (1985); C. Berger (Frejus Collaboration), in *Proceedings of the International Europhysics Conference on High Energy Physics*, Bari, Italy (1985) (to be published). See, however, E. Aprile *et al.*, (HPW Collaboration), in *Proceedings of the International Europhysics Conference on High Energy Physics*, Bari, Italy (1985) (to be published); Y. Oyama *et al.*, (KAMIOKANDE Collaboration), University of Tokyo preprint UT-ICEPP-85-03 (1985). See also L. E. Price, Argonne preprint ANL-HEP-CP-85-117, in *Proccedings of Annual Meeting of the Division of Particles*

and Fields of the American Physical Society, Eugene (1985), R. C. Hwa, ed., World Scientific, Singapore, (1986); Y. Totsuka, University of Tokyo preprint UT-ICEPP-85-05 (1985), in *Proceedings of 1985 International Symposium on Lepton and Photon Interactions at High Energies*, Kyoto (to be published).

31. N. Yamdagni, F. Halzen and P. Hoyer, Phys. Lett. B. 190:211 (1987).

32. A. M. Hillas, *19th International Cosmic Ray Conference*, La Jolla, 1985, NASA, Washington, D.C., (1985).

33. See *e.g.*, C. Cesarsky and P. O. Lagage, Astron. and Astrophysics 125:249 (1985).

34. T. K. Gaisser and F. Halzen, *XXIst Rencontre de Moriond*, Les Arcs (1986).

35. H. Sato, Prog. Theor. Phys. 58:549 (1977).

36. J. E. Gunn and J. P. Ostriker, Phys. Rev. Letters 22:728 (1969).

37. T. Nakamura, Y. Yamada and H. Sato, University of Kyoto preprint, KUNS 877, (1987).

Participants

Ahmed Ali
DESY
Notkestr. 85
D-2000, Hamburg 52, FRG

Richard Batley
Queen Mary College London
Mile End Road
London, E14NS, UK

Bradley Cox
Fermilab
P.O. Box 500
Batavia, IL 60510, USA

Andrea Contin
CERN
EP Division
1211 Geneva, Switzerland

Richard H. Dalitz
Oxford University
1 Keble Road
Oxford, OX1 3NP, UK

Alvaro De Rujula
CERN
TH Division
1211 Geneva 23, Switzerland

Luigi DiLella
CERN
EP Division
1211 Geneva 23, Switzerland

R. Keith Ellis
Fermilab
MS 106
P.O. Box 500
Batavia, IL 60510, USA

Michel Fontannaz
Lab de Physique Théorique
et Hautes Energies
Bât. 211
Univ. Paris-Sud
F-91405 Orsay, France

Alfred Fridman
CERN
1211 Geneva 23, Switzerland

Harald Fritzsch
Universität München
Theresienstr. 37a
D-8000 München, FRG

Thomas Gaisser
Bartol Research Institute
University of Delaware
Newark, DE, USA

Michel Gourdin
Université Pierre et Marie Curie
4, Place Jussieu
F-75230 Paris, France

Mario Greco
INFN
Laboratori Nazionali di Frascati
P.O. Box 13
I-00044 Frascati, Italy

Dieter Haidt
DESY
Notkestr. 85
D-2000 Hamburg 52, FRG

Francis Halzen
University of Wisconsin
Madison, WI 53705, USA

Gunnar Ingelman
DESY
Notkestr. 85
D-2000 Hamburg 52, FRG

Ken-Ichi Hikasa
DESY
Notkestr. 85
D-2000 Hamburg 52, FRG

Jasper Kirkby
CERN
EP Division
1211 Geneva, Switzerland

Zoltan Kunszt
ETH
Honggerberg
CH-8093 Zurich, Switzerland

Robert Meunier
CERN
EP Division
1211 Geneva 23, Switzerland

Nello Paver
Università di Trieste
Dipartamento di Fisica Teorica
Strada Costiera, 11
34014 Miramare-Grignano (TS), Italy

Herwig Schopper
CERN
DG Division
1211 Geneva 23, Switzerland

Barbara Schrempp
Universität München
Theoretische Physik
Theresienstr. 37
D-8000 München 2, FRG

John Sculli
New York University
Physics Department
4 Washington Place
New York, NY 10012, USA

Volker Soergel
DESY
Notkestr. 85
D-2000 Hamburg 52, FRG

Manfred Steuer
CERN
EP/L3
1211 Geneva 23, Switzerland

James Stirling
Physics Department
Durham University
Soth Road
Durham DH1 3LE, UK

John Thresher
CERN
EP Division
1211 Geneva 23, Switzerland

Bob van Eijk
CERN
EP Division
1211 Geneva 23, Switzerland

Bryan Webber
University of Cambridge
Cavendish Laboratory
Madigly Road
Cambridge CB3 0HE, UK

Gunter Wolf
DESY
Notkestr. 85
D-2000 Hamburg 52, FRG

Antonino Zichichi
CERN
EP Division
1211 Geneva 23, Switzerland

Ren-Yuan Zhu
CALTECH
256-48 HEP
Pasadena, CA 91125, USA

Index